PROGRESS IN BRAIN RESEARCH

VOLUME 160

GABA AND THE BASAL GANGLIA: FROM MOLECULES TO SYSTEMS

PROGRESS IN BRAIN RESEARCH

VOLUME 160

GABA AND THE BASAL GANGLIA: FROM MOLECULES TO SYSTEMS

EDITED BY

JAMES M. TEPPER
ELIZABETH D. ABERCROMBIE
Center for Molecular and Behavioral Neuroscience, Rutgers University, Newark, NJ 07102, USA

J. PAUL BOLAM
Medical Research Council, Anatomical Neuropharmacology Unit, Department of Pharmacology, University of Oxford, Oxford OX1 3TH, UK

ELSEVIER

AMSTERDAM – BOSTON – HEIDELBERG – LONDON – NEW YORK – OXFORD
PARIS – SAN DIEGO – SAN FRANCISCO – SINGAPORE – SYDNEY – TOKYO

Elsevier
Radarweg 29, PO Box 211, 1000 AE Amsterdam, The Netherlands
Linacre House, Jordan Hill, Oxford OX2 8DP, UK

First edition 2007

Library of Congress Cataloging-in-Publication Data
A catalog record for this book is available from the Library of Congress

British Library Cataloguing in Publication Data
A catalogue record for this book is available from the British Library

ISBN: 978-0-444-52184-2 (this volume)
ISSN: 0079-6123 (Series)

For information on all Elsevier publications
visit our website at books.elsevier.com

Printed and bound in The Netherlands

07 08 09 10   11   10 9 8 7 6 5 4 3 2 1

Working together to grow
libraries in developing countries
www.elsevier.com | www.bookaid.org | www.sabre.org
ELSEVIER   BOOK AID International   Sabre Foundation

List of Contributors

E.D. Abercrombie, Center for Molecular and Behavioral Neuroscience, Rutgers University, 197 University Avenue, Newark, NJ 07102, USA

G.W. Arbuthnott, Basal Ganglia Research Group, Department of Anatomy and Structural Biology, School of Medical Sciences, University of Otago, P.O. Box 913, 270 Great King Street, Dunedin, New Zealand

A. Arslan, Department of Clinical Neurobiology, University of Heidelberg, Im Neuenheimer Feld 364, 69120 Heidelberg, Germany

J. Baufreton, Laboratoire de physiologie et physiopathologie de la signalisation cellulaire, UMR CNRS 5543, Université de Bordeaux 2, 146 Rue Léo Saignat, 33076 Bordeaux Cedex, France

M.D. Bevan, Northwestern University, Department of Physiology, Feinberg School of Medicine, 303 E. Chicago Avenue, Chicago, IL 60611, USA

J.P. Bolam, Medical Research Council, Anatomical Neuropharmacology Unit, Department of Pharmacology, University of Oxford, Mansfield Road, Oxford, OX1 3TH, UK

J. Boyes, School of Life and Health Sciences, Aston University, Birmingham, UK

S. Charpier, Dynamique et Physiopathologie des Réseaux Neuronaux, INSERM U667, UPMC, Collège de France, 11 Place Marcelin Berthelot, 75231 Paris, Cedex 05, France

M.-F. Chesselet, Departments of Neurology and Neurobiology, The David Geffen School of Medicine at UCLA, 710 Westwood Plaza, Los Angeles, CA 90095-1769, USA

J.M. Deniau, Dynamique et Physiopathologie des Réseaux Neuronaux, INSERM U667, UPMC, Collège de France, 11 Place Marcelin Berthelot, 75231 Paris, Cedex 05, France

G. Drew, Institute of Physiology and Pathophysiology and Interdisciplinary Center for Neurosciences (IZN), University of Heidelberg, Im Neuenheimer Feld 326, D-69120 Heidelberg, Germany

P.C. Emson, The Babraham Institute, Babraham Research Campus, Cambridge, CB2 4AT, UK

M. Farrant, Department of Pharmacology, UCL (University College London), Gower Street, London WC1E 6BT, UK

A. Galvan, Department of Neurology, School of Medicine and Yerkes National Primate Research Center, Emory University, 954 Gatewood Road NE, Atlanta, GA 30322, USA

T. Goetz, Department of Clinical Neurobiology, University of Heidelberg, Im Neuenheimer Feld 364, 69120 Heidelberg, Germany

N.E. Hallworth, Northwestern University, Department of Physiology, Feinberg School of Medicine, 303 E. Chicago Avenue, Chicago, IL 60611, USA

O. Hikosaka, Laboratory of Sensorimotor Research, National Eye Institute, National Institute of Health, 49 Convent Drive, Bldg. 49, Rm. 2A50, Bethesda, MD 20892-4435, USA

K. Kaila, Department of Biological and Environmental Sciences and Neuroscience Center, Laboratory of Neurobiology, University of Helsinki, P.O. Box 65 (Viikinkaari 1), FIN-00014 Helsinki, Finland

H. Kita, Department of Anatomy and Neurobiology, The University of Tennessee Memphis, 855 Monroe Avenue, Memphis, TN 38163, USA

C.R. Lee, Center for Molecular and Behavioral Neuroscience, Rutgers University, 197 University Avenue, Newark, NJ 07102, USA

M.S. Levine, Mental Retardation Research Center, The David Geffen School of Medicine at UCLA, 710 Westwood Plaza, 58-258 NPI, Los Angeles, CA 90095-1769, USA

P. Mailly, Neurobiologie des Signaux Intercellulaires, CNRS UMR 7101, UPMC, 4 Place Jussieu, 75005 Paris, France

N. Maurice, Dynamique et Physiopathologie des Réseaux Neuronaux, INSERM U667, UPMC, Collège de France, 11 Place Marcelin Berthelot, 75231 Paris, Cedex 05, France

J.F. McGinty, Department of Neurosciences, Medical University of South Carolina, Charleston, SC 29425, USA

U. Misgeld, Institute of Physiology and Pathophysiology and Interdisciplinary Center for Neurosciences, (IZN), University of Heidelberg, Im Neuenheimer Feld 326, D-69120 Heidelberg, Germany

A. Nambu, Division of System Neurophysiology, National Institute for Physiological Sciences, 38 Nishigo-naka, Myodaiji, Okazaki, 444-8585, Japan

J.L. Plotkin, Departments of Neurology and Neurobiology, The David Geffen School of Medicine at UCLA, 710 Westwood Plaza, 58-258 NPI, Los Angeles, CA 90095-1769, USA

A. Schousboe, Department of Pharmacology and Pharmacotherapy, The Faculty of Pharmaceutical Sciences, University of Copenhagen, DK-2100 Copenhagen, Denmark

T. Shindou, Basal Ganglia Research Group, Department of Anatomy and Structural Biology, School of Medical Sciences, University of Otago, P.O. Box 913, Dunedin, New Zealand

J.M. Tepper, Center for Molecular and Behavioral Neuroscience, Rutgers University, 197 University Avenue, Newark, NJ 07102, USA

H.S. Waagepetersen, Department of Pharmacology and Pharmacotherapy, The Faculty of Pharmaceutical Sciences, University of Copenhagen, DK-2100 Copenhagen, Denmark

T. Wichmann, Department of Neurology, School of Medicine and Yerkes National Primate Research Center, Emory University, 954 Gatewood Road NE, Atlanta, GA 30322, USA

J.R. Wickens, Basal Ganglia Research Group, Department of Anatomy and Structural Biology, School of Medical Sciences, University of Otago, P.O. Box 913, 270 Great King Street, Dunedin, New Zealand

C.J. Wilson, Department of Biology, University of Texas at San Antonio, 6900 N. Loop 1604 W, San Antonio, TX 78249, USA

W. Wisden, Institute of Medical Sciences, University of Aberdeen, Foresterhill, Aberdeen, AB25 2ZD, Scotland, UK

N. Wu, Mental Retardation Research Center, The David Geffen School of Medicine at UCLA, 710 Westwood Plaza, 58-258 NPI, Los Angeles, CA 90095-1769, USA

P. Wulff, Department of Clinical Neurobiology, University of Heidelberg, Im Neuenheimer Feld 364, 69120 Heidelberg, Germany

Y. Yanovsky, Institute of Physiology and Pathophysiology and Interdisciplinary Center for Neurosciences (IZN), University of Heidelberg, Im Neuenheimer Feld 326, D-69120 Heidelberg, Germany

Preface

The genesis of this volume occurred during a car ride from Heathrow Airport to Oxford in the Spring of 2003 when two of us (JPB and JMT) were discussing that, for a couple of neuroscientists who were mostly interested in dopaminergic systems in the basal ganglia, we both found ourselves spending what seemed to us an inordinate amount of time doing experiments on GABAergic basal ganglia neurons and pathways. One of us mentioned to the other a little red book called "*GABA and the Basal Ganglia*" that was published by Raven Press in 1981 as Volume 30 of the *Advances in Biochemical Psychopharmacology* series. Both of us remembered this little camera-ready book edited by Gaetano DiChiara and Gian Luigi Gessa with great fondness, as it contained many chapters by some of the top names in basal ganglia research. Both of us had bought that book with our own money. As we were both of modest means in 1980, to buy a book with our own money indicated that there was something special about the book. We both thought that it had been a great investment and that little book proved very influential to us as it summarized much of the state-of-the-art of knowledge about the GABAergic mechanisms in the basal ganglia as it existed around 1980.

In the Spring of 2004 Elizabeth Abercrombie and I were fortunate enough to be able to spend our sabbatical leave in Paul Bolam's laboratory. One day, the three of us were having lunch (at one of the many great pubs around Oxford) and the subject of "*GABA and the Basal Ganglia*" came up again. It turned out that Elizabeth also was quite familiar with the book and felt similarly about it. We began to discuss the current state of knowledge concerning the biochemistry, anatomy and physiology of GABA in the basal ganglia and realized that in the 25 years since the original "*GABA and the Basal Ganglia*" was published, our knowledge of virtually all aspects of GABAergic function had taken giant leaps forward. Back in 1980 there was only a single GABA receptor known and the internal microstructure of the neostriatum was completely unknown. The GABAergic nature of some of the basal ganglia pathways was uncertain. Whole-cell recording did not exist, nor did in vivo microdialysis, biocytin or neurobiotin labeling, PHA-L tract tracing, or much of the huge host of modern neurobiological techniques that have enabled us to make so much progress in understanding the structural organization of the basal ganglia over the past 25 years.

Thus the three of us decided that the time was right for another volume on the subject of GABA and the Basal Ganglia. We had several additional meetings where we hashed out what chapters ought to be included and which subjects should be covered in those chapters. Finally we made a wish list of the authors we would like to write each chapter, and selected one or two back-up authors in case our first choice declined. In each case, the author we selected was a widely cited, internationally renowned authority on the nucleus or pathway to which we wished to assign them. We also decided that, to make the volume as self-contained as possible, we should also invite experts on GABA itself and GABA receptors. We then approached Elsevier with our idea and waited. We were thrilled when Elsevier gave approval to our outline and told us that the book would be published as part of their excellent *Progress in Brain Research* series. We then began contacting the authors and to our amazement and delight the response was unprecedented. Of the 16 authors we contacted, 15 wrote back that they would be delighted to contribute. The result is the volume you hold in your hand. We hope it has as great an impact and proves to be as useful for the current generation of neuroscientists studying the basal ganglia as the little red book was for us at the beginning of our careers 25 years ago.

James M. Tepper
Elizabeth D. Abercrombie
J. Paul Bolam

Contents

Section I. Fundamentals of GABA in the Basal Ganglia

Section II. GABAergic Microcircuits of the Basal Ganglia

Fundamentals of GABA in the Basal Ganglia

Tepper, Abercrombie & Bolam (Eds.)
Progress in Brain Research, Vol. 160
ISSN 0079-6123

CHAPTER 1

Basal ganglia macrocircuits

J.M. Tepper[1,*], E.D. Abercrombie[1] and J.P. Bolam[2]

[1]*Center for Molecular and Behavioral Neuroscience, Rutgers University, Newark, NJ 07102, USA*
[2]*Medical Research Council, Anatomical Neuropharmacology Unit, Department of Pharmacology, University of Oxford, Oxford OX1 3TH, UK*

Abstract: This is the introductory chapter to an edited volume comprising 18 chapters written by 38 specially selected authors covering the anatomy, physiology, biochemistry/pharmacology and behavioral aspects of GABA in the basal ganglia. In this chapter the various nuclei of the basal ganglia are defined and their cellular structure, connections and function reviewed in brief in order to provide an orientation for the subsequent 17 chapters.

Keywords: neostriatum; globus pallidus; substantia nigra; subthalamic nucleus

The purpose of this short introductory chapter is to give a brief overview of the functional organization of the basal ganglia. The aim is to help orient the reader, especially one who is not so familiar with the basal ganglia, to the eighteen chapters that follow. Chapters 2 through 5 deal with the basic aspects of the biochemistry (Chapter 2), receptor structure and pharmacology (Chapters 3 and 4) and molecular and ionic bases of receptor signalling (Chapter 5) of γ-aminobutyric acid (GABA) as a neurotransmitter in the basal ganglia. Chapters 6 through 18 review our understanding of the state-of-the-art functioning of GABAergic circuits in the basal ganglia from a systems perspective comprising anatomy, physiology and function.

The basal ganglia are traditionally seen to be composed of four major nuclei: the neostriatum, the globus pallidus (GP), the substantia nigra (SN) and the subthalamic nucleus (STN). The neostriatum is a single nucleus in rodents but is divided by the internal capsule into the caudate nucleus and putamen in higher vertebrates. The GP consists of two major parts, the external segment (GPe) and the internal segment (GPi). The external segment is simply referred to as the GP in rodents and the internal segment is equivalent (in terms of inputs and outputs) to the entopeduncular nucleus (EP) in rodents. The two divisions of the pallidal complex have different inputs and outputs and are functionally distinct. Similarly, the SN consists of two major sub-nuclei, the pars compacta (SNc) and the pars reticulata (SNr). These two parts of the SN share similar inputs but have mostly different outputs and are composed of neurochemically distinct neuron types (Gerfen and Wilson, 1996). These divisions are referred to as the dorsal aspects of the basal ganglia; the ventral division consists of the nucleus accumbens, the ventral pallidum (which probably includes a ventral equivalent of the GPi) and the medial aspects of the STN and SN. The dorsal division of the basal ganglia is primarily associated with motor and associative functions whereas the ventral division is more related to limbic functions.

---

*Corresponding author. Tel.: +973-353-1080 × 3151;
Fax: +973-353-1588; E-mail: jtepper@andromeda.rutgers.edu

DOI: 10.1016/S0079-6123(06)60001-0

3

4

The overwhelming majority of neurons in the basal ganglia are GABAergic and most are projection neurons. The neostriatum (Chapter 6), both segments of the GP (Chapters 7 and 8) and the SNr (Chapter 9) are each principally composed of GABAergic projection neurons, whereas the STN (Chapter 10) contains glutamatergic projection neurons and the SNc is made up almost exclusively of dopaminergic projection neurons. The neostriatum also contains clearly defined populations of interneurons (Chapter 15), all but one of which (the cholinergic interneuron) are GABAergic. Not surprisingly, each of these nuclei expresses high levels of $GABA_A$ and $GABA_B$ receptors, both pre- and post-synaptically (Chapters 13 and 14). From the quantitative analyses by Oorschot (1996), we know that the dorsal components of the basal ganglia in rats contain a total of 2886.3×10^3 neurons but only 32.8×10^3 are not GABAergic (i.e., the neurons of the STN, SNc and the cholinergic interneurons in the neostriatum). Thus 98.86% of neurons in the basal ganglia are GABAergic. (Perhaps the title of this book should have been "GABA *is* the Basal Ganglia".)

The principal pathways for information flow through the basal ganglia, the *macrocircuitry*, are illustrated in Fig. 1, which is also, of course, a simplification of the true state of affairs. The principal afferents of the basal ganglia arise from the cerebral cortex (both ipsi- and contralateral), the intralaminar thalamic cell nuclei (e.g., centromedian and parafascicular nucleus), the dorsal raphé nucleus and the amygdala. The densest innervation, and by far the greatest in quantitative terms, is from the cerebral cortex and thalamus, and the major recipient of these afferents is the neostriatum. Together, the cortical and thalamic inputs form about 85% of all the synapses in neostriatum. These afferents are glutamatergic, which contain small round vesicles and form Gray's Type I (asymmetric) synapses. The major target of these afferents, accounting for nearly 90% of corticostriatal synapses, is the heads of spines of the principal neurons of the neostriatum, the medium-sized, densely-spiny neurons (Wilson, 2004). These are GABAergic neurons, account for as much as 95% of the neurons in neostriatum in rodents (and 75–80% in primates) and are the

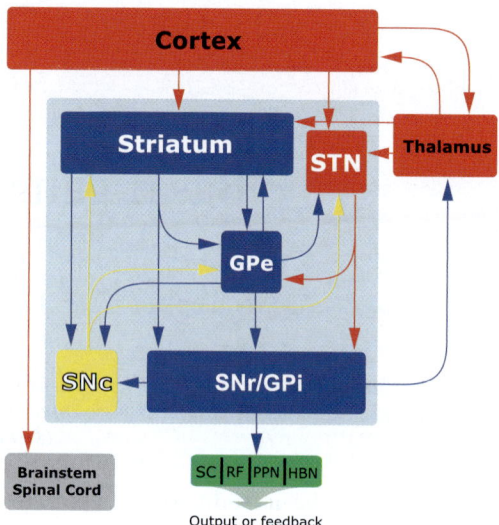

Fig. 1. Simplified diagram of the macrocircuits of the dorsal components of the basal ganglia. The nuclei of the basal ganglia are included in the light blue box and consist of the neostriatum (Striatum), the external segment of the globus pallidus (GPe), the subthalamic nucleus (STN), the substantia nigra pars reticulata and the internal segment of the globus pallidus (SNr/GPi) which together constitute the output nuclei of the basal ganglia and the substantia nigra pars compacta (SNc). The two major inputs to the basal ganglia are from the neocortex and the thalamus (mainly the intralaminar nuclei). The basal ganglia influence behaviour by the output nuclei projecting to the thalamus (mainly the ventral) and thence back to the cortex, and projections to the superior colliculus (SC), the reticular formation (RF), the pedunculopontine nucleus (PPN) and the lateral habenula (HBN). Dopamine neurons of the substantia nigra pars compacta (SNc) provide a massive feedback to the neostriatum and also the GPe and STN that modulates the flow of cortical and thalamic information through the basal ganglia. Dark blue indicates structures that are principally GABAergic; red indicates structures that are principally glutamatergic, yellow indicates structures that are dopaminergic and green indicates structures with variable neurochemistry.

principal projection neurons of the neostriatum, innervating the SN and pallidal complex. The spiny neurons emit a dense local axon collateral system that innervates other spiny cells and striatal interneurons (Chapter 6), influencing them in a complex way that has been subject to computational modelling (Chapter 18). Cortical and thalamic terminals also innervate the dendrites of the aspiny cholinergic interneuron and the GABAergic interneurons and in the latter neurons at least, cortical afferents may account for the majority of their afferent synapses (Bolam et al., 2000).

Projections arising from discrete areas of cortex have terminal fields that extend for great distances in neostriatum particularly in the rostro-caudal plane. They tend to cross over the dendrites of the spiny neuron at right angles ("cruciform axodendritic arrangement") with the result that each spiny neuron gets very few, at most only 1 or 2 synapses from a single corticostriatal neuron (Wilson, 2004). Similarly, thalamostriatal neurons extend their axons over wide regions of the neostriatum although the precise configuration varies with the thalamic nucleus of origin. The fact that the spiny neurons possess in the region of 15,000 spines implies that there is a massive degree of convergence of cortical and thalamic neurons at the level of individual spiny neurons in the neostriatum. The other major input to the neostriatum, from the SNc, is dopaminergic, and forms symmetric synapses with dendritic shafts or necks of spines of the spiny projection neurons, and, to a lesser extent, with other neostriatal cell types including the giant aspiny cholinergic neuron. Dopamine acts principally as a neuromodulator in neostriatum to powerfully modulate voltage-gated sodium, potassium and calcium channels in medium spiny neurons and cholinergic interneurons. This modulation leads directly to complex and state-dependent changes in neostriatal neuronal excitability (Surmeier, 2006). Dopamine also acts to modulate GABA release in the SN presynaptically (Chapter 14).

For the last 20 years or so, the macrocircuitry of the basal ganglia has been considered to be dominated by two principal pathways by which neocortical information is transmitted to the output nuclei of the basal ganglia, the SNr and GPi (Alexander and Crutcher, 1990; Delong, 1990; Smith et al., 1998). The so-called *direct* pathway comprises neostriatal GABAergic neurons that project directly to the SNr and GPi where they make direct synaptic contact with the GABAergic output neurons. Neurons giving rise to the direct pathway also give rise to collaterals to the GPe (Kawaguchi et al., 1990). The *indirect* pathway comprises neostriatal GABAergic neurons that project almost exclusively to the GABAergic

neurons of the GPe. These neurons, in turn, innervate the GABAergic output neurons in the SNr/GPi and also do so indirectly by innervating the glutamatergic neurons of the STN that then innervate the GABAergic output neurons in the SNr/GPi.

The neostriatal neurons giving rise to the direct and indirect pathways have very similar electrophysiological and morphological properties, but they have distinguishing neurochemical features: the direct pathway neurons express the dopamine D1 receptor and substance P and dynorphin whereas the indirect pathway neurons express the dopamine D2 receptor and enkephalin (Chapter 16). There is also a small proportion of neurons that expresses both sets of markers (Surmeier, 2006).

The GABAergic neurons of the GPe are in a unique position in that their extensive axon collaterals enable them to influence activity at every level of the basal ganglia (Chapter 7). Data from single-cell labelling studies (Kita and Kitai, 1994; Bevan et al., 1998; Sato et al., 2000) have demonstrated that all GPe projection neurons give rise to local axon collaterals and innervate the STN nucleus, the basal ganglia output nuclei and probably also the SNc. In each of these regions they make synapses with the cell bodies and proximal dendrites of their target neurons. In addition, about one-quarter to one-third of GPe neurons also innervate the neostriatum (Kita and Kitai, 1994; Bevan et al., 1998) where they are in a position to influence the activity of all neostriatal neurons by selectively innervating GABAergic interneurons (Bevan et al., 1998) which, in turn, innervate the spiny projection neurons (Kita, 1993; Bennett and Bolam, 1994; Koos and Tepper, 1999). Thus the role of GPe neurons is not to simply invert and transmit neostriatal information to the STN or basal ganglia output nuclei but rather, they are in a position to provide some sort of spatiotemporal selection of neurons at every level of the basal ganglia.

The second major port of entry of neocortical and thalamic information into the basal ganglia is the STN. This nucleus receives dense excitatory inputs from the motor and dorsal prefrontal cortices and the intralaminar thalamic nuclei. By virtue of the STN projection to the SNr/GPi, the

6

corticosubthalamic pathway (and probably the thalamosubthalamic pathway) is the fastest route by which cortical (and thalamic) information can influence activity in the output nuclei, and it has been referred to as the *hyperdirect* pathway (Kita, 1994). This pathway can be considered as a critical driving force in the basal ganglia output nuclei and in the GPe.

The output nuclei of the basal ganglia, the SNr and the GPi, are also composed essentially entirely of GABAergic projection neurons. These neurons project principally to the ventral tier of the dorsal thalamus and/or to the tectum where they exert powerful inhibition on their thalamic and tectal targets (Chapter 12). Many or most of the SNr/GPi neurons also co-express parvalbumin, calretinin or calbindin, but as in the neostriatum these neurochemically heterogeneous neurons exhibit very similar morphological and physiological properties. The SNr neurons also emit local axon collaterals that innervate the dopaminergic neurons of the SNc, thereby providing an important regulatory input to the nigrostriatal dopamine system (Tepper et al., 2002; Chapter 11).

It is thus evident that although the principal driving force of the basal ganglia, i.e., that derived from the cortex and thalamus, is glutamatergic, the majority of principal neurons in each division of the basal ganglia are GABAergic. Furthermore, every neuron in the basal ganglia is regulated by GABAergic inputs from multiple sources. The basal ganglia can thus be seen as a series or chain of GABAergic neurons that is regulated to a large extent by GABAergic inputs and that ultimately provide a GABAergic output to the basal ganglia targets.

Through these circuits the basal ganglia exert powerful control over voluntary movements and disease and/or damage to the basal ganglia result in well-defined movement disorders (Chapter 17). There is also a growing appreciation for the role of the basal ganglia in non-motor, higher-order cognitive functions such as learning (Graybiel, 2005). In a broader sense, it has been argued that the basal ganglia are part of a loop that supports thalamocortical interactions and thus cortical function by positive feedback (Wilson, 2004). Regardless of how one chooses to describe the basal ganglia's function, it is critical for our understanding of the basal ganglia to understand the organization, connectivity and properties of GABAergic neurons in each component of the system.

Acknowledgements

Supported in part by a grant from the National Institute for Neurological Diseases and Stroke (NS-34865), U.S.A. (JMT) and the Medical Research Council, U.K. (JPB). Many thanks to Ben Micklem for the preparation of the figure.

References

Alexander, G.E. and Crutcher, M.E. (1990) Functional architecture of basal ganglia circuits: neural substrates of parallel processing. Trends Neurosci., 13: 266–271.

Bennett, B.D. and Bolam, J.P. (1994) Synaptic input and output of parvalbumin-immunoreactive neurons in the neostriatum of the rat. Neurosci., 62: 707–719.

Bevan, M.D., Booth, P.A.C., Eaton, S.A. and Bolam, J.P. (1998) Selective innervation of neostriatal interneurons by a subclass of neuron in the globus pallidus of the rat. J. Neurosci., 18: 9438–9452.

Bolam, J.P., Hanley, J.J., Booth, P.A. and Bevan, M.D. (2000) Synaptic organisation of the basal ganglia. J. Anat., 196: 527–542.

Delong, M.R. (1990) Primate models of movement disorders of basal ganglia origin. Trends Neurosci., 13: 281–285.

Gerfen, C.R. and Wilson, C.J. (1996) The basal ganglia. In: Swanson L.W., Bjorklund A. and Hökfelt T. (Eds.), Handbook of Chemical Neuroanatomy, Vol. 12: Integrated Systems of the CNS, Part III. Elsevier Science BV, Amsterdam, pp. 371–468.

Graybiel, A.M. (2005) The basal ganglia: learning new tricks and loving it. Curr. Opin. Neurobiol., 15: 638–644.

Kawaguchi, Y., Wilson, C.J. and Emson, P.C. (1990) Projection subtypes of rat neostriatal matrix cells revealed by intracellular injection of biocytin. J. Neurosci., 10: 3421–3438.

Kita, H. (1993) GABAergic circuits of the striatum. Prog. Brain Res., 99: 51–72.

Kita, H. (1994) Physiology of two disynaptic pathways from the sensorimotor cortex to the basal ganglia output nuclei. In: Percheron G. (Ed.), The Basal Ganglia IV Adv. Behav. Biol., Vol. 41. Plenum Press, New York, pp. 263–276.

Kita, H. and Kitai, S.T. (1994) The morphology of globus pallidus projection neurons in the rat: an intracellular staining study. Brain Res., 636: 308–319.

Koos, T. and Tepper, J.M. (1999) Inhibitory control of neostriatal projection neurons by GABAergic interneurons. Nat. Neurosci., 2: 467–472.

Oorschot, D.E. (1996) Total number of neurons in the neostriatal, pallidal, subthalamic, and substantia nigral nuclei of the rat basal ganglia: A stereological study using the cavalieri and optical disector methods. J. Comp. Neurol., 366: 580–599.

Sato, F., Lavallee, P., Levesque, M. and Parent, A. (2000) Single-axon tracing study of neurons of the external segment of the globus pallidus in primate. J. Comp. Neurol., 417: 17–31.

Smith, Y., Bevan, M.D., Shink, E. and Bolam, J.P. (1998) Microcircuitry of the direct and indirect pathways of the basal ganglia. Neuroscience, 86: 353–387.

Surmeier, D.J. (2006) Microcircuits in the striatum: Cell types, intrinsic properties and Neuromodulation. In: Grillner S. and Graybiel A.M. (Eds.), Microcircuits — The Interface Between Neurons and Global Brain Function. MIT Press, Cambridge, pp. 105–112.

Tepper, J.M., Celada, P., Iribe, Y. and Paladini, C.A. (2002) Afferent control of nigral dopaminergic neurons — The role of GABAergic inputs. In: Graybiel A.M., DeLong M.R. and Kitai S.T. (Eds.), The Basal Ganglia VI, Adv. Behav. Biol., Vol. 54. Kluwer, Norwell, pp. 641–651.

Wilson, C.J. (2004) The basal ganglia. In: Shepherd G.M. (Ed.), The Synaptic Organization of the Brain. Oxford University Press, Oxford, pp. 361–414.

Tepper, Abercrombie & Bolam (Eds.)
Progress in Brain Research, Vol. 160
ISSN 0079-6123

CHAPTER 2

GABA: Homeostatic and pharmacological aspects

Arne Schousboe* and Helle S. Waagepetersen

Department of Pharmacology and Pharmacotherapy, The Faculty of Pharmaceutical Sciences, University of Copenhagen, DK-2100 Copenhagen, Denmark

Abstract: The central nervous system (CNS) operates by a fine-tuned balance between excitatory and inhibitory signalling. In this context, the inhibitory neurotransmission may be of particular interest as it has been suggested that such neuronal pathways may constitute 'command pathways' and the principle of 'disinhibition' leading ultimately to excitation may play a fundamental role (Roberts, E. (1974). Adv. Neurol., 5: 127–143). The neurotransmitter responsible for this signalling is γ-aminobutyrate (GABA) which was first discovered in the CNS as a curious amino acid (Roberts, E., Frankel, S. (1950). J. Biol. Chem., 187: 55–63) and later proposed as an inhibitory neurotransmitter (Curtis, D.R., Watkins, J.C. (1960). J. Neurochem., 6: 117–141; Krnjevic, K., Schwartz, S. (1967). Exp. Brain Res., 3: 320–336). The present review will describe aspects of GABAergic neurotransmission related to homeostatic mechanisms such as biosynthesis, metabolism, release and inactivation. Additionally, pharmacological and therapeutic aspects of this will be discussed.

Keywords: GAD; GABA-T; transport; inhibitors; epilepsy

GABA metabolism

Biosynthesis

In the first report (Roberts and Frankel, 1950) describing GABA as an interesting amino acid present in relatively large amounts in the brain it was demonstrated that the biosynthetic pathway consisted of a decarboxylation of L-glutamate catalysed by the enzyme L-glutamate decarboxylase (GAD). Subsequent studies led to a detailed characterization and purification of this enzyme (Roberts and Simonsen, 1963; Wu et al., 1973; Wu and Roberts, 1974). Further studies including cloning have revealed the presence of two distinct isoforms of GAD referred to as GAD_{65} and

GAD_{67} reflecting their molecular weights of 65 and 67 kDa, respectively (Martin and Rimval, 1993; Soghomonian and Martin, 1998). The two isoforms of GAD, which are encoded for by different genes, exhibit distinct differences with regard to regulation and subcellular localization in neurons. It appears that GAD_{65} mainly exists as a dormant apoenzyme, which may be rapidly activated by binding of the coenzyme, pyridoxal phosphate. This together with the demonstration that it is preferentially associated with nerve endings in GABAergic pathways makes it highly interesting in relation to biosynthesis of neurotransmitter GABA (Martin and Rimval, 1993; Waagepetersen et al., 2001). GAD_{67}, on the other hand, is present in the cytosol throughout the GABAergic neurons and it appears to exist mainly as the catalytically active holoenzyme having the coenzyme tightly bound (Martin and Rimval, 1993). Both isoforms

*Corresponding author. Tel.: +45 3530 6330;
Fax: +45 3530 6021; E-mail: as@dfuni.dk

DOI: 10.1016/S0079-6123(06)60002-2

9

are regulated by protein kinase-mediated phosphorylation, GAD_{65} being activated and GAD_{67} being inhibited (Martin and Rimval, 1993). Since the first immunocytochemical demonstrations of a selective localization of GAD in GABAergic neurons (Saito et al., 1974; McLaughlin et al., 1974; Ribak et al., 1976) it has been the general notion that GAD is only expressed in such neurons and hence, it has served as a marker for GABAergic neurons. Recent studies of the distribution of GAD particularly in hippocampal structures have, however, presented evidence that GAD is present also in certain glutamtergic neurons (Sloviter et al., 1996; Gutierrez, 2003). The functional implications of this are not well understood but it may be speculated that it could be related to the neurotrophic actions of GABA during early neuronal development as demonstrated in a number of studies (Belhage et al., 1998; Waagepetersen et al., 1999; Fiszman and Schousboe, 2004). In this context it may be of interest that it has recently been demonstrated that glutamatergic cerebellar granule neurons in culture are able to transport and concentrate GABA which is synthesized via GAD in and subsequently released by a small population of GABAergic neurons present in these cultures of dissociated cerebellum from 7-day-old mice (Sonnewald et al., 2004, 2006).

Although glutamate is the immediate precursor for GABA biosynthesis it has been demonstrated in different brain tissue preparations that glutamine is a more efficient exogenously supplied substrate (Reubi et al., 1978; Westergaard et al., 1995). This is in keeping with the concept that replenishment of GABA in vivo is accomplished by a combination of its reuptake in the presynaptic nerve ending (see below) and supply of glutamine from surrounding astrocytes by the operation of the GABA–glutamine–glutamate cycle (Waagepetersen et al., 2003). The synthesis of GABA from glutamine via glutamate requires the presence of phosphate-activated glutaminase (PAG) in GABAergic neurons, a prerequisite fulfilled by the demonstration of considerable PAG activity in such neurons (Larsson et al., 1985). It may, however, be noted that the two enzymes have different subcellular localizations, PAG being mitochondrial and GAD cytosolic (Balazs et al., 1966). Due to this mitochondria could play a functional role in GABA biosynthesis. Although for decades this was not investigated seriously, it has recently been demonstrated using ^{13}C-labelled glutamine as a precursor for GABA synthesis in GABAergic neurons that a large fraction ($>60\%$) of newly synthesized vesicular GABA originates from glutamate which has been produced in a fashion involving mitochondria and the tricarboxylic acid (TCA) cycle as illustrated in Fig. 1 (Waagepetersen et al., 2001, 2003). This may add regulatory mechanisms to the biosynthetic machinery involved in replenishment of vesicular GABA.

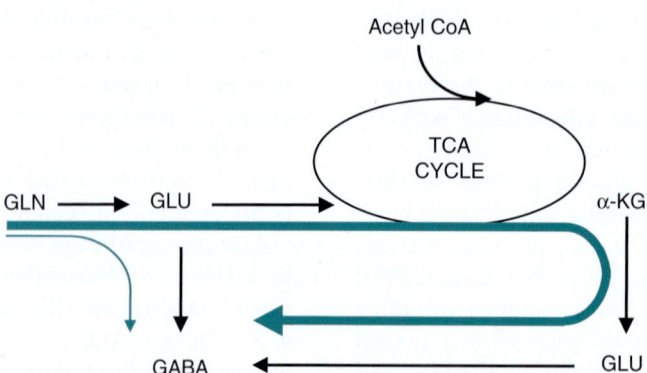

Fig. 1. An illustration of two pathways for GABA synthesis from glutamine. The bold arrow illustrates the predominant pathway where the carbon skeleton of glutamine is metabolized via the TCA cycle prior to synthesis of GABA. The thin arrow illustrates the direct synthesis of GABA from glutamine without involvement of the TCA cycle.

Degradation

The carbon skeleton of GABA is channelled into the TCA cycle as succinate by the concerted action of GABA-transaminase (GABA-T) and succinic acid semialdehyde dehydrogenase (Schousboe and Waagepetersen, 2007). Together with GAD this constitutes the GABA-shunt (Fig. 2), which represents an alternative route for converting α-ketoglutarate to succinate in the TCA cycle circumventing the succinyl-CoA step (Machiyama et al., 1970). This shunt is considered to account for about 10% of the flux through the TCA cycle in the brain (Balazs et al., 1970; Machiyama et al., 1970). It should be noted that this version of the TCA cycle yields slightly less ATP than the normally functioning TCA cycle due to the lack of succinyl-CoA production. The key enzyme of the

degradative part of the shunt is GABA-T, which is present in both neurons and astrocytes and in many other organs than the brain (Schousboe et al., 1977a; Wu et al., 1978; Larsson et al., 1985; Schousboe and Waagepetersen, 2007). This enzyme has been purified to homogeneity by several investigators and characterized with regard to K_m values for the substrates GABA and α-ketoglutarate (Schousboe et al., 1973, 1974; Cash et al., 1974; Bloch-Tardy et al., 1974; Maitre et al., 1975; John and Fowler, 1976). A high metabolic rate of conversion of GABA to CO_2 in situ is compatible with the low K_m values for both substrates (Machiyama et al., 1970; Yu and Hertz, 1983) and the fact that GABA-T is a mitochondrial enzyme associated with matrix (Schousboe et al., 1977b). It should be noted, however, that CO_2 production per se does not represent a net degradation of the C-4 unit derived from GABA (Fig. 2). Complete oxidation of the carbon skeleton of GABA requires pyruvate recycling, i.e., conversion of malate to pyruvate via malic enzyme and subsequently conversion of pyruvate to acetyl CoA and CO_2 by pyruvate dehydrogenase (Waagepetersen et al., 2003). This process, however, is only marginally present in neurons but does take place in astrocytes, making the latter cell type likely to be the major site for complete oxidative metabolism of GABA (Schousboe and Waagepetersen, 2006). This has some fundamental consequences for maintenance of optimal GABA-ergic activity in the CNS as discussed in detail recently (Schousboe et al., 2004a,b,c).

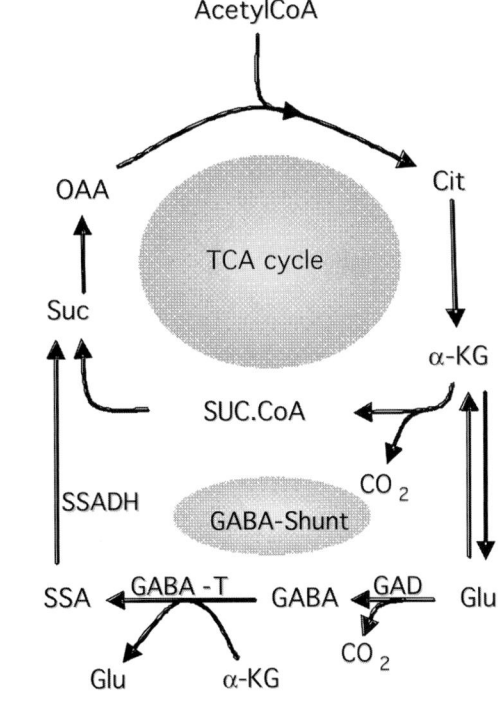

Fig. 2. Schematic representation of the TCA cycle and the GABA-shunt, which provides an alternative pathway for conversion of αKG to succinate. Abbreviations: αKG: α-ketoglutarate; SSADH: succinate-semialdehyde; GABA-T: GABA-transaminase; GABA: γ-aminobutyrate; GAD: glutamate decarboxylase; Glu: glutamate.

In vivo and ex vivo monitoring of GABA metabolism using NMR technology

GABA metabolism has been monitored using proton and ^{13}C NMR spectroscopy and different labelled substrates such as [1-^{13}C]glucose, [1,6-^{13}C]glucose and [2-^{13}C]acetate or [1-^{13}C]glucose in combination with [1,2-^{13}C]acetate in both normal animals and in various disease models (Cerdan et al., 1990; Haberg et al., 2001; Sonnewald and Kondziella, 2003; Patel et al., 2005). ^{13}C-labelled GABA was detected from [1-^{13}C]glucose in an in vivo ^{13}C NMR study in humans suggesting substantial GABA turnover (Gruetter et al., 1998). Glucose is metabolized in both neurons and astrocytes while acetate is a marker of astrocytic metabolism due to selective uptake into astrocytes (Waniewski and Martin, 1998). Thus, employing these substrates information regarding compartmentalized metabolism at the level of astrocytes and neurons can be obtained. Simultaneous injection of [1-^{13}C]glucose and [1,2-^{13}C]acetate into the same animal, glial metabolism is resolved from [4,5-^{13}C]glutamate and glutamine and [1,2-^{13}C]GABA is formed from [4,5-^{13}C]glutamine via transfer from the astrocytes to the GABAergic neurons and thus the GABA–glutamate–glutamine cycle activity can be estimated. On the other hand, [2-^{13}C]GABA derived from [2-^{13}C]acetyl CoA (formed from [1-^{13}C]glucose) reflects metabolism in the neuronal TCA-cycle (Haberg et al., 2001).

In an ex vivo study using [1-^{13}C]glucose or [2-^{13}C]acetate as substrates it was suggested that 3-NPA, which inhibits succinate dehydrogenase, somewhat selectively inhibits the TCA cycle of GABAergic neurons (Hassel and Sonnewald, 1995). Also vigabatrin which inhibits GABA-T in the neuronal compartment due to selective neuronal uptake (Schousboe et al., 1986; Pow et al., 1996) has been used in the monitoring of GABA metabolism in both in vivo and ex vivo experiments (Hassel et al., 1998; de Graaf et al., 2006; Patel et al., 2005, 2006). Evidence has been provided supporting the notion that the non-inhibited fraction of GABA-T, primarily localized in the astrocytes, is up-regulated due to a rise in the cellular GABA content upon acute vigabatrin treatment. Such explanation requires that GABA-T is not saturated in the astrocytic compartment. These considerations would explain an observed unchanged total GABA level, rate of GABA synthesis from glutamine and total GABA-T activity in vigabatrin treated animals (de Graaf et al., 2006).

The rate of the GAD-mediated GABA synthesis has been estimated in the range of 0.2–0.3 µmol/g per min (Hassel et al., 1998; Patel et al., 2005). It should be mentioned that in vivo NMR studies are performed under conditions of anaesthesia and the choice of drug has a significant impact on the results. For instance using α-chloralose the rate of

GAD was almost two-fold higher than that observed in rats exposed to halothane (Mason et al., 2001; Patel et al., 2005). GABA-shunt activity was estimated to be approximately 17% of total cerebral TCA cycle activity using [1-^{13}C]glucose in an ex vivo NMR spectroscopy study in mice (Hassel et al., 1998), a value being significantly higher than those discussed above.

The fraction of GABA which is taken up by the presynaptic neuron compared to that taken up by surrounding astrocytes has only been estimated from indirect measurements. From NMR studies it has been estimated that the GABA–glutamate–glutamine cycle accounts for approximately 23% of the total cycling of glutamine from astrocytes to neurons (Patel et al., 2005). The rate of the GABA–glutamate–glutamine cycle and the glucose oxidation rate of GABAergic neurons increase with cortical activity in line, although to a smaller extent, with that observed for glutamatergic neurons and the glutamate-glutamine cycle (Patel et al., 2005). It has previously been suggested that particularly GAD_{65} and not GAD_{67} may be associated with synthesis of neurotransmitter GABA during neuronal activity, mainly due to its localization in the nerve terminal (Kaufman et al., 1991). The importance of GAD_{65} for catalysing GABA synthesis during seizure was investigated in an experimental design using in vivo ^{1}H NMR spectroscopy in vigabatrin treated rats, a treatment which exhibits a selective inhibitory effect on GAD_{67} expression (Sheikh and Martin, 1998; Mason et al., 2001) and GAD_{65} was shown to be responsible for the majority of GABA synthesis during seizures (Patel et al., 2006). It was furthermore suggested that an increase in P_i, a known activator of $apoGAD_{65}$, during seizures might transform the large amount of dormant $apoGAD_{65}$ into the active holoform (de Graaf et al., 2006).

GABA release and uptake

Vesicular and non-vesicular release

In keeping with its role as a neurotransmitter, GABA is released from nervous tissue upon depolarisation in a Ca^{2+}-dependent manner (Curtis and Johnston, 1974; Schousboe et al., 1976; Otsuka, 1996). This represents vesicular release (Otsuka, 1996) which is the combined result of vesicular filling by transport of cytosolic GABA into vesicles via vesicular GABA transporters and subsequent depolarisation-coupled, Ca^{2+}-dependent fusion of the vesicular and plasma membranes (Otsuka, 1996).

It has been demonstrated in a number of neuronal preparations that depending upon the nature of the depolarising signal release of GABA may occur not only from vesicles but also by reversal of the GABA transporters (see below), i.e., release of GABA originating from the cytoplasmic pool of GABA (Pin and Bockaert, 1989; Bernath, 1992; Belhage et al., 1993). An important feature of this non-vesicular release is that it can be blocked by non-transportable GABA transport inhibitors such as diphenyl-butenyl-nipecotic acid and tiagabine di(methyl-thienyl)-butenyl-nipecotic acid (Belhage et al., 1993; Waagepetersen et al., 2001). An analogous release of GABA may also occur from glial cells (Minchin and Iversen, 1974). Altogether, such non-vesicular release of GABA occurring preferentially at extrasynaptic sites could be of considerable functional importance considering the recent demonstration that extrasynaptic GABA receptors may mediate tonic GABAergic inhibition (Mody, 2001; Krogsgaard-Larsen et al., 2004; Schousboe et al., 2004b).

Neuronal and glial transport

GABA uptake into inhibitory nerve terminals as well as glial elements was first reported using autoradiographic analysis of [3H]GABA uptake into brain slices (Hokfelt and Ljungdahl, 1970; Bloom and Iversen, 1971; Iversen and Bloom, 1972; Hösli and Hösli, 1976). Kinetic studies of GABA transport in a number of nervous tissue preparations including slices, homogenates, synaptosomes and bulk-prepared glial cells demonstrated high-affinity GABA transport systems (Iversen and Johnston, 1971; Beart and Johnston, 1973; Balcar and Johnston, 1973; Henn and Hamberger, 1971;

Levi and Raiteri, 1973). Subsequent kinetic studies in primary cultures of astrocytes and GABAergic neurons clearly confirmed that both of these cell types express high-affinity GABA transporters with K_m values around 10–20 μM (Schousboe et al., 1977a; Larsson et al., 1981). Numerous recent immunohistochemical studies using antibodies directed against the cloned GABA transporters (see below) have likewise demonstrated that both neurons (synapticly and extrasynapticly) as well as astroglial cells express a variety of GABA transporters (for references see Conti et al., 2004; Madsen et al., 2007).

Cloned transporters

The advent of the purification of a sodium and chloride dependent GABA transporting 80 kDa glycoprotein from rat brain synaptosomes (Radian et al., 1986) was instrumental in the subsequent first cloning of a rat brain GABA transporter (GAT1) having a K_m for GABA of 7 μM and exhibiting a Na^+ and Cl^- dependence (Guastella et al., 1990). Subsequently, two other GABA transporters with K_m values of 8 and 12 μM were cloned from rat brain (Borden et al., 1992) and referred to as GAT2 and GAT3. Four GABA transporters were almost simultaneously cloned from mouse brain and these were given the names GAT1, GAT2, GAT3 and GAT4 (Liu et al., 1992, 1993). One of these (GAT2) is homologous to the betaine transporter cloned from rat brain (BGT-1) by Yamauchi et al. (1992) and hence the nomenclature of the GABA transporters is somewhat confusing since rat GAT2 and rat GAT3 corresponds to mouse GAT3 and GAT4, respectively (Schousboe and Kanner, 2002). In the following, the mouse nomenclature will be used and therefore GAT2 is synonymous with BGT-1 (Schousboe and Kanner, 2002). It should be noted that the K_m for GABA of GAT2 is somewhat higher than that for the other three transporters (Bolvig et al., 1999). Nevertheless, as will be discussed below, this transporter may well be of considerable interest in relation to fine-tuning of GABAergic neurotransmission (Schousboe et al., 2004b,c).

Pharmacology of metabolism and uptake

Enzyme inhibitors

Inhibitors of the GABA synthesizing enzyme GAD mostly acting through the formation of covalent attachment to the carbonyl group of the coenzyme pyridoxal phosphate (e.g., formation of hydrazones) are always acting as convulsants since ultimately such action results in decreased GABA levels (Tapia, 1975). Since, so far no strategy has been found by which GAD activity can be enhanced, pharmacological intervention with GABA synthesis has not been a promising avenue to facilitate GABA neurotransmission. It should, however, be kept in mind that the GABA catabolic enzyme GABA-T, like GAD, is a pyridoxal-phosphate requiring enzyme (Schousboe et al., 1973). Therefore, inhibitors of GAD are generally also inhibitors of GABA-T (Tapia, 1975). In contrast to the problem with inhibitors of GAD acting as chemical convulsants (Tapia, 1975), inhibitors of GABA-T should theoretically increase GABA levels thereby acting as anti-convulsants (Tapia, 1975). Interestingly, aminooxyacetic acid has been shown to inhibit both GAD and GABA-T being 10-fold more potent as a GABA-T inhibitor (Wu and Roberts, 1974; Schousboe et al., 1974). Thus, the compound can act either as a convulsant or as an anti-convulsant drug depending on concentration (Tapia, 1975). It is therefore clear that if one could develop a carbonyl trapping agent which specifically would inhibit GABA-T leaving GAD activity intact, it might be possible to develop an anti-convulsant drug. Since GABA is a substrate for GABA-T with no affinity for GAD, it was used as a lead structure to develop an active site directed inhibitor of GABA-T by introducing an alkylating entity into the GABA molecule (Lippert et al., 1977). This led to development of γ-vinyl GABA and GABAculline, two highly specific suicide inhibitors of GABA-T (Lippert et al., 1977). These compounds when administered to animals or added to cultured GABAergic neurons are able to increase the synaptic GABA level significantly (Iadarola and Gale, 1980; Wood et al., 1981; Gram et al., 1988) which is likely to explain the anti-convulsive efficiency (Schechter et al., 1977). Gamma-vinyl GABA was subsequently developed into a clinically active drug (vigabatrin) used to treat certain types of epilepsy (Krämer, 2004).

Transport inhibitors

Inactivation of GABA as a neurotransmitter occurs exclusively by diffusion in the synaptic cleft and active transport into presynaptic nerve endings and astroglial cells ensheathing the synapse (for references see Schousboe and Waagepetersen, 2007). It is therefore clear that inhibition of such transport could lead to increased GABA levels in the synapse and thus to enhanced efficacy of inhibitory neurotransmission (e.g., Schousboe et al., 1983). As proposed by Schousboe et al. (1983) it would appear most attractive to selectively inhibit astroglial GABA transport as this would allow presynaptic transport to facilitate replenishment of vesicular GABA levels. A recent study using a series of astroglial GABA transport inhibitors as anti-convulsants in audiogenic seizure prone Fring's mice has demonstrated that this principle may well be correct emphasizing the important role of astroglial GABA transport in the regulation of GABAergic efficacy (White et al., 2002; Schousboe et al., 2004a).

Development of GABA transport inhibitors has been the topic of pharmacological characterization of the GABAergic system for more than 30 years. Hence, early studies led to the notion that β-alanine and diaminobutyric acid would act as selective inhibitors of glial and neuronal GABA transport, respectively (see Iversen and Kelly, 1975). However, particularly the concept of β-alanine being specific for glial GABA transport has been questioned partly because it acts as a substrate for the taurine carrier, which is predominantly expressed in glia (Larsson et al., 1986). Subsequent studies using the cloned GABA transporters have actually provided evidence that β-alanine preferentially inhibits GAT3 and GAT4 both of which are mainly expressed in glial cells (Borden, 1996).

An extensive characterization of the pharmacology of neuronal and glial GABA transport was initiated by the advent of the demonstration of THPO (4,5,6,7-tetrahydroisoxazolo[4,5-c]pyridin-3-ol) and its isostere nipecotic acid being active as GABA transport inhibitors with no activity on GABA receptors (Krogsgaard-Larsen and Johnston, 1975). Thus, in a series of studies using cultured astrocytes and neurons as well as the cloned transporters (GAT1–4) expressed in human embryonic kidney cells (HEK cells) a large number of GABA analogues of restricted conformation using nipecotic acid, THPO and *exo*-THPO (3-hydroxy-4-amino-4,5,6,7-tetrahydro-1,2-benzisoxazol) as the GABA mimetic entity have been characterized leading to the identification of drug candidates (for overview see Schousboe et al., 2004a; Clausen et al., 2006a,b; Madsen et al., 2007). Among these EF 1502 (*N*-[4,4-bis (3-methyl-2-thienyl)-3-butenyl]-4-(methyl-amino)-4,5,6,7-tetrahydrobenzo[*d*]isoxazol-3-ol) first described by Clausen et al. (2005) may be of particular interest as will be discussed below. As seen in Table 1, it represents an interesting GABA analogue among the series of compounds developed using the bicyclic isoxazole moity of THPO as a GABA molecular scaffold. It is the only compound of this very large library of GABA analogues (Falch et al., 1999; Clausen et al., 2005, 2006a,b), which potently inhibits both GAT1 and GAT2. All other compounds are GAT1 selective. The importance of this in relation to development of a second-generation clinically active anti-epileptic drugs is discussed in the following section.

Clinical applications

The development of the clinically active anti-epileptic drug tiagabine the action of which specifically relates to its inhibition of GABA transport (Suzdak and Jansen, 1995) has provided proof of principle that GABA transporters are indeed relevant drug targets for development of anti-epileptic drugs. Likewise, the application of vigabatrin in treatment of epilepsy shows that GABA-T is a suitable drug target. Recent attempts to develop new GABA analogues aimed at inhibiting GABA-T (Choi and Silverman, 2002) have, however, not led to development of clinically active drug candidates.

The demonstration that the glial selective GABA transport inhibitor *N*-methyl-*exo*THPO, being more than 100-fold less potent than tiagabine with regard to inhibition of GABA transport (Table 1) but essentially equipotent with tiagabine as an anti-convulsant (Table 1 and White et al., 2002) has provided evidence that glial GABA uptake may be an interesting drug target. Moreover, the finding that EF1502 which inhibits GAT2 and GAT1 equipotently not only is as potent as tiagabine as an anti-convulsant (White et al., 2005) but actually acts synergistically with tiagabine when administered in combination with this drug (White et al., 2005) provides new insights into the possibility that non-GAT1 inhibitors may be of clinical interest (Clausen et al., 2006a,b). This notion is further substantiated by the demonstration that

Table 1. Inhibitory parameters and anti-convulsant activity of selected GABA analogues

	Inhibition of uptake $IC_{50}/K_m/K_i$ (µM)[a]						Anti-convulsant activity ED_{50} (nmol icv.) (mg/kg i.p.)
	Neurons	Glia	GAT1	GAT2	GAT3	GAT4	
GABA	8	32	17	51	15	17	–
Nipecotic acid	12	16	24	>1000	113	159	–
THPO	501	262	1300	3000	800	5000	–
Exo-THPO	780	250	1000	3000	>3000	>3000	136
N-methyl-*exo*-THPO	405	48	450	>3000	>3000	>3000	59
EF 1500	4	2	3	>130	>100	>100	–
EF 1502	2	2	7	26	>300	>300	4.4
Tiagabine	0.5	0.2	0.1	>100	>100	>100	22

[a]Values for GABA uptake (K_m) and its inhibition (K_i/IC_{50}). Anti-convulsant activity of some of the analogues in Fring's audiogenic seizure susceptible mice.
Values cited from Larsson et al. (1981), Bolvig et al. (1999), White et al. (2002, 2005) and Sarup et al. (2003).

EF1502 is less toxic than tiagabine, i.e., it exhibits a more favourable therapeutic index (White et al., 2005).

Concluding remarks

Although the basic metabolic and homeostatic mechanisms governing functional aspects of GABA-mediated neurotransmission have been worked out decades ago, modern analytical technology such as NMR spectroscopy as well as cloning of enzymes, transporters and receptors have provided a wealth of information allowing a much more sophisticated and detailed knowledge about these matters to be obtained. Among other things it has become clear that the astrocytic entity of synapses has a profound importance for the regulation of GABAergic function. This has led to development of new therapeutic strategies by which neurological disorders involving disturbances of GABAergic activity may be treated.

Acknowledgements

The expert secretarial assistance of Ms. Hanne Danø is highly appreciated. The experimental work has been supported by grants from the Lundbeck, Hørslev and Benzon Foundations as well as the Danish Medical Research Council (22-03-0250 and 22-04-0314).

References

Balazs, R., Dahl, D. and Harwood, J.R. (1966) Subcellular distribution of enzymes of glutamate metabolism in rat brain. J. Neurochem., 13: 897–905.

Balazs, R., Machiyama, Y., Hammond, B.J., Julian, T. and Richter, D. (1970) The operation of the gamma-aminobutyrate bypath of the tricarboxylic acid cycle in brain tissue in vitro. Biochem. J., 116: 445–461.

Balcar, V.J. and Johnston, G.A. (1973) High affinity uptake of transmitters: studies on the uptake of L-aspartate, GABA, L-glutamate and glycine in cat spinal cord. J. Neurochem., 20: 529–539.

Beart, P.M. and Johnston, G.A. (1973) GABA uptake in rat brain slices: inhibition by GABA analogues and by various drugs. J. Neurochem., 20: 319–324.

Belhage, B., Hansen, G.H., Elster, E. and Schousboe, A. (1998) Effects of (γ-aminobutyric acid (GABA) on synaptogenesis and synaptic function. Perspec. Devel. Neurobiol., 5: 235–246.

Belhage, B., Hansen, G.H. and Schousboe, A. (1993) Depolarization by K^+ and glutamate activates different neurotransmitter release mechanisms in GABAergic neurons: vesicular versus non-vesicular release of GABA. Neuroscience, 54: 1019–1034.

Bernath, S. (1992) Calcium-independent release of amino acid neurotransmitters: fact or artifact? Prog. Neurobiol., 38: 57–91.

Bloch-Tardy, M., Rolland, B. and Gonnard, P. (1974) Pig brain 4-aminobutyrate 2-ketoglutarate transaminase. Purification, kinetics and physical properties. Biochimie., 56: 823–832.

Bloom, F.E. and Iversen, L.L. (1971) Localizing 3H-GABA in nerve terminals of rat cerebral cortex by electron microscopic autoradiography. Nature, 229: 628–630.

Bolvig, T., Larsson, O.M., Pickering, D.S., Nelson, N., Falch, E., Krogsgaard-Larsen, P. and Schousboe, A. (1999) Action of bicyclic isoxazole GABA analogues on GABA transporters and its relation to anticonvulsant activity. Eur. J. Pharmacol., 375: 367–374.

Borden, L.A. (1996) GABA transporter heterogeneity: pharmacology and cellular localization. Neurochem. Int., 29: 335–356.

Borden, L.A., Smith, K.E., Hartig, P.R., Branchek, T.A. and Weinshank, R.L. (1992) Molecular heterogeneity of the gamma-aminobutyric acid (GABA) transport system. Cloning of two novel high affinity GABA transporters from rat brain. J. Biol. Chem., 267: 21098–21104.

Cash, C., Maitre, M., Ciesielski, L. and Mandel, P. (1974) Purification and partial characterisatiohn of 4-aminobutyrate 2-ketoglutarate transaminase from human brain. FEBS Lett., 47: 199–203.

Cerdan, S., Kunnecke, B. and Seelig, J. (1990) Cerebral metabolism of [1,2–13C2]acetate as detected by in vivo and in vitro 13C NMR. J. Biol. Chem., 265: 12916–12926.

Choi, S. and Silverman, R.B. (2002) Inactivation and inhibition of gamma-aminobutyric acid aminotransferase by conformationally restricted vigabatrin analogues. J. Med. Chem., 45: 4531–4539.

Clausen, R.P., Madsen, K., Larsson, O.M., Frølund, B., Krogsgaard-Larsen, P. and Schousboe, A. (2006a) Structure-activity relationship and pharmacology of γ-aminobutyric acid (GABA) transport inhibitors. Adv. Pharmacol., 54: 265–284.

Clausen, R.P., Frølund, B., Larsson, O.M., Schousboe, A., Krogsgaard-Larsen, P. and White, H.S. (2006b) A novel selective γ-aminobutyric acid transport inhibitor demonstrates a functional role for GABA transporter subtype GAT2/BGT-1 in the CNS. Neurochem. Int., 48: 637–642.

Clausen, R.P., Moltzen, E.K., Perregaard, J., Lenz, S.M., Sanchez, C., Falch, E., Frølund, B., Sarup, A., Larsson, O.M., Schousboe, A. and Krogsgaard-Larsen, P. (2005) Selective inhibitors of GABA uptake: Synthesis and Molecular Pharmacology of 3-hydroxy-4-*N*-methylamino-4,5,6, 7-tetrahydro-1,2-benzo[*d*]isoxazole analogues. Bioorg. Med. Chem., 13: 895–908.

Conti, F., Minelli, A. and Melone, M. (2004) GABA transporters in the mammalian cerebral cortex: localization, development and pathological implications. Brain Res. Rev., 45: 196–212.

Curtis, D.R. and Johnston, G.A.R. (1974) Amino acid transmitters in the mammalian central nervous system. Rev. Physiol., 69: 97–188.

de Graaf, R.A., Patel, A.B., Rothman, D.L. and Behar, K.L. (2006) Acute regulation of steady-state GABA levels following GABA-transaminase inhibition in rat cerebral cortex. Neurochem. Int., 48: 508–514.

Falch, E., Perregaard, J., Frølund, B., Søkilde, B., Buur, A., Hansen, L.M., Frydenvang, K., Brehm, L., Bolvig, T., Larsson, O.M., Sanchez, C., White, H.S., Schousboe, A. and Krogsgaard-Larsen, P. (1999) Selective inhibitors of glial GABA uptake: Synthesis, absolute stereochemistry and pharmacology of the enantiomers of 3-hydroxy-4-amino-4,5,6,7-tetrahydro-1,2-benzisoxazole (Exo-THPO) and analogues. J. Med. Chem., 42: 5402–5414.

Fiszman, M.L. and Schousboe, A. (2004) Role of calcium and kinases on the neurotrophic effect induced by γ-aminobutyric acid. J. Neurosci. Res., 76: 435–441.

Gram, L., Larsson, O.M., Johnsen, A.H. and Schousboe, A. (1998) Effects of valproate, vigabatrin and aminooxyacetic acid on release of endogenous and exogenous GABA from cultured neurons. Epilepsy Res., 2: 87–95.

Gruetter, R., Seaquist, E.R., Kim, S. and Ugurbil, K. (1998) Localized in vivo 13C-NMR of glutamate metabolism in the human brain: initial results at 4 tesla. Dev. Neurosci., 20: 380–388.

Guastella, J., Nelson, N., Nelson, H., Czyzyk, L., Keynan, S., Miedel, M.C., Davidson, N., Lester, H.A. and Kanner, B.I. (1990) Cloning and expression of a rat brain GABA transporter. Science, 249: 1303–1306.

Gutierrez, R. (2003) The GABAergic phenotype of the glutamatergic granule cells of the dentate gyrus. Prog. Neurobiol., 71: 337–358.

Haberg, A., Qu, H., Saether, O., Unsgard, G., Haraldseth, O. and Sonnewald, U. (2001) Differences in neurotransmitter synthesis and intermediary metabolism between glutamatergic and GABAergic neurons during 4 hours of middle cerebral artery occlusion in the rat: the role of astrocytes in neuronal survival. J. Cereb. Blood Flow Metab, 21: 1451–1463.

Hassel, B., Johannessen, C.U., Sonnewald, U. and Fonnum, F. (1998) Quantification of the GABA shunt and the importance of the GABA shunt versus the 2-oxoglutarate dehydrogenase pathway in GABAergic neurons. J. Neurochem., 71: 1511–1518.

Hassel, B. and Sonnewald, U. (1995) Selective inhibition of the tricarboxylic acid cycle of GABAergic neurons with 3-nitropropionic acid in vivo. J. Neurochem., 65: 1184–1191.

Henn, F.A. and Hamberger, A. (1971) Glial cell function: Uptake of transmitter substances. Proc. Natl. Acad. Sci. USA, 68: 2686–2690.

Hokfelt, T. and Ljungdahl, A. (1970) Cellular localization of labeled gamma-aminobutyric acid (3H-GABA) in rat cerebellar cortex: an autoradiographic study. Brain Res., 22: 391–396.

Hösli, E. and Hösli, L. (1976) Autoradiographic studies on the uptake of ^{3}H-Noradrenaline and ^{3}H-GABA in cultured rat cerebellum. Exp. Brain Res., 26: 319–324.

Iadarola, M.J. and Gale, K. (1980) Evaluation of increases in nerve terminal-dependent vs nerve terminal-independent compartments of GABA in vivo. Brain Res. Bull., 5(Suppl 2): 13–19.

Iversen, L.L. and Bloom, F.E. (1972) Studies of the uptake of ^{3}H-GABA and (^{3}H)glycine in slices and homogenates of rat brain and spinal cord by electron microscopic autoradiography. Brain Res., 41: 131–143.

Iversen, L.L. and Johnston, G.A. (1971) GABA uptake in rat central nervous system: comparison of uptake in slices and homogenates and the effects of some inhibitors. J. Neurochem., 18: 1939–1950.

Iversen, L.L. and Kelly, J.S. (1975) Uptake and metabolism of gamma-aminobutyric acid by neurones and glial cells. Biochem. Pharmacol., 24: 933–938.

John, R.A. and Fowler, L.J. (1976) Kinetic and spectral properties of rabbit brain 4-aminobutyrate aminotransferase. Biochem. J., 155: 645–651.

Kaufman, D.L., Houser, C.R. and Tobin, A.J. (1991) Two forms of the gamma-aminobutyric acid synthetic enzyme glutamate decarboxylase have distinct intraneuronal distributions and cofactor interactions. J. Neurochem., 56: 720–723.

Krämer, G. (2004) Vigabatrin. In: Shorvon S., Perucca E., Fish D. and Dodson E. (Eds.), Treatment of Epilepsy (Sec Edn). Blackwell Science, Oxford, UK, pp. 540–547.

Krogsgaard-Larsen, P., Frolund, B., Liljefors, T. and Ebert, B. (2004) GABA(A) agonists and partial agonists: THIP (Gaboxadol) as a non-opioid analgesic and a novel type of hypnotic. Biochem. Pharmacol., 68: 1573–1580.

Krogsgaard-Larsen, P. and Johnston, G.A. (1975) Inhibition of GABA uptake in rat brain slices by nipecotic acid, various isoxazoles and related compounds. J. Neurochem., 25: 797–802.

Larsson, O.M., Drejer, J., Kvamme, E., Svenneby, G., Hertz, L. and Schousboe, A. (1985) Ontogenetic development of glutamate and GABA metabolizing enzymes in cultured cerebral cortex interneurons and in cerebral cortex in vivo. Int. J. Devl. Neurosci., 3: 177–185.

Larsson, O.M., Griffiths, R., Allen, I.C. and Schousboe, A. (1986) Mutual inhibition kinetic analysis of (γ-aminobutyric acid, taurine, taurine and β-alanine high affinity transport into neurons and astrocytes: Evidence for similarity between the taurine and β-alanine carriers in both cell types. J. Neurochem., 47: 426–432.

Larsson, O.M., Thorbek, P., Krogsgaard-Larsen, P. and Schousboe, A. (1981) Effect of homo-β-proline and other heterocyclic GABA analogues on GABA uptake in neurons and astroglial cells and on GABA receptor binding. J. Neurochem., 37: 1509–1516.

Levi, G. and Raiteri, M. (1973) GABA and glutamate uptake by subcellular fractions enriched in synaptosomes: critical

18

evaluation of some methodological aspects. Brain Res., 57: 165–185.

Lippert, B., Metcalf, B.W., Jung, M.J. and Casara, P. (1977) 4-Amino-Hex-5-Enoic Acid, a Selective Catalytic Inhibitor of 4-Aminobutyric-Acid Aminotransferase in Mammalian Brain. Eur. J. Biochem., 74: 441–445.

Liu, Q.R., Lopez-Corcuera, B., Mandiyan, S., Nelson, H. and Nelson, N. (1993) Molecular characterization of four pharmacologically distinct gamma-aminobutyric acid transporters in mouse brain (corrected). J. Biol. Chem., 268: 2106–2112.

Liu, Q.R., Mandiyan, S., Nelson, H. and Nelson, N. (1992) A family of genes encoding neurotransmitter transporters. Proc. Natl. Acad. Sci. USA, 89: 6639–6643.

Machiyama, Y., Balázs, R., Hammond, B.J., Julian, T. and Richter, D. (1970) The metabolism of γ-aminobutyrate and glucose in potassium ion-stimulated brain tissue in vitro. Biochem. J., 116: 469–481.

Madsen, K.B., Larsson, O.M. and Schousboe, A. (2007) Regulation of excitation by GABA neurotransmission: Focus on metabolism and transport. In: Darlison M. (Ed.), Inhibitory Regulation of Excitatory Neurotransmission. Springer Verlag, Berlin, in press.

Maitre, M., Ciesielski, L., Cash, C. and Mandel, P. (1975) Purification and studies on some properties of the 4-aminobutyrate: 2-oxoglutarate transaminase from rat brain. Eur. J. Biochem., 52: 157–169.

Martin, D.L. and Rimval, K. (1993) Regulation of γ-aminobutyric acid synthesis in the brain. J. Neurochem., 60: 395–407.

Mason, G.F., Martin, D.L., Martin, S.B., Manor, D., Sibson, N.R., Patel, A., Rothman, D.L. and Behar, K.L. (2001) Decrease in GABA synthesis rate in rat cortex following GABA-transaminase inhibition correlates with the decrease in GAD(67) protein. Brain Res., 914: 81–91.

McLaughlin, B.J., Wood, J.G., Saito, K., Barber, R., Vaughn, J.E., Roberts, E. and Wu, J.Y. (1974) The fine structural localization of glutamate decarboxylase in synaptic terminals of rodent cerebellum. Brain Res., 76: 377–391.

Minchin, M.C.W. and Iversen, L.L. (1974) Release of [^{3}H]gamma-aminobutyric acid from glial cells in rat dorsal root ganglia. J. Neurochem., 23: 535–540.

Mody, I. (2001) Distinguishing between GABA(A) receptors responsible for tonic and phasic conductances. Neurochem. Res., 26: 907–913.

Otsuka, M. (1996) Establishment of GABA as an inhibitory neurotransmitter at Crustacean neuromuscular junction and in the mammalian central nervous system. In: Tanaka C. and Bowery N.G. (Eds.), GABA: Receptors, transporters and Metabolism. Birkhäuser, Basel, Switzerland, pp. 1–6.

Patel, A.B., de Graaf, R.A., Martin, D.L., Battaglioli, G. and Behar, K.L. (2006) Evidence that GAD65 mediates increased GABA synthesis during intense neuronal activity in vivo. J. Neurochem., 97: 385–396.

Patel, A.B., de Graaf, R.A., Mason, G.F., Rothman, D.L., Shulman, R.G. and Behar, K.L. (2005) The contribution of GABA to glutamate/glutamine cycling and energy

metabolism in the rat cortex in vivo. Proc. Natl. Acad. Sci. U.S.A, 102: 5588–5593.

Pin, J.P. and Bockaert, J. (1989) Two distinct mechanisms, differentially affected by excitatory amino acids, trigger GABA release from fetal mouse striatal neurons in primary culture. J. Neurosci., 9: 648–656.

Pow, D.V., Baldridge, W. and Crook, D.K. (1996) Activity-dependent transport of GABA analogues into specific cell types demonstrated at high resolution using a novel immunocytochemical strategy. Neuroscience, 73: 1129–1143.

Radian, R., Bendahan, A. and Kanner, B.I. (1986) Purification and identification of the functional sodium- and chloride-coupled gamma-aminobutyric acid transport glycoprotein from rat brain. J. Biol. Chem., 261: 15437–15441.

Reubi, J.-C., Van der Berg, C. and Cuénod, M. (1978) Glutamine as precursor for the GABA and glutamate transmitter pools. Neurosci. Lett., 10: 171–174.

Ribak, C.E., Vaughn, J.E., Saito, K., Barber, R. and Roberts, E. (1976) Immunocytochemical localization of glutamate decarboxylase in rat substantia nigra. Brain Res., 116: 287–298.

Roberts, E. and Frankel, S. (1950) γ-Aminobutyric acid in brain: its formation from glutamic acid. J. Biol. Chem., 187: 55–63.

Roberts, E. and Simonsen, D.G. (1963) Some properties of L-glutamic decarboxylase in mouse brain. Biochem. Pharmacol., 12: 113–134.

Saito, K., Barber, R., Wu, J., Matsuda, T., Roberts, E. and Vaughn, J.E. (1974) Immunohistochemical localization of glutamate decarboxylase in rat cerebellum. Proc. Natl. Acad. Sci. USA, 71: 269–273.

Sarup, A., Larsson, O.M. and Schousboe, A. (2003) GABA transporters and GABA-transaminase as drug targets. Curr. Drug Targ. CNS Neurol. Dis., 2: 269–277.

Schechter, P.J., Tranier, Y., Jung, M.J. and Bohlen, P. (1977) Audiogenic seizure protection by elevated brain GABA concentration in mice: effects of gamma-acetylenic gaba and gamma-vinyl GABA, two irreversible GABA-T inhibitors. Eur. J. Pharmacol., 45: 319–328.

Schousboe, A. and Kanner, B. (2002) GABA transporters: functional and pharmacological properties. In: Egebjerg J., Schousboe A. and Krogsgaard-Larsen P. (Eds.), Glutamate and GABA Receptors and Transporters. Taylor & Francis Publisher, London, UK, pp. 337–349.

Schousboe, I., Bro, B. and Schousboe, A. (1977b) Intramitochondrial localization of the 4-aminobutyrate-2-ketoglutarate transaminase from ox brain. Biochem. J., 162: 303–307.

Schousboe, A., Larsson, O.M., Sarup, A. and White, H.S. (2004c) Role of the betaine/GABA transporter (BGT-1/GAT2) for the control of epilepsy. Eur. J. Pharmacol., 500: 281–287.

Schousboe, A., Larsson, O.M. and Seiler, N. (1986) Stereoselective uptake of the GABA-transaminase inhibitors gamma-vinyl GABA and gamma-acetylenic GABA into neurons and astrocytes. Neurochem. Res., 11: 1497–1505.

Schousboe, A., Larsson, O.M., Wood, J.D. and Krogsgaard-Larsen, P. (1983) Transport and metabolism of GABA in neurons and glia: Implications for epilepsy. Epilepsia, 24: 531–538.

Schousboe, A., Lisy, V. and Hertz, L. (1976) Postnatal alterations in effects of potassium on uptake and release of glutamate and GABA in rat brain cortex slices. J. Neurochem., 26: 1023–1027.

Schousboe, A., Sarup, A., Bak, L.K., Waagepetersen, H.S. and Larsson, O.M. (2004a) Role of astrocytic transport processes in glutamatergic and GABAergic neurotransmission. Neurochem. Int., 45: 512–527.

Schousboe, A., Sarup, A., Larsson, O.M. and White, H.S. (2004b) GABA transporters as drug targets for modulation of GABAergic activity. Biochem. Pharmacol., 68: 1557–1563.

Schousboe, A., Svenneby, G. and Hertz, L. (1977a) Uptake and metabolism of glutamate in astrocytes cultured from dissociated mouse brain hemispheres. J. Neurochem., 29: 999–1005.

Schousboe, A. and Waagepetersen, H.S. (2007) GABA neurotransmission: An overview. In: Lajtha A. et al. (Eds.), Handbook of Neurochem. and Molec. Neurobiol. 3rd ed., Springer, New York, in press.

Schousboe, A., Wu, J.-Y. and Roberts, E. (1973) Purification and characterization of the 4-aminobutyrate-2-ketogluterate transaminase from mouse brain. Biochemistry, 12: 2868–2873.

Schousboe, A., Wu, J.-Y. and Roberts, E. (1974) Subunit structure and kinetic properties of 4-aminobutyrate-2-ketoglutarate transaminase from mouse brain. J. Neurochem., 23: 1189–1195.

Sheikh, S.N. and Martin, D.L. (1998) Elevation of brain GABA levels with vigabatrin (gamma-vinylGABA) differentially affects GAD65 and GAD67 expression in various regions of rat brain. J. Neurosci. Res., 52: 736–741.

Sloviter, M., Dichter, M.A., Rachinsky, T.L., Dean, E., Goodman, J.H., Sollas, A.L. and Martin, D.L. (1996) Basal expression and induction of glutamate decarboxylase and GABA in excitatory granule cells of the rat and monkey hippocampal dentate gyrus. J. Comp. Neurol., 373: 593–618.

Soghomonian, J.J. and Martin, D.L. (1998) Two isoforms of glutamate decarboxylase: why? Trends Pharmacol. Sci., 19: 500–505 (Review).

Sonnewald, U. and Kondziella, D. (2003) Neuronal glial interaction in different neurological diseases studied by ex vivo 13C NMR spectroscopy. NMR Biomed., 16: 424–429.

Sonnewald, U., Kortner, T.M., Qu, H., Olstad, E., Sunol, C., Bak, L.K., Schousboe, A. and Waagepetersen, H. (2006) Demonstration of extensive GABA synthesis in a small population of GAD positive neurons in cerebellar cultures by the use of pharmacological tools. Neurochem. Int., 48: 572–578.

Sonnewald, U., Olstad, E., Qu, H., Babot, Z., Cristòfol, R., Sunol, C., Schousboe, A. and Waagepetersen, H.S. (2004) First direct demonstration of extensive GABA synthesis in mouse cerebellar neuronal cultures. J. Neurochem., 91: 796–803.

Suzdak, P.D. and Jansen, J.A. (1995) A review of the preclinical pharmacology of tiagabine: a potent and selective anticonvulsant GABA uptake inhibitor. Epilepsia, 36: 612–626.

Tapia, R. (1975) Biochemical pharmacology of GABA in CNS. Handbook Psychopharmacol., 4: 1–58.

Waagepetersen, H.S., Sonnewald, U., Gegelashvili, G., Larsson, O.M. and Schousboe, A. (2001) Metabolic distinction between vesicular and cytosolic GABA in cultured GABAergic neurons using ^{13}C MRS. J. Neurosci. Res., 63: 347–355.

Waagepetersen, H.S., Sonnewald, U. and Schousboe, A. (1999) The GABA paradox: Multiple roles as metabolite, neurotransmitter, and neurodifferentiative agent. J. Neurochem., 73: 1335–1342.

Waagepetersen, H.S., Sonnewald, U. and Schousboe, A. (2003) Compartmentation of glutamine, glutamate and GABA metabolism in neurons and astrocytes: functional implications. Neuroscientist, 9: 398–403.

Waniewski, R.A. and Martin, D.L. (1998) Preferential utilization of acetate by astrocytes is attributable to transport. J. Neurosci., 18: 5225–5233.

Westergaard, N., Sonnewald, U., Petersen, S.B. and Schousboe, A. (1995) Glutamate and glutamine metabolism in cultured GABAergic neurons studied by ^{13}C NMR spectroscopy: Evidence for compartmentation and mitochondrial heterogeneity. Neurosci. Lett., 185: 24–28.

White, H.S., Sarup, A., Bolvig, T., Kristensen, A.S., Petersen, G., Nelson, N., Pickering, D.S., Larsson, O.M., Frølund, B., Krogsgaard-Larsen, P. and Schousboe, A. (2002) Correlation between anticonvulsant activity and inhibitory action on glial GABA uptake of the highly selective mouse GAT1 inhibitor 3-hydroxy-4-amino-4,5,6,7-tetrahydro-1,2-benzisoxazole (exo-THPO) and its N-alkylated analogs. J. Pharmacol. Exp. Therap., 302: 636–644.

White, H.S., Watson, W.P., Hansen, S., Slough, S., Sarup, A., Bolvig, T., Petersen, G., Larsson, O.M., Clausen, R.P., Frølund, B., Krogsgaard-Larsen, P. and Schousboe, A. (2005) First demonstration of a functional role for CNS betaine/GABA transporter (mGAT2) based on synergistic anticonvulsant action among inhibitors of mGAT1 and mGAT2. J. Pharmacol. Exp. Therap., 312: 866–874.

Wood, J.D., Kurylo, E. and Tsui, S.K. (1981) Interactions of di-n-propylacetate, gabaculine, and aminooxyacetic acid: anticonvulsant activity and the gamma-aminobutyrate system. J. Neurochem., 37: 1440–1447.

Wu, J.Y., Matsuda, T. and Roberts, E. (1973) Purification and characterization of glutamate decarboxylase from mouse brain. J. Biol. Chem., 248: 3029–3034.

Wu, J.Y., Moss, L.G. and Chude, O. (1978) Distribution and tissue specificity of 4-aminobutyrate-2-oxoglutarate aminotransferase. Neurochem. Res., 3: 207–219.

Wu, J.Y. and Roberts, E. (1974) Properties of brain L-glutamate decarboxylase: inhibition studies. J. Neurochem., 23: 759–767.

Yamauchi, A., Uchida, S., Kwon, H.M., Preston, A.S., Robey, R.B., Garcia-Perez, A., Burg, M.B. and Handler, J.S. (1992) Cloning of a Na(+)- and Cl(−)-dependent betaine transporter that is regulated by hypertonicity. J. Biol. Chem., 267: 649–652.

Yu, A.C.H. and Hertz, L. (1983) Metabolic sources of energy in astrocytes. In: Hertz L., Kvamme E., McGeer E.G. and Schousboe A. (Eds.), Glutamine, Glutamate and GABA in the Central Nervous System. Alan R. Liss, Inc, New York, pp. 431–438.

Tepper, Abercrombie & Bolam (Eds.)
Progress in Brain Research, Vol. 160
ISSN 0079-6123

CHAPTER 3

GABA$_A$ receptors: structure and function in the basal ganglia

T. Goetz[1], A. Arslan[1], W. Wisden[2] and P. Wulff[1,*]

[1]*Department of Clinical Neurobiology, University of Heidelberg, Im Neuenheimer Feld 364, 69120 Heidelberg, Germany*
[2]*Institute of Medical Sciences, University of Aberdeen, Foresterhill, Aberdeen AB25 2ZD, Scotland, UK*

Abstract: γ-Aminobutyric acid type A (GABA$_A$) receptors, the major inhibitory neurotransmitter receptors responsible for fast inhibition in the basal ganglia, belong to the superfamily of "cys–cys loop" ligand-gated ion channels. GABA$_A$ receptors form as pentameric assemblies of subunits, with a central Cl$^-$ permeable pore. On binding of two GABA molecules to the extracellular receptor domain, a conformational change is induced in the oligomer and Cl$^-$, in most adult neurons, moves into the cell leading to an inhibitory hyperpolarization. Nineteen mammalian subunit genes have been identified, each showing distinct regional and cell-type-specific expression. The combinatorial assembly of the subunits generates considerable functional diversity. Here we place the focus on GABA$_A$ receptor expression in the basal ganglia: striatum, globus pallidus, substantia nigra and subthalamic nucleus, where, in addition to the standard $\alpha1\beta2/3\gamma2$ receptor subtype, significant levels of other subunits ($\alpha2$, $\alpha3$, $\alpha4$, $\gamma1$, $\gamma3$ and δ) are expressed in some nuclei.

Keywords: GABA; GABA$_A$ receptor; basal ganglia; striatum; globus pallidus; substantia nigra; benzodiazepines

GABA$_A$ receptors are essential for the function of the entire basal ganglia network, providing fast (millisecond) synaptic as well as tonic extrasynaptic inhibition within and between the various basal ganglia nuclei (reviewed in Smith et al., 1998; Misgeld, 2004; Tepper and Bolam, 2004). As for other brain regions, different neuronal subtypes within the basal ganglia employ different GABA$_A$ receptor subtypes. Here we first review the genetics and structure of GABA$_A$ receptor subtypes; we then consider key drugs that act on the receptors to enhance or decrease GABAs actions; finally, we summarize which receptor subunit combinations are expressed in the various nuclei of the basal ganglia.

GABA$_A$ receptors are GABA-gated anion channels responsible (together with ligand-gated glycine receptors) for most fast inhibitory synaptic transmission in the vertebrate central nervous system. GABA$_A$ receptors are permeable to HCO$_3^-$ and Cl$^-$ ions; the permeability ratio of HCO$_3^-$/Cl$^-$ is approximately 0.2 to 0.4 (reviewed in Kaila et al., 1997). HCO$_3^-$ moves out of the cell causing a mild depolarization (the reversal potential for HCO$_3^-$ is $-12\,mV$). In mature neurons Cl$^-$ usually moves into the cell overriding this mild depolarization, causing a strong inhibitory hyperpolarization, as the Cl$^-$ reversal potential is 15–20 mV more negative than the resting membrane potential. The Cl$^-$ gradient is maintained by K–Cl co-transporters (Rivera et al., 2005). Depending on the intracellular Cl$^-$ concentration, GABA$_A$ receptor activation

---

*Corresponding author. Tel.: +0044-1224-551941; Fax: +0044-1224-555719; E-mail: peer.wulff@urz.uni-heidelberg.de

DOI: 10.1016/S0079-6123(06)60003-4

21

can also lead to Cl⁻ efflux and depolarization. This is the case, for example, during embryonic and early postnatal development when K–Cl co-transporters are not expressed at sufficient levels to efficiently transport Cl⁻ out of the cell (Rivera et al., 2005). There are interesting caveats: adult dopaminergic neurons in the substantia nigra pars compacta have little KCC2 expression (Gulasci et al., 2003), possibly explaining the relatively low efficacy of GABA$_A$ receptor-mediated inhibition in nigral dopaminergic neurons (Gulasci et al., 2003). Further, KCC2 expression can vary in subdomains of neurons, thus affecting local Cl⁻ gradients. KCC2 is absent from the axon initial segments of neocortical pyramidal cells (Szabadics et al., 2006). Thus GABAergic terminals arriving at this location may produce depolarization via GABA$_A$ receptors in this context. In hippocampal neurons the dendritic KCC2 channels can also be transiently inhibited by Ca^{2+} entry through voltage-gated Ca^{2+} channels, thus producing local changes in the dendritic Cl⁻ gradient and affecting the efficacy of GABA$_A$ receptor inhibition, and possibly inducing plasticity at GABAergic synapses (Fiumelli et al., 2005). This is a potential mechanism to bear in mind when considering GABAergic function in the basal ganglia.

GABA$_A$ receptors: genes

In mammals, GABA$_A$ receptors form as hetero-pentameric assemblies from a family of 19 subunits encoded by distinct genes (α1–α6, β1–β3, γ1–γ3, δ, ε, θ, π and ρ1–ρ3) (Korpi et al., 2002a; Rudolph and Moehler, 2006; Whiting, 2006). Depending on the subunit composition GABA$_A$ receptors differ in their biophysical properties and affinity for GABA (see Section "GABA$_A$ receptors: how subunit combinations affect synaptic and extrasynaptic transmission" below), their pharmacology (see Section "GABA$_A$ receptor agonists, antagonists and allosteric modulators" below) and location on the cell (see Section "Extrasynaptic GABA$_A$ receptors: α4βδ subtype" below). Along with the closely related glycine receptors, GABA$_A$ receptors were originally cloned by the classical tour-de-force method: peptide sequences obtained from purified

(bovine brain) receptors were used to construct synthetic DNA probes to screen brain cDNA libraries (Grenningloh et al., 1987; Schofield et al., 1987). This was the starting point. Within a few years, this now historical technique of screening cDNA libraries had revealed most of the gene family, all the α1–α6, β1–β3, γ1–γ3 subunits and one δ subunit (Seeburg et al., 1990); over the remaining decade, a few more subunits, such as ε, θ and π were characterized (Davies et al., 1997; Hedblom and Kirkness, 1997; Bonnert et al., 1999; Sinkkonen et al., 2000). With the completion of the human genome database, Simon et al. (2004) did an in silico hybridization screen, searching for further undescribed mammalian GABA$_A$ receptor genes but found none. Most of the subunit gene family members are in clusters (Simon et al., 2004), suggesting gene and then cluster duplication during the evolutionary origin of vertebrates: β2, α6, α1, γ2 form a cluster in that order on human chromosome 5q34; the β3, α5, γ3 genes cluster in that order on human chromosome 15q13; the γ1, α2, α4, β1 genes cluster in that order on chromosome 4p12; the ε, α3, θ genes cluster in that order on Xq28; the ρ1 and ρ2 genes are 40 Kb apart on 6q15; the π, ρ3 and δ subunit genes are isolated on human chromosomes 5q35.1, 3q12.1 and 1p36.3 respectively (Simon et al., 2004). The complete genome data makes it an easy task to see the gene cluster organizations at a few keyboard strokes (http://www.ensembl.org/Homo_sapiens/index.html).

As determined by both in situ hybridization (mRNA localization) with gene-specific probes and immunocytochemistry (protein localization) with subunit-specific antibodies, the expression of the individual subunit genes is age- and region-specific (Laurie et al., 1992a, b; Wisden et al., 1992; Fritschy and Mohler, 1995; Schwarzer et al., 2001). Some GABA$_A$ receptor subunit genes have extremely restricted expression patterns; the α6 subunit gene expresses only in cerebellar and cochlear nucleus granule cells (Luddens et al., 1990), the ρ subunit genes are mainly expressed in retina with low transcript levels in the hippocampus and colliculi — these receptors, because of their unique pharmacology used to be termed as "GABA$_C$"; the π gene is expressed in non-neural tissues (Hedblom and Kirkness, 1997). The ε and θ subunit genes are mainly transcribed in the locus ceruleus (the adrenergic nucleus in the brainstem), dorsal raphe (serotonergic cells) and cholinergic cells (Sinkkonen et al., 2000; Moragues et al., 2002).

The clustering of the GABA$_A$ receptor subunit genes raises the question of whether the clustered genes are co-regulated. The α1 and β2 genes do indeed share identical transcription patterns, nucleus for nucleus and even have the same RNA levels in each area (Wisden et al., 1992; Duncan et al., 1995); thus these two genes may share regulatory elements. All the other subunit genes have sometimes common, sometimes divergent expression patterns, with no correlation with which gene is in which cluster. The expression of the GABA$_A$ receptor genes in the basal ganglia is reviewed in the Section "Expression of GABA$_A$ receptor subunit genes in the basal ganglia" below.

GABA$_A$ receptor structure

The GABA$_A$ receptor belongs to a superfamily of ligand-gated ion channels ("Cys–loop receptors") that in vertebrates include the nicotinic acetylcholine receptors (nAChR), the 5-hydroxytryptamine type 3 (5-HT3) receptors, the zinc-activated ion channel (ZAC) and the glycine receptors (reviewed in Cromer et al., 2002; Lester et al., 2004; Peters et al., 2005; Unwin, 2005). In imagining how the GABA$_A$ receptor must look, we can do no better than quote Unwin (2005) for his empirical observations on the *Torpedo* nicotinic acetylcholine receptor: "The receptor (a large 290 kDa glycoprotein) is composed of elongated subunits, which associate with their long axes approximately normal to the membrane, creating a continuous wall around the central ion-conducting path. The whole assembly presents a rounded, nearly fivefold symmetric assembly when viewed from the synaptic cleft, but is wedge-shaped when viewed parallel with the membrane plane. All the subunits of the receptor have a similar size 30 Å × 40 Å × 160 Å and the same three-dimensional fold. Each subunit is a three-domain protein and so portions the channel naturally into its ligand-binding, membrane-spanning and intracellular parts" (Unwin, 2005).

In GABA$_A$ receptors, the arrangement of subunits around the channel is probably γβαβα counter-clockwise when viewed from the extracellular space (Baumann et al., 2002). Current thinking is that for those cells in which they are expressed, ϵ and π subunits can replace the γ and δ subunit within the pentamer, whereas the θ subunit might replace a β subunit (Sieghart and Sperk, 2002). As for all members of the nicotinic receptor superfamily, all GABA$_A$ receptor subunits contain a large extracellular N-terminal domain of approximately 200 amino acids shaped by a cysteine disulfide bridge (the so-called "Cys–loop"). For GABA$_A$ subunits, the amino acid consensus sequence of the Cys–loop is C******F/YP*D***C*****S (where * is a degenerate residue; Simon et al., 2004).

Each subunit contains four predicted transmembrane spanning domains (TM1 to TM4) of about 20 amino acids and a large intracellular loop between TM3 and TM4 (TM3–TM4 loop) (Fig. 1) (Macdonald and Haas, 2000). Many GABA$_A$ receptor subunits have the amino acid sequence (TTVLTMTT) in the TM2 domain (Seeburg et al., 1990). Five of these eight amino acids have been proposed to line the ion channel. TM1, TM3 and TM4 segregate TM2 from membrane lipid (Unwin, 2005). The amino acids specifying that the nicotinic receptors gate cations have been identified. The selectivity filter and gate lies at the intracellular end

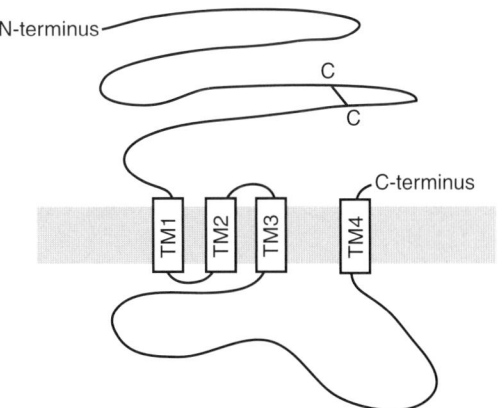

Fig. 1. Predicted topology of a GABA$_A$ receptor subunit. The cysteine-disulfide bridge in the N-terminus is indicated by a black bar. Transmembrane domains are shown as open boxes labelled TM1-4.

of the TM2 domains and includes part of the TM1-TM2 loop. Mutating these amino acids in the α7 subunit of homomeric nicotinic receptors produced acetylcholine-gated anion channels (Galzi et al., 1992). The converse can also be done: mutation of five amino acids in the TM1-TM2 loop of the GABA$_A$ receptor β3 subunit to the corresponding amino acids of the α7 nicotinic acetylcholine subunit produces cation-selective GABA$_A$ receptors (Jensen et al., 2002). Similar mutations in the α2 or γ2 subunits did not change ion selectivity. Thus the β subunits predominantly determine the ion selectivity of the GABA$_A$ receptor (Jensen et al., 2002).

In the 5-HT3 and nicotinic receptors, residues in the TM3–TM4 loop region influence single channel conductance (Peters et al., 2005; Hales et al., 2006). The TM3–TM4 loop, which may be relatively unstructured in the nAChR (Kukhtina et al., 2006), contributes key sites for attaching anchor and regulatory proteins involved in locating the receptor at synapses and in governing the activity of GABA$_A$ receptors (Kittler and Moss, 2003) (see Sections "Synaptic GABA$_A$ receptors: αβγ subunit combinations and anchoring role of the γ2 subunit and gephyrin" and "Regulation of GABA$_A$ receptor function by neuromodulators: the role of kinases and phosphatases"). But the TM4 region of the γ2 subunit is necessary and sufficient to confer a synaptic localization on the receptor (Alldred et al., 2005).

The atomic structure of a GABA$_A$ receptor subunit complex has not so far been solved directly. Instead, realistic models have used the empirically determined structural coordinates of the muscle nicotinic acetylcholine receptor from the electric organ of the *Torpedo* ray fish and a related snail acetylcholine receptor binding protein (AChBP) (Brejc et al., 2001; Cromer et al., 2002; Ernst et al., 2003; Unwin, 2003, 2005). The AChBP shows sequence similarity with the N-terminus of the nAChR at regions that build the agonist-binding sites; the AChBP contains a Cys–loop but lacks the transmembrane domains. It assembles as soluble homopentamers (Brejc et al., 2001). The crystal structure of the AChBP, with bound ligand, provided a template for comparative modelling of the N-terminal extracellular domain of GABA$_A$ receptors (Ernst et al., 2003), whereas Unwin's most recent structure of the *Torpedo* nAChR at 4 Å resolution obtained by cryo-EM, and incorporating insights from the AChBP, has given a full-scale atomic model (Protein Data Bank Code 2BG9). According to Xiu and colleagues, 2BG9 represents a substantial advance for the field, and all modern attempts to obtain molecular scale information on the structure and function of Cys–loop receptors must consider this as a starting point (Xiu et al., 2005; Unwin, 2005). 2BG9 provides us with a view of how the entire GABA$_A$ receptor must look, including the transmembrane and large cytoplasmic loops (Unwin, 2003, 2005).

Before 2BG9, modellers used family conservation patterns and fold predictions to estimate that 60–75% of the amino acid residues of the GABA$_A$ receptor subunits have structural equivalents in the AChBP template (Ernst et al., 2003). The accuracy of the GABA$_A$ receptor model will be limited in regions where alignment is unclear (e.g., due to low sequence identity) or in regions where the AChBP differs from other family members due to its soluble, non-membrane-bound nature (Ernst et al., 2003). A model of the extracellular domain of a pentameric GABA$_A$ receptor consisting of two α, two β and one γ2 subunit is shown in Fig. 2. In this model the amino acids known to contribute to ligand-binding sites and interfaces are correctly positioned and the interface-forming segments and the solvent accessibility of individual residues correlate well with experimental data (Ernst et al., 2003). Six "loops" (loop A, B, C for the plus side and D, E, F for the minus side) at the interface between neighbouring subunits form the ligand-binding sites (Sigel and Buhr, 1997; Olsen et al., 2004). The binding pocket for GABA forms at the interface between the α and the β subunit (Figs. 2a, c, d), the binding pocket for benzodiazepines lies at the interface of the α and the γ subunit (Figs. 2a, b). The predicted space for agonist binding is formed by loops A, B, C, D and E (blue volume in Fig. 2d) and correlates with experimental data from photo-labeling of α1F64 by [³H]muscimol and substituted cysteine accessibility mapping (Ernst et al., 2003; Olsen et al., 2004). Amino acid residues on loops A, B, C, D and E at the interface of the α and the γ subunit influence binding, potency and efficacy of

25

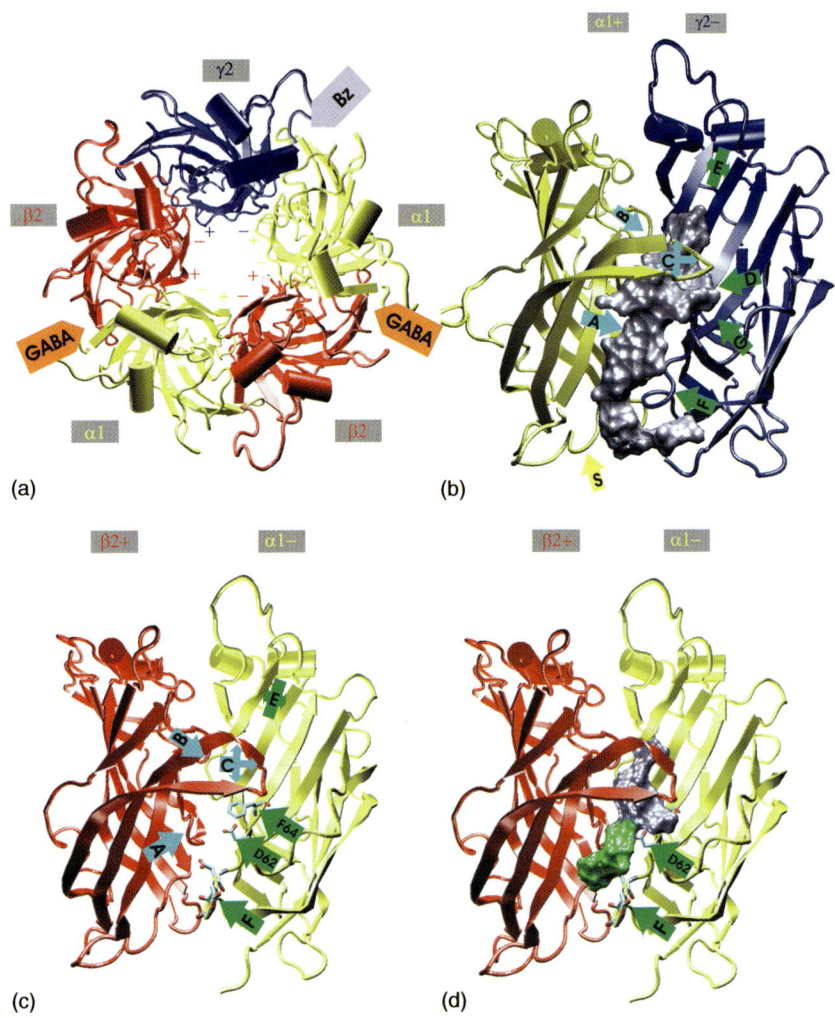

Fig. 2. Model of the extracellular domains of a pentameric GABA$_A$ receptor consisting of two α, two β and one γ2 subunit. (a) View from the extracellular space. GABA binds to the interface between the α and the β subunit, benzodiazepines bind to the interface between the α and the γ2 subunit. (b) Predicted benzodiazepine-binding pocket between the α and the γ2 subunit, viewed from the side. The binding site loops are labelled A to G. (c) and (d) The α and β subunit viewed from the side. Loops A, B, C, D and E form the predicted GABA-binding pocket (blue volume in (d)). The volume shown in green might be used in antagonist-bound states. (Adapted from Ernst et al., 2003 used with permission.)

benzodiazepines (see Section "GABA$_A$ receptors: allosteric modulation by benzodiazepines and related ligands") (Fig. 2b) (Ernst et al., 2003). The predicted benzodiazepine pocket is larger than the GABA pocket. It communicates with the Cys–loop of the α subunit and extends down to the membrane-near part, which possibly contains side chains from the linker between transmembrane region 2 and 3 of the α subunit (Ernst et al., 2003).

GABA$_A$ receptor gating by agonist

As for all other ligand-gated channels, GABA$_A$ receptors convert chemical messages into electrical signals. In less than a millisecond, the binding of two (tiny) molecules of GABA between the α and β subunits induces a conformational change in the (giant) receptor oligomer that opens the central ion channel (see Baumann et al., 2003). This remarkable process is called "gating". In the opinion of Xiu et al. (2005), "the gating mechanism for the Cys-loop superfamily is one of the most challenging questions in molecular neuroscience". The full 2BG9 model of the nAChR suggests ways in which the agonist-binding site couples to the transmembrane region and initiates gating; principles applying to the nAChR are likely to apply, with minor variations, to other members of the superfamily (Unwin, 2005).

The basis for a model of how gating works is that specific ion pairs exert precise control over gating (Kash et al., 2003; Xiu et al., 2005). The membrane-near location of two flexible loops, loop 2 and loop 7 (the Cys–loop) in the crystal structure of the AChBP suggested an involvement in gating of the Cys–loop in the $GABA_A$ receptor. Indeed, Kash et al. (2003) using an "ion pair model", found by site-directed mutagenesis in the $\alpha 1$ subunit that optimal gating needs electrostatic interactions between negatively charged residues in loops 2 and 7 (Asp57 and Asp 149) and a positively charged residue in the region linking transmembrane domains 2 and 3 (Lys 279). For the $\beta 2$ subunit of the $GABA_A$ receptor the interaction between an acidic residue in loop 7 (Asp 146) and a basic residue in pre-transmembrane domain-1 (Lys 215) helps couple agonist binding to channel gating (Kash et al., 2004). Studies on other members of the "Cys–loop" family found residues at corresponding regions in the nicotinic acetylcholine receptor and the serotonin 5-HT3 receptor as critical coupling elements for gating (Lee and Sine, 2005: Lummis et al., 2005). Nevertheless, building on the results of Kash and colleagues in a detailed and broad examination of electrostatic interactions in the subunits, Xiu et al. (2005) concluded that no specific ion pair interaction in fact influences gating, but instead a cluster of charges is important; specific ion pair interactions are non-essential and it is misleading to focus only on specific residues: "Receptors have evolved to create a compatible collection of charged residues that allows the receptor to assemble and also facilitates the existence of and interconversions among multiple states" (Xiu et al., 2005).

Subunit assembly rules for $GABA_A$ receptors

The $GABA_A$ receptor subunit combinations found in brain are partly governed by which cell types express which genes (e.g., Wisden et al., 1992) and partly by preferential partnering of subunits within a given cell (e.g., Jones et al., 1997); for example, the $\alpha 4$ and $\alpha 6$ subunits assemble preferentially with the δ subunit (Jones et al., 1997; Peng et al., 2002). The majority of mammalian brain $GABA_A$ receptors are probably $\alpha\beta\gamma 2$ combinations. The subunit ratio is probably $2\alpha/2\beta/1\gamma$ (Ernst et al., 2003). Some receptors also contain different α and β subunits, e.g., $\alpha 1\alpha 2\beta 2\gamma 2$ (Benke et al., 2004). According to Benke et al., 2004, who analysed whole mouse brain samples, the $\alpha 1\alpha 1\beta\gamma 2$ combination is the most abundant $GABA_A$ receptor subtype in the brain (61% of total). Other combinations were found in smaller quantities: $\alpha 1\alpha 2\beta\gamma 2$ (13%), $\alpha 1\alpha 3\beta\gamma 2$ (15%), $\alpha 2\alpha 2\beta\gamma 2$ (12%), $\alpha 2\alpha 3\beta\gamma 2$(2%) and $\alpha 3\alpha 3\beta\gamma 2$ (4%). Within the $\alpha 1$-containing receptor population, most receptors are $\alpha 1\alpha 1\beta\gamma 2$, whereas in the $\alpha 2$- and $\alpha 3$-containing receptor populations, receptors with two different α subunit types predominate (Benke et al., 2004). Of course, these percentages are from homogenized brain; within particular cell types, some of these rare subtypes will be the most important receptor subtype. Other receptor subtypes relevant for the basal ganglia are *predicted* from subunit expression patterns as $\alpha 4\beta\delta$ (or possibly $\alpha 4\beta$ Bencsits et al., 1999) in the striatum and $\alpha 1\beta 2\gamma 1$ in the globus pallidus.

Synaptic $GABA_A$ receptors: $\alpha\beta\gamma$ subunit combinations and anchoring role of the $\gamma 2$ subunit and gephyrin

Placing $GABA_A$ receptors at synapses requires specific proteins that interact directly or indirectly with the γ subunits. For example, targeting some $GABA_A$ receptor subtypes to GABAergic terminals involves the widely expressed microtubule-binding protein gephyrin (Ramming et al., 2000). The best studied brain area for this has been the hippocampus, but splice forms of gephyrin are found throughout the basal ganglia (Ramming et al., 2000) and so the principles of $GABA_A$ receptor targeting would be expected to be similar there. In hippocampal neurons in vitro and in vivo, gephyrin either helps convey some $GABA_A$ receptor subtypes to the synapse or anchors them there — this requires the $\gamma 2$ subunit (Kneussel et al., 1999;

Brunig et al., 2002). Without the γ2 subunit, no GABA$_A$ receptors are found in synapses in the developing or adult hippocampus (Gunther et al., 1995; Essrich et al., 1998; Schweizer et al., 2003), and without gephyrin, much reduced numbers of some synaptic GABA$_A$ receptor subtypes, especially α2-containing, are found; some receptor clusters, especially those containing the α1 subunit, persist in hippocampal gephyrin knockout neurons (Levi et al., 2004). Other γ subunits can replace synaptic targeting function of γ2; in γ2 knockout mice, GABA$_A$ receptors can be restored to hippocampal synapses by expressing the γ3 subunit by transgenic rescue (Baer et al., 1999; Luscher and Keller, 2004). Some conserved sequence identity in the large TM3–TM4 intracellular loops of the γ subunits may indicate binding sites for parts of the synapse-anchoring mechanism. Distributed cysteine residues are conserved in the γ subunit large intracellular loops, but are absent from the α, β and δ subunits. Palmitoylation of these cysteine residues via a thio-ester bond plays some role in targeting γ subunit-containing receptors to the synapse (Luscher and Keller, 2004); in cultured hippocampal neurons, cysteine–alanine substitutions in the γ2 subunit loop region interfere with expression and clustering of receptors (Rathenberg et al., 2004). A surprising finding is that the TM4 region of the γ2 subunit is also involved in synaptic targeting, possibly by interacting with lipid rafts occurring in the synapse or by other membrane proteins (Alldred et al., 2005). In transfected hippocampal cultures, analyses of chimeric γ2/α2 subunit constructs showed that γ2 TM4 is necessary and sufficient for postsynaptic clustering of GABA$_A$ receptors, whereas the cytoplasmic γ2 subunit domains are dispensable (Alldred et al., 2005). In contrast, both the TM3–TM4 loop and the TM4 domain of the γ2 subunit contribute to efficient recruitment of gephyrin to postsynaptic receptor clusters and are essential for restoration of miniature inhibitory postsynaptic currents (IPSCs) (Alldred et al., 2005). Thus the γ2 subunit TM3–TM4 cytoplasmic loop might be needed for inserting receptors into the plasma membrane but is dispensable for delivery of receptors to subsynaptic dendritic sites (Alldred et al., 2005). Gephyrin does not bind the γ2 receptor subunit directly. The identity of the missing link(s)

between gephyrin and GABA$_A$ receptor subunits is unknown. As mentioned earlier, the targeting of γ2-containing receptors to hippocampal synapses must depend on both the α subunit and the γ2 subunit. According to some investigators α5βγ2 receptors seem largely extrasynaptic and non-colocalized with gephyrin (Crestani et al., 2002) and when α6βγ2 receptors (normally only found in cerebellar granule cells) are ectopically expressed in pyramidal cells these receptors remain extrasynaptic (Wisden et al., 2002). So it is not simply that a γ2 subunit (or even gephyrin) guarantees a stable synaptic placement of the GABA$_A$ receptor. In addition to gephyrin other clustering proteins must contribute to the synaptic localization of selected GABA$_A$ receptor subtypes (Kneussel et al., 2001).

GABA$_A$ receptor occupancy at synapses is dynamic

It is important to keep in mind that GABA$_A$ receptor expression on the surface of neurons is dynamic; receptors rapidly recycle and leave from or insert into the synapse by rapid lateral diffusion and/or endo/exocytosis; a static crystalline scaffold of GABA$_A$ receptors anchored at the synapse would seem to be the wrong view (Kittler and Moss, 2003; Thomas et al., 2005); GABA$_A$ receptors diffuse into a synaptic zone and are transiently "captured" by the anchoring complex. However, for some inhibitory hippocampal synapses, a direct relationship exists between the number of synaptic GABA$_A$ receptors and the strength of the synapse, but it is not clear what mechanisms maintain fixed numbers of GABA$_A$ receptors long-term at specific synapses (reviewed in Nusser, 1999). As for glutamate receptors at excitatory synapses, neurons probably recycle GABA$_A$ receptors as a strategy for setting their degree of excitability (Kittler and Moss, 2003). GABA$_A$ receptors constitutively internalize by clathrin-dependent endocytosis; this requires interactions between the β and γ2 subunits and the AP2 adaptin complex (Kittler and Moss, 2003).

Extrasynaptic GABA$_A$ receptors: α4βδ subtype

Besides mediating precisely timed synaptic point to point inhibition (phasic inhibition) via γ2 subunit-containing receptors, GABA$_A$ receptors can also convey less time-locked signals. Low GABA concentrations in the extracellular space, resulting from GABA diffusing from the synapse, can tonically activate extrasynaptic GABA$_A$ receptors (Fig. 3) (Brickley et al., 2001; Farrant and Nusser, 2005; Staley and Scharfmann, 2005). This "tonic inhibition" is temporally uncoupled from the fast synaptic events, causing a continually present background inhibitory conductance. Such conductances alter the input resistance of the cell and thus influence synaptic efficacy and integration; tonic extrasynaptic conductances, by increasing the electrical leakiness of the dendritic membrane, substantially and indiscriminately diminish the size of excitatory signals in dendrites (reviewed in Farrant and Nusser, 2005; Staley and Scharfmann, 2005).

Receptors with the δ subunit, α4βδ in forebrain and α6βδ in cerebellar granule cells, are extrasynaptic; δ subunits are perisynaptic (annular), localized around the edge of synapses in hippocampal dentate granule cells and totally extrasynaptic on cerebellar granule cells (Nusser et al., 1998; Wei et al., 2003). In all regions so far tested (cerebellar granule cells, hippocampal dentate granule cells, thalamic relay nuclei), δ subunits contribute to GABA$_A$ receptors that provide an extrasynaptic tonic conductance (Brickley et al., 2001; Stell et al., 2003; Cope et al., 2005) For GABA$_A$ receptors containing α4βδ subunits in the basal ganglia, for example in the striatum, it is predicted that the receptors are extrasynaptic and that their key properties are high affinity for neurotransmitter and

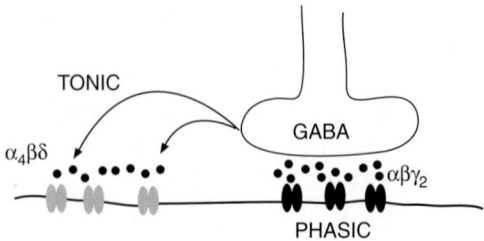

Fig. 3. Phasic and tonic GABAergic inhibition. Fast synaptic (phasic) inhibition is mediated mainly via γ2 subunit-containing receptors (shown in black). δ subunit-containing receptors (shown in grey) are located peri- or extrasynaptically and are tonically activated by GABA diffusing out of the synaptic cleft.

limited desensitization, enabling them to contribute to tonic background conductances (see above) (Brickley et al., 2001; Semyanov et al., 2004).

GABA$_A$ receptors: how subunit combinations affect synaptic and extrasynaptic transmission

Many factors will influence the type of IPSCs mediated by GABA$_A$ receptors: the number of GABA$_A$ receptors at the synapse; the subunit composition of the receptors which influences the kinetics; the phosphorylation state of the receptor; a differential modulation of synaptic and nonsynaptic receptors; the Cl$^-$ reversal potential; the GABA transient in the synaptic cleft and regulation by neuromodulators (Mody and Pearce, 2004).

A typical synaptic pulse of GABA is often cited as 0.3–1.0 mM lasting less than 1 ms (Mody and Pearce, 2004). Under these conditions, all synaptic GABA$_A$ receptors will be saturated and give maximal responses. Nevertheless, receptor subunit composition affects the single channel conductance, how fast the receptors gate, how fast they switch off (deactivate and desensitize) and how they respond to allosteric modulators. GABA also diffuses out of the synaptic cleft, where μM GABA concentrations are typically present (Nusser and Farrant, 2005); at these concentrations the GABA sensitivity of the receptor is critical. The sensitivity of α1β3- and α1β3δ-containing receptors is significantly higher (mean EC$_{50}$ of approx. 2 and 3.5 μM, respectively) than that of α1β3γ2-containing receptors (mean EC$_{50}$ of approx. 13 μM) (Fisher and Macdonald, 1997). The single channel conductance of recombinant αβγ2 or αβδ receptors lies in the range of 25–30 pS. The single channel conductance of αβ-heterodimeric channels is 11–15 pS (Fisher and Macdonald, 1997).

The transient kinetic properties of GABA$_A$ receptors depend on the subunit composition. The time course of the GABA$_A$ receptor current is governed by three different kinetic processes: activation, desensitization and deactivation. During activation the current shows a rapid rise to the maximum. During this time the agonist binds to the receptor and the channel opens. Desensitization describes the unresponsiveness of the receptor and

current decline in the continued presence of agonist. Deactivation describes the current decline
after removal of the agonist. Current activation for
recombinant α1β3 and α1β3δ receptors is slower
than for α1β3γ2 receptors, with mean 10–90% rise
time varying from 1.7 to 2.4 ms for α1β3 and α1β3δ
receptors and only 0.5 ms for α1β3γ2 receptors
(Haas and Macdonald, 1999). The α subunit also
influences the activation rate. Recombinant α2β1γ2
receptors have a more rapid activation (10–90%
rise time of 0.5 ms) than α1β1γ2 receptors (10–90%
rise time of 1 ms) (Lavoie et al., 1997). Desensitization is also influenced by the subunit composition. Currents of recombinant α1β3 and α1β3γ2
receptors desensitize quicker and more completely
than currents of α1β3δ receptors (Haas and
Macdonald, 1999). Thus the δ subunit reduces
speed and extent of receptor desensitization. As
receptors with the δ subunit are primarily extrasynaptically and perisynaptically located (Nusser
et al., 1998; Wei et al., 2003), their limited desensitization and high sensitivity to GABA might be
important for their roles in tonic background
conductance (see Section "Extrasynaptic GABA$_A$
receptors: α4βδ subtype", above). α1β3γ2 currents
deactivate more slowly than α1β3 or α1β3δ currents, mainly due to a significantly longer slow
decay component (Haas and Macdonald, 1999).
Again also the α subunit composition influences
the kinetics. Deactivation of α2β1γ2 containing receptors is six to seven times slower than deactivation of α1β1γ2 receptors (Lavoie et al., 1997).

**GABA$_A$ receptor agonists, antagonists and
allosteric modulators**

GABA$_A$ receptors display a rich pharmacology
(Korpi et al., 2002a; Sieghart and Sperk, 2002;
Rudolph and Moehler, 2006; Whiting, 2006).
Generic GABA$_A$ receptors are selectively activated
by the GABA agonist muscimol and blocked com-
petitively by the GABA antagonists bicuculline and
SR95531 (receptors assembled with ρ subunits are
bicuculline- and barbiturate-insensitive, having
their own unique pharmacology). Picrotoxin blocks
GABA$_A$ receptors non-competitively, probably by
binding to a site in the channel (Korpi et al., 2002a).

Many drugs bind at sites on the GABA$_A$ receptor
distinct from the GABA-binding site; these drugs
change the shape of the receptor oligomer so that
the efficacy of GABA at opening the channel is
either increased (positive allosteric agonists, e.g.,
diazepam) or decreased (negative allosteric agonists, e.g., the β-carboline, DMCM). A few allosteric modulators occur naturally in the brain
(e.g., Zn^{2+}, neurosteroids). Generally, positive
allosteric agonists are used widely in medicine
(e.g., for the induction and maintenance of general
anaesthesia or to treat anxiety disorders, states of
agitation, epilepsy or sleep disorders) and there is
scope to further develop these drugs to produce receptor subtype-selective drugs with fewer side-
effects (Whiting, 2006; Rudolph and Moehler,
2006); however, negative allosteric agonists also
have potential clinical applications; for example,
the drug L-655 708 works selectively at α5βγ2 receptors (a subtype mainly expressed in the hippocampus) and by decreasing GABAs action there
it acts as a cognition enhancer (Rudolph and
Moehler, 2006). A feature of all allosteric modulators is that they usually only work when GABA is
at submaximal activating concentrations (below
1 mM) and they do not work in the absence of
GABA (with the exception of some intravenous
anaesthetics). Nevertheless, some modulators (e.g.,
benzodiazepines) also strongly influence the deactivation rate of the receptors even at peak synaptic
GABA concentrations and this maybe how some of
their in vivo effects originate (Mellor and Randall,
1997). In the following sections, we briefly consider
the drugs that could act on GABA$_A$ receptor subunit combinations relevant for the basal ganglia
(e.g., α1β2γ2, α2β2/3γ2, α4βγ2, α4βδ) (see Section
"Expression of GABA$_A$ receptor subunit genes in
the basal ganglia" below).

*GABA$_A$ receptors: allosteric modulation by
benzodiazepines and related ligands*

The main effects of benzodiazepines are sedation,
anxiolysis, suppression of seizures and muscle relaxation. These drugs require αβγ2-type receptors
(e.g., α1β2γ2 or α2β2γ2 or α1β3γ2) with the drug-
binding site located between the α and γ2 subunits

(Ernst et al., 2003). Note α4β2-type receptors, which could potentially form in some nuclei of the basal ganglia, are insensitive to most BZ drugs, as are any receptors that contain the δ subunit. The substances that act at the benzodiazepine-binding site include the classical benzodiazepines like diazepam or flunitrazepam as well as chemically different substances like the imidazopyridine zolpidem (relatively selective for α1β2-type receptors). Depending on the ligand the benzodiazepine site can mediate different effects. Benzodiazepine antagonists like flumazenil (Ro 15–1788) inhibit the effects of both agonists (positive allosteric modulators) and inverse agonists (negative allosteric modulators). In clinics, flumazenil is used in cases of benzodiazepine intoxication.

The most abundant receptor subtype in the brain, α1β2γ2 or α1β3γ2, corresponds to the pharmacologically defined BZ1 site with high affinity for flumazenil (Ro 15–1788), Ro 15–4513 and flunitrazepam and selective affinity for ligands like zolpidem (Niddam et al., 1987; Pritchett et al., 1989). Receptor subtypes containing the α2 or α3 subunit along with β and γ2 correspond to the BZ2 site with high affinity for flumazenil, Ro 15–4513 and flunitrazepam but lower affinity to zolpidem (Pritchett et al., 1989; Hadingham et al., 1993) (see also the Section "Autoradiography of GABA$_A$ receptors"). The type of β subunit has no effect on benzodiazepine pharmacology.

The BZ site is situated at the interface between the α and the γ subunit (see Section "GABA$_A$ receptor structure") (Ernst et al., 2003; Ogris et al., 2004). In the α1, α2, α3 and α5 subunits a mutation from histidine to arginine at position 101 abolishes binding of classic agonists like diazepam (Wieland et al., 1992; Korpi et al., 2002a). The diazepam-insensitive α4 (or α6) subunits naturally contain an arginine residue at the homologous position and so α4β2 or α6β2 receptors are insensitive to most BZ ligands (Luddens et al., 1990; Wisden et al., 1991; Korpi et al., 2002a). In the γ2 subunit, a replacement of phenylalanine by isoleucine at position 77 abolishes binding of zolpidem, DMCM and flumazenil whereas flunitrazepam still shows high-affinity binding (Buhr et al., 1997; Wingrove et al., 1997; Cope et al., 2004; Ogris et al., 2004). Methionine at position 130 in the γ2 subunit is

required for high-affinity binding of flunitrazepam but not flumazenil (Ro 15–1788) (Wingrove et al., 1997). The residues are distributed over the N-terminal domains of the α and γ subunits. In the assembled receptor the residues that form the benzodiazepine-binding pocket are brought into close physical proximity by the so-called binding site "loops" (see Section "GABA$_A$ receptor structure"). It is not clear, however, whether each of the above-mentioned amino acid residues really participates in the lining of the BZ-binding site or whether the inserted mutations have allosteric effects.

GABA$_A$ receptors: allosteric modulation by intravenous anaesthetics

At clinically relevant concentrations, general anaesthetics modulate the activity of various ion channels (Krasowski and Harrison, 1999; Thompson and Wafford, 2001). Whereas volatile anaesthetics (e.g., halothane, enflurane or isoflurane) are positive modulators of recombinant GABA$_A$ receptors, the main targets of these drugs in vivo are probably two pore domain (K2P) potassium channels (Franks and Honore, 2004). The intravenous anaesthetics (e.g., barbiturates, steroidal anaesthetics, propofol and etomidate) can modulate GABAs action at the receptor but can also activate the receptor directly in the absence of GABA at higher concentrations (Korpi et al., 2002a). Based on the analysis of knock-in mouse lines with propofol- and etomidate-insensitive β subunits (see below), propofol and etomidate exert nearly all of their anaesthetic actions entirely through GABA$_A$ receptors (Rudolph and Mohler, 2004).

The action of etomidate and propofol absolutely requires residues in TM2 and TM3 in the β2 or β3 subunits (Jurd et al., 2003). A mutation of asparagine to methionine at position 265 (N265 M) in the 2nd transmembrane domain of the β3 subunit abolishes the modulatory and direct effects of etomidate and propofol in recombinant receptors (Jurd et al., 2003). A mutation of aspargine at the same position in the β2 subunit also abolishes the action of etomidate on the GABA$_A$ receptor (Reynolds et al., 2003). In β3(N265 M) mice propofol and etomidate do not suppress noxious-evoked movements and show a strongly decreased duration of the loss of righting reflex, two different endpoints of anaesthesia. These results suggest that propofol and etomidate act mainly via the GABA$_A$ receptor and the β3 subunit in particular to induce deep anaesthesia. The remaining effects of propofol and etomidate could be mediated by β2 subunit-containing receptors. Studies on β2(N265S) mice suggested that the β2 subunit mediates the sedative effects of etomidate whereas the β3 subunit is required for etomidate to induce a loss of consciousness (Reynolds et al., 2003). A highly interesting issue is the location in the brain where etomidate and propofol exert their anaesthetic effects. Is the modulation of GABA$_A$ receptors in specific nuclei required to induce anaesthesia or do these drugs produce global effects at many GABA$_A$ receptors in all brain circuits? In any case, as GABA$_A$ receptors with both β2 and β3 are found throughout the basal ganglia (see Section "Expression of GABA$_A$ receptor subunit genes in the basal ganglia"), the operation of these nuclei will be profoundly affected by propofol and etomidate.

GABA$_A$ receptors: allosteric modulation by neurosteroids

Neuroactive steroids modulate GABA$_A$ receptor function in many brain regions (Belelli and Lambert, 2005; Farrant and Nusser, 2005). Naturally occurring steroid metabolites form locally in the brain: 5α-reductase transforms progesterone to 5α-DPH, which in turn is reduced by 3α-hydroxysteroid oxidoreductase to allopregnanolone. Allopregnanolone potently activates GABA$_A$ receptors. No absolute specificity of neurosteroids for particular GABA$_A$ receptor subunit combinations exits. Many GABA$_A$ receptors are sensitive to the steroid tetrahydrodeoxycorticosterone (THDOC), but receptors with the δ subunit are particularly sensitive — 30 nM THDOC enhances the peak currents of α1β3δ GABA$_A$ receptors (with 1 μM GABA) by up to 800%; other receptor isoform currents, e.g., from α1β3γ2 are enhanced to a smaller degree 1(5–50%) (Mihalek et al., 1999; Wohlfarth et al., 2002; Stell et al., 2003). Thus

endogenous allopregnanolone may act on extrasynaptic αβδ GABA$_A$ receptors to increase basal levels of inhibition. Mice without functional δ subunits have decreased sensitivity to the sedative/hypnotic, anxiolytic and pro-absence effects of neuroactive steroids (Mihalek et al., 1999).

GABA$_A$ receptors: allosteric modulation by Zn^{2+}

Zn^{2+} inhibits GABA$_A$ receptors (Hosie et al., 2003). In various brain regions, Zn^{2+} is synaptically released together with other neurotransmitters, both GABA and glutamate (reviewed by Mathie et al., 2006). Hippocampal mossy fibres have actually been the main area where synaptic Zn^{2+} actions have been investigated, but Zn^{2+} is worth bearing in mind as a *potential* modulator of GABAergic function in the basal ganglia. Zn^{2+} can reduce the amplitude, slow the rise time and accelerate the decay of mIPSCs. On recombinant GABA$_A$ receptors, Zn^{2+} has an inhibitory potency 3400 times higher on αβ receptors than on αβγ2 receptors (reviewed in Hosie et al., 2003). Thus the γ2 subunit lowers the sensitivity of the GABA$_A$ receptor complex to Zn^{2+}. Hosie and colleagues hypothesize that in addition to its role in promoting synaptic targeting and single channel conductance, the γ2 subunit evolved to retain the fidelity of GABAergic inhibition in the presence of Zn^{2+} (Hosie et al., 2003). Nevertheless, the potency of Zn^{2+} is also α subunit- and δ subunit-dependent.

Extrasynaptic δ subunit-containing GABA$_A$ receptors: allosteric potentiation by ethanol

Blood alcohol levels of 1–3 mM can result from drinking half a glass of wine or less. Ethanol influences many channels, including the *N*-methyl-D-aspartate (NMDA) glutamate receptor (Hanchar et al., 2005). But amongst GABA$_A$ receptor subunit combinations, low concentrations of ethanol (about 3 mM, a concentration six times lower than the legal blood-alcohol limit for driving in most States in the USA) specifically potentiate GABA responses of cloned α4βδ and α6βδ receptors expressed in *Xenopus* oocytes (Sundstrom-Poromaa et al., 2002; Wallner et al., 2003; Hanchar et al., 2004, 2005). This effect is β subunit-dependent; β3 subunits provide maximal sensitivity to ethanol (Wallner et al., 2003). Thus a glass of wine might, via α4β3δ and α6β3δ receptors, enhance GABAergic tonic (extrasynaptic) inhibition in the striatum and cerebellum respectively (Hanchar et al., 2005). On the other hand, the ethanol sensitivity of α6/δ KO mice is not different from wild-type mice (Korpi et al., 1999).

New subtype-selective drugs for GABA_A receptors

GABA$_A$ receptors have always been fertile ground for drug companies. Benzodiazepines, although for many years the main stay of clinical treatments for anxiety disorders, fell out of favour to selective serotonin reuptake inhibitors (SSRIs) due to side-effects like sedation, cognitive impairment and abuse liability. But SSRIs are too slow acting for some situations, requiring several weeks to work. Thus there is a medical need for fast-acting anxiolytics with few/no side-effects (Whiting, 2006). To dissociate the wanted anxiolytic effects from the unwanted side-effects of GABA$_A$ agonists, two different strategies were pursued: The development of partial agonists and the development of receptor subtype specific compounds. Despite promising preclinical assays, partial agonists (e.g., bretazenil) so far did not meet the expectations in clinical trials. One compound (ocinaplon) though was reported to show anxiolysis and strongly-reduced sedative side-effects in Phase II clinical trials (Whiting, 2006). Since different effects of benzodiazepines are mediated by different receptor subtypes, the second strategy has focussed on the development of compounds with selectivity for those subtypes proposed to mediate the anxiolytic effects of benzodiazepines, that is α2 and α3 subunit containing receptors (see Section "Function and physiological significance of GABA$_A$ receptor diversity for the basal ganglia", below). Some of these compounds (e.g., L-838417, TP003, SL 651498) showed a promising separation of anxiolytic effects and side-effects when tested in rodents and primates (McKernan et al., 2000; Griebel et al., 2001, 2003; Dias et al., 2005; Rowlett et al., 2005;

Whiting, 2006). The usefulness of this second approach now has to be evaluated in clinical trials.

Regulation of GABA_A receptor function by neuromodulators: the role of kinases and phosphatases

Given that dopamine is a key neurotransmitter/neuromodulator in many parts of the basal ganglia and that dopamine receptors couple to G-protein-linked second messenger systems to alter kinase and phosphatase activity, it is appropriate to consider how GABA$_A$ receptor function is regulated by phosphorylation initiated by dopamine receptor activation (e.g., Chen et al., 2006). There has been one intriguing report that dopamine D5 receptors directly crosslink with GABA$_A$ receptors by using the TM3-TM4 loop of the γ2 subunit (Liu et al., 2000); so far this finding has not been followed up, but it should not be dismissed too prematurely. But more conventionally, phosphorylation is the common way to regulate ion channels (Kittler and Moss, 2003). This is underlined by studies on, for example, PKC ∈ knockout mice; these mice show increased anxiety and have impaired GABA$_A$ receptor function (Hodge et al., 1999). For the GABA$_A$ receptor, the intracellular loops of the β and γ2 subunits in particular are phosphorylation targets. Studies of recombinant receptors have shown phosphorylation of these subunits by PKA, PKC, Src and PKB (Kittler and Moss, 2003; Wang et al., 2003). Depending on the subunit phosphorylation can have different functional effects, which might contribute to the diversity of GABA$_A$ receptor function. For example PKA mediated phosphorylation of the β1 subunit (serine 409) leads to negative modulation of the receptor, whereas phosphorylation of the β3 subunit (serine 408 and serine 409) enhances the activity of GABA$_A$ receptors (Kittler and Moss, 2003). Neuromodulators that influence GABA$_A$ receptor function via PKC include M1 muscarinic acetylcholine receptors, serotonin (5-HT) type 4 receptors and TrKB receptor stimulation via BDNF (Kittler and Moss, 2003; Jovanovic et al., 2004). Modulation of GABA$_A$ receptor function after dopamine D4 and D3 receptor activation depends on PKA (Wang et al., 2002; Chen et al., 2006). Protein kinase B (Akt) can phosphorylate the β2 subunit at serine 410, which promotes rapid insertion of the GABA$_A$ receptor into the membrane, resulting in increased sIPSC amplitudes after stimulation with insulin (Wang et al., 2003).

Function and physiological significance of GABA$_A$ receptor diversity for the basal ganglia

An important consequence of differences in subunit expression between cells or differential subcellular localization of subunits within a cell is that GABA$_A$ receptor kinetics might differ between different cells and different synapses (Thomson et al., 2000; Nyiri et al., 2001; Freund, 2003). In the hippocampus, for example, functionally distinct interneurons might signal via distinct GABA$_A$ receptor subtypes (Freund, 2003). A similar situation could occur in the basal ganglia.

Knockout mouse lines have been generated for many of the GABA$_A$ receptor subunit genes (reviewed by Vicini and Ortinski, 2004), but it would be hard to make observations specifically about altered basal ganglia function in these lines. In a beautiful series of papers, mice with specific mutations in the key H101 coding position affecting BZ sensitivity were generated in the α1, α2, α3 and α5 subunit genes and the behavioural effects of diazepam were tested (Rudolph and Mohler, 2004). These mice have normal GABA$_A$ receptors, but in α1H101R mice for example, only the α2β2, α3β2 and α5β2-type GABA$_A$ receptors are diazepam-sensitive. Thus by a process of subtraction, it can be deduced how different α1β2, α2β2, α3β2 and α5β2 subtypes contribute to the diverse in vivo pharmacological effects of diazepam and other ligands requiring the H101 site. Thus, α1H101R mice no longer become sleepy when given diazepam and so the α1β2 receptors are required for the sedative effects of diazepam (Rudolph and Mohler, 2004), whereas the α2 subunit mediates diazepam's anxiolytic effects (under the influence of diazepam α2H101R mice do not venture more into threatening areas, whereas their wild-type littermates do) (Rudolph and Mohler, 2004). A different set of studies using α3 selective

inverse agonists and agonists also showed a significant contribution of the α3 subunit in anxiogenesis and anxiolysis (Atack et al., 2005; Dias et al., 2005). The muscle relaxant activity of diazepam is mediated by the α2 and α3 subunits, probably because these subunits are expressed in spinal motor neurons (Rudolph and Mohler, 2004). Nevertheless, because of the widespread expression of the α1, α2 and α3 subunits, it is not easy to use results obtained from these mouse lines to make observations specifically about which GABA$_A$ receptor subtypes respond to diazepam and influence basal ganglia function at the whole animal level; stereotactic injection of ligands into the various lines might be possible.

Expression of GABA$_A$ receptor subunit genes in the basal ganglia

GABA$_A$ receptor expression in the basal ganglia has been mapped by ligand autoradiography (e.g., Niddam et al., 1987; Faull and Villiger, 1988; Olsen et al., 1990; Duncan et al., 1995; Waldvogel et al., 1999; Korpi et al., 2002a), in situ hybridization with gene-specific probes (e.g., Laurie et al., 1992a; Persohn et al., 1992; Petri et al., 2002; Wisden et al., 1992), single-cell polymerase chain reaction (PCR) (e.g., Criswell et al., 1997; Guyon et al., 1999; Okada et al., 2004) and immunocytochemistry with subunit-specific antibodies at the light and electron microscopic level (e.g., Fritschy and Mohler, 1995; Somogyi et al., 1996; Pirker et al., 2000; Fujiyama et al., 2000, 2002; Schwarzer et al., 2001).

Autoradiography of GABA$_A$ receptors

Ligand autoradiography, although usually lacking cellular resolution, serves as a highly-useful indicator of which αβγ2 subunit combination is present in a brain region (Korpi et al., 2002a) and is also a superbly quantitative technique (Korpi et al., 2002a). For example, the ligand [³H]flunitrazepam is incubated with a brain section in the absence or presence of the discriminating ligand CL 218 872. If the binding signal is reduced or even completely vanishes by co-incubation with CL 218 872 (which has high affinity for α1βγ2 receptors and so displaces [³H]flunitrazepam), then this is a "BZ1 region". If CL 218 872 fails to displace [³H]flunitrazepam, then this is a "BZ2 -region" (reviewed in Niddam et al., 1987). Thus BZ1-type binding marks α1βγ2 receptors; ³H-zolpidem autoradiography on brain sections also selectively highlights α1βγ2 receptors directly (Duncan et al., 1995; Korpi et al., 2002a). BZ2-type binding marks α2, α3 and α5βγ2 type receptors (the α4βγ2 receptors are not picked up by this method, as they do not bind [³H]flunitrazepam — see Section "GABA$_A$ receptors: allosteric modulation by benzodiazepines and related ligands"). The αβγ-type receptors with γ1 or γ3 are not picked up by BZ1 and BZ2 screening, neither are α4βδ-type receptors. However, many novel ligands are available these days, which could be used as autoradiographic probes.

Autoradiography with [³H]muscimol (high-affinity site), commonly assumed to mark all GABA$_A$ receptors, selectively marks only α4βδ and α6βδ receptors; mouse brain sections with no δ subunit no longer give detectable [³H]muscimol signals (Korpi et al., 2002b). Thus the [³H]muscimol autoradiographic signals detected in many basal ganglia areas such as the striatum (Olsen et al., 1990; Jones et al., 1997; Korpi et al., 2002b) will originate from α4δ -containing receptors.

In situ hybridization and immunocytochemistry of GABA$_A$ receptors

In situ hybridization is used to localize mRNA in brain regions or cell types. The radioactive version is a sensitive assay for which cell type expresses which gene; the use of digoxygenin-labelled probes is less sensitive but permits double labelling for multiple gene expression. There is good agreement between different labs on the results with GABA$_A$ receptor gene expression, probably because hybridization of nucleic acids is reasonably standardized between labs. The disadvantage of in situ hybridization is that no information is obtained on protein localization on the cell, or even if the mRNA is translated. Thus immunocytochemistry with subunit-specific antibodies gives the ultimate biological information. However, antibodies can

be problematic; Saper and Sawchenko (2003) point out that antibodies are biological agents, not standard chemical reagents: antibodies may bind to a wide variety of antigens other than the one that they were raised to recognize and there is no way to be sure that the pattern they stain really represents that antigen. If investigating localization in the rodent nervous system, a knockout mouse for the particular antigen is the best control of antibody specificity (Saper and Sawchenko, 2003; Aller et al., 2005). This has been done for some (e.g., α1, α3, α6 and δ) (Jones et al., 1997; Tretter et al., 2001; Yee et al., 2005; Kralic et al., 2006), but not all GABA$_A$ receptor antibodies used in published papers.

Occurrence of α1/β2/3γ2 receptors throughout the basal ganglia

The α1/β2/γ32 GABA$_A$ receptor subtype is the most abundant subtype in the brain (Benke et al., 2004). Many cell types in the basal ganglia use the "standard" BZ1-type α1β2γ2 receptor subtype: neurons in the caudate-putamen (striatum), substantia nigra pars reticulata; globus pallidus and subthalamic nucleus, all use this receptor (see Niddam et al., 1987; Wisden et al., 1989, 1992; Criswell et al., 1997; Chen et al., 2004). The subthalamic nucleus, substantia nigra pars reticulata, globus pallidus and ventral pallidum are classic BZ1 sites, having some of the highest densities of [³H]zolpidem binding (Niddam et al., 1987; Duncan et al., 1995).

Mixed GABA$_A$ receptors in the striatum: synaptic α1β2γ2, α2β3γ2, α3β2/3γ2 and extrasynaptic α4β3δ

Consistent with the diversity of neuronal cell types in the striatum (Tepper and Bolam, 2004), a diverse mixture of GABA$_A$ receptor subtypes exists in this structure. By in situ hybridization, this region expresses moderately α1, strongly α2, moderately α3, strongly α4 (about the same as α2), moderately β2 (about the same as α1), strongly β3 (about the same as α2 and α4), moderately γ1, γ2 and γ3 and moderately δ mRNAs (Wisden et al.,

1991, 1992; Herb et al., 1992). In the rat, $\gamma 3$ mRNA is slightly more abundant in the caudate-putamen than $\gamma 2$ and $\gamma 1$ mRNAs (see Figs. 2 and 3 in Herb et al., 1992). The significance of receptors containing the $\gamma 1$ and $\gamma 3$ type receptors has not been investigated, but we think it would be rewarding to examine.

In the rat striatum, based on assembly rules from other brain regions, we might expect $\alpha 1\beta 2\gamma 2$, $\alpha 2\beta 3\gamma 2$, $\alpha 3\beta 2\gamma 2$ (all synaptic, or $\alpha 1\alpha X$ mixtures), and $\alpha 4\beta 3\delta$ (or possibly $\alpha 4\beta 3$) receptors, distributed within and between different cell types, and sometimes co-expressed on the same cell type (Fujiyama et al., 2000). Just by the sheer abundance and rather uniform distribution in the striatum of the $\alpha 2$ and $\beta 3$ mRNAs, we predict that $GABA_A$ receptors containing $\alpha 2$ and $\beta 3$ will be on GABAergic projection neurons.

Strong developmental switch in GABA_A receptor subunit expression in the striatum from neonate to adult

There is much interest in identifying the role(s) GABA and its (generic) receptors may have in shaping the development of the postnatal nervous system (Ben-Ari et al., 2004). A marked switch in α subunit gene expression takes place during rat striatal development: the $\alpha 2$, and $\alpha 5$ mRNAs are present in the neonatal period at E19 to P6; but in the adult, the $\alpha 2$ and $\alpha 4$ mRNAs predominate (Laurie et al., 1992b). The $\alpha 5$ subunit gene is strongly expressed in the neonatal period in the caudate; and then the gene switches off. Any significance of this very clear subunit switch in $GABA_A$ receptor expression during striatal development has remained unstudied.

Occurrence of α1β2γ1 or α2β2γ1 receptors in the rodent globus pallidus

The globus pallidus expresses a lot of $\alpha 1$ mRNA, some $\alpha 2$, a lot of $\beta 2$ mRNA, and some of both $\gamma 1$ and some $\gamma 2$, with more $\gamma 1$ than $\gamma 2$ mRNA; $\gamma 3$ mRNA is not present (Wisden et al., 1992). The subunits in the globus pallidus (examined with $\alpha 1$, $\beta 2/3$ and $\gamma 2$ antibodies) are concentrated in

GABAergic (symmetrical, type II) synapses as assessed by postembedding, immuno-electron-microscopy (Somogyi et al., 1996). The mIPSCs recorded in the rat globus pallidus are zolpidem-sensitive, fitting with the expected $\alpha 1\beta 2\gamma 2$ type receptors (Chen et al., 2004). But the globus pallidus is one of the few brain areas that have elevated levels of the otherwise rarely expressed $\gamma 1$ subunit, although its levels in the neighbouring bed nucleus, stria terminalis and medial preoptic nucleus are even higher (Pirker et al., 2000; Schwarzer et al., 2001; Herb et al., 1992; Laurie et al., 1992b; Wisden et al., 1992). The brain areas that express $\gamma 1$ tend to also have $\alpha 2$ mRNA (Wisden et al., 1992) and so $\alpha 2\beta 2\gamma 1$ might be the combination. It is not known if the occurrence of $\alpha 1$, $\alpha 2$, $\beta 2$, $\gamma 1$ and $\gamma 2$ produce subtypes of receptor on the same cell type or if the $\gamma 1$ and $\gamma 2$ are in different cell types in the globus pallidus. The function of receptors with the $\gamma 1$ subunit has been relatively ignored by $GABA_A$ researchers, but one would expect that, for example, $\alpha 1\beta 2\gamma 1$ receptors would be synaptically located, with a 25–30 pS single channel conductance, but with an unknown pharmacological profile.

Strong developmental changes in GABA_A receptor subunit expression in the developing globus pallidus

During postnatal development, the expression of $\gamma 1$ mRNA in the globus pallidus is even higher (Laurie et al., 1992b). As for the striatum, a very marked switch in subunit gene expression takes place during rat globus pallidus development: the $\alpha 2$, ($\alpha 3$), $\beta 2$ and $\gamma 1$ genes are expressed in the neonatal (E19–P6) globus pallidus, whereas the $\alpha 1$, ($\alpha 2$), $\beta 2$, $\gamma 1$ and $\gamma 2$ mRNAs are found in the adult structure (Laurie et al., 1992b).

Substantia nigra reticulata

Substantia nigra neurons express mainly $\alpha 1$, $\beta 2$ and $\gamma 2$ mRNA and protein (Wisden et al., 1992; Fujiyama et al., 2002); post-embedding immuno-gold labelling showed that these subunits are enriched in symmetrical (GABAergic) synapses (Fujiyama et al., 2002). The main receptor subtype in all cell types in the reticulata will thus be the "standard" BZ1-version (α1/β2/3γ2) (see also Niddam et al., 1987; Duncan et al., 1995).

Substantia nigra compacta

The dopaminergic compacta cells receive GABA from GABAergic interneurons in the pars reticulata and in addition, a direct GABAergic input from the striatum and the pallidum, although controversy exists about the involvement of postsynaptic GABA$_A$ and/or GABA$_B$ receptors in the striato-nigral and pallido-nigral paths (Misgeld, 2004). GABA$_A$ receptors could be on the cell soma and dendrites of the compacta cells; some GABA$_A$ receptors could be located on the compacta axon terminals far away in the striatum, where they might regulate locally release of dopamine. Compacta cells express a complex mix of GABA$_A$ receptor subunits, which may also vary with species: by in situ hybridization, rodent cells express significant α4, less α3, no δ, β3, and possibly all three γ subunits (Wisden et al., 2002). Note this is one of the few brain areas where the α4 subunit occurs without the δ subunit, hippocampal pyramidal cells being the other example; thus in nigra compacta cells, α4 either forms α4β complexes or α4βγ complexes; this is unknown. Single-cell PCR on rodent dopaminergic compacta cells gives contradictory results: α3, α4, β2, β3 and γ3 were detected in one study (Guyon et al., 1999), α2, α3, α4, β1, β3 and γ2 in another (Okada et al., 2004). Both studies agree on α3 and α4, and that there is no δ subunit mRNA, fitting with the in situ hybridization data. In contrast, immunocytochemical analysis showed immunoreactivity mainly for α3, γ3 and δ on potentially dopaminergic neurons (Schwarzer et al., 2001); but here the detection of the δ immunoreactivity is at odds with the RNA data. The relevance of the α3 subunit in dopaminergic neurons is supported by electrophysiological analysis of α3 KO animals. Whole-cell GABA currents were reduced to less than 50% after bath application of GABA to compacta neurons (Yee et al., 2005).

Species differences in expression are likely: an in situ hybridization study on human substantia nigra compacta found strong expression of α1 subunit mRNA along with β2 and γ2 in about 25% of the compacta cells, suggesting an inhibitory input via the "standard" α1β2γ2 receptor subtype in a subpopulation of dopaminergic neurons in humans (Petri et al., 2002).

Acknowledgements

Research in our lab is supported by the Deutsche Forschungsgemeinschaft (Wi 1951/2–1) (Wulff and Wisden), the DFG Graduate College (GK 791) (Arslan), University of Heidelberg Young Investigator Award (Wulff) and Fonds der Chemischen Industrie (Wisden).

References

Alldred, M.J., Mulder-Rosi, J., Lingenfelter, S.E., Chen, G. and Luscher, B. (2005) Distinct gamma2 subunit domains mediate clustering and synaptic function of postsynaptic GABA$_A$ receptors and gephyrin. J. Neurosci., 25: 594–603.

Aller, M.I., Veale, E.L., Linden, A.M., Sandu, C., Schwaninger, M., Evans, L.J., Korpi, E.R., Mathie, A., Wisden, W. and Brickley, S.G. (2005) Modifying the subunit composition of TASK channels alters the modulation of a leak conductance in cerebellar granule neurons. J. Neurosci., 25: 11455–11467.

Atack, J.R., Hutson, P.H., Collinson, N., Marshall, G., Bentley, G., Moyes, C., Cook, S.M., Collins, I., Wafford, K., McKernan, R.M. and Dawson, G.R. (2005) Anxiogenic properties of an inverse agonist selective for alpha3 subunit-containing GABA$_A$ receptors. Br. J. Pharmacol., 144: 357–366.

Baer, K., Essrich, C., Benson, J.A., Benke, D., Bluethmann, H., Fritschy, J.M. and Luscher, B. (1999) Postsynaptic clustering of GABA$_A$ receptors by the γ3 subunit in vivo. Proc. Natl. Acad. Sci. USA., 96: 12860–12865.

Baumann, S.W., Baur, R. and Sigel, E. (2002) Forced subunit assembly in alpha1beta2gamma2 GABA$_A$ receptors. Insight into the absolute arrangement. J. Biol. Chem., 277: 46020–46025.

Baumann, S.W., Baur, R. and Sigel, E. (2003) Individual properties of the two functional agonist sites in GABA$_A$ receptors. J. Neurosci., 23: 11158–11166.

Belelli, D. and Lambert, J.J. (2005) Neurosteroids: endogenous regulators of the GABA(A) receptor. Nat. Rev. Neurosci., 6: 565–755.

Ben-Ari, Y., Khalilov, I., Represa, A. and Gozlan, H. (2004) Interneurons set the tune of developing networks. Trends Neurosci., 27: 422–427.

Bencsits, E., Ebert, V., Tretter, V. and Sieghart, W. (1999) A significant part of native gamma-aminobutyric AcidA receptors containing alpha4 subunits do not contain gamma or delta subunits. J. Biol. Chem., 274: 19613–19616.

Benke, D., Fakitsas, P., Roggenmoser, C., Michel, C., Rudolph, U. and Mohler, H. (2004) Analysis of the presence and abundance of GABA$_A$ receptors containing two different types of alpha subunits in murine brain using point-mutated alpha subunits. J. Biol. Chem., 279: 43654–43660.

Bonnert, T.P., McKernan, R.M., Farrar, S., le Bourdelles, B., Heavens, R.P., Smith, D.W., Hewson, L., Rigby, M.R., Sirinathsinghji, D.J., Brown, N., Wafford, K.A. and Whiting, P.J. (1999) Theta, a novel gamma-aminobutyric acid type A receptor subunit. Proc. Natl. Acad. Sci. USA, 96: 9891–9896.

Brejc, K., van Dijk, W.J., Klaassen, R.V., Schuurmans, M., van Der Oost, J., Smit, A.B. and Sixma, T.K. (2001) Crystal structure of an ACh-binding protein reveals the ligand-binding domain of nicotinic receptors. Nature, 411: 269–276.

Brickley, S.G., Revilla, V., Cull-Candy, S.G., Wisden, W. and Farrant, M. (2001) Adaptive regulation of neuronal excitability by a voltage-independent potassium conductance. Nature, 409: 88–92.

Brunig, I., Suter, A., Knuesel, I., Luscher, B. and Fritschy, J.M. (2002) GABAergic terminals are required for postsynaptic clustering of dystrophin but not of GABA(A) receptors and gephyrin. J. Neurosci., 22: 4805–4813.

Buhr, A., Baur, R. and Sigel, E. (1997) Subtle changes in residue 77 of the gamma subunit of alpha1beta2gamma2 GABA$_A$ receptors drastically alter the affinity for ligands of the benzodiazepine binding site. J. Biol. Chem., 272: 11799–11804.

Chen, G., Kittler, J.T., Moss, S.J. and Yan, Z. (2006) Dopamine D3 receptors regulate GABAA receptor function through a phospho-dependent endocytosis mechanism in nucleus accumbens. J. Neurosci., 26: 2513–2521.

Chen, L., Savio Chan, C. and Yung, W.H. (2004) Electrophysiological and behavioral effects of zolpidem in rat globus pallidus. Exp. Neurol, 186: 212–220.

Cope, D.W., Hughes, S.W. and Crunelli, V. (2005) GABA$_A$ receptor-mediated tonic inhibition in thalamic neurons. J. Neurosci., 25: 11553–11563.

Cope, D.W., Wulff, P., Oberto, A., Aller, M.I., Capogna, M., Ferraguti, F., Halbsguth, C., Hoeger, H., Jolin, H.E., Jones, A., McKenzie, A.N., Ogris, W., Poeltl, A., Sinkkonen, S.T., Vekovischeva, O.Y., Korpi, E.R., Sieghart, W., Sigel, E., Somogyi, P. and Wisden, W. (2004) Abolition of zolpidem sensitivity in mice with a point mutation in the GABA$_A$ receptor gamma2 subunit. Neuropharmacology, 47: 17–34.

Crestani, F., Keist, R., Fritschy, J.M., Benke, D., Vogt, K., Prut, L., Bluthmann, H., Mohler, H. and Rudolph, U. (2002) Trace fear conditioning involves hippocampal alpha5 GABA(A) receptors. Proc. Natl. Acad. Sci. USA., 99: 8980–8985.

Criswell, H.E., McCown, T.J., Moy, S.S., Oxford, G.S., Mueller, R.A., Morrow, A.L. and Breese, G.R. (1997) Action of zolpidem on responses to GABA in relation to mRNAs for GABA(A) receptor alpha subunits within single cells: evidence for multiple functional GABA(A) isoreceptors on individual neurons. Neuropharmacology, 36: 1641–1652.

Cromer, B.A., Morton, C.J. and Parker, M.W. (2002) Anxiety over GABA$_A$ receptor structure relieved by AChBP. Trends Biochem. Sci., 27: 280–287.

Davies, P.A., Hanna, M.C., Hales, T.G. and Kirkness, E.F. (1997) Insensitivity to anaesthetic agents conferred by a class of GABA(A) receptor subunit. Nature, 385: 820–823.

Dias, R., Sheppard, W.F., Fradley, R.L., Garrett, E.M., Stanley, J.L., Tye, S.J., Goodacre, S., Lincoln, R.J., Cook, S.M., Conley, R., Hallett, D., Humphries, A.C., Thompson, S.A., Wafford, K.A., Street, L.J., Castro, J.L., Whiting, P.J., Rosahl, T.W., Atack, J.R., McKernan, R.M., Dawson, G.R. and Reynolds, D.S. (2005) Evidence for a significant role of alpha 3-containing GABA$_A$ receptors in mediating the anxiolytic effects of benzodiazepines. J. Neurosci., 25: 10682–10688.

Duncan, G.E., Breese, G.R., Criswell, H.E., McCown, T.J., Herbert, J.S., Devaud, L.L. and Morrow, A.L. (1995) Distribution of [^{3}H]zolpidem binding sites in relation to messenger RNA encoding the alpha 1, beta 2 and gamma 2 subunits of GABA$_A$ receptors in rat brain. Neuroscience, 64: 1113–1128.

Ernst, M., Brauchart, D., Boresch, S. and Sieghart, W. (2003) Comparative modeling of GABA(A) receptors: limits, insights, future developments. Neuroscience, 119: 933–943.

Essrich, C., Lorez, M., Benson, J.A., Fritschy, J.M. and Luscher, B. (1998) Postsynaptic clustering of major GABA$_A$ receptor subtypes requires the gamma 2 subunit and gephyrin. Nat. Neurosci., 1: 563–571.

Farrant, M. and Nusser, Z. (2005) Variations on an inhibitory theme: phasic and tonic activation of GABA(A) receptors. Nat. Rev. Neurosci., 6: 215–229.

Faull, R.L. and Villiger, J.W. (1988) Multiple benzodiazepine receptors in the human basal ganglia: a detailed pharmacological and anatomical study. Neuroscience, 24: 433–451.

Fisher, J.L. and Macdonald, R.L. (1997) Single channel properties of recombinant GABA$_A$ receptors containing gamma 2 or delta subtypes expressed with alpha 1 and beta 3 subtypes in mouse L929 cells. J. Physiol., 505: 283–297.

Fiumelli, H., Cancedda, L. and Poo, M.M. (2005) Modulation of GABAergic transmission by activity via postsynaptic Ca^{2+}-dependent regulation of KCC2 function. Neuron, 48: 773–786.

Franks, N.P. and Honore, E. (2004) The TREK K2P channels and their role in general anaesthesia and neuroprotection. Trends Pharmacol. Sci., 25: 601–608.

Freund, T.F. (2003) Interneuron Diversity series: rhythm and mood in perisomatic inhibition. Trends Neurosci, 26: 489–495.

Fritschy, J.M. and Mohler, H. (1995) GABA$_A$-receptor heterogeneity in the adult rat brain: differential regional and cellular distribution of seven major subunits. J. Comp. Neurol., 359: 154–194.

Fujiyama, F., Fritschy, J.M., Stephenson, F.A. and Bolam, J.P. (2000) Synaptic localization of GABA$_A$ receptor subunits in the striatum of the rat. J. Comp. Neurol., 416: 158–172.

Fujiyama, F., Stephenson, F.A. and Bolam, J.P. (2002) Synaptic localization of GABA$_A$ receptor subunits in the substantia nigra of the rat: effects of quinolinic acid lesions of the striatum. Eur. J. Neurosci., 15: 1961–1975.

Galzi, J.L., Devillers-Thiery, A., Hussy, N., Bertrand, S., Changeux, J.P. and Bertrand, D. (1992) Mutations in the channel domain of a neuronal nicotinic receptor convert ion selectivity from cationic to anionic. Nature, 359: 500–505.

38

Grenningloh, G., Rienitz, A., Schmitt, B., Methfessel, C., Zensen, M., Beyreuther, K., Gundelfinger, E.D. and Betz, H. (1987) The strychnine-binding subunit of the glycine receptor shows homology with nicotinic acetylcholine receptors. Nature, 328: 215–220.

Griebel, G., Perrault, G., Simiand, J., Cohen, C., Granger, P., Decobert, M., Francon, D., Avenet, P., Depoortere, H., Tan, S., Oblin, A., Schoemaker, H., Evanno, Y., Sevrin, M., George, P. and Scatton, B. (2001) SL651498: an anxioselective compound with functional selectivity for alpha2- and alpha3-containing gamma-aminobutyric acid(A) (GABA(A)) receptors. J. Pharmacol. Exp. Ther., 298: 753–768.

Griebel, G., Perrault, G., Simiand, J., Cohen, C., Granger, P., Depoortere, H., Francon, D., Avenet, P., Schoemaker, H., Evanno, Y., Sevrin, M., George, P. and Scatton, B. (2003) SL651498, a GABAA receptor agonist with subtype-selective efficacy, as a potential treatment for generalized anxiety disorder and muscle spasms. CNS Drug Rev, 9: 3–20.

Gulasci, A., Lee, C.R., Sik, A., Viitanen, T., Kaila, K., Tepper, J.M. and Freund, T.F. (2003) Cell-type specific differences in chloride-regulatory mechanisms and GABA$_A$ receptor-mediated inhibition in rat substantia nigra. J. Neurosci., 23: 8237–8246.

Gunther, U., Benson, J., Benke, D., Fritschy, J.M., Reyes, G., Knoflach, F., Crestani, F., Aguzzi, A., Arigoni, M., Lang, Y., Bluethmann, H., Mohler, H. and Luscher, B. (1995) Benzodiazepine-insensitive mice generated by targeted disruption of the gamma 2 subunit gene of gamma-aminobutyric acid type A receptors. Proc. Natl. Acad. Sci. USA., 92: 7749–7753.

Guyon, A., Laurent, S., Paupardin-Tritsch, D., Rossier, J. and Eugene, D. (1999) Incremental conductance levels of GABAA receptors in dopaminergic neurones of the rat substantia nigra pars compacta. J. Physiol., 516: 719–737.

Haas, K.F. and Macdonald, R.L. (1999) GABA$_A$ receptor subunit gamma2 and delta subtypes confer unique kinetic properties on recombinant GABA$_A$ receptor currents in mouse fibroblasts. J. Physiol., 514: 27–45.

Hadingham, K.L., Wingrove, P., Le Bourdelles, B., Palmer, K.J., Ragan, C.I. and Whiting, P.J. (1993) Cloning of cDNA sequences encoding human alpha 2 and alpha 3 gamma-aminobutyric acidA receptor subunits and characterization of the benzodiazepine pharmacology of recombinant alpha 1-, alpha 2-, alpha 3-, and alpha 5-containing human gamma-aminobutyric acidA receptors. Mol. Pharmacol., 43: 970–975.

Hales, T.G., Dunlop, J.I., Deeb, T.Z., Carland, J.E., Kelley, S.P., Lambert, J.J. and Peters, J.A. (2006) Common determinants of single channel conductance within the large cytoplasmic loop of 5-HT3 and alpha 4beta 2 nicotinic acetylcholine receptors. J. Biol. Chem., in press.

Hanchar, H.J., Dodson, P.D., Olsen, R.W., Otis, T.S. and Wallner, M. (2005) Alcohol-induced motor impairment caused by increased extrasynaptic GABA(A) receptor activity. Nat. Neurosci., 8: 339–345.

Hanchar, H.J., Wallner, M. and Olsen, R.W. (2004) Alcohol effects on γ-aminobutyric acid type A receptors: are extrasynaptic receptors the answer? Life Sci, 76: 1–8.

Hedblom, E. and Kirkness, E.F. (1997) A novel class of GABA$_A$ receptor subunit in tissues of the reproductive system. J. Biol. Chem., 272: 15346–15350.

Herb, A., Wisden, W., Luddens, H., Puia, G., Vicini, S. and Seeburg, P.H. (1992) The third gamma subunit of the gamma-aminobutyric acid type A receptor family. Proc. Natl. Acad. Sci. USA., 89: 1433–1437.

Hodge, C.W., Mehmert, K.K., Kelley, S.P., McMahon, T., Haywood, A., Olive, M.F., Wang, D., Sanchez-Perez, A.M. and Messing, R.O. (1999) Supersensitivity to allosteric GABA(A) receptor modulators and alcohol in mice lacking PKCepsilon. Nat. Neurosci., 2: 997–1002.

Hosie, A.M., Dunne, E.L., Harvey, R.J. and Smart, T.G. (2003) Zinc-mediated inhibition of GABA(A) receptors: discrete binding sites underlie subtype specificity. Nat. Neurosci., 6: 362–369.

Jensen, M.L., Timmermann, D.B., Johansen, T.H., Schousboe, A., Varming, T. and Ahring, P.K. (2002) The β subunit determines the ion selectivity of the GABA$_A$ receptor. J. Biol. Chem., 277: 41438–41447.

Jones, A., Korpi, E.R., McKernan, R.M., Pelz, R., Nusser, Z., Makela, R., Mellor, J.R., Pollard, S., Bahn, S., Stephenson, F.A., Randall, A.D., Sieghart, W., Somogyi, P., Smith, A.J. and Wisden, W. (1997) Ligand-gated ion channel subunit partnerships: GABA$_A$ receptor alpha6 subunit gene inactivation inhibits delta subunit expression. J. Neurosci., 17: 1350–1362.

Jovanovic, J.N., Thomas, P., Kittler, J.T., Smart, T.G. and Moss, S.J. (2004) Brain-derived neurotrophic factor modulates fast synaptic inhibition by regulating GABA(A) receptor phosphorylation, activity, and cell-surface stability. J. Neurosci., 24: 522–530.

Jurd, R., Arras, M., Lambert, S., Drexler, B., Siegwart, R., Crestani, F., Zaugg, M., Vogt, K.E., Ledermann, B., Antkowiak, B. and Rudolph, U. (2003) General anesthetic actions in vivo strongly attenuated by a point mutation in the GABA(A) receptor beta3 subunit. FASEB J, 17: 250–252.

Kaila, K., Lamsa, K., Smirnov, S., Taira, T. and Voipio, J. (1997) Long-lasting GABA-mediated depolarization evoked by high-frequency stimulation in pyramidal neurons of rat hippocampal slice is attributable to a network-driven, bicarbonate-dependent K$^+$ transient. J. Neurosci., 17: 7662–7672.

Kash, T.L., Dizon, M.J., Trudell, J.R. and Harrison, N.L. (2004) Charged residues in the beta2 subunit involved in GABA$_A$ receptor activation. J. Biol. Chem., 279: 4887–4893.

Kash, T.L., Jenkins, A., Kelley, J.C., Trudell, J.R. and Harrison, N.L. (2003) Coupling of agonist binding to channel gating in the GABA(A) receptor. Nature, 421: 272–275.

Kittler, J.T. and Moss, S.J. (2003) Modulation of GABA$_A$ receptor activity by phosphorylation and receptor trafficking: implications for the efficacy of synaptic transmission. Curr. Opin. Neurobiol., 13: 341–347.

Kneussel, M., Brandstatter, J.H., Gasnier, B., Feng, G., Sanes, J.R. and Betz, H. (2001) Gephyrin-independent clustering of postsynaptic GABA(A) receptor subtypes. Mol. Cell Neurosci., 17: 973–982.

Kneussel, M., Brandstatter, J.H., Laube, B., Stahl, S., Muller, U. and Betz, H. (1999) Loss of postsynaptic GABA(A) receptor clustering in gephyrin-deficient mice. J. Neurosci., 19: 9289–9297.

Korpi, E.R., Grunder, G. and Luddens, H. (2002a) Drug interactions at GABA(A) receptors. Prog. Neurobiol., 67: 113–159.

Korpi, E.R., Koikkalainen, P., Vekovischeva, O.Y., Makela, R., Kleinz, R., Uusi-Oukari, M. and Wisden, W. (1999) Cerebellar granule-cell-specific GABA$_A$ receptors attenuate benzodiazepine-induced ataxia: evidence from alpha 6-subunit-deficient mice. Eur. J. Neurosci., 11: 233–240.

Korpi, E.R., Mihalek, R.M., Sinkkonen, S.T., Hauer, B., Hevers, W., Homanics, G.E., Sieghart, W. and Luddens, H. (2002b) Altered receptor subtypes in the forebrain of GABA(A) receptor delta subunit-deficient mice: recruitment of gamma 2 subunits. Neuroscience, 109: 733–743.

Kralic, J.E., Sidler, C., Parpan, F., Homanics, G.E., Morrow, A.L. and Fritschy, J.M. (2006) Compensatory alteration of inhibitory synaptic circuits in cerebellum and thalamus of gamma-aminobutyric acid type A receptor alpha1 subunit knockout mice. J. Comp. Neurol., 495: 408–421.

Krasowski, M.D. and Harrison, N.L. (1999) General anaesthetic actions on ligand-gated ion channels. Cell Mol. Life Sci., 55: 1278–1303.

Kukhtina, V., Kottwitz, D., Strauss, H., Heise, B., Chebotareva, N., Tsetlin, V. and Hucho, F. (2006) Intracellular domain of nicotinic acetylcholine receptor: the importance of being unfolded. J. Neurochem., 97: 63–67.

Laurie, D.J., Seeburg, P.H. and Wisden, W. (1992a) The distribution of 13 GABA$_A$ receptor subunit mRNAs in the rat brain. II. Olfactory bulb and cerebellum. J. Neurosci., 12: 1063–1076.

Laurie, D.J., Wisden, W. and Seeburg, P.H. (1992b) The distribution of thirteen GABA$_A$ receptor subunit mRNAs in the rat brain. III. Embryonic and postnatal development. J. Neurosci., 12: 4151–4172.

Lavoie, A.M., Tingey, J.J., Harrison, N.L., Pritchett, D.B. and Twyman, R.E. (1997) Activation and deactivation rates of recombinant GABA(A) receptor channels are dependent on alpha-subunit isoform. Biophys. J., 73: 2518–2526.

Lee, W.Y. and Sine, S.M. (2005) Principal pathway coupling agonist binding to channel gating in nicotinic receptors. Nature, 438: 243–247.

Lester, H.A., Dibas, M.I., Dahan, D.S., Leite, J.F. and Dougherty, D.A. (2004) Cys-loop receptors: new twists and turns. Trends Neurosci, 27: 329–336.

Levi, S., Logan, S.M., Tovar, K.R. and Craig, A.M. (2004) Gephyrin is critical for glycine receptor clustering but not for the formation of functional GABAergic synapses in hippocampal neurons. J. Neurosci., 24: 207–217.

Liu, F., Wan, Q., Pristupa, Z.B., Yu, X.M., Wang, Y.T. and Niznik, H.B. (2000) Direct protein–protein coupling enables cross-talk between dopamine D5 and gamma-aminobutyric acid A receptors. Nature, 403: 274–280.

Luddens, H., Pritchett, D.B., Kohler, M., Killisch, I., Keinanen, K., Monyer, H., Sprengel, R. and Seeburg, P.H. (1990)

Cerebellar GABA$_A$ receptor selective for a behavioural alcohol antagonist. Nature, 346: 648–651.

Lummis, S.C., Beene, D.L., Lee, L.W., Lester, H.A., Broadhurst, R.W. and Dougherty, D.A. (2005) Cis–trans isomerization at a proline opens the pore of a neurotransmitter-gated ion channel. Nature, 438: 248–252.

Luscher, B. and Keller, C.A. (2004) Regulation of GABA$_A$ receptor trafficking, channel activity, and functional plasticity of inhibitory synapses. Pharmacol. Ther., 102: 195–221.

Macdonald, R.L. and Haas, K.F. (2000) Kinetic Properties of GABA$_A$ Receptor Channels. In: Martin D.L. and Olsen R.W. (Eds.), GABA in the Nervous System: The View at Fifty Years. Lippincott Williams and Wilkins, Philadelphia, pp. 141–165.

Mathie, A., Sutton, G.L., Clarke, C.E. and Veale, E.L. (2006) Zinc and copper: pharmacological probes and endogenous modulators of neuronal excitability. Pharmacol Ther., 111: 567–583.

McKernan, R.M., Rosahl, T.W., Reynolds, D.S., Sur, C., Wafford, K.A., Atack, J.R., Farrar, S., Myers, J., Cook, G., Ferris, P., Garrett, L., Bristow, L., Marshall, G., Macaulay, A., Brown, N., Howell, O., Moore, K.W., Carling, R.W., Street, L.J., Castro, J.L., Ragan, C.I., Dawson, G.R. and Whiting, P.J. (2000) Sedative but not anxiolytic properties of benzodiazepines are mediated by the GABA(A) receptor alpha1 subtype. Nat. Neurosci., 3: 587–592.

Mellor, J.R. and Randall, A.D. (1997) Frequency-dependent actions of benzodiazepines on GABA$_A$ receptors in cultured murine cerebellar granule cells. J. Physiol., 503: 353–369.

Mihalek, R.M., Banerjee, P.K., Korpi, E.R., Quinlan, J.J., Firestone, L.L., Mi, Z.P., Lagenaur, C., Tretter, V., Sieghart, W., Anagnostaras, S.G., Sage, J.R., Fanselow, M.S., Guidotti, A., Spigelman, I., Li, Z., DeLorey, T.M., Olsen, R.W. and Homanics, G.E. (1999) Attenuated sensitivity to neuroactive steroids in gamma-aminobutyrate type A receptor delta subunit knockout mice. Proc. Natl. Acad. Sci. USA, 96: 12905–12910.

Misgeld, U. (2004) Innervation of the substantia nigra. Cell Tissue Res, 318: 107–114.

Mody, I. and Pearce, R.A. (2004) Diversity of inhibitory neurotransmission through GABA$_A$ receptors. Trends Neurosci, 27: 569–575.

Moragues, N., Ciofi, P., Tramu, G. and Garret, M. (2002) Localisation of GABA(A) receptor epsilon-subunit in cholinergic and aminergic neurones and evidence for co-distribution with the theta-subunit in rat brain. Neuroscience, 111: 657–669.

Niddam, R., Dubois, A., Scatton, B., Arbilla, S. and Langer, S.Z. (1987) Autoradiographic localization of [^{3}H]zolpidem binding sites in the rat CNS: comparison with the distribution of [^{3}H]flunitrazepam binding sites. J. Neurochem., 49: 890–899.

Nusser, Z. (1999) A new approach to estimate the number, density and variability of receptors at central synapses. Eur. J. Neurosci., 11: 745–752.

Nusser, Z., Sieghart, W. and Somogyi, P. (1998) Segregation of different GABA$_A$ receptors to synaptic and extrasynaptic membranes of cerebellar granule cells. J. Neurosci., 18: 1693–1703.

Nyiri, G., Freund, T.F. and Somogyi, P. (2001) Input-dependent synaptic targeting of alpha(2)-subunit-containing GABA(A) receptors in synapses of hippocampal pyramidal cells of the rat. Eur. J. Neurosci., 13: 428–442.

Ogris, W., Poltl, A., Hauer, B., Ernst, M., Oberto, A., Wulff, P., Hoger, H., Wisden, W. and Sieghart, W. (2004) Affinity of various benzodiazepine site ligands in mice with a point mutation in the GABA(A) receptor gamma2 subunit. Biochem. Pharmacol., 68: 1621–1629.

Okada, H., Matsushita, N., Kobayashi, K. and Kobayashi, K. (2004) Identification of GABA$_A$ receptor subunit variants in midbrain dopaminergic neurons. J. Neurochem., 89: 7–14.

Olsen, R.W., Chang, C.S., Li, G., Hanchar, H.J. and Wallner, M. (2004) Fishing for allosteric sites on GABA(A) receptors. Biochem. Pharmacol., 68: 1675–1684.

Olsen, R.W., McCabe, R.T. and Wamsley, J.K. (1990) GABA$_A$ receptor subtypes: autoradiographic comparison of GABA, benzodiazepine, and convulsant binding sites in the rat central nervous system. J. Chem. Neuroanat., 3: 59–76.

Peng, Z., Hauer, B., Mihalek, R.M., Homanics, G.E., Sieghart, W., Olsen, R.W. and Houser, C.R. (2002) GABA(A) receptor changes in delta subunit-deficient mice: altered expression of alpha4 and gamma2 subunits in the forebrain. J. Comp. Neurol., 446: 179–197.

Persohn, E., Malherbe, P. and Richards, J.G. (1992) Comparative molecular neuroanatomy of cloned GABA$_A$ receptor subunits in the rat CNS. J. Comp. Neurol., 326: 193–216.

Peters, J.A., Hales, T.G. and Lambert, J.J. (2005) Molecular determinants of single channel conductance and ion selectivity in the Cys-loop family: insights from the 5-HT$_3$ receptor. Trends Pharmacol. Sci., 26: 587–594.

Petri, S., Krampfl, K., Dengler, R., Bufler, J., Weindl, A. and Arzberger, T. (2002) Human GABA$_A$ receptors on dopaminergic neurons in the pars compacta of the substantia nigra. J. Comp. Neurol., 452: 360–366.

Pirker, S., Schwarzer, C., Wieselthaler, A., Sieghart, W. and Sperk, G. (2000) GABA(A) receptors: immunocytochemical distribution of 13 subunits in the adult rat brain. Neuroscience, 101: 815–850.

Pritchett, D.B., Luddens, H. and Seeburg, P.H. (1989) Type I and type II GABA$_A$-benzodiazepine receptors produced in transfected cells. Science, 245: 1389–1392.

Ramming, M., Kins, S., Werner, N., Hermann, A., Betz, H. and Kirsch, J. (2000) Diversity and phylogeny of gephyrin: tissue-specific splice variants, gene structure, and sequence similarities to molybdenum cofactor-synthesizing and cytoskeleton-associated proteins. Proc. Natl. Acad. Sci. USA., 97: 10266–10271.

Rathenberg, J., Kittler, J.T. and Moss, S.J. (2004) Palmitoylation regulates the clustering and cell surface stability of GABA$_A$ receptors. Mol. Cell Neurosci., 26: 251–257.

Reynolds, D.S., Rosahl, T.W., Cirone, J., O'Meara, G.F., Haythornthwaite, A., Newman, R.J., Myers, J., Sur, C., Howell, O., Rutter, A.R., Atack, J., Macaulay, A.J., Hadingham, K.L., Hutson, P.H., Belelli, D., Lambert, J.J.,

Dawson, G.R., McKernan, R., Whiting, P.J. and Wafford, K.A. (2003) Sedation and anesthesia mediated by distinct GABA(A) receptor isoforms. J. Neurosci., 23: 8608–8617.

Rivera, C., Voipio, J. and Kaila, K. (2005) Two developmental switches in GABAergic signalling: the K$^+$–Cl$^-$ cotransporter KCC2 and carbonic anhydrase CAVII. J. Physiol., 562: 27–36.

Rowlett, J.K., Platt, D.M., Lelas, S., Atack, J.R. and Dawson, G.R. (2005) Different GABA$_A$ receptor subtypes mediate the anxiolytic, abuse-related, and motor effects of benzodiazepine-like drugs in primates. Proc. Natl. Acad. Sci. USA., 102: 915–920.

Rudolph, U. and Moehler, H. (2006) GABA-based therapeutic approaches: GABA$_A$ receptor subtype functions. Curr. Opin. Pharmacol., 6: 18–23.

Rudolph, U. and Mohler, H. (2004) Analysis of GABA$_A$ receptor function and dissection of the pharmacology of benzodiazepines and general anesthetics through mouse genetics. Annu. Rev. Pharmacol. Toxicol., 44: 475–498.

Saper, C.B. and Sawchenko, P.E. (2003) Magic peptides, magic antibodies: guidelines for appropriate controls for immunohistochemistry. J. Comp. Neurol., 465: 161–163.

Schofield, P.R., Darlison, M.G., Fujita, N., Burt, D.R., Stephenson, F.A., Rodriguez, H., Rhee, L.M., Ramachandran, J., Reale, V., Glencorse, T.A., et al. (1987) Sequence and functional expression of the GABA$_A$ receptor shows a ligand-gated receptor super-family. Nature, 328: 221–227.

Schwarzer, C., Berresheim, U., Pirker, S., Wieselthaler, A., Fuchs, K., Sieghart, W. and Sperk, G. (2001) Distribution of the major gamma-aminobutyric acid(A) receptor subunits in the basal ganglia and associated limbic brain areas of the adult rat. J. Comp. Neurol., 433: 526–549.

Schweizer, C., Balsiger, S., Bluethmann, H., Mansuy, I.M., Fritschy, J.M., Mohler, H. and Luscher, B. (2003) The gamma 2 subunit of GABA(A) receptors is required for maintenance of receptors at mature synapses. Mol. Cell. Neurosci., 24: 442–450.

Seeburg, P.H., Wisden, W., Verdoorn, T.A., Pritchett, D.B., Werner, P., Herb, A., Luddens, H., Sprengel, R. and Sakmann, B. (1990) The GABA$_A$ receptor family: molecular and functional diversity. Cold Spring Harb. Symp. Quant. Biol., 55: 29–40.

Semyanov, A., Walker, M.C., Kullmann, D.M. and Silver, R.A. (2004) Tonically active GABA$_A$ receptors: modulating gain and maintaining the tone. Trends Neurosci, 27: 262–269.

Sieghart, W. and Sperk, G. (2002) Subunit composition, distribution and function of GABA(A) receptor subtypes. Curr. Top. Med. Chem., 2: 795–816.

Sigel, E. and Buhr, A. (1997) The benzodiazepine binding site of GABA$_A$ receptors. Trends Pharmacol. Sci., 18: 425–429.

Simon, J., Wakimoto, H., Fujita, N., Lalande, M. and Barnard, E.A. (2004) Analysis of the set of GABA$_A$ receptor genes in the human genome. J. Biol. Chem., 279: 41422–41435.

Sinkkonen, S.T., Hanna, M.C., Kirkness, E.F. and Korpi, E.R. (2000) GABA(A) receptor epsilon and theta subunits display unusual structural variation between species and are enriched in the rat locus ceruleus. J. Neurosci., 20: 3588–3595.

Smith, Y., Bevan, M.D., Shink, E. and Bolam, J.P. (1998) Microcircuitry of the direct and indirect pathways of the basal ganglia. Neuroscience, 86: 353–387.

Somogyi, P., Fritschy, J.M., Benke, D., Roberts, J.D. and Sieghart, W. (1996) The gamma2 subunit of the $GABA_A$ receptor is concentrated in synaptic junctions containing the alpha1 and beta2/3 subunits in hippocampus, cerebellum and globus pallidus. Neuropharmacology, 35: 1425–1444.

Staley, K. and Scharfmann, H. (2005) A woman's prerogative. Nat. Neurosci., 8: 697–699.

Stell, B.M., Brickley, S.G., Tang, C.Y., Farrant, M. and Mody, I. (2003) Neuroactive steroids reduce neuronal excitability by selectively enhancing tonic inhibition mediated by delta subunit-containing $GABA_A$ receptors. Proc. Natl. Acad. Sci. USA., 100: 14439–14444.

Sundstrom-Poromaa, I., Smith, D.H., Gong, Q.H., Sabado, T.N., Li, X., Light, A., Wiedmann, M., Williams, K. and Smith, S.S. (2002) Hormonally regulated alpha(4)beta(2)delta GABA(A) receptors are a target for alcohol. Nat. Neurosci., 5: 721–722.

Szabadics, J., Varga, C., Molnar, G., Olah, S., Barzo, P. and Tamas, G. (2006) Excitatory effect of GABAergic axo-axonic cells in cortical microcircuits. Science, 311: 233–235.

Tepper, J.M. and Bolam, J.P. (2004) Functional diversity and specificity of neostriatal interneurons. Curr. Opin. Neurobiol., 14: 685–692.

Thomas, P., Martensen, M., Hosie, A.M. and Smart, T.G. (2005) Dynamic mobility of functional $GABA_A$ receptors at inhibitory synapses. Nat. Neurosci., 8: 889–897.

Thompson, S.A. and Wafford, K. (2001) Mechanism of action of general anaesthetics—new information from molecular pharmacology. Curr. Opin. Pharmacol., 1: 78–83.

Thomson, A.M., Bannister, A.P., Hughes, D.I. and Pawelzik, H. (2000) Differential sensitivity to Zolpidem of IPSPs activated by morphologically identified CA1 interneurons in slices of rat hippocampus. Eur. J. Neurosci., 12: 425–436.

Tretter, V., Hauer, B., Nusser, Z., Mihalek, R.M., Hoger, H., Homanics, G.E., Somogyi, P. and Sieghart, W. (2001) Targeted disruption of the GABA(A) receptor delta subunit gene leads to an up-regulation of gamma 2 subunit-containing receptors in cerebellar granule cells. J. Biol. Chem., 276: 10532–10538.

Unwin, N. (2003) Structure and action of the nicotinic acetylcholine receptor explored by electron microscopy. FEBS Lett, 555: 91–95.

Unwin, N. (2005) Refined structure of the nicotinic acetylcholine receptor at 4A resolution. J. Mol. Biol., 346: 967–989.

Vicini, S. and Ortinski, P. (2004) Genetic manipulations of $GABA_A$ receptor in mice make inhibition exciting. Pharmacol. Ther., 103: 109–120.

Waldvogel, H.J., Kubota, Y., Fritschy, J., Mohler, H. and Faull, R.L. (1999) Regional and cellular localisation of GABA(A) receptor subunits in the human basal ganglia: An autoradiographic and immunohistochemical study. J. Comp. Neurol., 415: 313–340.

Wallner, M., Hanchar, H.J. and Olsen, R.W. (2003) Ethanol enhances alpha 4 beta 3 delta and alpha 6 beta 3 delta gamma-aminobutyric acid type A receptors at low concentrations known to affect humans. Proc. Natl. Acad. Sci. USA, 100: 5218–15223.

Wang, Q., Liu, L., Pei, L., Ju, W., Ahmadian, G., Lu, J., Wang, Y., Liu, F. and Wang, Y.T. (2003) Control of synaptic strength, a novel function of Akt. Neuron, 38: 915–928.

Wang, X., Zhong, P. and Yan, Z. (2002) Dopamine D4 receptors modulate GABAergic signaling in pyramidal neurons of prefrontal cortex. J. Neurosci., 22: 9185–9193.

Wei, W., Zhang, N., Peng, Z., Houser, C.R. and Mody, I. (2003) Perisynaptic localization of delta subunit-containing GABA(A) receptors and their activation by GABA spillover in the mouse dentate gyrus. J. Neurosci., 23: 10650–10661.

Whiting, P.J. (2006) $GABA_A$ receptors: a viable target for novel anxiolytics? Curr. Opin. Pharmacol., 6: 24–29.

Wieland, H.A., Luddens, H. and Seeburg, P.H. (1992) A single histidine in $GABA_A$ receptors is essential for benzodiazepine agonist binding. J. Biol. Chem., 267: 1426–1429.

Wingrove, P.B., Thompson, S.A., Wafford, K.A. and Whiting, P.J. (1997) Key amino acids in the gamma subunit of the gamma-aminobutyric acidA receptor that determine ligand binding and modulation at the benzodiazepine site. Mol Pharmacol, 52: 874–881.

Wisden, W., Cope, D., Klausberger, T., Hauer, B., Sinkkonen, S.T., Tretter, V., Lujan, R., Jones, A., Korpi, E.R., Mody, I., Sieghart, W. and Somogyi, P. (2002) Ectopic expression of the GABA(A) receptor alpha6 subunit in hippocampal pyramidal neurons produces extrasynaptic receptors and an increased tonic inhibition. Neuropharmacology, 43: 530–549.

Wisden, W., Herb, A., Wieland, H., Keinanen, K., Luddens, H. and Seeburg, P.H. (1991) Cloning, pharmacological characteristics and expression pattern of the rat $GABA_A$ receptor alpha 4 subunit. FEBS Lett, 289: 227–230.

Wisden, W., Laurie, D.J., Monyer, H. and Seeburg, P.H. (1992) The distribution of 13 $GABA_A$ receptor subunit mRNAs in the rat brain. I. Telencephalon, diencephalon, mesencephalon. J. Neurosci., 12: 1040–1062.

Wisden, W., Morris, B.J., Darlison, M.G., Hunt, S.P. and Barnard, E.A. (1989) Localization of $GABA_A$ receptor alpha-subunit mRNAs in relation to receptor subtypes. Mol. Brain Res., 5: 305–310.

Wohlfarth, K.M., Bianchi, M.T. and Macdonald, R.L. (2002) Enhanced neurosteroid potentiation of ternary GABA(A) receptors containing the delta subunit. J. Neurosci., 22: 1541–1549.

Xiu, X., Hanek, A.P., Wang, J., Lester, H.A. and Dougherty, D.A. (2005) A unified view of the role of electrostatic interactions in modulating the gating of Cys loop receptors. J. Biol. Chem., 280: 41655–41666.

Yee, B.K., Keist, R., von Boehmer, L., Studer, R., Benke, D., Hagenbuch, N., Dong, Y., Malenka, R.C., Fritschy, J.M., Bluethmann, H., Feldon, J., Mohler, H. and Rudolph, U. (2005) A schizophrenia-related sensorimotor deficit links alpha 3-containing $GABA_A$ receptors to a dopamine hyperfunction. Proc. Natl. Acad. Sci. USA, 102: 17154–17159.

Tepper, Abercrombie & Bolam (Eds.)
Progress in Brain Research, Vol. 160
ISSN 0079-6123

CHAPTER 4

GABA$_B$ receptors: structure and function

Piers C. Emson*

The Babraham Institute, Babraham Research Campus, Cambridge CB2 4AT, UK

Abstract: In the basal ganglia the effects of γ-aminobutyrate (GABA) are mediated by both ionotropic (GABA$_A$) and metabotropic (GABA$_B$) receptors. Although the existence and widespread distribution in the CNS of the GABA$_B$ receptor had been established in the 1980s the field of GABA$_B$ research was revolutionized with the discovery that two related G-protein-coupled receptors (GPCRs) needed to dimerize to form the functional GABA$_B$ receptor at the cell surface. This finding lead to a number of studies of oligomerization in GPCRs and detailed pharmacological studies of the cloned receptors and their splice variants. Particular interest has focused on the proteins interacting with the receptor which may be important in mediating the longer term signalling effects of the receptor and modifying its cellular localization or physiology. The cloning of the GABA$_B$ receptors also lead to the identification of the first compounds interacting in an allosteric fashion with the receptor some of which may have therapeutic value. Most recently "knockouts" of both the GABA$_B$ subunits have been produced where in general as expected there is a loss of the majority of the inhibitory effects of the GABA$_B$ receptor

Keywords: GABA$_B$ receptor; G-protein coupled receptor; allosteric regulator; signalling properties; dimerization

Introduction

The actions of γ-aminobutyrate (GABA) in the basal ganglia and throughout the mammalian CNS are mediated by two classes of GABA receptors, the GABA$_A$ (see Wisden and Kaila chapters in this volume) and GABA$_B$ receptors. The first class described was the ligand-gated chloride GABA$_A$ channels (see early reviews, e.g., Curtis and Johnston, 1974; Krnjevic, 1984) that could be detected by binding assays using ^{3}H-GABA where binding was blocked by the antagonist, bicuculline (Curtis et al., 1974) and that were dependent on chloride (Zukin et al., 1974). In contrast, in 1979 Bowery and colleagues studying the release of ^{3}H-noradrenaline

from the rat atria reported that GABA would reduce the release of ^{3}H-noradrenaline, but the process was not chloride dependent or could not be blocked by bicuculline suggesting the presence of a separate pharmacologically distinct GABA receptor (Bowery et al., 1979, 1980, 1981a, b), although this was disputed at the time (Bowery, 1993). Further studies identified β-*p*-chlorophenyl GABA (baclofen) as an active agonist at this putative GABA receptor and the use of the ^{3}H-labelled form of baclofen allowed Bowery and colleagues to reveal a Ca^{2+}-dependent binding of ^{3}H-baclofen and ^{3}H-GABA in the CNS at the site they had earlier termed the GABA$_B$ site (Hill and Bowery, 1981), establishing GABA$_B$ as a CNS, as well as a peripheral receptor (Bowery et al., 1983; Curtis and Johnston, 1974). Baclofen was already known to decrease transmitter release and depress neuronal

*Corresponding author. Tel.: 01223 496502;
Fax: 01223 496022; E-mail: piers.emson@bbsrc.ac.uk

activity probably through an action on calcium channels. Subsequent work established a physiological role for a GABA$_B$ postsynaptic receptor where activation produced an increase in membrane potassium conductance and neuronal hyperpolarization (Dutar and Nicoll, 1988; Nicoll, 2004). Thus, activation of the presynaptic GABA$_B$ receptors will inhibit the release of other neurotransmitters (e.g., noradrenaline) through a decrease in membrane Ca^{2+} conductance whereas postsynaptic receptors induce an increase in membrane potassium conductance through G-protein coupling to inwardly rectifying GIRK or Kir3 potassium channels (Nicoll, 2004) (see schematic Fig. 1). The GABA$_B$ receptor, in contrast to the GABA$_A$ channel, is coupled to G-proteins and is a metabotropic receptor (Hill, 1985; Karbon and Enna, 1985). Despite the relatively early recognition (1981) of

the GABA$_B$ receptor and the identification of ^{3}H-baclofen as a ligand, it was not until 1997 that the first GABA$_B$ G-protein-coupled receptor was cloned (now termed GABA$_{B(1)}$) by Kaupmann et al. (1997). Subsequently a number of groups established that the functional GABA$_B$ receptor exists as a heterodimer of two components, subunits GABA$_{B(1)}$ and GABA$_{B(2)}$ (Jones et al., 1998; Kaupmann et al., 1998a; White et al., 1998; Kuner et al., 1999; Ng et al., 1999a). The cloning of these two GABA$_B$ receptor subunits and the realization that G-protein-coupled receptors (GPCRs) can exist as heterodimers led to a large number of ongoing studies of oligmerization between GPCRs and the realization that the properties of oligomeric GPCRs may differ substantially from their monomers (Bulenger et al., 2005). Thus for example heterodimerization of the κ and δ opiod receptors

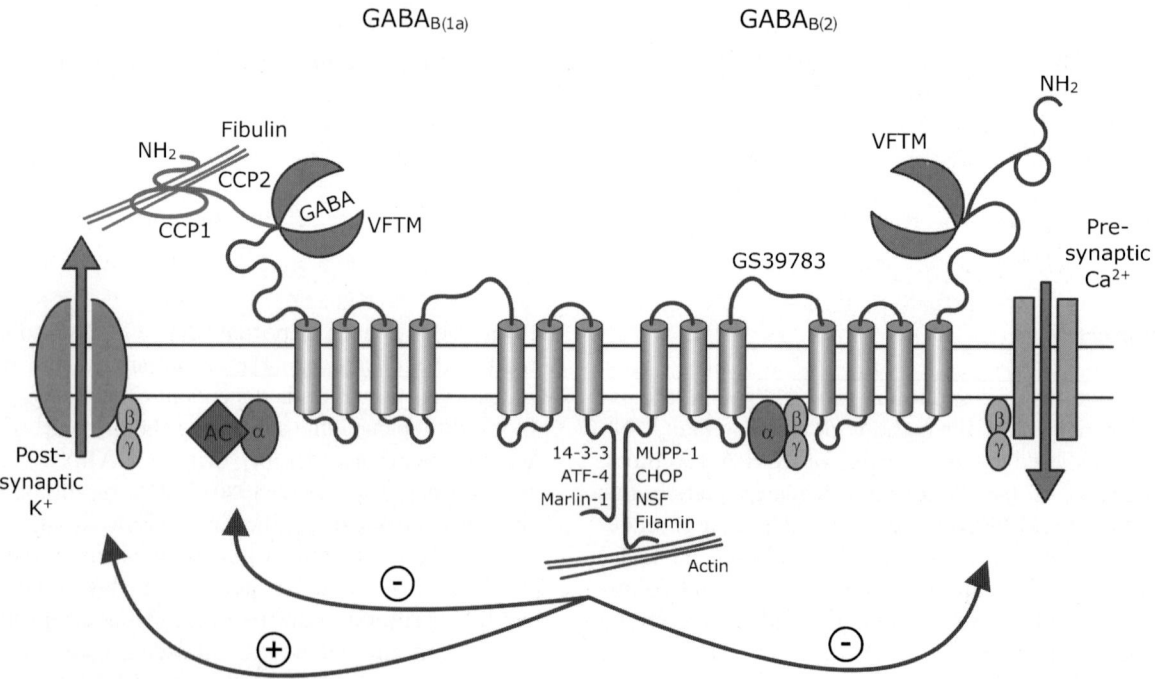

Fig. 1. Diagrammatic illustration of the GABA$_B$ receptor and its components in a 'signalling' complex. The amino-terminal domains GABA$_{B(1)}$ and GABA$_{B(2)}$ contain one venus fly trap motif (VFTM) each, but only the one in the GABA$_{B(1)}$ subunit binds GABA. Also in the amino terminal of GABA$_{B(1a)}$ are two complement control protein (CCP) or sushi domains. A number of proteins are known to bind to the carboxy-terminal coiled-coil domains of the heterodimers (see text). Allosteric effectors such as CGS 39783 bind to the heptahelical domain of GABA$_{B(2)}$ subunit. Activation of the heterodimer opens presynaptic calcium channels and post-synaptic potassium channels (GIRKs) through the G-proteins G$_{\beta/\gamma}$. A number of proteins (see text for details) have been shown to bind to the coiled-coil domains of the two subunits GABA$_{B(1)}$ and GABA$_{B(2)}$.

results in a novel receptor with a distinct pharmacology (Jordan and Devi, 1999) and the μ opioid receptor can heterodimerize with the α_{2a} adrenergic receptor (Jordan et al., 2003). To summarize briefly at this point, the overwhelming evidence is that $GABA_B$ receptors function primarily as heterodimers in which the $GABA_{B(1)}$ subunit binds GABA and the $GABA_{B(2)}$ subunit is responsible for G-protein coupling and signalling (for up to date reviews see Bettler et al., 2004 and Couve et al., 2004).

Molecular characterization/structure

Cloning

Despite the realization that $GABA_B$ sites could be recognized by labelled agonists such as baclofen it proved very difficult to purify the protein. Ligands such as ^{3}H-baclofen bound only with low affinity and the binding activity disappeared on membrane solubilization. The success in 1997 of Kaupmann et al. (1997) depended on the use of labelled ligands such as $[^{125}I]$-CGP 64213 or $[^{125}I]$-CGP 71872 with nanomolar affinity for the $GABA_B$ binding site to detect surface expression of a cDNA corresponding to a $GABA_B$ receptor. These ligands used in conjunction with expression cloning in COS-1 cells identified several $GABA_{B(1)}$ clones with many of the properties of the endogenous $GABA_B$ receptor, except that they did not couple well to signalling pathways. The originally identified cDNAs (now termed $GABA_{B(1a)}$ and $GABA_{B(1b)}$) were GPCRs with the characteristic seven trans-membrane domains with marked sequence similarities to the metabotropic glutamate receptors and belonging to the Family 3 (or Family C) GPCRs which include the metabotropic, calcium sensing and vomeronasal receptors. Subsequent work by a number of other groups (Jones et al. 1998; White et al., 1998; Kuner, et al. 1999; Ng et al., 1999a, as well as Kaupman et al., 1998a) showed that the reason for the poor coupling of the $GABA_{B(1)}$ isoforms was due to requirement for a second $GABA_B$ receptor subunit, $GABA_{B(2)}$. $GABA_{B(2)}$ must be co-expressed with a $GABA_{B(1)}$ subunit to form a functional heterodimeric $GABA_B$ receptor as

expression of the $GABA_{B(2)}$ subunit on its own did not give a fully functional receptor. The evidence for heterodimerization included the demonstration of the co-expression of $GABA_{B(1)}$ and $GABA_{B(2)}$ mRNAs in many CNS areas (although the striatum is relatively low in $GABA_{B(2)}$ mRNA signal), and principally that co-expression of $GABA_{B(1)}$ and $GABA_{B(2)}$ in cell lines yielded a functional receptor with a 10-fold increase in agonist potency compared to $GABA_{B(1)}$ on its own that coupled well to signalling pathways (Jones et al., 1998, 2000; Kaupmann et al., 1998a; Kuner et al., 1999; Ng et al., 1999b; White et al., 1998). Work by White et al. (1998) using yeast two-hybrid methods (White et al., 2002) revealed that the C-terminal coiled-coil domain of the $GABA_{B(1)}$ subunit would bind the C-terminal of $GABA_{B(2)}$ thus the two proteins interact preferentially as heterodimers and in general do not form functional receptors as monomers or homodimers. Although homodimers are formed, they are retained in the endoplasmic reticulum (ER) or Golgi compartments (Villemure et al., 2005).

Following the recognition of the $GABA_{B(1a)}$ and $GABA_{B(1b)}$ isoforms further studies indicated a number of splice variants existed of the $GABA_{B(1a)}$ isoform (Kaupmann et al., 1997, 1998b; Isomoto et al., 1998; Pfaff et al., 1999; Schwarz et al., 2000; Billinton et al., 2001; Martin et al., 2001; Wei et al., 2001; Bettler et al., 2004) (see Fig. 2). There has been a certain amount of confusion over the nomenclature of these variants, but in agreement with Bettler et al. (2004) the rat variants with an extra 31 amino acids in the second extracellular domain are perhaps best described as $GABA_{B(1c-a)}$ and $GABA_{B(1c-b)}$ to distinguish then from the human variant $GABA_{B(1c)}$ which lacks one sushi domain (Fig. 2). Most interesting is perhaps the $GABA_{B(1e)}$ isoform (Schwarz et al., 2000), which expressed in cell cultures can be secreted as a soluble protein and in the plasma membrane of transfected cells can heterodimerize with the $GABA_{B(2)}$ subunit indicating that the coiled-coil interaction is not essential for membrane targeting. It would be very interesting if this shortened protein were produced in vivo as it could act in a dominant-negative manner reducing the efficacy of the normal full length $GABA_{B(1)}$. Developmentally

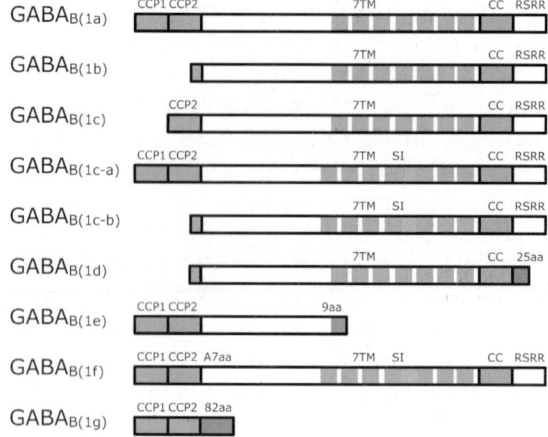

GABA_B(1a)

GABA_B(1b)

GABA_B(1c)

GABA_B(1c-a)

GABA_B(1c-b)

GABA_B(1d)

GABA_B(1e)

GABA_B(1f)

GABA_B(1g)

Fig. 2. Representation of the various variants of $GABA_{B(1)}$. The two main isoforms are $GABA_{B(1a)}$ and $GABA_{B(1b)}$ originally cloned by Kaupman et al. (1997). Other variants include $GABA_{(1c-a)}$ and $GABA_{B(1c-b)}$, which are rat variants with an additional insertion spliced into the second extracellular loop (SI). $GABA_{B(1c)}$ is a human variant without the first CCP domain. $GABA_{B(1c)}$ would be soluble form if produced in vivo. Other abbreviations: CC coiled-coil domain and RSRR C-terminal endoplasmic reticulum (ER) retention signal. (Adapted from Bettler et al., 2004, used with permission.)

the human $GABA_{B(1c)}$ variant has been reported to be up-regulated Calver et al. (2000) and Blein et al. (2004) have examined the structure of the two sushi or complement control protein (CCP) domains in $GABA_{B(1a)}$. They showed that the two sushi/CCP modules differ strikingly in structure with the second CCP module (CCP2) having a structure very similar to the CCP module in other proteins whilst CCP1 in $GABA_{B(1a)}$ has a more disordered structure although like CCP2 it could bind the extracellular protein, fibulin-2 (Blein et al., 2004).

In contrast to the variety of transcripts found for the $GABA_{B(1)}$ subunit, the $GABA_{B(2)}$ transcripts reported in man may be artefacts as consideration of the genomic sequence does not indicate the presence of appropriate splice sites to generate these alternative transcripts and $GABA_{B(2)}$ does not seem to occur in alternative forms (Martin et al., 2001). Regional distribution studies indicate that there are significant differences between the distribution of $GABA_{B(1a)}$ and $GABA_{B(1b)}$ mRNA and protein, that is the $GABA_{B(1)}$ subunits

$GABA_{B(1a)}$ with the sushi or CCP domains, and $GABA_{B(1b)}$ without the sushi domains, with variable amounts of 1a and 1b protein or being found in cerebral cortex hippocampus and brain stem (see for example Kaupmann et al., 1998b; Benke et al., 1999; Bischoff et al., 1999; Poorkhalkali et al., 2000; Princivalle et al., 2001). In these regions the 1b form is more prominent but in the striatum (Fig. 3) the 1a form (with the sushi domains) is more abundant. The significance of this difference is so far unclear but it may allow possible interactions of the NH_2 terminal extracellular sequence with other extracellular/surface proteins (such as fibulin-2) on neighbouring striatal neurones or influence the subcellular localization of the receptor.

Surprisingly the two subunits 1a and 1b arise not from alternative splicing, but from the use of an alternative promoter where the short $GABA_{B(1b)}$ NH_2 terminal arises from transcription within a $GABA_{B(1a)}$ intron resulting in extension of exon 6 at its 5' end (Pfaff et al., 1999; Martin et al., 2001). Of more interest with respect to the striatum is that the $GABA_{B(2)}$ mRNA is only poorly expressed in this region (Bischoff et al., 1999; Kuner et al., 1999; Margeta-Mitrovic et al., 1999; Martin et al., 1999; Clark et al., 2000) so it had been suggested that in this region the $GABA_{B(1)}$ protein might interact with another partner. However, in the $GABA_{B(2)}$ knockout described by Thuault et al. (2004) there was no evidence for functional $GABA_{B(1)}$ receptors on striatal neurones. This does not completely resolve the issue as the $GABA_{B(2)}$ knockout produced by Thuault et al. (2004) probably expressed a mutated $GABA_{B(2)}$ protein expression albeit without function, so that in contrast to the full $GABA_{B(2)}$ knockout described by Gassmann et al. (2004) the presence of the $GABA_{B(2)}$ mutated protein may have prevented the $GABA_{B(1)}$ receptor protein from trafficking to the surface on its own, or with another partner. However, what is clear is that the presence of the functional $GABA_{B(2)}$ protein is required for the established $GABA_B$ response in striatal neurones. For basal ganglia neuronal function it may be important that the $GABA_{B(1)}$ subunit interacts with the common $GABA_A$ receptor subunit $\gamma 2S$, which is highly expressed in the basal

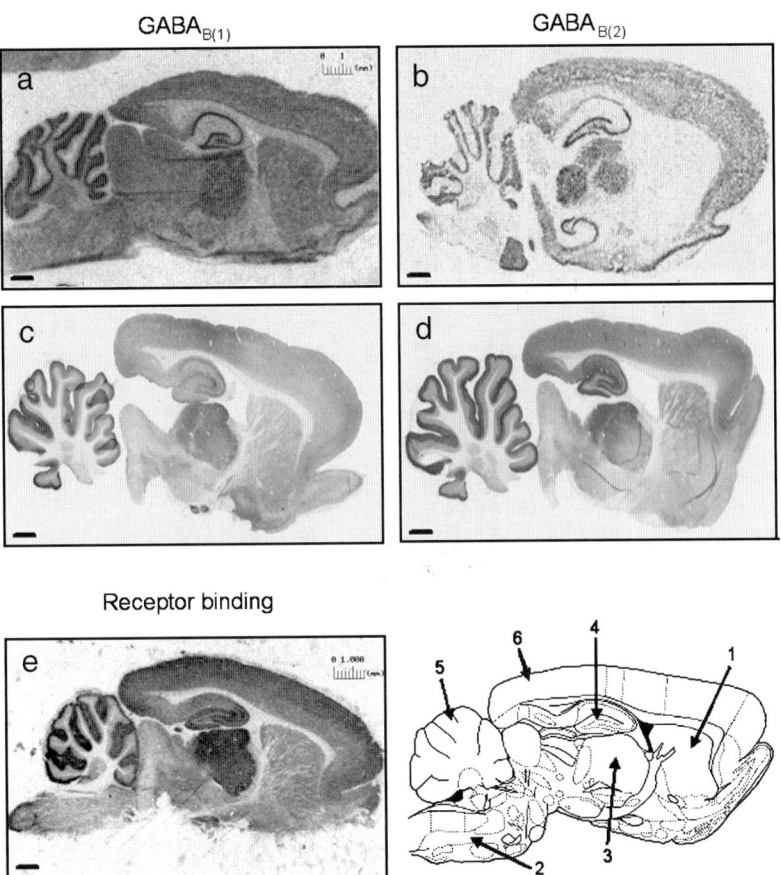

Fig. 3. Distribution of GABA$_{B(1)}$ and GABA$_{B(2)}$ mRNA by in situ hybridization using variant non-specific oligonucleotides (a), (b), GABA$_{B(1)}$ and GABA$_{B(2)}$ proteins by immunohistochemistry using variant non-specific antibodies (c), (d) and GABA$_B$ binding sites by receptor autoradiography (e). The mismatch of mRNA signals in the striatum is clear in (a) and (b). A comprehensive overlap of GABA$_{B(1)}$ and GABA$_{B(2)}$ protein localization is evident from immunohistochemistry with polyclonal antibodies (c) and (d) with cortex, hippocampus thalamus and cerebellum showing high levels of receptor subunit expression. Part (e) shows the binding of 0.5 nm [^{3}H]-CGP62349, a high-affinity antagonist radioligand for GABA$_B$ receptors, which is in good agreement with immunohistochemical localization (c) and (d) [(e) image courtesy of Professor N. Bowery, University of Birmingham, UK]. Labels: 1, striatum; 2, brainstem; 3, thalamus; 4, hippocampus; 5, cerebellum; 6, cortex. Scale bar, 1 mm.(Adapted from Billinton et al., 2001, Copyright Elsevier with permission.)

ganglia (Fig. 4). Although the interaction between the GABA$_A$ receptor subunit and GABA$_{B(1)}$ did not form a functional receptor, the γ2S subunit association resulted in an enhancement of the normal GABA$_B$ heterodimeric receptors' response to GABA (Balasubramanian et al., 2004). Given the close association between the GABA$_A$ and GABA$_B$ receptors in the basal ganglia (Fig. 4) this may enhance the efficacy of GABA at GABA$_B$ receptors.

Protein trafficking and GABA$_B$ associated proteins

It had been expected from earlier pharmacology studies that a number of GABA$_B$ receptor subtypes would be found, for example (Bonanno and Raiteri, 1993; Bowery, 1993; Bonanno et al., 1998) but expression studies of GABA$_{1(a)}$ or GABA$_{1(b)}$ with GABA$_{B(2)}$ do not detect significant functional differences between the expressed subunits.

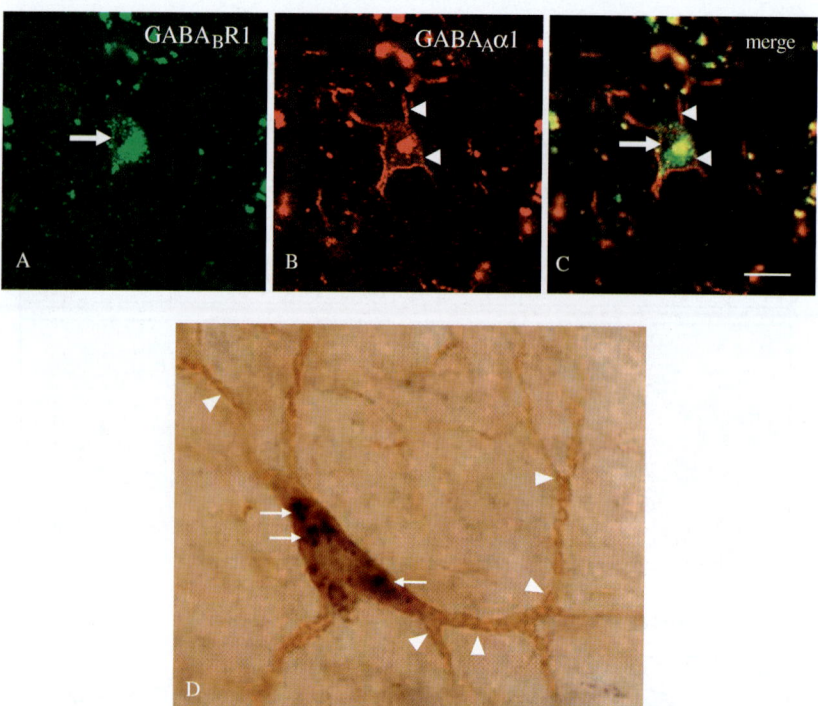

Fig. 4. Colour confocal laser scanning images showing double-labelled neurons in the human striatum double-labelled for GABA$_{B(1)}$ (A) and GABA$_A\alpha$1(B) and the merged images (C). Scale bar 20 μm (D). Double-labelled human striatal neuron showing DAB GABA$_A\alpha$1(brown, white arrowheads on cell surface) and GABA$_{B(1)}$ intracellular (DAB nickel black, white arrows). (*Source*: Images courtesy of Dr. H. Waldvogel, University of Auckland New Zealand with permission.)

Pharmacological differences may arise from interactions with extracellular proteins or with proteins interacting with the cytoplasmic tails of the heterodimer, especially the C-terminal coiled-coil domains, and the study of interacting proteins has revealed a number of proteins which may associate and interact with GABA$_B$ or its subunits. Thus, White et al. (2000) identified a number of proteins interacting with the coiled-coil domain of GABA$_{B(1)}$ including MUPP-1, CREB2, and ATF4. In primary neuronal cultures or in the cerebellum, CREB2 and GABA$_{B(1)}$ immunoreactivity co-distribute (Fig. 5). Exploration of this interaction in CHO cells showed that upon an activation of these cells with baclofen CREB2 moved from the cytoplasm into the nucleus where it could initiate transcription (White et al., 2000). These results and earlier studies suggest GABA$_B$ receptors may signal through CREB proteins to the nucleus to mediate long-term effects. Another transcription factor that

has recently been shown to interact with the coiled-coil domain of GABA$_{B(1a)}$ is CHOP (transcription factor CCAAT/enhancer binding protein C/EBP), which like CREB can co-localize and co-immunoprecipitate with GABA$_B$ in neurones (Sauter et al., 2005). Further work established that the leucine zipper of CHOP interacts with the leucine zipper in the C-terminal of GABA$_{B(2)}$. Interestingly, CHOP did not bind to the C-terminal of GABA$_{B(1)}$ isoforms but could interact with the GABA$_{B(1a)}$ subtype selectively via its amino-terminal domain (Sauter et al., 2005). The physiological significance of the interaction between CHOP with GABA$_B$ remains to be elucidated.

Other proteins that may influence cellular levels of expression of GABA$_B$ include 14-3-3 protein which competes with GABA$_{B(2)}$ for binding to GABA$_{B(1)}$ (Couve et al., 2001) and Marlin-1. Marlin-1 that binds to the coiled-coil domain of GABA$_{B(1)}$ is a novel brain-specific RNA binding

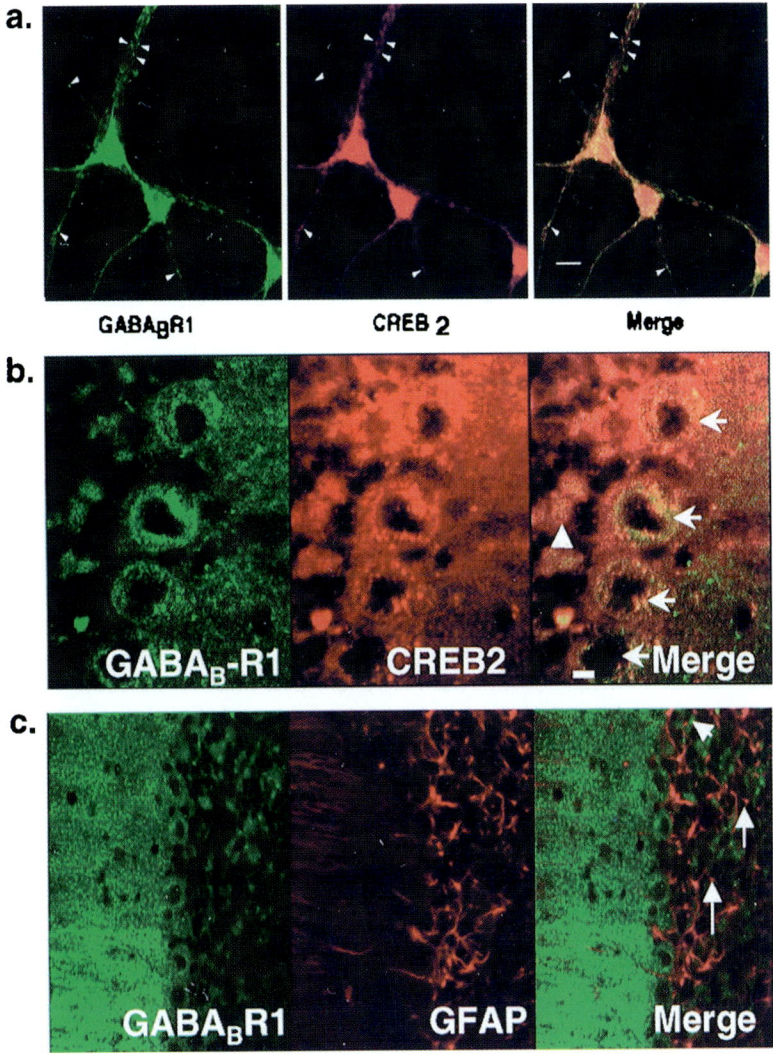

Fig. 5. Codistribution of GABA$_B$-R1 and CREB2 in neurons. (a) Codistribution of GABA$_B$-R1 and CREB2 in rat primary cortical cultures. Fixed rat-cortical neurons (E17) were stained with primary antibodies for GABA$_B$-R1 and CREB2. Images were viewed for GABA$_B$-R1 (green) and CREB2 (red), and the images were merged [bar = 10 μm]. Arrowheads indicate areas of distinct codistribution for the two proteins in the distal dendrites. (b) Codistribution of GABA$_B$-R1 and CREB2 in rat-cerebellar cortex. Fluorescence was detected by confocal microscopy for GABA$_B$R1 (green) and CREB2 (red) and the images were merged. Arrows indicate Purkinje cells, and the arrowhead shows granule cells [bar = 15 μm]. (c) Distinct noncodistribution of GABA$_B$-R1 and GFAP in rat-cerebellar cortex. Flourescence was detected by confocal microscopy for GABA$_B$-R1 (green) and GFAP (red), and the images were merged [bar = 15 μm]. (Taken from White et al., 2000. Copyright National Academy of Sciences, USA with permission.)

protein (Couve et al., 2004). Marlin-1 like CREB, 14-3-3 and MUPP1 was identified by yeast two-hybrid studies. Marlin-1 binds to various RNA species including the 3′ untranslated regions of GABA$_{B(1)}$ and GABA$_{B(2)}$. Use of an antisense marlin-1 oligonucleotide sequence in an appropriate vector reduced marlin-1 expression in transfected COS cells and neurones and produced an increase in content of GABA$_{B(2)}$ the significance of which remains to be established. It is clear, however, that the carboxy-terminal of the GABA$_B$ heterodimer provides a complex and changeable scaffold for the

50

assembly of a signalling complex. Other proteins established as part of this complex although not binding through to the coiled-coil domain are the G-protein receptor kinase GRK4 and a cAMP-dependent kinase (see Couve et al., 2004).

As noted the C-terminals of the $GABA_{B(1)}$ and $GABA_{B(2)}$ interact through a coiled-coil domain which is essential for membrane trafficking and it is suggested that the coiled-coil domain masks the ER retention signal (RSR) in $GABA_{B(1)}$. Recent expression studies by Restituto et al. (2005) and Gassmann et al. (2005), have generated $GABA_{B(1)}$ mutants where inactivation of the RSR C-terminal retention motif by its modification to ASA, allows as predicted, the $GABA_{B(1)}$ protein to be expressed at the plasma membrane and to form functional receptors with $GABA_{B(2)}$. Mutation studies revealed inactivation of the RSR motif is not completely critical for release of the receptor complex from the ER as its effects can be reduced or abolished by moving the RSR sequence closer to the heptaheical domain, but even when it can reach the membrane surface $GABA_{B(1)}$ requires $GABA_{B(2)}$ to form a functional receptor. Thus, Restituito et al. (2005) found that mutation of the RSR sequence to ASA did not result in 100% trafficking of the receptor heterodimer to the surface and they showed that inactivation of an LL motif in the cytoplasmic domain of $GABA_{B(1)}$ substantially improved surface expression of $GABA_{B(1)}$ ASA. As part of this study Restituito et al. (2005) also identified msec7-1 a guanidine exchange factor that interacted with the LL motif. In the context of trafficking and the RSR retention sequence it had been proposed that 14-3-3 or coat protein complex 1 (COPI) might act as trafficking control proteins for $GABA_B$. COPI is known to mediate retrieval of RXR proteins from the cis-golgi into the ER in COPI-coated vesicles. Recent work by Brock et al. (2005) has shown that both COPI and 14-3-3 proteins bind the RSR sequence of $GABA_{B(1)}$ and COPI is involved in the intracellular retention of $GABA_{B(1)}$, but that 14-3-3 proteins (overexpressed) did not influence the kinetics of $GABA_B$ receptor trafficking to the surface or influence the ability of the $GABA_B$ receptors to inhibit forskolin-stimulated cAMP formation. The precise role of 14-3-3, msec7-1 and

Marlin-1 proteins as with the other proteins associated with $GABA_B$ is unknown.

Considering the function of the heterodimer in more detail, both the $GABA_{B(1)}$ and $GABA_{B(2)}$ subunits have NH_2 terminal sequences that include sequences homologous to the bacterial periplasmic leucine-rich binding protein (LBP) corresponding to the lobes of a venus fly trap motif (VFTM) which could close on binding of ligand (Fig. 1). Mutagenesis and modelling studies by Pin and colleagues identified cysteine residues as important in folding the VFTM of $GABA_{B(1)}$ (Galvez et al., 1999). Consistent with this suggestion $GABA_{B(1)}$ could be made constitutively active by introducing a Cys–Cys bond in between the lobes "locking" the structure in the closed constituitively active form (Pin et al., 2004b). Modification of the amino acids in the VFTM (Galvez et al., 1999) had already suggested that a structure where Ser-246 and Gln-465 bind GABA in the $GABA_{B(1)}$ subunit VFTM. The constitutively active receptor could not be inhibited by antagonists but normal functioning could be restored by the use of the reducing agent dithiothreitol (Kniazeff et al., 2004a, b). In contrast, the VFTM is not well conserved in the $GABA_{B(2)}$ sequence and neither are the putative GABA-binding domains (Ser-246 and Gln-465) consistent with the evidence which indicates that it is only the $GABA_{B(1)}$ sequence which binds GABA and as noted earlier the $GABA_{B(2)}$ subunit couples the G-proteins and increases the affinity of the $GABA_{B(1)}$ receptor subunit for GABA. This interaction can happen at the level of the VFTM domains as using fluorescence resonance energy transfer (FRET). Pin et al. (2004a) could demonstrate a direct interaction between the two VFTM domains in the absence of the 7 transmembrane domains of $GABA_{B(2)}$. So far there is no evidence that the VFTM of the subunit recognizes a natural ligand as would be required for receptor activation (Pin et al., 2005), although it is the site of action of novel low molecular weight allosteric effectors (Urwyler et al., 2005). Once GABA is bound to the $GABA_{B(1)}$ VTFM there is a stabilization of the closed state as noted, which activates the heptahelical domain of $GABA_{B(2)}$ required for G-protein activation and signalling and a chimaeric receptor with the $GABA_{B(1)}$ VFTM and the

GABA$_{B(2)}$ heptahelical domain does not signal indicating the importance of the dimer formation for function (Gassmann et al., 2005). However, a dimeric receptor with the VFTM domain from GABA$_{B(1)}$ with two heptaheical domains from GABA$_{B(2)}$ will activate G-proteins upon agonist application so that the GABA$_{B(2)}$ heptahelical domain is sufficient for G-protein coupling.

Early studies had established that GABA$_B$ was coupled to G-proteins based on sensitivity to GTP analogues, islet-activated protein and pertussis toxin (Hill et al., 1984; Asano et al., 1985). The sensitivity to pertussis toxin suggested the involvement of G$_{\alpha i}$ or G$_{\alpha o}$ G-proteins. Following the cloning of the GABA$_B$ subunits chimaeric G-proteins were used to show that the heterodimers recognize G$_{qi\alpha}$ or G$_{qo\alpha}$ proteins (Franek et al., 1999). This study has recently been extended by Uezono et al. (2005) who have investigated the G-protein coupling of GABA$_B$ receptors in more detail. Using BHK cells or *Xenopus* oocytes and GABA$_{B(1)}$ or GABA$_{B(2)}$ with attached fluorescent reporters they showed using FRET that the reporter-tagged GABA$_{B(1)}$ and GABA$_{B(2)}$ constructs would form heterodimers. Functionally expression in oocytes depended on co-expression of GIRKs and to determine which subunit coupled to the signalling mechanism. Uezono et al. (2005) used GABA$_{B(1)}$ or GABA$_{B(2)}$ constructs fused with G$_{\gamma i2}$. In a similar fashion they used GABA$_{B(1)}$ or GABA$_{B(2)}$ constructs with a fused G$_{\gamma qi5}$ that activates phospholipase C in BHK cells. They found as predicted that only GABA$_{B(2)}$ with either G$_{\gamma i2}$ or G$_{\gamma qi5}$ would signal through oocytes or BHK cells, respectively, elegantly confirming that signal transducing G-proteins couple to GABA$_{B(2)}$ and not GABA$_{B(1)}$. Finally, it has recently been shown that the GABA$_B$ heterodimer in neurones or in CHO cells is associated with lipid rafts (Becher et al., 2004) where the receptor at least in CHO cells was less active with an increased EC$_{50}$.

Pharmacology

As noted, evidence for GABA$_B$ receptor subtypes is limited and when expressed in CHO cells the pharmacology of the recombinant GABA$_{B(1a \text{ or } 1b)}$

and GABA$_{B(2)}$ subunits is indistinguishable (Green et al., 2000). Similarly, the discovery of the original active agonist, baclofen (Bowery et al., 1980, 1981b), raised hopes that a range of selective agonists might be developed with selective effects in terms of antispasticity, antinociception or anti-drug craving. However, although a number of structures based on the phosphinic analogues, 3 aminopropyl phosphinic (2APPA) or its methyl homologue were synthesized (Froestl et al., 1995) which are more potent than baclofen, there is no evidence for any selectivity in physiological effects of these compounds (for overview see Bowery et al., 2002). The structure of baclofen also provided the basis for the production of the first GABA$_B$ antagonists including phaclofen and saclofen (Kerr et al., 1987; Ong et al., 1987) (Fig. 6), which were important in defining the CNS physiological effects of GABA$_B$ receptor (Dutar and Nicholl, 1988). Dramatic improvement in the affinity of antagonists came with the attachment of 3,4-dichlorophenyl or 3-carboxybenzyl substituent to the 'basic' baclofen/phaclofen structure (Ong et al., 1998). These compounds with affinities in the nanomolar range were also produced as radioligands and [^{125}I]-CGP71872 and [^{125}I]-CGP64213 (Fig. 7) were critical to the first expression cloning of GABA$_{1A}$ and GABA$_{1B}$ receptor (for detailed review of structures see Bowery et al., 2002).

The range of drugs interacting with the GABA$_B$ receptor has been considerably extended by the recent discovery of novel allosteric effectors for GABA$_B$ (Urwyler et al., 2001), which have

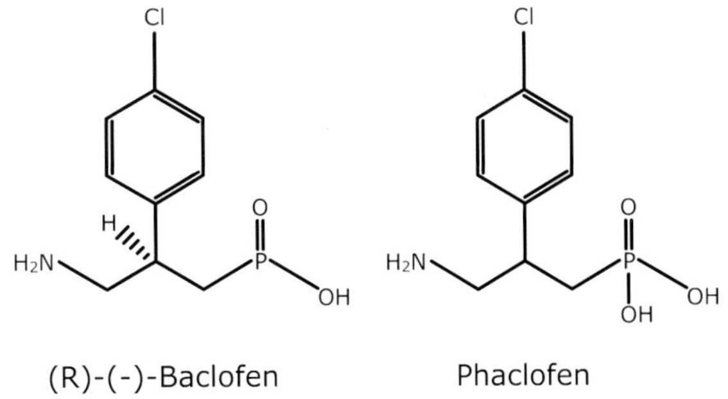

(R)-(-)-Baclofen   Phaclofen

Fig. 6. Structures of baclofen (agonist) and phaclofen (antagonist).

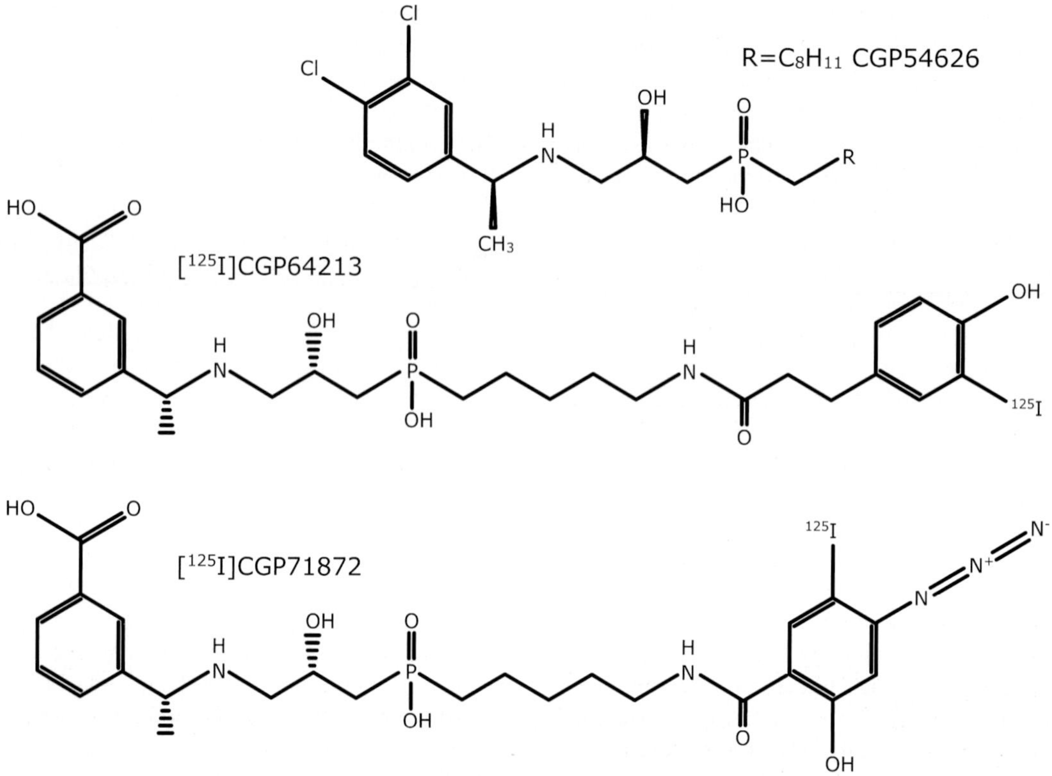

Fig. 7. Structures of high-affinity GABA$_B$ antagonists. [125I] CGP71872 and [125I]CGP 64213 which were important in the initial cloning of GABA$_{B(1a)}$ and GABA$_{B(1b)}$.

aroused considerable interest as they may have therapeutic advantages. The first group reported by Urwyler et al. (2001) including CGP7930 and CGP13501 (see Fig. 8) were positive modulators of GABA binding increasing the efficacy of GABA on the GABA$_B$ hetero-dimer. Further work by Urwyler et al. (2003) identified another class of compounds exemplified by GS39783 (Fig. 8) which like CGP7930 and CGP13501 were also positive modulators of GABA$_B$, increasing the potency and efficacy of GABA on the GABA$_B$ receptor but without any effects themselves and modulating responses which were blockable by GABA$_B$ antagonists. Clarification of the site of action of these compounds came from the studies of Binet et al. (2004a, b) who demonstrated that CGP7930 activates the GABA$_B$ heterodimer through the GABA$_{B(2)}$ subunit. This effect is retained in the recombinant receptor where the GABA$_{B(2)}$ subunit has had the extracellular domains deleted showing these low molecular weight allosteric modulators act through the heptahelical domain of GABA$_{B(2)}$. Behavioural studies with GS 39783 indicate it is devoid of the sedative and cognitive effects of baclofen (Cryan et al., 2004) and these compounds may have potential as anxiolytics especially as GABA$_B$-deficient mice have 'panic' attacks and are more anxious in anxiety related behavioural screens (Mombereau et al., 2005).

In considering the 'classical' effects of GABA$_B$ receptors, the original observations of Bowery and others were based on the ability of GABA$_B$ to inhibit (autoreceptors) or increase neurotransmitter release (heteroreceptors). These effects are likely mediated through inhibition of Ca^{2+} channels of N, P or Q types by G$_{\beta/\gamma}$ (see Fig. 1) and a recent study (Richman et al., 2004) has shown that N-type calcium channels can act as scaffold proteins in assembling signalling molecules such as members of the Ras pathway which modulate

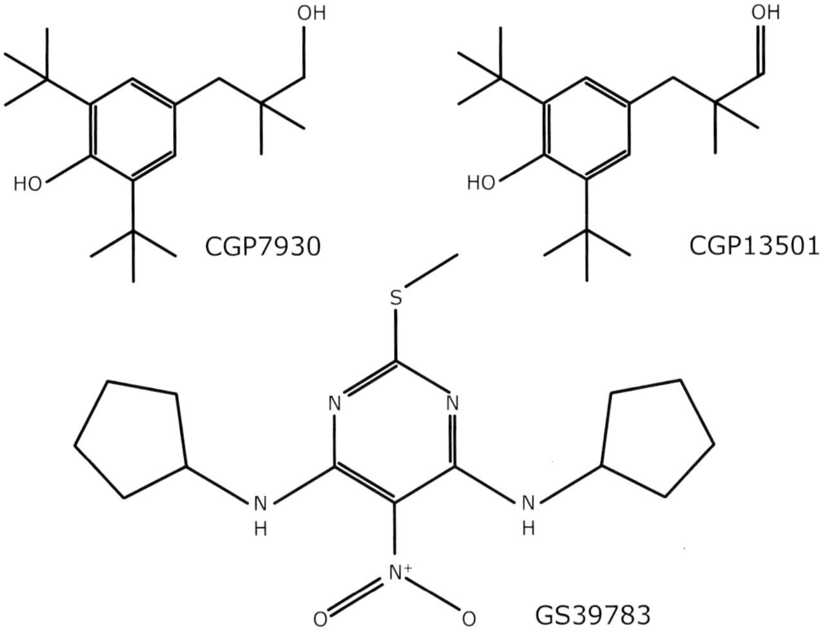

Fig. 8. Structures of some positive allosteric modulators of GABA$_B$ receptors.

GABA$_B$ effects. In a similar fashion G$_{\beta/\gamma}$ subunits open K$^+$ channels (principally the inward rectifying Kir3 channels) as well as activating or inhibiting adenylate cyclase isoforms. The G-proteins associated with the GABA$_B$ receptor are believed as discussed earlier on the basis pertussis sensitivity, antisense knockdown and knockout experiments to be G$_{\alpha i}$ or G$_{\alpha o}$ type G-protein (Morishita et al., 1990; Campbell et al., 1993; Greif et al., 2000) and G$_{\alpha i}$ and G$_{\alpha o}$ are known to inhibit adenyl cyclase types I, III, I and VI while G$_{\beta\gamma}$ can stimulate adenyl cyclase types II, IV and VII. The ability of GABA$_B$ receptors to stimulate adenyl cyclase is considered to be due to the presence of G$_{s\alpha}$ coupled GPCRs in the vicinity. Hashimoto and Kuriyama (1997), for example, could demonstrate in the rat striatum that baclofen would potentiate the generation of cAMP by isoprenaline in microdialysis experiments in freely moving animals. In general terms, the ability of the GABA$_B$ receptor to couple to Ca^{2+} channels makes them strong inhibitors of neurotransmitter release although the precise Ca^{2+} channels implicated in these effects is likely to vary depending on the CNS region. In mice lacking GABA$_{B(1)}$ or

GABA$_{B(2)}$ receptors, baclofen-induced outward currents are absent, a feature also found in mice lacking the inwardly rectifying K$^+$ channel, Kir3 (Nicoll, 2004) consistent with the important role of this channel in GABA$_B$ receptor function.

Knockouts (KO) of both the GABA$_{B(1)}$ and GABA$_{B(2)}$ receptors have been produced, the B2 KO most recently (Schuler et al., 2001, Gassmann et al., 2004; Thuault et al., 2004). All the knockout mice have similar behavioural phenotypes showing spontaneous seizures, hyper-algesia, hyper-locomotion and memory impairments consistent with the loss of GABA$_B$ receptors. The first KO mice produced were knockouts of the GABA$_{B(1)}$ receptor and as expected there was no loss of GABA$_{B(2)}$ mRNA, but there was in both strains of mice a dramatic loss of GABA$_{B(2)}$ protein consistent with the need for the heterodimer partner GABA$_{B(1)}$ for stable expression of the GABA$_{B(2)}$ protein. In the GABA$_{B(1)}$ KOs as expected there was a complete loss of ligand binding (^{3}H-CGP54626 and ^{125}I-CGP64213) consistent with the loss of the GABA/ligand binding site on the NH$_2$-terminal of GABA$_{B(1)}$. Electrophysiological analysis of hippocampal slices from these mice revealed the loss of all GABA$_B$ receptors whether pre- or post-synaptic (Prosser et al., 2001; Schuler et al., 2001) providing further evidence against the existence of multiple GABA$_B$ receptors. The GABA$_{B(2)}$ KO mice as noted had a similar behavioural phenotype as the GABA$_{B(1)}$ KO mice and detectable expression of the GABA$_{B(1)}$ mRNA. The GABA$_{B(1)}$ protein was still produced albeit at substantially lower levels in the full KO (Gassmann et al., 2004). In this mouse although ligand ([^{125}I]-CGP64213) binding or [^{35}S]-GTP$_{\gamma s}$ binding was not detectable immunohistochemistry, however, revealed that some GABA$_{B(1)}$ protein was trafficked to the surface of neurones and was detectable on synaptic membranes so that some GABA$_{B(1)}$ protein could exit the ER but remained in the cell body and dendrites of the hippocampus. However, although electrophysiological experiments were unable to demonstrate the presence of interneuronal autoreceptors in these GABA$_{B(2)}$ KO mice there was a baclofen-induced inward current in the CA2 pyramidal cells which was blocked by GABA$_B$ antagonists whereas in the equivalent wild-type controls the Kir3 current is outward (see Fig. 1). This response was G-protein mediated and represents what may be a functional response of the GABA$_{B(1)}$ subunit. In contrast, in the KO mouse of Thuault et al. (2004), which had only a C-terminal deletion of the GABA$_{B(2)}$, the GABA$_{B(1)}$ was not trafficked to the surface. As the striatum expresses far more GABA$_{B(1)}$ mRNA than GABA$_{B(2)}$ mRNA it may be that certain striatal neurones would express this GABA$_{B(1)}$ receptor subunit without the GABA$_{B(2)}$ subunit producing what would be a functionally very distinct somatic GABA$_B$ receptor. This possibility needs to be borne in mind in considering the role(s) of GABA$_B$ in the basal ganglia. Important tools in investigating GABA$_B$ function will be further transgenic mice including conditional KOs and a floxed allele for inactivation of the GABA$_{B(1)}$ gene has recently been produced (Haller et al., 2004). Such an approach will allow regional- and time-specific deletion of GABA$_B$ receptors and further our understanding of the role of GABA$_B$ in the CNS which may provide clues to the use of GABA$_B$ in conditions such as anxiety and depression (Cryan and Kaupmann, 2005).

References

Asano, T., Ui, M. and Ogasawara, N. (1985) Prevention of the agonist binding to gamma-aminobutyric acid B receptors by guanine nucleotides and islet-activating protein, pertussis toxin, in bovine cerebral cortex. Possible coupling of the toxin-sensitive GTP-binding proteins to receptors. J. Biol. Chem., 260: 12653–12658.

Balasubramanian, S., Teissere, J.A., Raju, D.V. and Hall, R.A. (2004) Hetero-oligomerization between GABA$_A$ and GABA$_B$ receptors regulates GABA$_B$ receptor trafficking. J. Biol. Chem., 279: 18840–18850.

Becher, A., Green, A., Ige, A.O., Wise, A., White, J.H. and McIlhinney, R.A. (2004) Ectopically expressed gamma-aminobutyric acid receptor B is functionally down-regulated in isolated lipid raft-enriched membranes. Biochem. Biophys. Res. Commun., 321: 981–987.

Benke, D., Honer, M., Michel, C., Bettler, B. and Mohler, H. (1999) gamma-aminobutyric acid type B receptor splice variant proteins GBR1a and GBR1b are both associated with GBR2 in situ and display differential regional and subcellular distribution. J. Biol. Chem., 274: 27323–27330.

Bettler, B., Kaupmann, K., Mosbacher, J. and Gassmann, M. (2004) Molecular structure and physiological functions of GABA(B) receptors. Physiol. Rev., 84: 835–867.

Billinton, A., Ige, A.O., Bolam, J.P., White, J.H., Marshall, F.H. and Emson, P.C. (2001) Advances in the molecular understanding of GABA(B) receptors. Trends Neurosci., 24: 277–282.

Binet, V., Brajon, C., Le Corre, L., Acher, F., Pin, J.P. and Prezeau, L. (2004a) The heptahelical domain of GABA(B2) is activated directly by CGP7930, a positive allosteric modulator of the GABA(B) receptor. J. Biol. Chem., 279: 29085–29091.

Binet, V., Goudet, C., Brajon, C., Le Corre, L., Acher, F., Pin, J.P. and Prezeau, L. (2004b) Molecular mechanisms of GABA(B) receptor activation: new insights from the mechanism of action of CGP7930, a positive allosteric modulator. Biochem. Soc. Trans, 32: 871–872.

Bischoff, S., Leonhard, S., Reymann, N., Schuler, V., Shigemoto, R., Kaupmann, K. and Bettler, B. (1999) Spatial distribution of GABA(B)R1 receptor mRNA and binding sites in the rat brain. J. Comp. Neurol., 412: 1–16.

Blein, S., Ginham, R., Uhrin, D., Smith, B.O., Soares, D.C., Veltel, S., McIlhinney, R.A., White, J.H. and Barlow, P.N. (2004) Structural analysis of the complement control protein (CCP) modules of GABA(B) receptor 1a: only one of the two CCP modules is compactly folded. J. Biol. Chem., 279: 48292–48306.

Bonanno, G., Fassio, A., Sala, R., Schmid, G. and Raiteri, M. (1998) GABA(B) receptors as potential targets for drugs able to prevent excessive excitatory amino acid transmission in the spinal cord. Eur. J. Pharmacol., 362: 143–148.

Bonanno, G. and Raiteri, M. (1993) Multiple GABA$_B$ receptors. Trends Pharmacol. Sci., 14: 259–261.

Bowery, N.G. (1993) GABA$_B$ receptor pharmacology. Annu. Rev. Pharmacol. Toxicol., 33: 109–147.

Bowery, N.G., Bettler, B., Froestl, W., Gallagher, J.P., Marshall, F., Raiteri, M., Bonner, T.I. and Enna, S.J. (2002) International Union of Pharmacology. XXXIII. Mammalian gamma-aminobutyric acid(B) receptors: structure and function. Pharmacol. Rev., 54: 247–264.

Bowery, N.G., Doble, A., Hill, D.R., Hudson, A.L., Shaw, J.S. and Turnbull, M.J. (1979) Baclofen: a selective agonist for a novel type of GABA receptor[proceedings]. Br. J. Pharmacol., 67: 444P–445P.

Bowery, N.G., Doble, A., Hill, D.R., Hudson, A.L., Shaw, J.S., Turnbull, M.J. and Warrington, R. (1981a) Bicuculline-insensitive GABA receptors on peripheral autonomic nerve terminals. Eur. J. Pharmacol., 71: 53–70.

Bowery, N.G., Doble, A., Hill, D.R., Hudson, A.L., Turnbull, M.J. and Warrington, R. (1981b) Structure/activity studies at a baclofen-sensitive, bicuculline-insensitive GABA receptor. Adv. Biochem. Psychopharmacol., 29: 333–341.

Bowery, N.G., Hill, D.R. and Hudson, A.L. (1983) Characteristics of $GABA_B$ receptor binding sites on rat whole brain synaptic membranes. Br. J. Pharmacol., 78: 191–206.

Bowery, N.G., Hill, D.R., Hudson, A.L., Doble, A., Middlemiss, D.N., Shaw, J. and Turnbull, M. (1980) (−)Baclofen decreases neurotransmitter release in the mammalian CNS by an action at a novel GABA receptor. Nature, 283: 92–94.

Brock, C., Boudier, L., Maurel, D., Blahos, J. and Pin, J.P. (2005) Assembly-dependent surface targeting of the heterodimeric $GABA_B$ receptor is controlled by COPI, but not 14-3-3. Mol. Biol. Cell, 16: 5572–5578.

Bulenger, S., Marullo, S. and Bouvier, M. (2005) Emerging role of homo- and heterodimerization in G-protein-coupled receptor biosynthesis and maturation. Trends Pharmacol. Sci., 26: 131–137.

Calver, A.R., Medhurst, A.D., Robbins, M.J., Charles, K.J., Evans, M.L., Harrison, D.C., Stammers, M., Hughes, S.A., Hervieu, G., Couve, A., Moss, S.J., Middlemiss, D.N. and Pangalos, M.N. (2000) The expression of GABA(B1) and GABA(B2) receptor subunits in the cNS differs from that in peripheral tissues. Neuroscience, 100: 155–170.

Campbell, V., Berrow, N. and Dolphin, A.C. (1993) $GABA_B$ receptor modulation of Ca^{2+} currents in rat sensory neurones by the G protein G(0): antisense oligonucleotide studies. J. Physiol., 470: 1–11.

Clark, J.A., Mezey, E., Lam, A.S. and Bonner, T.I. (2000) Distribution of the GABA(B) receptor subunit gb2 in rat CNS. Brain Res., 860: 41–52.

Couve, A., Calver, A.R., Fairfax, B., Moss, S.J. and Pangalos, M.N. (2004) Unravelling the unusual signalling properties of the GABA(B) receptor. Biochem. Pharmacol., 68: 1527–1536.

Couve, A., Kittler, J.T., Uren, J.M., Calver, A.R., Pangalos, M.N., Walsh, F.S. and Moss, S.J. (2001) Association of GABA(B) receptors and members of the 14-3-3 family of signaling proteins. Mol. Cell Neurosci., 17: 317–328.

Cryan, J.F. and Kaupmann, K. (2005) Don't worry B' happy!: a role for GABA(B) receptors in anxiety and depression. Trends Pharmacol. Sci., 26: 36–43.

Cryan, J.F., Kelly, P.H., Chaperon, F., Gentsch, C., Mombereau, C., Lingenhoehl, K., Froestl, W., Bettler, B., Kaupmann, K. and Spooren, W.P. (2004) Behavioral characterization of the novel $GABA_B$ receptor-positive modulator GS39783 (N,N'-dicyclopentyl-2-methylsulfanyl-5-nitro-pyrimidine-4,6-diamine): anxiolytic-like activity without side effects associated with baclofen or benzodiazepines. J. Pharmacol. Exp. Ther., 310: 952–963.

Curtis, D.R. and Johnston, G.A. (1974) Amino acid transmitters in the mammalian central nervous system. Ergeb. Physiol., 69: 97–188.

Curtis, D.R., Johnston, G.A., Game, C.J. and McCulloch, R.M. (1974) Central action of bicuculline. J. Neurochem., 23: 605–606.

Dutar, P. and Nicoll, R.A. (1988) A physiological role for $GABA_B$ receptors in the central nervous system. Nature, 332: 156–158.

Franek, M., Pagano, A., Kaupmann, K., Bettler, B., Pin, J.P. and Blahos, J. (1999) The heteromeric $GABA_B$ receptor recognizes G-protein alpha subunit C-termini. Neuropharmacology, 38: 1657–1666.

Froestl, W., Mickel, S.J., von Sprecher, G., Diel, P.J., Hall, R.G., Maier, L., Strub, D., Melillo, V., Baumann, P.A., Bernasconi, R., et al. (1995) Phosphinic acid analogues of GABA. 2. Selective, orally active $GABA_B$ antagonists. J. Med. Chem., 38: 3313–3331.

Galvez, T., Parmentier, M.L., Joly, C., Malitschek, B., Kaupmann, K., Kuhn, R., Bittiger, H., Froestl, W., Bettler, B. and Pin, J.P. (1999) Mutagenesis and modeling of the $GABA_B$ receptor extracellular domain support a venus flytrap mechanism for ligand binding. J. Biol. Chem., 274: 13362–13369.

Gassmann, M., Haller, C., Stoll, Y., Aziz, S.A., Biermann, B., Mosbacher, J., Kaupmann, K. and Bettler, B. (2005) The RXR-type endoplasmic reticulum-retention/retrieval signal of $GABA_{B1}$ requires distant spacing from the membrane to function. Mol. Pharmacol., 68: 137–144.

Gassmann, M., Shaban, H., Vigot, R., Sansig, G., Haller, C., Barbieri, S., Humeau, Y., Schuler, V., Muller, M., Kinzel, B., Klebs, K., Schmutz, M., Froestl, W., Heid, J., Kelly, P.H., Gentry, C., Jaton, A.L., Van der Putten, H., Mombereau, C., Lecourtier, L., Mosbacher, J., Cryan, J.F., Fritschy, J.M., Luthi, A., Kaupmann, K. and Bettler, B. (2004) Redistribution of GABAB(1) protein and atypical $GABA_B$ responses in GABAB(2)-deficient mice. J. Neurosci., 24: 6086–6097.

Green, A., Walls, S., Wise, A., Green, R.H., Martin, A.K. and Marshall, F.H. (2000) Characterization of [(3)H]-CGP54626A binding to heterodimeric GABA(B) receptors stably expressed in mammalian cells. Br. J. Pharmacol., 131: 1766–1774.

Greif, G.J., Sodickson, D.L., Bean, B.P., Neer, E.J. and Mende, U. (2000) Altered regulation of potassium and calcium channels by GABA(B) and adenosine receptors in hippocampal neurons from mice lacking Galpha(o). J. Neurophysiol., 83: 1010–1018.

Haller, C., Casanova, E., Muller, M., Vacher, C.M., Vigot, R., Doll, T., Barbieri, S., Gassmann, M. and Bettler, B. (2004)

Floxed allele for conditional inactivation of the GABA$_{B(1)}$ gene. Genesis, 40: 125–130.

Hashimoto, T. and Kuriyama, K. (1997) In vivo evidence that GABA(B) receptors are negatively coupled to adenylate cyclase in rat striatum. J. Neurochem., 69: 365–370.

Hill, D.R. (1985) GABA$_B$ receptor modulation of adenylate cyclase activity in rat brain slices. Br. J. Pharmacol., 84: 249–257.

Hill, D.R. and Bowery, N.G. (1981) ^{3}H-baclofen and ^{3}H-GABA bind to bicuculline-insensitive GABA$_B$ sites in rat brain. Nature, 290: 149–152.

Hill, D.R., Bowery, N.G. and Hudson, A.L. (1984) Inhibition of GABA$_B$ receptor binding by guanyl nucleotides. J. Neurochem., 42: 652–657.

Isomoto, S., Kaibara, M., Sakurai-Yamashita, Y., Nagayama, Y., Uezono, Y., Yano, K. and Taniyama, K. (1998) Cloning and tissue distribution of novel splice variants of the rat GABA$_B$ receptor. Biochem. Biophys. Res. Commun., 253: 10–15.

Jones, K.A., Borowsky, B., Tamm, J.A., Craig, D.A., Durkin, M.M., Dai, M., Yao, W.J., Johnson, M., Gunwaldsen, C., Huang, L.Y., Tang, C., Shen, Q., Salon, J.A., Morse, K., Laz, T., Smith, K.E., Nagarathnam, D., Noble, S.A., Branchek, T.A. and Gerald, C. (1998) GABA(B) receptors function as a heteromeric assembly of the subunits GABA(B)R1 and GABA(B)R2. Nature, 396: 674–679.

Jones, K.A., Tamm, J.A., Craig, D.A., Yao, W. and Panico, R. (2000) Signal transduction by GABA(B) receptor heterodimers. Neuropsychopharmacology, 23: S41–S49.

Jordan, B.A. and Devi, L.A. (1999) G-protein-coupled receptor heterodimerization modulates receptor function. Nature, 399: 697–700.

Jordan, B.A., Gomes, I., Rios, C., Filipovska, J. and Devi, L.A. (2003) Functional interactions between mu opioid and alpha 2A-adrenergic receptors. Mol. Pharmacol., 64: 1317–1324.

Karbon, E.W. and Enna, S.J. (1985) Characterization of the relationship between gamma-aminobutyric acid B agonists and transmitter-coupled cyclic nucleotide-generating systems in rat brain. Mol. Pharmacol., 27: 53–59.

Kaupmann, K., Huggel, K., Heid, J., Flor, P.J., Bischoff, S., Mickel, S.J., McMaster, G., Angst, C., Bittiger, H., Froestl, W. and Bettler, B. (1997) Expression cloning of GABA(B) receptors uncovers similarity to metabotropic glutamate receptors. Nature, 386: 239–246.

Kaupmann, K., Malitschek, B., Schuler, V., Heid, J., Froestl, W., Beck, P., Mosbacher, J., Bischoff, S., Kulik, A., Shigemoto, R., Karschin, A. and Bettler, B. (1998a) GABA(B)-receptor subtypes assemble into functional heteromeric complexes. Nature, 396: 683–687.

Kaupmann, K., Schuler, V., Mosbacher, J., Bischoff, S., Bittiger, H., Heid, J., Froestl, W., Leonhard, S., Pfaff, T., Karschin, A. and Bettler, B. (1998b) Human gamma-aminobutyric acid type B receptors are differentially expressed and regulate inwardly rectifying K$^+$ channels. Proc. Natl. Acad. Sci. USA, 95: 14991–14996.

Kerr, D.I., Ong, J., Prager, R.H., Gynther, B.D. and Curtis, D.R. (1987) Phaclofen: a peripheral and central baclofen antagonist. Brain Res., 405: 150–154.

Kniazeff, J., Bessis, A.S., Maurel, D., Ansanay, H., Prezeau, L. and Pin, J.P. (2004a) Closed state of both binding domains of homodimeric mGlu receptors is required for full activity. Nat. Struct. Mol. Biol., 11: 706–713.

Kniazeff, J., Saintot, P.P., Goudet, C., Liu, J., Charnet, A., Guillon, G. and Pin, J.P. (2004b) Locking the dimeric GABA(B) G-protein-coupled receptor in its active state. J. Neurosci., 24: 370–377.

Krnjevic, K. (1984) Neurochemistry of synaptic transmission in the cerebral cortex. Neurochem. Res., 9: 445–448.

Kuner, R., Kohr, G., Grunewald, S., Eisenhardt, G., Bach, A. and Kornau, H.C. (1999) Role of heteromer formation in GABA$_B$ receptor function. Science, 283: 74–77.

Margeta-Mitrovic, M., Mitrovic, I., Riley, R.C., Jan, L.Y. and Basbaum, A.I. (1999) Immunohistochemical localization of GABA(B) receptors in the rat central nervous system. J. Comp. Neurol., 405: 299–321.

Martin, S.C., Russek, S.J. and Farb, D.H. (1999) Molecular identification of the human GABA$_{BR2}$: cell surface expression and coupling to adenylyl cyclase in the absence of GABA$_{BR1}$. Mol. Cell Neurosci., 13: 180–191.

Martin, S.C., Russek, S.J. and Farb, D.H. (2001) Human GABA(B)R genomic structure: evidence for splice variants in GABA(B)R1 but not GABA(B)R2. Gene, 278: 63–79.

Mombereau, C., Kaupmann, K., Gassmann, M., Bettler, B., van der Putten, H. and Cryan, J.F. (2005) Altered anxiety and depression-related behaviour in mice lacking GABA$_{B(2)}$ receptor subunits. Neuroreport, 16: 307–310.

Morishita, R., Kato, K. and Asano, T. (1990) GABA$_B$ receptors couple to G proteins Go, Go* and Gi1 but not to Gi2. FEBS Lett., 271: 231–325.

Ng, G.Y., Clark, J., Coulombe, N., Ethier, N., Hebert, T.E., Sullivan, R., Kargman, S., Chateauneuf, A., Tsukamoto, N., McDonald, T., Whiting, P., Mezey, E., Johnson, M.P., Liu, Q., Kolakowski Jr., L.F., Evans, J.F., Bonner, T.I. and O'Neill, G.P. (1999a) Identification of a GABA$_B$ receptor subunit, gb2, required for functional GABA$_B$ receptor activity. J. Biol. Chem., 274: 7607–7610.

Ng, G.Y., McDonald, T., Bonnert, T., Rigby, M., Heavens, R., Whiting, P., Chateauneuf, A., Coulombe, N., Kargman, S., Caskey, T., Evans, J., O'Neill, G.P. and Liu, Q. (1999b) Cloning of a novel G-protein-coupled receptor GPR 51 resembling GABA$_B$ receptors expressed predominantly in nervous tissues and mapped proximal to the hereditary sensory neuropathy type 1 locus on chromosome 9. Genomics, 56: 288–295.

Nicoll, R.A. (2004) My close encounter with GABA(B) receptors. Biochem. Pharmacol., 68: 1667–1674.

Ong, J., Kerr, D.I., Bittiger, H., Waldmeier, P.C., Baumann, P.A., Cooke, N.G., Mickel, S.J. and Froestl, W. (1998) Morpholin-2-yl-phosphinic acids are potent GABA(B) receptor antagonists in rat brain. Eur. J. Pharmacol., 362: 27–34.

Ong, J., Kerr, D.I. and Johnston, A.R. (1987) Differing actions of beta-phenyl-GABA and baclofen in the guinea pig isolated ileum. Neurosci. Lett., 77: 109–112.

Pfaff, T., Malitschek, B., Kaupmann, K., Prezeau, L., Pin, J.P., Bettler, B. and Karschin, A. (1999) Alternative splicing generates a novel isoform of the rat metabotropic GABA(B)R1 receptor. Eur. J. Neurosci., 11: 2874–2882.

Pin, J.P., Kniazeff, J., Binet, V., Liu, J., Maurel, D., Galvez, T., Duthey, B., Havlickova, M., Blahos, J., Prezeau, L. and Rondard, P. (2004a) Activation mechanism of the heterodimeric GABA(B) receptor. Biochem. Pharmacol., 68: 1565–1572.

Pin, J.P., Kniazeff, J., Goudet, C., Bessis, A.S., Liu, J., Galvez, T., Acher, F., Rondard, P. and Prezeau, L. (2004b) The activation mechanism of class-C G-protein coupled receptors. Biol. Cell, 96: 335–342.

Pin, J.P., Kniazeff, J., Liu, J., Binet, V., Goudet, C., Rondard, P. and Prezeau, L. (2005) Allosteric functioning of dimeric class C G-protein-coupled receptors. FEBS J., 272: 2947–2955.

Poorkhalkali, N., Juneblad, K., Jonsson, A.C., Lindberg, M., Karlsson, O., Wallbrandt, P., Ekstrand, J. and Lehmann, A. (2000) Immunocytochemical distribution of the GABA(B) receptor splice variants GABA(B) R1a and R1b in the rat CNS and dorsal root ganglia. Anat. Embryol. (Berl.), 201: 1–13.

Princivalle, A.P., Pangalos, M.N., Bowery, N.G. and Spreafico, R. (2001) Distribution of GABA(B(1a)), GABA(B(1b)) and GABA(B2) receptor protein in cerebral cortex and thalamus of adult rats. Neuroreport, 12: 591–595.

Prosser, H.M., Gill, C.H., Hirst, W.D., Grau, E., Robbins, M., Calver, A., Soffin, E.M., Farmer, C.E., Lanneau, C., Gray, J., Schenck, E., Warmerdam, B.S., Clapham, C., Reavill, C., Rogers, D.C., Stean, T., Upton, N., Humphreys, K., Randall, A., Geppert, M., Davies, C.H. and Pangalos, M.N. (2001) Epileptogenesis and enhanced prepulse inhibition in GABA(B1)-deficient mice. Mol. Cell. Neurosci., 17: 1059–1070.

Restituito, S., Couve, A., Bawagan, H., Jourdain, S., Pangalos, M.N., Calver, A.R., Freeman, K.B. and Moss, S.J. (2005) Multiple motifs regulate the trafficking of GABA(B) receptors at distinct checkpoints within the secretory pathway. Mol. Cell. Neurosci., 28: 747–756.

Richman, R.W., Tombler, E., Lau, K.K., Anantharam, A., Rodriguez, J., O'Bryan, J.P. and Diverse-Pierluissi, M.A. (2004) N-type Ca^{2+} channels as scaffold proteins in the assembly of signaling molecules for $GABA_B$ receptor effects. J. Biol. Chem., 279: 24649–24658.

Sauter, K., Grampp, T., Fritschy, J.M., Kaupmann, K., Bettler, B., Mohler, H. and Benke, D. (2005) Subtype-selective interaction with the transcription factor CCAAT/enhancer-binding protein (C/EBP) homologous protein (CHOP) regulates cell surface expression of $GABA_B$ receptors. J. Biol. Chem., 280: 33566–33572.

Schwarz, D.A., Barry, G., Eliasof, S.D., Petroski, R.E., Conlon, P.J. and Maki, R.A. (2000) Characterization of gamma-aminobutyric acid receptor $GABA_{B(1e)}$, a $GABA_{B(1)}$ splice variant encoding a truncated receptor. J. Biol. Chem., 275: 32174–32181.

Thuault, S.J., Brown, J.T., Sheardown, S.A., Jourdain, S., Fairfax, B., Spencer, J.P., Restituito, S., Nation, J.H., Topps,

S., Medhurst, A.D., Randall, A.D., Couve, A., Moss, S.J., Collingridge, G.L., Pangalos, M.N., Davies, C.H. and Calver, A.R. (2004) The GABA(B2) subunit is critical for the trafficking and function of native GABA(B) receptors. Biochem. Pharmacol., 68: 1655–1666.

Uezono, Y., Kanaide, M., Kaibara, M., Barzilai, R., Dascal, N., Sumikawa, K. and Taniyama, K. (2005) Coupling of the $GABA_B$ receptor $GABA_{B(2)}$ subunit to G proteins: evidence from the Xenopus oocyte and baby kidney hamster cell expression system. Am. J. Physiol. Cell Physiol.

Urwyler, S., Gjoni, T., Koljatic, J. and Dupuis, D.S. (2005) Mechanisms of allosteric modulation at $GABA_B$ receptors by CGP7930 and GS39783: effects on affinities and efficacies of orthosteric ligands with distinct intrinsic properties. Neuropharmacology, 48: 343–353.

Urwyler, S., Mosbacher, J., Lingenhoehl, K., Heid, J., Hofstetter, K., Froestl, W., Bettler, B. and Kaupmann, K. (2001) Positive allosteric modulation of native and recombinant gamma-aminobutyric acid(B) receptors by 2,6-Di-tert-butyl-4-(3-hydroxy-2,2-dimethyl-propyl)-phenol (CGP7930) and its aldehyde analog CGP13501. Mol. Pharmacol., 60: 963–971.

Urwyler, S., Pozza, M.F., Lingenhoehl, K., Mosbacher, J., Lampert, C., Froestl, W., Koller, M. and Kaupmann, K. (2003) N,N'-Dicyclopentyl-2-methylsulfanyl-5-nitro-pyrimidine-4,6-diamine (GS39783) and structurally related compounds: novel allosteric enhancers of gamma-aminobutyric acid B receptor function. J. Pharmacol. Exp. Ther., 307: 322–330.

Villemure, J.F., Adam, L., Bevan, N.J., Gearing, K., Chenier, S. and Bouvier, M. (2005) Subcellular distribution of GABA(B) receptor homo- and hetero-dimers. Biochem. J., 388: 47–55.

Wei, K., Eubanks, J.H., Francis, J., Jia, Z. and Snead Jr., O.C. (2001) Cloning and tissue distribution of a novel isoform of the rat GABA(B)R1 receptor subunit. Neuroreport, 12: 833–837.

White, J.H., McIllhinney, R.A., Wise, A., Ciruela, F., Chan, W.Y., Emson, P.C., Billinton, A. and Marshall, F.H. (2000) The $GABA_B$ receptor interacts directly with the related transcription factors CREB2 and ATFx. Proc. Natl. Acad. Sci. USA, 97: 13967–13972.

White, J.H., Wise, A., Main, M.J., Green, A., Fraser, N.J., Disney, G.H., Barnes, A.A., Emson, P., Foord, S.M. and Marshall, F.H. (1998) Heterodimerization is required for the formation of a functional GABA(B) receptor. Nature, 396: 679–682.

White, J.H., Wise, A. and Marshall, F.H. (2002) Heterodimerization of gamma-aminobutyric acid B receptor subunits as revealed by the yeast two-hybrid system. Methods, 27: 301–310.

Zukin, S.R., Young, A.B. and Snyder, S.H. (1974) Gamma-aminobutyric acid binding to receptor sites in the rat central nervous system. Proc. Natl. Acad. Sci. USA, 71: 4802–4807.

Tepper, Abercrombie & Bolam (Eds.)
Progress in Brain Research, Vol. 160
ISSN 0079-6123

CHAPTER 5

The cellular, molecular and ionic basis of GABA$_A$ receptor signalling

Mark Farrant[1,*] and Kai Kaila[2,*]

[1]*Department of Pharmacology, UCL (University College London), Gower Street, London WC1E 6BT, UK*
[2]*Department of Biological and Environmental Sciences and Neuroscience Center, P.O. Box 65 (Viikinkaari 1), University of Helsinki, FIN-00014 Helsinki, Finland*

Abstract: GABA$_A$ receptors mediate fast synaptic inhibition in the CNS. Whilst this is undoubtedly true, it is a gross oversimplification of their actions. The receptors themselves are diverse, being formed from a variety of subunits, each with a different temporal and spatial pattern of expression. This diversity is reflected in differences in subcellular targetting and in the subtleties of their response to GABA. While activation of the receptors leads to an inevitable increase in membrane conductance, the voltage response is dictated by the distribution of the permeant Cl$^-$ and HCO$_3^-$ ions, which is established by anion transporters. Similar to GABA$_A$ receptors, the expression of these transporters is not only developmentally regulated but shows cell-specific and subcellular variation. Untangling all these complexities allows us to appreciate the variety of GABA-mediated signalling, a diverse set of phenomena encompassing both synaptic and non-synaptic functions that can be overtly excitatory as well as inhibitory.

Keywords: bicarbonate; chloride; CLC-2; excitation; gating; inhibition; KCC3; NKCC1; NKCC2; shunting inhibition; synaptic; tonic

Introduction

The spectacular progress of biosciences during recent decades has led to a vast accumulation of knowledge concerning mechanisms that govern the development and functions of living beings. It has also shown that much of the variation among species and organ systems, as well as among distinct brain regions and nerve cells, is largely due to what can be thought of as "combinatorics" of basic molecular elements and signalling pathways. In fact, given the large number of possible conformational states and allosteric interactions that proteins and complexes thereof can assume, any potential signalling molecule can bind to an enormous variety of biologically relevant binding sites or receptors in diverse target cells. In the case of a neurotransmitter substance such as GABA, these ligand-binding site interactions take place at ionotropic and metabotropic receptors, at vesicular and plasmalemmal transporters and with metabolic enzymes. In the nervous system, such interactions of GABA culminate in its most important action — inhibition of neuronal firing (Kuffler and Edwards, 1958; Krnjevic, 1974).

*Corresponding authors. Tel.: +44(0)20 7679 4121;
Fax: +44(0)20 7679 7298; E-mail: m.farrant@ucl.ac.uk, or
Tel.: +358 9 19159860; Fax: +358 9 19159810;
E-mail: kai.kaila@helsinki.fi

DOI: 10.1016/S0079-6123(06)60005-8
59

The role of GABA as the main inhibitory transmitter in the mammalian brain is undoubtedly fundamental, but it should be noted that GABA also has excitatory actions (and "dual" effects — both inhibitory and excitatory) in a number of cell types (Cherubini et al., 1991; Voipio and Kaila, 2000; Ben-Ari, 2001; Marty and Llano, 2005). This is, only to be expected, because the properties of the ligand-binding site of a transmitter receptor channel do not dictate or even influence the biophysical properties of the channel's ionic filter or the transmembrane ion gradients that set the polarity and driving force of the channel-mediated current. As inhibition and excitation are central issues in the present review, we want to note here that postsynaptic *inhibition* is best defined as a transient decrease in the probability of firing in the target cell, while postsynaptic *excitation* has the opposite effect. Very often, *hyperpolarization* and *depolarization* are taken as synonyms of inhibition and excitation, which is an error, as is evident from the above definition (and from what is described below). Importantly, the actions of GABA can be manifest as conventional synaptic transmission, involving the transient or "phasic" activation of receptors, or they can reflect the persistent or "tonic" activation of receptors in a manner temporally divorced form identifiable presynaptic events (Mody, 2001; Semyanov et al., 2004; Farrant and Nusser, 2005). Such tonic receptor activation can also have varied functional consequences.

GABA's role as a neurotransmitter goes far beyond its immediate effects on the excitability of an individual target cell. Extensive work on cortical structures has shown that the activity of GABAergic neurons is crucially involved in the assembly of neurons into functional networks, and in shaping oscillations and transient population events at the network level (McBain and Fisahn, 2001; Buzsaki, 2002; Freund, 2003; Whittington and Traub, 2003; Buzsaki and Draguhn, 2004; Jonas et al., 2004; Vida et al., 2006). Many aspects of these network-level phenomena are based on GABA-mediated changes in the integrative properties of principal neurons that depend critically on synapse location and the timing of activity. Thus, numerous examples illustrate the importance of spatially distinct GABAergic inputs (axonic, perisomatic, proximal or distal dendritic) in enabling or regulating important neuronal

behaviours in the cortex (Spruston et al., 1995; Miles et al., 1996; Pouille and Scanziani, 2001; Somogyi and Klausberger, 2005; Szabadics et al., 2006) and basal ganglia (Plenz, 2003; Tepper and Bolam, 2004; see also *Chapters 6–11 in this volume*). While a great deal of work has been carried out on GABAergic transmission in the basal ganglia, it is quite obvious that the amount of information available is not comparable to that for GABA actions in cortical neurons and networks. The aim of the present review is to provide a general overview of $GABA_A$ receptor-mediated actions in an attempt to facilitate future work on $GABA_A$ receptor-mediated transmission at the cellular and network level in the basal ganglia. Our main focus is on the electrophysiological, biophysical and ion-regulatory mechanisms that shape GABA-mediated postsynaptic potentials and tonic signalling.

Molecular basis of $GABA_A$ receptor function

Ionotropic GABA receptors are members of the "Cys-loop" superfamily of ligand-gated ion channels (cl-LGIC), so named for a conserved motif in the amino-terminal domain in which a pair of cysteines forms a disulphide bridge (Simon et al., 2004). Other family members include the nicotinic acetylcholine receptors, glycine receptors and 5-HT3 receptors. In each case, the receptors are pentameric assemblies of subunits (each with four transmembrane domains; M1–4) that form a central ion channel, which is gated by the binding of a small neurotransmitter molecule (Lester et al., 2004; Peters et al., 2005; Unwin, 2005; Sine and Engel, 2006). This remarkable gating reaction is the key to cl-LGIC function. Somehow, the binding of transmitter molecules to extracellular domains of the receptor triggers an extremely rapid conformational change that propagates through the protein to the transmembrane region and results in opening of the ion channel. In the case of $GABA_A$ receptors, two GABA molecules bind at the extracellular interfaces between α and β subunits (see below). Investigations into the molecular basis of this phenomenon have been greatly facilitated by recent models of receptor structure, combining the atomic-scale model of the nicotinic acetylcholine receptor (nAChR) from

Torpedo marmorata electric organ (Unwin, 2005) and the crystallographic structure of the soluble acetylcholine-binding protein (AChBP) from *Lymnaea stagnalis* (Brejc et al., 2001; Celie et al., 2004). This has allowed earlier studies, in which important ion-pair interactions were identified at the interface of the extracellular binding domain and the channel region (Hu et al., 2003; Kash et al., 2003; Kash et al., 2004), to be extended to a more complete atomic-scale picture of the gating process. This now incorporates translation of movement at the agonist-binding site, through electrostatic interactions within the binding domain-ion channel interface, to the M2 region, causing disruption of hydrophobic interactions within the channel and removal of the barrier to ion flow (Czajkowski, 2005; Lee and Sine, 2005; Lummis et al., 2005; Xiu et al., 2005; Corry, 2006; Sine and Engel, 2006).

The molecular diversity of ionotropic GABA receptors has been reviewed extensively (Barnard et al., 1998; Korpi et al., 2002; Sieghart and Sperk, 2002; Rudolph and Mohler, 2004; Sieghart and Ernst, 2005). In mammals, 19 GABA$_A$ receptor subunit genes are grouped into eight families according to their sequence similarity (α1–6, β1–3, γ1–3, δ, ϵ, θ, π and ρ1–3), with additional variation coming from alternative splicing (notably for α5, β2, β3 and γ2 subunits (Barnard et al., 1998; Simon et al., 2004)). It is often, and correctly, stated that such subunit diversity predicts enormous heterogeneity of receptor types. Indeed, without any form of constraint, 19 different subunits could form more than two million unique pentameric permutations. However, basic "rules" of assembly (Kittler et al., 2002; Luscher and Keller, 2004) and a differential distribution of subunit types among brain regions and neuronal populations (Wisden et al., 1992; Fritschy and Mohler, 1995; Pirker et al., 2000) greatly reduce the number of receptor subtypes that exist in the CNS. While some subunits such as the α1 and γ2 are ubiquitous, others are much more restricted in their distribution. For example, the α6 subunit is confined to granule cells of the cerebellum and inferior colliculus (Luddens et al., 1990), the ϵ and θ subunits are found principally in nuclei belonging to various diffuse modulatory systems (cholinergic, noradrenergic, serotonergic, dopaminergic,

histaminergic and peptidergic cell groups (Sinkkonen et al., 2000; Moragues et al., 2002, 2003; Sergeeva et al., 2002, 2005) and the π subunit is present at low levels (if at all) in brain, but is strongly expressed in various other organs, including uterus and breast (Hedblom and Kirkness, 1997; Zafrakas et al., 2006).

In general, combinations of α and β subunits are sufficient to form functional GABA$_A$ receptors. However, the vast majority of native receptors are known to contain a third subunit type. Based, primarily, on data from studies employing subunit-specific antibodies, the most abundant GABA$_A$ receptor subtype in brain is formed from α1, β2 and γ2 subunits (McKernan and Whiting, 1996; Sieghart and Sperk, 2002; Whiting, 2003; Benke et al., 2004). Data from quantitative immunoprecipitation (Tretter et al., 1997), fluorescence resonance energy transfer between epitope-tagged subunits (Farrar et al., 1999) and electrophysiology of receptors with concatenated subunits (Baumann et al., 2002; Boileau et al., 2005; Baur et al., 2006) suggest a stoichiometry of two α, two β and one γ subunit. Receptors formed from other α, β and γ combinations (e.g., $\alpha_2\beta_3\gamma_2$ or $\alpha_3\beta_3\gamma_2$) are also widely expressed. Less numerous, though no less significant for the specific neuronal populations in which they are expressed, are receptors in which the γ subunit is replaced by a δ subunit (e.g., $\alpha_4\beta_3\delta$ or $\alpha_6\beta_3\delta$). In yet other receptor subtypes, the γ subunit can be replaced by the ϵ or π subunits, while the π and θ subunits may be capable of co-assembling with α, β *and* γ subunits to form receptors containing representatives from four families (Bonnert et al., 1999; Neelands and Macdonald, 1999; Neelands et al., 1999; Sieghart and Sperk, 2002). Finally, additional variability comes from the fact that individual receptors may contain two different α or β subunit isoforms (Benke et al., 2004; Minier and Sigel, 2004; Boulineau et al., 2005).

While most subunits assemble as heteromers, the three ρ subunits form functional homo- or heteromeric assemblies that have sometimes been classed as GABA$_C$ receptors (Bormann, 2000; Chebib and Johnston, 2000; Zhang et al., 2001), based on their pharmacological similarity to bicuculline-insensitive GABA receptors originally identified in spinal cord (for review, see Johnston, 1996).

Nevertheless, they may be more appropriately considered a sub-class of GABA$_A$ receptor subunits (Kaila, 1994; Barnard et al., 1998) and, indeed, they appear capable of forming functional receptors with $\gamma2$ subunits (Qian and Ripps, 1999; Pan and Qian, 2005) or with both $\alpha1$ and $\gamma2$ subunits (Milligan et al., 2004).

Ionic permeability of GABA$_A$ receptors

Irrespective of their subunit composition, all GABA$_A$ receptors are permeable to the same ions. Based on measurements of the permeation of a series of large, weakly hydrated probe anions, the diameter of the narrowest part of the GABA$_A$ receptor channel has been estimated at about 0.55 nm (Inomata et al., 1986; Bormann et al., 1987; Akaike et al., 1989; Fatima-Shad and Barry, 1993; Kaila, 1994; Wotring et al., 1999). However, the GABA$_A$ channel is not a simple water-filled pore (Takeuchi and Takeuchi, 1971) and its ion selectivity sequence ($SCN^- > I^- > Br^- > Cl^- > F^-$) differs from the corresponding sequence of relative ion mobilities in free solution, indicating electrostatic interaction between permeant ions and binding sites within the channel (Takeuchi and Takeuchi, 1971; Bormann et al., 1987; Akaike et al., 1989; Fatima-Shad and Barry, 1993). The ionic filter appears to be highly conserved, since both the channel diameter and ionic selectivity sequences are similar in diverse species, ranging from crayfish to frogs to mammals (Kaila, 1994).

Studies of different cl-LGICs have shown that aligned residues within the α-helical M2 region form the lining of the channel pore (Xu and Akabas, 1996; Keramidas et al., 2004), which is at its narrowest towards the intracellular portion of the channel. For homomeric $\rho1$ receptors, point mutations introduced towards the cytoplasmic mouth of the channel (within M2 and the proximal region of the M1–M2 linker), which alter effective charges, profoundly affect ion selectivity (Wotring et al., 2001, 2003; Wotring and Weiss, 2002; Carland et al., 2004). In heteromeric $\alpha\beta\gamma$ receptors, comparable mutations indicate that selectivity is determined by the β subunit (Jensen et al., 2002, 2005a, b). Residues within more

intracellular sections of the M1–M2 and M3–M4 loops, together with some towards the cytoplasmic end of M1, have also been suggested to contribute to the pathway for ion permeation in various cl-LGICs (Kelley et al., 2003; Filippova et al., 2004; Hales et al., 2006) and these may form parts of intracellular "portals" (i.e., narrow openings potentially capable of allowing ion passage), identified between adjacent subunits (Peters et al., 2005; Unwin, 2005).

Under physiological conditions, Cl^- and HCO_3^- are the only anions that act as carriers of current through GABA$_A$ channels. Data from measurements of the GABA$_A$ reversal potential (E_{GABA}) entered into the Goldman-Hodgkin-Katz (GHK) equation have provided estimates for the relative HCO_3^-/Cl^- permeability that range from 0.18 in mouse cultured spinal neurons (Bormann et al., 1987) and 0.3 in crayfish muscle fibre (Kaila and Voipio, 1987; Kaila et al., 1989) to 0.6 in hippocampal pyramidal neurons (Fatima-Shad and Barry, 1993). While some differences are seen for recombinant receptors (e.g., homomeric $\rho1$ vs. $\alpha\beta\gamma$; Wotring et al., 1999), much of this large variation is likely to have a methodological rather than a biological basis, and novel estimates of $P_{HCO_3^-}/P_{Cl^-}$ from native mammalian receptors would be of much interest.

GABA$_A$ receptor channel conductance

Although the channels formed by GABA$_A$ receptors are similarly *ion selective*, they do display differences in single-channel *conductance*, i.e., the rate at which they allow ions to flow. Most receptors display a prominent main conductance state and one or more subconductance states. The absolute conductance depends on the concentrations of permeant ions present, but with near symmetrical $[Cl^-]$ (~130–160 mM; room temperature in outside-out patches) the main conductance state of homomeric $\rho1$ receptors is ~1–5 pS, that of dimeric $\alpha\beta$ receptors is ~15 pS, and that of ternary $\alpha\beta\gamma$ or $\alpha\beta\delta$ assemblies is ~25–28 pS (Puia et al., 1990; Verdoorn et al., 1990; Angelotti and Macdonald, 1993; Fisher and Macdonald, 1997; Pan et al., 1997; Brickley et al., 1999; Boileau et al., 2005)).

Changing the type of α or β has little or no effect on the conductance. These numbers show that GABA$_A$ receptor channels, while maintaining near-constant selectivity, can still allow ions to pass at high but variable rates. For a membrane voltage of −70 mV, the main conductance of αβγ combinations (the predominant conductance level seen in neurons Gunther et al., 1995; Brickley et al., 1999; Lorez et al., 2000) corresponds to the flow of ~1.1–1.2 × 10^7 ions/s, or 1 ion per 85 ns. The exact determinants of conductance remain to be established. The higher conductance of γ- or δ-containing receptors correlates with a greater net positive charge in the extracellular M2–M3 loop (Fisher and Macdonald, 1997; Brickley et al., 1999), but for the γ2 subunit, a polar serine residue within the M2 pore region has been shown to be important (Luu et al., 2005). This residue is also present in the δ subunit. As indicated above, in the case of other cl-LGICs (nACh and 5-HT3 receptors), residues within cytoplasmic loops, thought to form ion portals (see above), profoundly influence channel conductance (Peters et al., 2005; Hales et al., 2006), demonstrating that interactions beyond the narrow pore play a key role in ion conduction.

It is worth noting that while the behaviour of the vast majority of native or recombinant GABA$_A$ receptors fits the pattern just described, the single-channel conductance of some GABA$_A$ receptors has been reported to vary widely in apparent multiples of an elementary conductance. This variation depends on the concentration of agonist used and it is markedly influenced by allosteric modulators such as benzodiazepines, which produce an increase in conductance or a shift towards occupancy of pre-existing higher conductance states, as described for receptors in hippocampal neurons (Eghbali et al., 1997, 2003; Birnir et al., 2001; Lindquist and Birnir, 2006) and in dopaminergic (DA) neurons of the substantia nigra (SN) (Guyon et al., 1999).

Under physiological conditions, I–V plots of GABA$_A$ receptor-mediated currents typically show GHK-type outward rectification (Hille, 2001), which is attributable to the asymmetric transmembrane distribution of Cl$^-$ (Kaila, 1994). However, a genuine voltage dependence of GABA$_A$ receptor single-channel conductance has been reported in some studies (Gray and Johnston, 1985; Gage and Chung, 1994; Eghbali et al., 2003; Lindquist and Birnir, 2006). In most cases, I–V plots for single channels are linear in symmetrical Cl$^-$, and rectification of macroscopic responses (e.g., Bormann et al., 1987; Weiss et al., 1988; Fisher, 2002; Pytel et al., 2006) has been attributed to a voltage dependence of receptor binding, gating or desensitisation (see below).

With these fundamental features of channel function in mind, we will now consider the way in which the receptors are activated in the brain, and the consequences of their activation.

GABA concentration changes at the synapse

GABA is synthesized in the cytosol and accumulated into synaptic vesicles by a vesicular transporter (VGAT/VIAAT (McIntire et al., 1997; Chaudhry et al., 1998; Wojcik et al., 2006)) that is able to generate an intra-lumenal concentration thought to be in the range of several hundred millimolar (Axmacher et al., 2004). Thus, the fusion of a single vesicle liberates many thousands of GABA molecules into the synaptic cleft (~20 nm wide and 0.05–0.2 μm^2 in area), generating a GABA concentration waveform that peaks in the millimolar range (Mody et al., 1994; Mozrzymas et al., 2003; Mozrzymas, 2004).

The GABA transient to which the postsynaptic receptors are exposed is very brief. This is because diffusion of transmitter out of the synaptic cleft is predicted to be rapid (Overstreet et al., 2002; Mozrzymas, 2004), even allowing for the fact that diffusion within the cleft may be several-fold slower than in free solution (Nielsen et al., 2004). Indeed, experiments that have measured the displacement of low-affinity competitive antagonists such as SR-95103 (that dissociate on a timescale comparable to that of the synaptic concentration transient) or the effects of agents such as chlorpromazine or protons that slow down the binding of GABA to its receptors (and thus cause a greater reduction on inhibitory postsynaptic current (IPSC) amplitude when the exposure of synaptic receptors to GABA is brief) indicate a time constant of synaptic GABA clearance of 300–600 µs (Overstreet et al., 2002) or even as fast as ~100 µs (Mozrzymas et al., 2003; Mozrzymas, 2004). Once GABA has diffused beyond the synaptic cleft it is taken up by selective transport molecules into nerve terminals and/or glial cells.

In the simple case, where a population of receptors in the postsynaptic density is affected by transmitter molecules liberated from a single vesicle only, the transmitter concentration waveform experienced by those receptors will be influenced by a variety of synapse-specific factors. Importantly, it will be altered by variations in vesicle size and content, the nature of vesicle fusion, the geometry of the synaptic cleft and the number and arrangement of transporters and receptors in relation to the site of release (Liu, 2003). Such intrinsic factors are thought to contribute to the marked developmental increase in the speed of the synaptic concentration transient for both GABA and glutamate (Barberis et al., 2005; Cathala et al., 2005). Greater variation in the concentration waveform of GABA can arise when there is interaction between multiple vesicles released from a single synaptic specialization (multivesicular release), from neighbouring synapses, or following repeated synaptic activation. In each case, the postsynaptic receptors will be exposed to a GABA concentration transient different from that arising from the release of a single vesicle, and this will be particularly pronounced if vesicle release is temporally dispersed (asynchronous release).

GABA$_A$ receptor gating and the IPSC

GABA$_A$ receptors undergo considerable spontaneous motion while in the closed state (Bera and Akabas, 2005), suggesting a level of flexibility appropriate for rapid gating (Maconochie et al., 1994; Jones and Westbrook, 1995; McClellan and Twyman, 1999; Burkat et al., 2001; Chakrapani and Auerbach, 2005). In theory, the channel can open in the absence of agonist, albeit with an extremely low probability (Chang and Weiss, 1999a; Campo-Soria et al., 2006). Of note, some recombinant (Sigel et al., 1989; Maksay et al., 2003; Lindquist et al., 2004; Wagner et al., 2005; Ranna et al., 2006) and native (Birnir et al., 2000;

Jones et al., 2006; McCartney et al., 2007) GABA$_A$ receptors do exhibit measurable spontaneous activity, but most allow ion flux only after they have been "occupied" by agonist, two molecules of GABA being needed for efficient gating (Baumann et al., 2003). In such a scheme — shared with other LGICs, and based on an extension of the del Castillo-Katz (del Castillo and Katz, 1957) mechanism (Fig. 1) — the agonist can be seen as activating the receptor by producing a massive (~100 K–10 M fold) increase in the basal probability of opening (Campo-Soria et al., 2006; Sine and Engel, 2006). Thus, exposure to GABA drives receptors through mono- and di-liganded closed states to the open state.

In order to describe observed channel behaviour (and hopefully to reflect physical events (Colquhoun, 2006)), extensions of such reaction schemes have been developed by many groups (Macdonald et al., 1989; Weiss and Magleby, 1989; Twyman et al., 1990; Jones and Westbrook, 1995; Haas and Macdonald, 1999; Jayaraman et al., 1999; Burkat et al., 2001; Akk et al., 2004; Celentano and Hawkes, 2004; Lema and Auerbach, 2006). All these models contain multiple open and closed states (Figs. 1d, e). A key difference among them is the presence of additional agonist-bound closed states, which can be entered *instead* of the open states (desensitized states; (Jones and Westbrook, 1995)), or *prior* to the open states, so-called "pre-gateway non-conducting" (Lema and Auerbach, 2006) or "flip" states (Colquhoun, 2006).

At the synapse, following the release of a single vesicle, GABA reaches a peak concentration that would, at equilibrium, produce maximal receptor activation. However, because the on-rate of binding of GABA is slower than the rate of diffusion (Jones et al., 1998), the short exposure means that only a fraction of the ten to a few hundred receptors clustered opposite the single release site (Edwards et al., 1990; Mody et al., 1994; Nusser et al., 1997; Brickley et al., 1999) will be fully occupied. The degree of receptor occupancy varies between synapses on different neurons and even between those on a single cell (Nusser et al., 1997; Perrais and Ropert, 1999; Hajos et al., 2000; Overstreet et al., 2002; Mozrzymas et al., 2003), and will, of

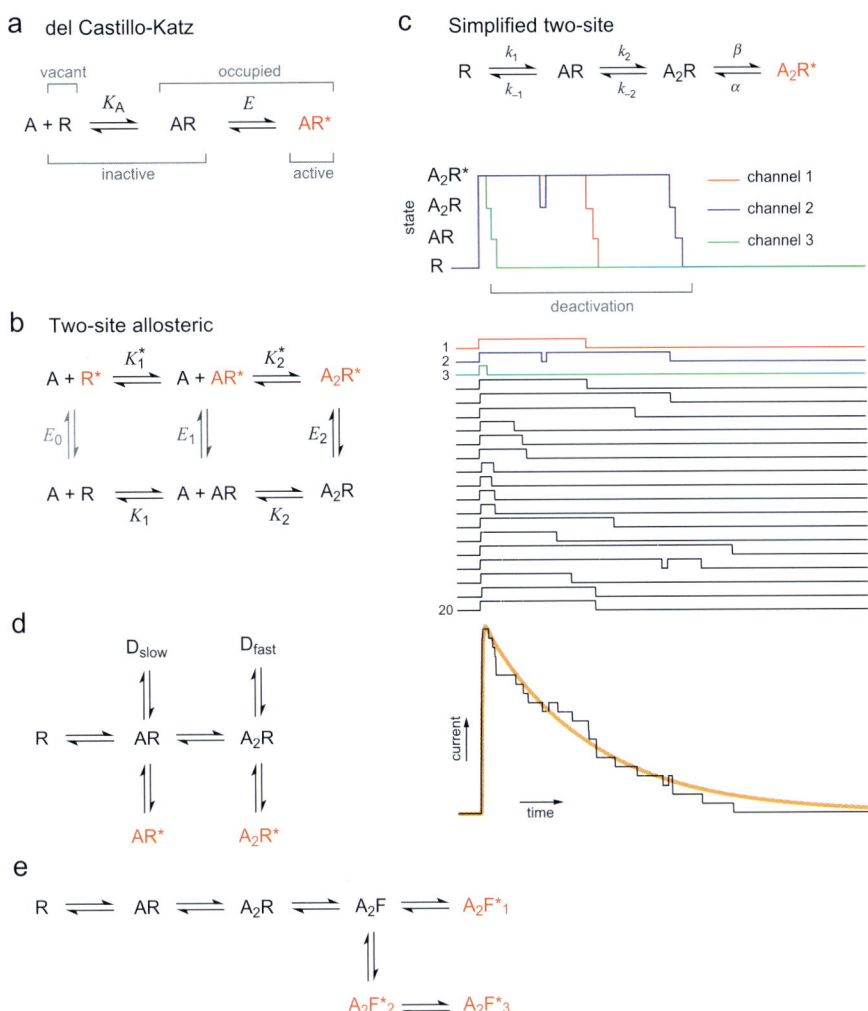

Fig. 1. Reaction mechanisms for cl-LGICs and the generation of a postsynaptic current. Panel (a) The del Castillo-Katz scheme (del Castillo and Katz, 1957), in which the agonist (A) binds to the receptor (R) forming a complex (AR) that changes conformation to yield a receptor with an open channel (AR*). K_A is the equilibrium dissociation constant for binding and E is the equilibrium constant for the gating step (opening rate constant/shutting rate constant; β/α, panel c). Panel (b), Extension of this reaction mechanism, in line with the Monod-Wyman-Changeux scheme (see Colquhoun, 1998 for discussion), to include two binding sites and the potential for constitutive (un-liganded) channel opening. E_0 is the equilibrium constant for gating in the absence of agonist. The progressively darker shading moving from E_0 to E_2 is designed to indicate the massively increased likelihood of conformational change in the presence of two agonist molecules (see also Downing et al., 2005; Campo-Soria et al., 2006; Sine and Engel, 2006). Panel (c) A simplified two-site model, in which un-liganded and mono-liganded openings are ignored. The microscopic rate constants are shown for binding (k_1 and k_2), unbinding (k_{-1} and k_{-2}), opening (β) and closing (α). Shown beneath the reaction scheme are idealized state transitions (after Sine and Engel, 2006) for three channels activated by a brief pulse of agonist. Rate constants k_1, k_2, k_{-1}, k_{-2} and β determine the rising phase of the response (activation), while k_{-2}, β and α govern the decay (deactivation). Open states of the same three events (1–3) are shown below, together with 17 other simulated openings. When summed, these produce the exponential decay characteristic of the synaptic current. The continuous orange line is the numerical integration of the same simulation. Panels (d) and (e) Examples of reaction schemes that have been used to describe the behaviour of GABA$_A$ receptors. The widely used model of Jones and Westbrook (Jones and Westbrook, 1995) (d) incorporates two liganded closed (desensitized) states (D$_{slow}$ and D$_{fast}$), while the model of Lema and Auerbach (Lema and Auerbach, 2006) (e) contains an additional "pre-open closed" or gateway state. In all schemes, the open (ion conducting) states of the receptors are shown in red.

course, be sensitive to factors that modify the time course of the GABA concentration transient. As outlined above, for those receptors that are occupied by GABA, a conformational change is elicited that may ultimately lead to channel opening. The behaviour of the receptors can be envisaged as a series of transitions through various closed, open and desensitized states. The time spent in each of the various states, and thus the time course of the postsynaptic current, is determined by the properties of the receptors and the profile of GABA exposure (Mozrzymas, 2004).

GABAergic miniature inhibitory postsynaptic currents (mIPSCs), resulting from the release of a single vesicle, have a rapid onset with rise times of only a few hundred microseconds (Bier et al., 1996; Nusser et al., 1997; Brickley et al., 1999). This reflects the proximity of the receptors to the site of GABA release, the rapid binding of GABA and the speed of the transition between closed and open states (i.e., a high rate constant for opening). As the GABA concentration transient is brief, the decay of the IPSC reflects the closure of channels following removal of ligand (deactivation). The duration of this process depends on the rate constant for channel closing, and also on various transitions of the receptor, notably entry into and exit from desensitized states that have been viewed as trapping GABA on the receptor prior to the final unbinding (Jones and Westbrook, 1995; Chang and Weiss, 1999b; Haas and Macdonald, 1999; Bianchi and Macdonald, 2001). Because the rates of these transitions differ for $GABA_A$ receptors of different subunit composition, the expression of different receptor subtypes contributes to differences in IPSC decay observed at different stages of development (Okada et al., 2000; Vicini et al., 2001) and in different cell types (Nusser et al., 1999; Bacci et al., 2003; Ramadan et al., 2003). Typically, IPSCs have a much longer duration in immature than in mature neurons.

As intimated above, discrete point-to-point communication, mediated by GABA release from an axon terminal onto closely apposed receptors, is not the only form of GABAergic synaptic signalling. The factors that determine the transmitter exposure of any particular receptor include its location relative to the GABA release site(s), the geometry and arrangement of the neighbouring cellular elements, the presence of diffusional barriers and the proximity of GABA transporters (see above; and Barbour and Hausser, 1997; Kullmann, 2000; Overstreet et al., 2002). Particularly under conditions favouring the release of multiple vesicles, GABA can escape the synaptic cleft at a sufficiently high concentration to activate additional receptors that may be located adjacent to the immediate postsynaptic receptors ("perisynaptic"), at other postsynaptic clusters within the same synaptic contact, outside the synapse ("extrasynaptic"), or at nearby synapses. Such "spillover" of GABA onto extrasynaptic or perisynaptic (α6- and α4-containing) receptors has been shown to contribute a slow component to IPSCs in granule cells of the cerebellum (Rossi and Hamann, 1998) and dentate gyrus (Wei et al., 2003). A more unusual form of spillover (possibly with activation of receptors *only* at sites remote from the point of release) is thought to contribute to the generation of slowly rising and decaying IPSCs ("$GABA_{A, slow}$") seen in hippocampal CA1 pyramidal neurons (Pearce, 1993; Banks et al., 2000). These events originate at distal dendritic sites (Banks et al., 1998) and appear to reflect the activity of a unique population of interneurons and the activation of receptors with a different subunit composition (α5-containing) to those that underlie fast IPSCs from somatic and dendritic sites (α2- and α1-containing, respectively (Prenosil et al., 2006)). Of note, similar slow IPSCs are also present in pyramidal neurons of the subiculum, which lack α5 subunits (Prenosil et al., 2006).

Tonic activity of $GABA_A$ receptors

$GABA_A$ receptors are not activated only during "phasic", i.e., synaptic signalling. Persistent or "tonic" activation of receptors, independent of any identified release event, occurs prior to synapse formation in embryonic (Valeyev et al., 1993; LoTurco et al., 1995; Owens et al., 1999; Demarque et al., 2002) and immature (Sipila et al., 2007) as well as newly derived postnatal neurons (Nguyen et al., 2003; Liu et al., 2005; Ge et al., 2006), as well as in a variety of mature (synaptically connected) neurons (Mody, 2001; Semyanov et al., 2004; Farrant and Nusser, 2005). Tonic GABAergic signalling in mature neurons was first identified in granule cells of the cerebellar cortex (Kaneda et al., 1995; Brickley et al., 1996; Tia et al., 1996; Wall and Usowicz, 1997), where, in addition to blocking IPSCs, GABA$_A$ receptor antagonists were shown to decrease resting membrane conductance (see Farrant and Nusser, 2005). To date, similar GABA-mediated tonic conductances have been demonstrated in granule cells of the dentate gyrus (Nusser and Mody, 2002; Stell and Mody, 2002; Stell et al., 2003; Mtchedlishvili and Kapur, 2006), pyramidal cells and inhibitory interneurons in the CA1 region of the hippocampus (Semyanov et al., 2003; Scimemi et al., 2005; Shen et al., 2005), pyramidal neurons and interneurons in the somatosensory cortex (Yamada et al., 2006; Keros and Hablitz, 2005), thalamocortical relay neurons of the ventral basal complex and dorsal lateral geniculate nucleus (Porcello et al., 2003; Belelli et al., 2005; Cope et al., 2005; Jia et al., 2005), magnocellular neurosecretory neurons of the supraoptic nucleus (Park et al., 2006), and at axon terminals of retinal bipolar cells (Hull et al., 2006).

While some GABA$_A$ receptors exhibit a low but measurable probability of opening in the absence of agonist (see above), a variety of evidence suggest that the widely-observed tonic activity of GABA$_A$ receptors reflects, in some neurons atleast, the persistent *ligand-induced activation* of specific receptor populations by low concentrations (nM to μM) of ambient GABA (see Cavelier et al., 2005; Farrant and Nusser, 2005; Santhakumar et al., 2006; but see McCartney et al., 2007). For different neurons, the concentration of ambient GABA to which they are continuously exposed may differ, but it is ultimately dictated by the effectiveness of Na$^+$- and Cl$^-$-dependent GABA uptake carriers. Often, such transporters are viewed as simply removing GABA from the extracellular space. In so doing, they act to limit spillover of GABA molecules from the synapse, alter the GABA concentration waveform in the synaptic cleft and ensure that "resting" GABA concentration remains sufficiently low to limit desensitization of synaptic receptors (Roepstorff and Lambert, 1994; Overstreet et al., 2002; Keros and Hablitz, 2005). However, transporters also have a

significant and dynamic influence on ambient GABA (Richerson and Wu, 2003). This is because the extracellular concentration of GABA at which they are at equilibrium is dictated by the membrane potential and the transmembrane gradients for the transporter substrates (GABA, Na$^+$, Cl$^-$), each of which can vary (Overstreet and Westbrook, 2001; Richerson and Wu, 2003; Wu et al., 2003, 2006; Allen et al., 2004). In the extreme case, transporters can operate in reverse, and could, therefore, act as a source of GABA (Attwell et al., 1993). In general, such reversal does not contribute to the generation of tonic GABA$_A$ receptor activity. Rather, blockade of transport increases the magnitude of the tonic current (Wall and Usowicz, 1997; Nusser and Mody, 2002; Jensen et al., 2003; Rossi et al., 2003; Semyanov et al., 2003; Sipila et al., 2004; Keros and Hablitz, 2005; Scimemi et al., 2005; Sipila et al., 2007), suggesting that, in some cases at least, normally functioning uptake fails to reduce the ambient GABA concentration below that capable of activating certain GABA$_A$ receptors.

The identity of GABA$_A$ receptors that generate tonic activity has been established in several neuronal populations (see Farrant and Nusser, 2005). Of central importance in many cases is the δ subunit. Unlike the α1, α2, α3, α6, β2/3 and γ2 subunits, which are found throughout the neuronal membrane but are highly enriched in GABAergic synapses (Craig et al., 1994; Nusser et al., 1995a, b, 1998; Somogyi et al., 1996; Fritschy et al., 1998; Brunig et al., 2002), the δ subunit has been shown to be present exclusively in extrasynaptic and perisynaptic locations (Nusser et al., 1998; Wei et al., 2003). This finding is consistent with the idea that a γ subunit (most commonly the γ2) is indispensable for synaptic clustering of GABA$_A$ receptors (Essrich et al., 1998; Baer et al., 1999; Schweizer et al., 2003; Alldred et al., 2005; see also, Fritschy and Brunig, 2003; Luscher and Keller, 2004). The δ subunit co-assembles with the α6 subunit in cerebellar granule cells and with the α4 subunit in several areas, including the thalamus and dentate gyrus (Barnard et al., 1998; Sur et al., 1999). The initial suggestion that extrasynaptic α6- and δ-containing receptors might mediate the tonic conductance in cerebellar granule cells (Brickley et al., 1996; Wall and Usowicz, 1997; Nusser et al., 1998) followed the recognition that the postnatal development of the conductance mirrored the delayed expression of these subunits, and that $\alpha 6$- and δ-containing receptors display a high affinity for GABA and desensitize slowly and less extensively than $\alpha\beta\gamma$ receptors (Saxena and Macdonald, 1996; see also Farrant and Nusser, 2005). Subsequent studies showed that the tonic conductance in cerebellar granule cells is reduced by the $\alpha 6$-selective antagonist furosemide (Hamann et al., 2002) and abolished in mice lacking either the $\alpha 6$ (Brickley et al., 2001) or δ subunits (Stell et al., 2003). Likewise, deletion of the δ subunit (and concomitant loss of $\alpha 4$ expression (Peng et al., 2002)) reduces the tonic receptor activation in granule cells of the dentate gyrus (Stell et al., 2003). The δ subunit (in combination with $\alpha 4$ and $\beta 2$ subunits) is also implicated in the generation of the tonic conductance seen in thalamocortical neurons of the dorsal lateral geniculate and ventral basal thalamus (Belelli et al., 2005; Cope et al., 2005; Jia et al., 2005).

Although δ-containing receptors, by virtue of their high affinity for GABA and limited desensitization, would appear to underlie most of the tonic GABA$_A$ receptor activity seen in mature neurons, other receptor subtypes have been suggested to sustain a similar role. Although it seems obvious, it should be remembered that all GABA$_A$ receptors (even those exhibiting significant desensitization) could contribute to the generation of a tonic conductance if the ambient concentration of GABA were high enough to activate them. As discussed by Scimemi and colleagues (Scimemi et al., 2005), this is an important consideration when comparing different in vitro studies, some of which have measured tonic receptor activity in the presence of added GABA (Stell and Mody, 2002; Stell et al., 2003; see also McCartney et al., 2007), blockers of GABA uptake (Semyanov et al., 2003), or blockers of GABA metabolism (Wu et al., 2003; Caraiscos et al., 2004). It is also important with regard to the interpretation of the effects on tonic receptor activity of (potentially selective) positive allosteric modulators. While a lack of effect may be used to rule out the contribution of specific receptor subtypes, enhancement of the tonic conductance could reflect either a selective increase in the affinity of those receptors responsible for the conductance or a recruitment of receptor populations of lower

affinity that do not ordinarily exhibit tonic activity. In hippocampal pyramidal neurons, for example, $\alpha 5$-containing receptors would appear to contribute to the tonic conductance only when the concentration of extracellular GABA is elevated, while in mature tissue under "normal" conditions (without blockade of GABA transaminase (Caraiscos et al., 2004)) the conductance is mediated by δ-containing receptors (Scimemi et al., 2005). These observations also point to a tonic activity that is potentially dynamic, with different GABA$_A$ receptor populations playing a role under different conditions, as might result from developmental changes, pharmacological intervention, changes in the efficacy of GABA uptake, physiologically relevant changes in network activity or pathological changes such as epilepsy (see Scimemi et al., 2005).

Neuronal ion regulation and the driving force for GABA$_A$ receptor-mediated currents

As noted above, GABA$_A$ receptors are permeable to two physiologically relevant anions, Cl$^-$ and HCO$_3^-$, with a HCO$_3^-$/Cl$^-$ permeability ratio ($P_{HCO_3^-}/P_{Cl^-}$) of around 0.2–0.4. A frequent and erroneous assumption in the GABA literature is that this permeability ratio translates directly into identical quantitative relations of the respective anion currents during synaptic and tonic responses. With an intracellular pH (pH$_i$) of about 7.1–7.2 and the above $P_{HCO_3^-}/P_{Cl^-}$ ratio, the intraneuronal HCO$_3^-$ concentration ([HCO$_3^-$]$_i$ ~ 15 mM) produces an influence on E_{GABA} that is equal to about 3–6 mM Cl$^-$. Hence, the role of HCO$_3^-$ in setting the value for IPSP reversal can be significant, not only quantitatively but also qualitatively. In neurons with a rather hyperpolarized resting membrane potential and low [Cl$^-$]$_i$, the HCO$_3^-$-mediated current component can, in fact, exceed the Cl$^-$-mediated current, which leads to HCO$_3^-$-dependent depolarizing IPSPs as shown in experiments on pyramidal neurons in adult neocortical slices (Kaila et al., 1993; see also Connors et al., 1982; Gulledge and Stuart, 2003).

The above observations have important consequences. They show that modestly depolarizing

IPSPs can be generated in certain neurons under resting, steady-state conditions (i.e., in the absence of activity-dependent redistribution of Cl^-; see below) despite the presence of a functional Cl^- extrusion mechanism and a Cl^- equilibrium potential (E_{Cl}) that is more negative than the resting membrane potential. More generally, it is obvious that E_{GABA} is always more positive than E_{Cl}, and this difference is accentuated when $[Cl^-]_i$ is low. A corollary of this is that in immature neurons and other neurons that have a high $[Cl^-]_i$, the contribution of HCO_3^- to E_{IPSP} is negligible (Fig. 2).

Hyperpolarization of the membrane potential is a key property of the "classical" IPSPs described in the original work by Eccles and colleagues on the glycinergic mechanisms in spinal cord motoneurons (Burke, 2006) and in most textbooks of neuroscience in relation to both glycinergic and $GABA_A$ receptor-mediated responses. However, from the above it is evident that robustly hyperpolarizing IPSPs in resting neurons are probably not as common as is generally thought. Because of the HCO_3^- permeability, the IPSP reversal potential can be very close to resting V_m in a number of cell types. This, however, does not imply an absence of an inhibitory effect of such IPSPs because the opening of the $GABA_A$ receptors leads to what is known as "shunting inhibition": a decrease in the input resistance of the neuron with a consequent decrease in the membrane time constant and space constant of the target cell. Here, it is important to note that shunting inhibition is local and lasts only for the duration of the synaptic conductance change, whereas hyperpolarizing or depolarizing synaptic potentials outlast the originating conductance change, with their duration and spread dictated by passive membrane properties and voltage-gated channels. Thus, the decrease in time constant caused by shunting inhibition will suppress the temporal summation of simultaneously occurring excitatory inputs, while the decrease in space constant will suppress the spatial summation of such inputs. In fact, slightly depolarizing IPSPs such as those seen in hippocampal dentate granule cells may exert a more effective inhibitory action than hyperpolarizing ones. This is because the $GABA_A$ receptor current shows outward rectification, and the slope conductance of the current-voltage (I–V)

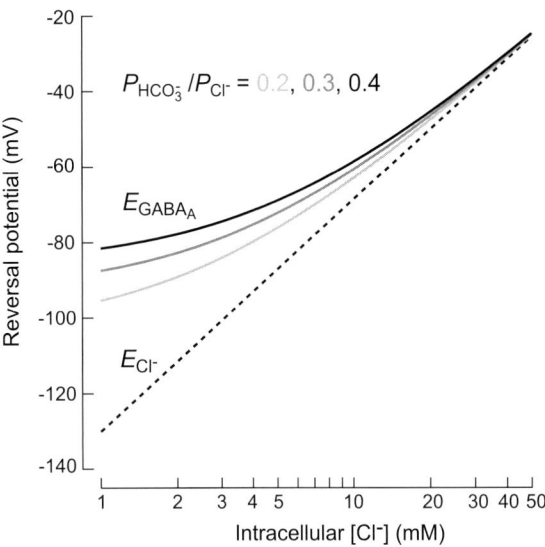

Fig. 2. The effect of bicarbonate on the $GABA_A$ reversal potential. The relationship between the intracellular concentration of Cl^- ($[Cl^-]_i$) and the reversal potential (E) for a purely Cl^--mediated event and one mediated by $GABA_A$ receptors (with permeability to both Cl^- and HCO_3^-) was calculated for different permeability ratios ($P_{HCO_3^-}/P_{Cl^-}$) according to:

$$E = \frac{RT}{F} \ln \frac{P_{Cl^-}[Cl^-]_i + P_{HCO_3^-}[HCO_3^-]_i}{P_{Cl^-}[Cl^-]_o + P_{HCO_3^-}[HCO_3^-]_o},$$

where R is the ideal gas constant, T is the absolute temperature at 37°C and F is Faraday's constant. The extracellular concentration of Cl^- ($[Cl^-]_o$) was assumed to be 130 mM, the extracellular concentration of HCO_3^- ($[HCO_3^-]_o$) was assumed to be 25 mM, and the intracellular concentration of HCO_3^- ($[HCO_3^-]_i$) was calculated to be 14.1 mM (assuming 5% ambient CO_2 and an intracellular pH of 7.15). *Note:* Light grey (0.2), mid grey (0.3) and black (0.4).

relation is enhanced at depolarized values of V_m. In addition, the absence of a hyperpolarizing component also prevents the generation of "rebound" excitation and the associated synchronization in the firing of cortical principal neurons that is often seen after hyperpolarizing IPSPs (Bevan et al., 2002; Somogyi and Klausberger, 2005). Interestingly, a recent study on subthalamic nucleus (STN) neurons (Baufreton et al., 2005) has shown that hyperpolarizing IPSPs increase the availability of voltage-gated Na^+ channels that leads to a negative shift in the action potential (AP) firing threshold and to a dramatic enhancement of EPSP-AP coupling. This implies that GABAergic inhibition can prime STN neurons to respond more efficiently to excitatory input, which adds a new dimension to the conventional view that GABAergic transmission from the globus pallidus (GP) to the STN restrains the activity of the STN (Albin et al., 1989).

Tonic GABA$_A$ receptor activity results in a conductance increase that (just like a synaptically evoked increase) reduces the magnitude and duration of the voltage response to an injected current and increases the decrement of voltage with distance. The obvious difference between the two is that while one is spatially and temporally discrete, the other is distributed and persistent and leads to a maintained shunting inhibition. Predictably, therefore, tonic GABA$_A$ receptor activation makes a cell less easily excited in response to a depolarizing input, and shifts the relationship between excitation current and output firing rate to the right (i.e., to higher levels of current) (Brickley et al., 1996; Hamann et al., 2002; Chadderton et al., 2004; Cope et al., 2005). In the case of excitatory synaptic conductances (as opposed to step current inputs), the effect of shunting inhibition is to decrease the slope of the input-output relationship, producing a change in "gain" (Chance et al., 2002; Mitchell and Silver, 2003). Again, just as for phasic events, the polarity of currents evoked by tonic receptor activation is not fixed and will depend on the membrane voltage and the net movement of Cl$^-$ and HCO$_3^-$, and thus on the Cl$^-$ and pH regulation of the cell (see below). In thalamocortical neurons, for example, persistent GABA$_A$ receptor activation has been shown to produce a hyperpolarization of 1–2 mV, and block of the tonic receptor activity causes a depolarization that promotes a shift from a burst firing towards a tonic firing mode (Cope et al., 2005). By contrast, in immature neurons, tonic GABA$_A$ receptor activity can cause depolarization, as described below.

Neuronal chloride regulation

Neuronal Cl$^-$ homeostasis is mainly controlled by the *SLC12A* family of cation-chloride co-transporters (CCCs; Fig. 3). The CCCs are composed of glycoproteins (MW of monomers in the range of 120–200 kDa) with 12 putative transmembrane segments flanked on one side by a small intracellular amino-terminus and by a large intracellular

carboxy-terminus on the other (Payne et al., 2003). CCCs are secondary active transporters that do not directly consume ATP, but derive the energy for Cl$^-$ translocation from the Na$^+$ and K$^+$ gradients generated by the Na-K-ATPase. The Na–K–2Cl co-transporters (NKCCs) mediate Cl$^-$ uptake fuelled by the plasmalemmal Na$^+$ gradient, and K–Cl co-transporters (KCCs) extrude Cl$^-$ under physiological conditions using the K$^+$ gradient. Both types of transporters are electrically neutral and hence do not have direct effects on the membrane potential. So far, two NKCC (NKCC1 and NKCC2) and four KCC isoforms (KCC1–4) have been identified (Delpire and Mount, 2002; Payne et al., 2003; Gamba, 2005). From these, NKCC1 as well as KCC2 and KCC3 have been shown to be expressed and functional in neurons (Payne et al., 1996; Plotkin et al., 1997; Rivera et al., 1999; Boettger et al., 2003). Some immature neurons express KCC4, at least at the mRNA level (Li et al., 2002).

NKCCs are kinetically activated by phosphorylation while KCCs are activated by dephosphorylation (Russell, 2000; Gamba, 2005). In neurons, much work remains to be done to identify the specific signalling cascades that modify the kinetics of Cl$^-$ transporters (Kelsch et al., 2001; Payne et al., 2003; Khirug et al., 2005). The phosphorylation cascade that targets NKCC is suppressed by intracellular Cl$^-$ that leads to a negative feedback loop that inhibits NKCC1 activity when a sufficiently high [Cl$^-$]$_i$ is achieved (Lytle and Forbush, 1996). As for KCC2, an increase in [Cl$^-$]$_i$ leads to an immediate increase in net Cl$^-$ efflux that is caused by the increase in the outward driving force for K-Cl co-transport and not by a change in the kinetics of the transporter (Payne et al., 2003; Khirug et al., 2005). Recent work points to a role for WNK3, a member of the WNK family of serine-threonine kinases, in the control of [Cl$^-$]$_i$ by upregulation of NKCC1 and down-regulation of KCC2 activity (Kahle et al., 2005).

The KCC2 molecule has gained much attention since it has an exclusively neuron-specific expression and is responsible for the generation of "classical" hyperpolarizing IPSPs (Rivera et al., 1999; Hubner et al., 2001). KCC2 is operational under isotonic conditions (Mercado et al., 2006) and in cortical principal cells it is constitutively active

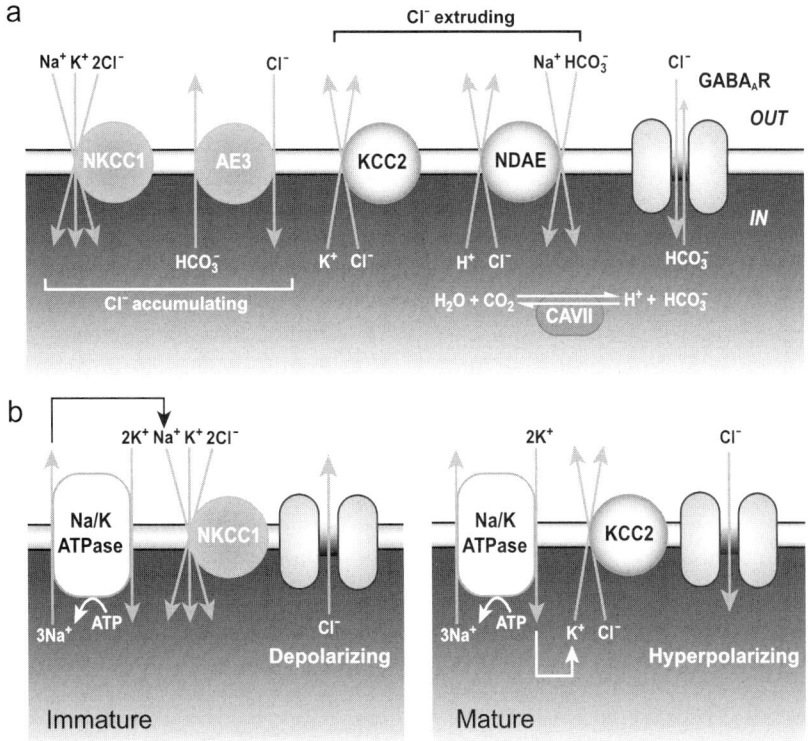

Fig. 3. Ion transporters set the value of E_{GABA}. Panel (a) The net effect of various Cl^- co-transporters and exchangers determines the reversal potential for $GABA_A$ receptor-mediated currents. On the left are shown those mechanisms that result in accumulation of Cl^- in neurons, in the middle are shown mechanisms that result in extrusion of Cl^-, and on the right is shown a Cl^- and HCO_3^- permeable GABA-gated channel. The Na–K–2Cl co-transporter isoform 1, NKCC1, mediates electrically neutral Cl^- uptake; this is a secondary active transport process driven by the Na^+ concentration gradient that is generated by the Na/K ATPase (see Panel (b)). The K-Cl co-transporter isoform 2, KCC2, mediates Cl^- extrusion driven by the K^+ gradient. The Na^+-dependent and -independent anion exchangers (NDAE and AE3, respectively) regulate pH_i, but they also influence neuronal Cl^- homeostasis, and provide a link between neuronal HCO_3^-/pH and Cl^- regulation. The carbonic anhydrase isoform VII (CAVII) is present in several types of neurons, and catalyzes the formation of HCO_3^- from CO_2. Panel (b) The "developmental switch" from depolarization to hyperpolarization. In immature neurons (left), depolarizing $GABA_A$ receptor-mediated responses are largely attributable to neuronal Cl^- accumulation via NKCC1. During maturation, the developmental up-regulation of functional KCC2 and down-regulation of NKCC1 result in a decrease in $[Cl^-]_i$ thereby providing the basis for conventional, hyperpolarizing IPSPs (right).

(Khirug et al., 2005; see also Lee et al., 2005). Unlike some other KCC isoforms (especially KCC4), KCC2 is not sensitive to changes in cellular volume as such. However, cell (and especially dendritic) swelling that is triggered in neurons during intense synaptic activity (Dietzel et al., 1980; Jefferys, 1995) is linked to an increase in $[Cl^-]_i$ that leads to an increase in the Cl^- driving force acting on KCC2, and hence to a "quasi volume-regulatory" enhancement of chloride extrusion. In principle, both NKCCs and KCCs are bidirectional and can carry out net Cl^- influx or efflux, depending on the concentration gradients of the relevant ions. However, a reversal of the direction of Cl^- transport is physiologically important only in the case of KCCs and in particular for KCC2, which often operates at close-equilibrium conditions. Hence, a modest increase in extracellular K^+ can lead to a KCC2-mediated net influx of Cl^- and to a consequent depolarizing shift in the IPSP reversal potential (Thompson and Gahwiler, 1989; Kakazu et al., 2000).

At the subcellular level, KCC2 is heavily expressed in dendritic spines both in cortical principal neurons as well as in interneurons (Gulyas et al., 2001), but "hot spots" of KCC2 expression occur also elsewhere, including the postsynaptic membranes in inhibitory synapses located in the dendrites of GABAergic neurons in the SN (Gulacsi et al., 2003). Work on hippocampal pyramidal cells has shown that in immature neurons, NKCC1 has a mainly somatic expression that thereafter develops into a predominantly dendritic one (Marty et al., 2002).

It has been recently shown that KCC3 is expressed in a large variety of neurons, including hippocampal pyramidal cells and cerebellar Purkinje neurons, where it has been shown to influence E_{GABA} (Boettger et al., 2003). Unlike KCC2, KCC3 is genuinely volume sensitive (Race et al., 1999). Clearly, more work is needed to gain an understanding of the role of this KCC isoform in the regulation of [Cl$^-$]$_i$ in various kinds of neurons.

Both functional and structural data indicate that transporter molecules taking up and extruding chloride, NKCC1 and KCCs, are co-expressed in certain types of mature neurons (Misgeld et al., 1986; Vardi et al., 2000; Martina et al., 2001; Duebel et al., 2006). In fact, various kinds of cells employ several distinct transport mechanisms, where the principal ionic substrate is transported in opposite directions. This seemingly paradoxical push-pull design permits a precise control of the set-point of the particular intracellular ion under various physiological conditions (Roos and Boron, 1981). Such a situation appears to be true for dentate gyrus granule cells, which co-express KCC2 plus NKCC1 (Kanaka et al., 2001) and exhibit powerful GABAergic inhibition in spite of their slightly depolarizing GABA$_A$ responses (Misgeld et al., 1986). Furthermore, discrete subcellular distributions of NKCC1 and KCC2–3 may produce intraneuronal Cl$^-$ gradients (even under "resting" conditions) that shape GABAergic responses in a cell-region dependent manner (Duebel et al., 2006; Szabadics et al., 2006). Additionally, the observation of large changes in monovalent cation concentrations (Rose and Konnerth, 2001) seen in neuronal spines during excitatory transmission, raises the possibility that spines form microdomains where Cl$^-$ levels undergo large activity-dependent changes. However, there is currently no direct evidence for or against this idea.

In addition to the CCCs, [Cl$^-$]$_i$ is also influenced by the Na$^+$-dependent and -independent anion (Cl$^-$/HCO$_3^-$) exchangers NDAE (or NDCBE) and AE3 that extrude and accumulate Cl$^-$, respectively. These transporters have a major role in the regulation of neuronal pH$_i$, as described below.

Neuronal pH regulation and E_{GABA}

In virtually all kinds of cells, including neurons, acid extrusion mechanisms (extrusion of H$^+$, uptake of HCO$_3^-$) maintain pH$_i$ at a significantly more alkaline level than what would be predicted on the basis of a passive distribution of H$^+$ and HCO$_3^-$ ions (Roos and Boron, 1981). Therefore, the equilibrium potential for HCO$_3^-$ (which equals the equilibrium potential of H$^+$) is much more positive than the resting membrane potential (typically about -10 to -15 mV). Unlike Cl$^-$ that can mediate either depolarizing or hyperpolarizing currents across GABA$_A$ receptor channels, HCO$_3^-$ invariably mediates a depolarizing current under physiological conditions (Kaila and Voipio, 1987; Kaila, 1994).

Neuronal acid extrusion is mainly achieved by Na$^+$-driven Cl$^-$/HCO$_3^-$ exchange ("NDAE", synonymous with "NDCBE") and Na$^+$/H$^+$ exchange (Chesler, 2003; Payne et al., 2003; Romero et al., 2004). Both types of transporters are activated by a fall in pH$_i$, which can be generated, e.g., by GABAergic transmission that leads to a channel-mediated net efflux of HCO$_3^-$ (Kaila et al., 1990) and by glutamatergic transmission and spiking that produce an increase in intraneuronal Ca^{2+} and a consequent acidosis (Ballanyi and Kaila, 1998). An excess alkalosis, which is most likely a rather exceptional event under physiological conditions, leads to activation of the Na$^+$-independent anion exchanger (AE3) that extrudes base equivalents in the form of HCO$_3^-$ in exchange for Cl$^-$ (Romero et al., 2004).

Carbonic anhydrases (CA) are enzymes that catalyze the reversible hydration of CO$_2$ into water and HCO$_3^-$. To date, 12 catalytically active CA isozymes are known, and five of these show a cytosolic localization (Pastorekova et al., 2004). During a large channel-mediated net efflux of

HCO_3^-, the intracellular HCO_3^- is quickly replenished by the activity of a cytosolic CA isoform (Kaila et al., 1990; Pasternack et al., 1993). In neonatal rat pyramidal neurons, CA activity is absent until around P10, and thereafter a steep increase in the expression of the CAVII isoform takes place (Ruusuvuori et al., 2004). Hence, CAVII acts as a second "developmental switch" (after KCC2 expression; see below) at the ion-regulatory level to modulate postsynaptic GABA-ergic responses in a qualitative manner.

The "developmental switch" from depolarization to hyperpolarization

In a pioneering study, Obata et al. (1978) showed that GABA has an excitatory action in co-cultures of muscle and spinal neurons taken from 6- to 8–day-old chick embryos, while an inhibitory effect was observed when culturing was started on day 10. Immature neurons are known to have very high input impedance that makes recordings with conventional sharp microelectrodes difficult and unreliable. However, the excitatory effect of GABA (and glycine) demonstrated by Obata and co-workers was clearly a genuine response of intact neurons, since recordings from the muscle cells showed that the GABA-induced end-plate potentials were blocked by TTX, which was indicative of motoneuronal spiking. In another hallmark study, Ben-Ari et al. (1989) demonstrated a similar shift in hippocampal pyramidal neurons from depolarizing to hyperpolarizing $GABA_A$ receptor-mediated responses. This "ontogenetic switch" in GABA action appears to be an almost ubiquitous feature of central neurons. In hippocampal and neocortical neurons, this switch is attributable to the developmental up-regulation of KCC2 that is paralleled by a down-regulation of NKCC1 (Rivera et al., 1999; Yamada et al., 2004; Rivera et al., 2005).

The cellular signalling cascades that trigger the expression of KCC2 and the postsynaptic switch from depolarizing to hyperpolarizing GABA action have not been identified. At least in neuronal cultures from both the hippocampus and the midbrain, neither spiking nor the transmitter actions of GABA or glutamate are absolute requirements for the induction of KCC2 expression and hyperpolarizing GABA responses (Ludwig et al., 2003; Titz et al., 2003). In cultures of cortical neurons as well as in neurons in the brain stem, KCC2 is initially expressed in an inactive form (Balakrishnan et al., 2003), but in cortical pyramidal neurons KCC2 becomes functionally active at the time of its expression (Khirug et al., 2005). These observations indicate that the mechanisms underlying the developmental up-regulation of functional KCC2 show some degree of variation among different neuronal types, and call for caution in any direct extrapolation from data on, e.g., cortical neurons to basal ganglia (Blaesse et al., 2006).

In immature pyramidal neurons, the specific NKCC1 antagonist bumetanide selectively blocks the depolarizing GABAergic response (Yamada et al., 2004; Sipila et al., 2006). Unfortunately, there is no selective blocker available for KCC2, and hence a gene knock-down strategy was used in the study of the role of this K-Cl co-transporter isoform in the maturation of GABAergic responses (Rivera et al., 1999). Furosemide is often used to block KCC2, but this drug is equally potent in inhibiting NKCC1 (Payne et al., 2003).

In immature neurons, depolarizing GABAergic responses have been shown to activate voltage-gated Ca^{2+} channels and to facilitate the activation of NMDA receptors, which leads to intracellular Ca^{2+} transients and to the activation of various intracellular signaling cascades (Yuste and Katz, 1991; Leinekugel et al., 1997; Fukuda et al., 1998). Such mechanisms are important for trophic actions of GABA that influence neuronal proliferation and migration, as well as the morphological maturation of individual neurons, and synaptogenesis (LoTurco et al., 1995; Owens and Kriegstein, 2002). However, the exact manner in which these processes contribute to normal development (or crucially perhaps, the nature of any redundancy in the system) remains to be determined, as synapse formation and early brain development appear essentially unaltered when GABA synthesis (Ji et al., 1999), vesicular transport (Wojcik et al., 2006) or vesicular release (Verhage et al., 2000; Varoqueaux et al., 2002) are eliminated (see also, Represa and Ben-Ari, 2005).

In the hippocampus, depolarizing GABAergic transmission appears to play a key role in the generation of endogenous network events that are thought to be crucial for the proper development of neuronal connections (Ben-Ari, 2001). Traditionally, these early events and particularly the so-called "giant depolarizing potentials" (GDPs) generated in the immature hippocampus were thought to be driven by depolarizing and excitatory GABAergic transmission. However, recent work has demonstrated that GDPs are not paced by GABAergic interneurons but rather by glutamatergic principal neurons, and that depolarization caused by $GABA_A$ receptor activation (brought about by both tonic and synaptically released GABA) has a *facilitatory* or *permissive* role in the generation of GDPs (Sipila and Kaila, In press; Sipila et al., 2005), and perhaps also of other kinds of early spontaneous events (Zheng et al., 2006).

Ionic plasticity of $GABA_A$ receptor-mediated signalling

Ionotropic glutamatergic transmission does not appear to be heavily modulated by activity-dependent changes in the driving force of the postsynaptic currents. In contrast to this, GABAergic signalling has the unique property of "ionic plasticity" which is based on both short-term and long-term shifts in the concentrations of Cl^- and HCO_3^- in postsynaptic neurons. This type of plasticity phenomena fall into two categories: short-term modulatory changes that are solely attributable to activity-dependent ionic shifts in the target cell (Voipio and Kaila, 2000) and long-term effects that involve changes in the kinetic properties or synthesis and trafficking of ion-regulatory proteins (Woodin et al., 2003; Rivera et al., 2004; Fiumelli et al., 2005).

The voltage response associated with $GABA_A$-mediated single IPSPs evoked under resting conditions can be robustly hyperpolarizing (such as in hippocampal pyramidal neurons and in SN GABAergic neurons), slightly hyperpolarizing or depolarizing (e.g., neocortical pyramidal neurons and SN DA neurons), slightly but consistently depolarizing (e.g., dentate granule cells). As discussed earlier, robustly depolarizing and sometimes even excitatory $GABA_A$-mediated PSPs occur not only in immature neurons but also in some interneurons that seem to actively accumulate Cl^- (Martina et al., 2001; Chavas and Marty, 2003; Marty and Llano, 2005). However, it has been known for a long time (Kaila and Voipio, 1987) that persistent activation of $GABA_A$ receptors can lead to changes in the distribution of Cl^- and HCO_3^- in target cells, and hence, to marked changes in the effects of $GABA_A$-mediated transmission (Kaila, 1994).

A well-known example of short-term ionic plasticity is the biphasic GABAergic responses in hippocampal pyramidal neurons that can be evoked by high-frequency stimulation (HFS) of hippocampal interneurons in the absence of ionotropic glutamatergic transmission (Grover et al., 1993; Kaila et al., 1997; Smirnov et al., 1999; Ruusuvuori et al., 2004). The biphasic response has a fast initial hyperpolarizing component that is followed by a depolarization with a duration (up to several seconds) that outlasts any possible action of synaptically released GABA. Importantly, the depolarization is strong enough to trigger spike bursts in the CA1 pyramids, which demonstrates that during intense interneuronal activity, GABA's signalling role can undergo a qualitative change from inhibitory to excitatory. A detailed analysis of the ionic bases of the biphasic response has shown that the initial part of the depolarization is caused by anionic redistribution, where the inwardly directed HCO_3^- current drives a depolarization that promotes the uptake of Cl^- and, hence, a positive shift in E_{GABA} (Kaila et al., 1997; Smirnov et al., 1999). However, the prolonged depolarization that is seen afterwards is caused by an increase in extracellular K^+, which is attributable to the recovery of the neuronal $[Cl^-]_i$ levels that by necessity requires a net efflux of Cl^- and K^+ in a 1:1 stoichiometry. Hence, during the prolonged depolarization, the V_m of the pyramidal neurons achieves a level that is much more depolarized than the simultaneous value of E_{GABA}. As expected, membrane-permeant CA inhibitors strongly suppress HFS-induced GABAergic excitation (Ruusuvuori et al., 2004).

While the above data have been obtained under rather extreme experimental conditions (similar to those used in conventional LTP-induction paradigms), it is likely that qualitatively similar activity-dependent ionic redistributions take place in various kinds of cells in the brain. A point worth noting here is that the GABAergic neurons in the SN have high firing rates. Also, the lack of KCC2 in DA neurons (see above) suggests that E_{GABA} is rather volatile in these neurons and hence likely to be subject to marked activity-induced changes.

As to long-term ionic plasticity, there are interesting observations suggesting that the efficacy of KCC2 is kinetically modulated during physiological activity (Woodin et al., 2003; Fiumelli et al., 2005). A recent study (Wang et al., 2006) has demonstrated that the reversal potential of IPSPs triggered by GP neurons in the STN is modulated in a bi-directional manner by rebound spiking: a long-lasting potentiation of IPSPs was seen in STN neurons that fired three or more spikes, whereas a long-lasting depression was seen in neurons that had 0–2 rebound spikes. Variations in the influx pathways of Ca^{2+} were shown to play a crucial role in this bi-directional IPSP plasticity, but the calcium-dependent, down-stream mechanisms — which apparently act on ion transporters — have not been identified (cf. Woodin et al., 2003; Fiumelli et al., 2005).

An example of ionic plasticity at the level of protein synthesis is provided by work on the effect of epileptiform activity on KCC2. Epileptic activity is known to produce an increase in the expression of brain-derived neurotrophic factor (BDNF) and its receptor TrkB (Binder et al., 2001; Huang and Reichardt, 2003). Hence, it was of much interest to find that following in vivo kindling, the expression of KCC2 shows a rapid fall in hippocampal regions where BNDF-TrkB up-regulation is most salient (Rivera et al., 2002). A direct causal link from TrkB activation to KCC2 down-regulation was established in experiments on hippocampal slices and slice cultures (Rivera et al., 2002, 2004). Work on transgenic mice with point mutations in their TrkB receptors (see Minichiello et al., 1998) showed that KCC2 down-regulation was triggered by the simultaneous activation of the two major TrkB-mediated signalling cascades, the PLCγ and Shc pathways. Interestingly, activation of the Shc cascade only led to an increase in KCC2 expression (Rivera et al., 2004), which may shed light on the mechanisms that underlie the developmental up-regulation of KCC2. The lifetime of membrane-associated KCC2 is very short, in the range of tens of minutes (Rivera et al., 2004), which makes KCC2 well suited for mediating GABAergic long-term ionic plasticity under various kinds of physiological and pathophysiological conditions.

Ionic mechanisms of GABAergic inhibition in substantia nigra

SN pars compacta

There is little information about the ionic mechanisms that shape GABAergic actions in basal ganglia. However, a rather exceptional finding was made in recent work (Gulacsi et al., 2003) on dopaminergic (DA) neurons in the substantia nigra (SN) pars compacta. In contrast to most of the neurons in the CNS, these DA neurons do not express KCC2, as shown by several kinds of immunohistochemical techniques. In contrast to this, GABAergic neurons in the SN were immunopositive for KCC2, with high levels of expression in the dendritic plasma membrane close to inhibitory synapses.

Consistent with the anatomical observations, gramicidin-perforated patch clamp recordings showed that GABAergic SN neurons had a significantly more hyperpolarized E_{IPSP} than the DA neurons (Gulacsi et al., 2003). Another interesting difference between these two cell types was that the DA neurons had an abundant expression of the chloride channel, ClC-2 (see also Smith et al., 1995), while no ClC-2 protein was detected in the GABAergic neurons. Inward rectification of the current carried by ClC-2 (Staley, 1994) suggests that this channel facilitates the clearance of intracellular Cl^- following a significant chloride load (acting in a somewhat valve-like manner), which may be especially important for a neuron devoid of KCC2. However, a channel-mediated ion flux is, of course, always "downhill" along the electrochemical ion gradient, and cannot produce any driving force, such as the slightly but significantly hyperpolarizing IPSP driving force that was observed in the DA neurons. The enigma of hyperpolarizing IPSPs was resolved by recordings in the absence of CO_2/HCO_3^- (in HEPES buffer) where a marked depolarizing shift in E_{IPSP} was observed, suggesting that the Cl^- extrusion by the Na^+-dependent anion exchanger (NDCBE, aka NDAE; see above) was responsible for the modestly hyperpolarizing IPSPs in the DA neurons (Gulacsi et al., 2003).

The DA neurons are known to exhibit less effective $GABA_A$ receptor-mediated inhibition than GABAergic neurons of the SN. This may have a number of underlying causes (cf Chapter 8 in this volume) but, nevertheless, is in agreement with the lack of KCC2 in the former. Since neuronal damage is often associated with hyperexcitability, it is tempting (but admittedly highly speculative) to raise the question whether the lack of KCC2 plays a role in DA neuron vulnerability in Parkinson's disease. Changes in neuronal volume are also known to have pathophysiological causes and effects (Jefferys, 1995), and even though KCC2 is not genuinely volume sensitive (see above), it responds to an increase in $[Cl]_i$ and thereby helps in maintaining a constant neuronal volume.

SN pars reticulata

Regional differences in KCC2 expression have been implicated in seizure control by the GABAergic neurons of the SN pars reticulata (SNr). The SNr is activated in various seizure models, especially during generalized seizures (reviewed by Veliskova and Moshe, 2006). This takes place in a region-specific pattern, where the anterior part of the SNr is active just before the onset and the posterior part during the seizure. Pharmacological manipulations that suppress the firing of GABAergic neurons in the anterior SNr lead to an attenuation of seizures. In addition to anterior-posterior SNr differences in the structure of the neuronal circuits, $GABA_A$ receptor subunit compositions and other features, a factor that may be of significance in the present context is the higher levels of expression of KCC2 mRNA in the anterior SNr (Galanopoulou, 2005). Consistent with a possible role for KCC2 in SNr seizure control, activation of $GABA_A$ receptors by local

injections of muscimol had a proconvulsant action in 2-week-old male pups that have a lower level of KCC2 expression in the SNr. Furthermore, there is evidence for sexual dimorphism in the SNr, based on a delayed increase in KCC2 expression in the male compared to the female, which correlates with a delayed switch to hyperpolarizing $GABA_A$ responses (Kyrozis et al., 2006). KCC2 expression in the SNr (and elsewhere, for that matter) may be under the control of sex hormones, as suggested by the effects of administration of sex hormones to rat pups (Galanopoulou, 2005; Veliskova and Moshe, 2006).

Conclusion

If there is one straightforward conclusion to be drawn from the preceding sections, it might be that extrapolation of observations regarding $GABA_A$ receptor-mediated signalling in one neuron or network to another cell type or brain region is not easy. Of course, this does not mean that it should not be attempted, but rather the opposite. A comparative approach — at the level of different species or brain regions — is a fundamental strategy of neurobiological research that has generated lots of novel experimental designs and theories and helped in the identification of basic principles in the design of living systems.

For $GABA_A$ receptors in different regions of the brain and in different neurons, profound heterogeneity is seen at the level of subunit expression. This contributes to a diversity of receptor subtypes that display differences in their affinity for GABA, their conductance, their kinetics of activation and deactivation and their onset and recovery from desensitization. Such factors determine the conditions under which the receptors may be activated, the magnitude and shape of the conductance change and responses to repeated activation. The functional significance of differences in biophysical properties among receptor subtypes will depend on their location and on any variations in their exposure to GABA — be those differences in the concentration waveform of GABA at or near the synapse or the concentration of ambient GABA to which they are persistently exposed.

Finally, the magnitude and polarity of current responses (dictated by the membrane voltage and the transmembrane distribution Cl^- and HCO_3^-) can vary widely, within individual neurons, between different neurons, during development and in response to physiological or pathological stimuli. Thus, despite common underlying principles of action, variations in each of the discrete steps that contribute to $GABA_A$ receptor-mediated signalling generate phenomena that are remarkably diverse and endlessly fascinating.

References

Akaike, N., Inomata, N. and Yakushiji, T. (1989) Differential effects of extra- and intracellular anions on GABA-activated currents in bullfrog sensory neurons. J. Neurophysiol., 62: 1388–1399.

Akk, G., Bracamontes, J. and Steinbach, J.H. (2004) Activation of $GABA_A$ receptors containing the $\alpha 4$ subunit by GABA and pentobarbital. J. Physiol., 556: 387–399.

Albin, R.L., Young, A.B. and Penney, J.B. (1989) The functional anatomy of basal ganglia disorders. Trends Neurosci., 12: 366–375.

Alldred, M.J., Mulder-Rosi, J., Lingenfelter, S.E., Chen, G. and Luscher, B. (2005) Distinct $\gamma 2$-subunit domains mediate clustering and synaptic function of postsynaptic $GABA_A$ receptors and gephyrin. J. Neurosci., 25: 594–603.

Allen, N.J., Rossi, D.J. and Attwell, D. (2004) Sequential release of GABA by exocytosis and reversed uptake leads to neuronal swelling in simulated ischemia of hippocampal slices. J. Neurosci., 24: 3837–3849.

Angelotti, T.P. and Macdonald, R.L. (1993) Assembly of $GABA_A$ receptor subunits: $\alpha 1\beta 1$ and $\alpha 1\beta 12S$ subunits produce unique ion channels with dissimilar single-channel properties. J. Neurosci., 13: 1429–1440.

Attwell, D., Barbour, B. and Szatkowski, M. (1993) Nonvesicular release of neurotransmitter. Neuron, 11: 401–407.

Axmacher, N., Stemmler, M., Engel, D., Draguhn, A. and Ritz, R. (2004) Transmitter metabolism as a mechanism of synaptic plasticity: A modeling study. J. Neurophysiol., 91: 25–39.

Bacci, A., Rudolph, U., Huguenard, J.R. and Prince, D.A. (2003) Major differences in inhibitory synaptic transmission onto two neocortical interneuron subclasses. J. Neurosci., 23: 9664–9674.

Baer, K., Essrich, C., Benson, J.A., Benke, D., Bluethmann, H., Fritschy, J.M. and Luscher, B. (1999) Postsynaptic clustering of γ-aminobutyric acid type A receptors by the $\gamma 3$-subunit in vivo. Proc. Natl. Acad. Sci. USA, 96: 12860–12865.

Balakrishnan, V., Becker, M., Lohrke, S., Nothwang, H.G., Guresir, E. and Friauf, E. (2003) Expression and function of chloride transporters during development of inhibitory neurotransmission in the auditory brainstem. J. Neurosci., 23: 4134–4145.

Ballanyi, K. and Kaila, K. (1998) Transmitter-evoked shifts in intracellular pH. In: Kaila K. and Ransom B.R. (Eds.), pH and Brain Function. Wiley-Liss, New York, pp. 291–308.

Banks, M.I., Li, T.B. and Pearce, R.A. (1998) The synaptic basis of $GABA_{A,slow}$. J. Neurosci., 18: 1305–1317.

Banks, M.I., White, J.A. and Pearce, R.A. (2000) Interactions between distinct $GABA_A$ circuits in hippocampus. Neuron, 25: 449–457.

Barberis, A., Lu, C., Vicini, S. and Mozrzymas, J.W. (2005) Developmental changes of GABA synaptic transient in cerebellar granule cells. Mol. Pharmacol., 67: 1221–1228.

Barbour, B. and Hausser, M. (1997) Intersynaptic diffusion of neurotransmitter. Trends Neurosci., 20: 377–384.

Barnard, E.A., Skolnick, P., Olsen, R.W., Mohler, H., Sieghart, W., Biggio, G., Braestrup, C., Bateson, A.N. and Langer, S.Z. (1998) International union of pharmacology. XV. Subtypes of γ-aminobutyric acid$_A$ receptors: classification on the basis of subunit structure and receptor function. Pharmacol. Rev., 50: 291–313.

Baufreton, J., Atherton, J.F., Surmeier, D.J. and Bevan, M.D. (2005) Enhancement of excitatory synaptic integration by GABAergic inhibition in the subthalamic nucleus. J. Neurosci., 25: 8505–8517.

Baumann, S.W., Baur, R. and Sigel, E. (2002) Forced subunit assembly in $\alpha 1\beta 2\gamma 2$ $GABA_A$ receptors. Insight into the absolute arrangement. J. Biol. Chem., 277: 46020–46025.

Baumann, S.W., Baur, R. and Sigel, E. (2003) Individual properties of the two functional agonist sites in $GABA_A$ receptors. J. Neurosci., 23: 11158–11166.

Baur, R., Minier, F. and Sigel, E. (2006) A $GABA_A$ receptor of defined subunit composition and positioning: concatenation of five subunits. FEBS Lett., 580: 1616–1620.

Belelli, D., Peden, D.R., Rosahl, T.W., Wafford, K.A. and Lambert, J.J. (2005) Extrasynaptic $GABA_A$ receptors of thalamocortical neurons: a molecular target for hypnotics. J. Neurosci., 25: 11513–11520.

Ben-Ari, Y. (2001) Developing networks play a similar melody. Trends Neurosci., 24: 353–360.

Ben-Ari, Y., Cherubini, E., Corradetti, R. and Gaiarsa, J.L. (1989) Giant synaptic potentials in immature rat CA3 hippocampal neurones. J. Physiol., 416: 303–325.

Benke, D., Fakitsas, P., Roggenmoser, C., Michel, C., Rudolph, U. and Mohler, H. (2004) Analysis of the presence and abundance of $GABA_A$ receptors containing two different types of α subunits in murine brain using point-mutated α−subunits. J. Biol. Chem., 279: 43654–43660.

Bera, A.K. and Akabas, M.H. (2005) Spontaneous thermal motion of the $GABA_A$ receptor M2 channel-lining segments. J. Biol. Chem., 280: 35506–35512.

Bevan, M.D., Magill, P.J., Hallworth, N.E., Bolam, J.P. and Wilson, C.J. (2002) Regulation of the timing and pattern of action potential generation in rat subthalamic neurons in vitro by GABA-A IPSPs. J. Neurophysiol., 87: 1348–1362.

Bianchi, M.T. and Macdonald, R.L. (2001) Agonist trapping by $GABA_A$ receptor channels. J. Neurosci., 21: 9083–9091.

78

Bier, M., Kits, K.S. and Borst, J.G. (1996) Relation between rise times and amplitudes of GABAergic postsynaptic currents. J. Neurophysiol., 75: 1008–1012.

Binder, D.K., Croll, S.D., Gall, C.M. and Scharfman, H.E. (2001) BDNF and epilepsy: too much of a good thing? Trends Neurosci., 24: 47–53.

Birnir, B., Eghbali, M., Cox, G.B. and Gage, P.W. (2001) GABA concentration sets the conductance of delayed $GABA_A$ channels in outside-out patches from rat hippocampal neurons. J. Membr. Biol., 181: 171–183.

Birnir, B., Everitt, A.B., Lim, M.S. and Gage, P.W. (2000) Spontaneously opening $GABA_A$ channels in CA1 pyramidal neurons of rat hippocampus. J. Membr. Biol., 174: 21–29.

Blaesse, P., Guillemin, I., Schindler, J., Schweizer, M., Delpire, E., Khiroug, L., Friauf, E. and Nothwang, H.G. (2006) Oligomerization of KCC2 correlates with development of inhibitory neurotransmission. J. Neurosci., 26: 10407–10419.

Boettger, T., Rust, M.B., Maier, H., Seidenbecher, T., Schweizer, M., Keating, D.J., Faulhaber, J., Ehmke, H., Pfeffer, C., Scheel, O., Lemcke, B., Horst, J., Leuwer, R., Pape, H.C., Volkl, H., Hubner, C.A. and Jentsch, T.J. (2003) Loss of K-Cl co-transporter KCC3 causes deafness, neurodegeneration and reduced seizure threshold. EMBO J., 22: 5422–5434.

Boileau, A.J., Pearce, R.A. and Czajkowski, C. (2005) Tandem subunits effectively constrain $GABA_A$ receptor stoichiometry and recapitulate receptor kinetics but are insensitive to $GABA_A$ receptor-associated protein. J. Neurosci., 25: 11219–11230.

Bonnert, T.P., McKernan, R.M., Farrar, S., le Bourdelles, B., Heavens, R.P., Smith, D.W., Hewson, L., Rigby, M.R., Sirinathsinghji, D.J., Brown, N., Wafford, K.A. and Whiting, P.J. (1999) θ, a novel γ-aminobutyric acid type A receptor subunit. Proc. Natl. Acad. Sci. USA, 96: 9891–9896.

Bormann, J. (2000) The "ABC" of GABA receptors. Trends Pharmacol. Sci., 21: 16–19.

Bormann, J., Hamill, O.P. and Sakmann, B. (1987) Mechanism of anion permeation through channels gated by glycine and γ-aminobutyric acid in mouse cultured spinal neurones. J. Physiol., 385: 243–286.

Boulineau, N., Baur, R., Minier, F. and Sigel, E. (2005) Consequence of the presence of two different β subunit isoforms in a $GABA_A$ receptor. J. Neurochem., 95: 1724–1731.

Brejc, K., van Dijk, W.J., Klaassen, R.V., Schuurmans, M., van Der Oost, J., Smit, A.B. and Sixma, T.K. (2001) Crystal structure of an ACh-binding protein reveals the ligand-binding domain of nicotinic receptors. Nature, 411: 269–276.

Brickley, S.G., Cull-Candy, S.G. and Farrant, M. (1996) Development of a tonic form of synaptic inhibition in rat cerebellar granule cells resulting from persistent activation of $GABA_A$ receptors. J. Physiol., 497: 753–759.

Brickley, S.G., Cull-Candy, S.G. and Farrant, M. (1999) Single-channel properties of synaptic and extrasynaptic $GABA_A$ receptors suggest differential targeting of receptor subtypes. J. Neurosci., 19: 2960–2973.

Brickley, S.G., Revilla, V., Cull-Candy, S.G., Wisden, W. and Farrant, M. (2001) Adaptive regulation of neuronal excitability by a voltage-independent potassium conductance. Nature, 409: 88–92.

Brunig, I., Scotti, E., Sidler, C. and Fritschy, J.M. (2002) Intact sorting, targeting, and clustering of γ-aminobutyric acid A receptor subtypes in hippocampal neurons in vitro. J. Comp. Neurol., 443: 43–55.

Burkat, P.M., Yang, J. and Gingrich, K.J. (2001) Dominant gating governing transient $GABA_A$ receptor activity: A first latency and P-o/o analysis. J. Neurosci., 21: 7026–7036.

Burke, R.E. (2006) John Eccles' pioneering role in understanding central synaptic transmission. Prog. Neurobiol., 78: 173–188.

Buzsaki, G. (2002) Theta oscillations in the hippocampus. Neuron, 33: 325–340.

Buzsaki, G. and Draguhn, A. (2004) Neuronal oscillations in cortical networks. Science, 304: 1926–1929.

Campo-Soria, C., Chang, Y. and Weiss, D.S. (2006) Mechanism of action of benzodiazepines on GABA(A) receptors. Br. J. Pharmacol., 148: 984–990.

Caraiscos, V.B., Elliott, E.M., You-Ten, K.E., Cheng, V.Y., Belelli, D., Newell, J.G., Jackson, M.F., Lambert, J.J., Rosahl, T.W., Wafford, K.A., MacDonald, J.F. and Orser, B.A. (2004) Tonic inhibition in mouse hippocampal CA1 pyramidal neurons is mediated by α5 subunit-containing γ-aminobutyric acid type A receptors. Proc. Natl. Acad. Sci. USA, 101: 3662–3667.

Carland, J.E., Moorhouse, A.J., Barry, P.H., Johnston, G.A. and Chebib, M. (2004) Charged residues at the 2' position of human $GABA_C$ ρ1 receptors invert ion selectivity and influence open state probability. J. Biol. Chem., 279: 54153–54160.

Cathala, L., Holderith, N.B., Nusser, Z., DiGregorio, D.A. and Cull-Candy, S.G. (2005) Changes in synaptic structure underlie the developmental speeding of AMPA receptor-mediated EPSCs. Nat. Neurosci., 8: 1310–1318.

Cavelier, P., Hamann, M., Rossi, D., Mobbs, P. and Attwell, D. (2005) Tonic excitation and inhibition of neurons: ambient transmitter sources and computational consequences. Prog. Biophys. Mol. Biol., 87: 3–16.

Celentano, J.J. and Hawkes, A.G. (2004) Use of the covariance matrix in directly fitting kinetic parameters: application to $GABA_A$ receptors. Biophys. J., 87: 276–294.

Celie, P.H., van Rossum-Fikkert, S.E., van Dijk, W.J., Brejc, K., Smit, A.B. and Sixma, T.K. (2004) Nicotine and carbamylcholine binding to nicotinic acetylcholine receptors as studied in AChBP crystal structures. Neuron, 41: 907–914.

Chadderton, P., Margrie, T.W. and Hausser, M. (2004) Integration of quanta in cerebellar granule cells during sensory processing. Nature, 428: 856–860.

Chakrapani, S. and Auerbach, A. (2005) A speed limit for conformational change of an allosteric membrane protein. Proc. Natl. Acad. Sci. USA, 102: 87–92.

Chance, F.S., Abbott, L.F. and Reyes, A.D. (2002) Gain modulation from background synaptic input. Neuron, 35: 773–782.

Chang, Y. and Weiss, D.S. (1999a) Allosteric activation mechanism of the α1β2γ2 γ-aminobutyric acid type A receptor revealed by mutation of the conserved M2 leucine. Biophys. J., 77: 2542–2551.

Chang, Y. and Weiss, D.S. (1999b) Channel opening locks agonist onto the $GABA_C$ receptor. Nat. Neurosci., 2: 219–225.

Chaudhry, F.A., Reimer, R.J., Bellocchio, E.E., Danbolt, N.C., Osen, K.K., Edwards, R.H. and Storm-Mathisen, J. (1998) The vesicular GABA transporter, VGAT, localizes to synaptic vesicles in sets of glycinergic as well as GABAergic neurons. J. Neurosci., 18: 9733–9750.

Chavas, J. and Marty, A. (2003) Coexistence of excitatory and inhibitory GABA synapses in the cerebellar interneuron network. J. Neurosci., 23: 2019–2031.

Chebib, M. and Johnston, G.A. (2000) GABA-activated ligand gated ion channels: medicinal chemistry and molecular biology. J. Med. Chem., 43: 1427–1447.

Cherubini, E., Gaiarsa, J.L. and Ben-Ari, Y. (1991) GABA: an excitatory transmitter in early postnatal life. Trends Neurosci., 14: 515–519.

Chesler, M. (2003) Regulation and modulation of pH in the brain. Physiol. Rev., 83: 1183–1221.

Colquhoun, D. (1998) Binding, gating, affinity and efficacy: the interpretation of structure-activity relationships for agonists and of the effects of mutating receptors. Br. J. Pharmacol., 125: 924–947.

Colquhoun, D. (2006) Agonist-activated ion channels. Br. J. Pharmacol., 147(Suppl 1): S17–S26.

Connors, B.W., Gutnick, M.J. and Prince, D.A. (1982) Electrophysiological properties of neocortical neurons in vitro. J. Neurophysiol., 48: 1302–1320.

Cope, D.W., Hughes, S.W. and Crunelli, V. (2005) $GABA_A$ receptor-mediated tonic inhibition in thalamic neurons. J. Neurosci., 25: 11553–11563.

Corry, B. (2006) An energy-efficient gating mechanism in the acetylcholine receptor channel suggested by molecular and Brownian dynamics. Biophys. J., 90: 799–810.

Craig, A.M., Blackstone, C.D., Huganir, R.L. and Banker, G. (1994) Selective clustering of glutamate and γ-aminobutyric acid receptors opposite terminals releasing the corresponding neurotransmitters. Proc. Natl. Acad. Sci. USA, 91: 12373–12377.

Czajkowski, C. (2005) Neurobiology: triggers for channel opening. Nature, 438: 167–168.

del Castillo, J. and Katz, B. (1957) Interaction at end-plate receptors between different choline derivatives. Proc. R. Soc. Lond. B. Biol. Sci., 146: 369–381.

Delpire, E. and Mount, D.B. (2002) Human and murine phenotypes associated with defects in cation-chloride cotransport. Annu. Rev. Physiol., 64: 803–843.

Demarque, M., Represa, A., Becq, H., Khalilov, I., Ben-Ari, Y. and Aniksztejn, L. (2002) Paracrine intercellular communication by a Ca-2 +- and SNARE- independent release of GABA and glutamate prior to synapse formation. Neuron, 36: 1051–1061.

Dietzel, I., Heinemann, U., Hofmeier, G. and Lux, H.D. (1980) Transient changes in the size of the extracellular space in the sensorimotor cortex of cats in relation to stimulus-induced changes in potassium concentration. Exp. Brain Res., 40: 432–439.

Downing, S.S., Lee, Y.T., Farb, D.H. and Gibbs, T.T. (2005) Benzodiazepine modulation of partial agonist efficacy and spontaneously active $GABA_A$ receptors supports an allosteric model of modulation. Br. J. Pharmacol., 145: 894–906.

Duebel, J., Haverkamp, S., Schleich, W., Feng, G., Augustine, G.J., Kuner, T. and Euler, T. (2006) Two-photon imaging reveals somatodendritic chloride gradient in retinal ON-type bipolar cells expressing the biosensor Clomeleon. Neuron, 49: 81–94.

Edwards, F.A., Konnerth, A. and Sakmann, B. (1990) Quantal analysis of inhibitory synaptic transmission in the dentate gyrus of rat hippocampal slices: a patch-clamp study. J. Physiol., 430: 213–249.

Eghbali, M., Birnir, B. and Gage, P.W. (2003) Conductance of $GABA_A$ channels activated by pentobarbitone in hippocampal neurons from newborn rats. J. Physiol., 552: 13–22.

Eghbali, M., Curmi, J.P., Birnir, B. and Gage, P.W. (1997) Hippocampal $GABA_A$ channel conductance increased by diazepam. Nature, 388: 71–75.

Essrich, C., Lorez, M., Benson, J.A., Fritschy, J.M. and Luscher, B. (1998) Postsynaptic clustering of major $GABA_A$ receptor subtypes requires the γ2 subunit and gephyrin. Nat. Neurosci., 1: 563–571.

Farrant, M. and Nusser, Z. (2005) Variations on an inhibitory theme: phasic and tonic activation of $GABA_A$ receptors. Nat. Rev. Neurosci., 6: 215–229.

Farrar, S.J., Whiting, P.J., Bonnert, T.P. and McKernan, R.M. (1999) Stoichiometry of a ligand-gated ion channel determined by fluorescence energy transfer. J. Biol. Chem., 274: 10100–10104.

Fatima-Shad, K. and Barry, P.H. (1993) Anion permeation in GABA- and glycine-gated channels of mammalian cultured hippocampal neurons. Proc. Biol. Sci., 253: 69–75.

Filippova, N., Wotring, V.E. and Weiss, D.S. (2004) Evidence that the TM1-TM2 loop contributes to the ρ1 GABA receptor pore. J. Biol. Chem., 279: 20906–20914.

Fisher, J.L. (2002) A lysine residue in the β3-subunit contributes to the regulation of GABAA receptor activity by voltage. Mol. Cell. Neurosci., 20: 683–694.

Fisher, J.L. and Macdonald, R.L. (1997) Single channel properties of recombinant $GABA_A$ receptors containing γ2-or δ-subtypes expressed with α1- and β3-subtypes in mouse L929 cells. J. Physiol., 505: 283–297.

Fiumelli, H., Cancedda, L. and Poo, M.M. (2005) Modulation of GABAergic transmission by activity via postsynaptic Ca^{2+}-dependent regulation of KCC2 function. Neuron, 48: 773–786.

Freund, T.F. (2003) Rhythm and mood in perisomatic inhibition. Trends Neurosci., 26: 489–495.

Fritschy, J.M. and Brunig, I. (2003) Formation and plasticity of GABAergic synapses: physiological mechanisms and pathophysiological implications. Pharmacol. Ther., 98: 299–323.

Fritschy, J.M., Johnson, D.K., Mohler, H. and Rudolph, U. (1998) Independent assembly and subcellular targeting of

GABA$_A$-receptor subtypes demonstrated in mouse hippocampal and olfactory neurons in vivo. Neurosci. Lett., 249: 99–102.

Fritschy, J.M. and Mohler, H. (1995) GABA$_A$-receptor heterogeneity in the adult-rat brain: differential regional and cellular-distribution of 7 major subunits. J. Comp. Neurol., 359: 154–194.

Fukuda, A., Muramatsu, K., Okabe, A., Shimano, Y., Hida, H., Fujimoto, I. and Nishino, H. (1998) Changes in intracellular Ca^{2+} induced by GABA$_A$ receptor activation and reduction in Cl$^-$ gradient in neonatal rat neocortex. J. Neurophysiol., 79: 439–446.

Gage, P.W. and Chung, S.H. (1994) Influence of membrane potential on conductance sublevels of chloride channels activated by GABA. Proc. Biol. Sci., 255: 167–172.

Galanopoulou, A.S. (2005) GABA receptors as broadcasters of sexually differentiating signals in the brain. Epilepsia, 46(Suppl 5): 107–112.

Gamba, G. (2005) Molecular physiology and pathophysiology of electroneutral cation-chloride cotransporters. Physiol. Rev., 85: 423–493.

Ge, S., Goh, E.L., Sailor, K.A., Kitabatake, Y., Ming, G.L. and Song, H. (2006) GABA regulates synaptic integration of newly generated neurons in the adult brain. Nature, 439: 589–593.

Gray, R. and Johnston, D. (1985) Rectification of single GABA-gated chloride channels in adult hippocampal neurons. J. Neurophysiol., 54: 134–142.

Grover, L.M., Lambert, N.A., Schwartzkroin, P.A. and Teyler, T.J. (1993) Role of HCO$_3^-$ ions in depolarizing GABA$_A$ receptor-mediated responses in pyramidal cells of rat hippocampus. J. Neurophysiol., 69: 1541–1555.

Gulacsi, A., Lee, C.R., Sik, A., Viitanen, T., Kaila, K., Tepper, J.M. and Freund, T.F. (2003) Cell type-specific differences in chloride-regulatory mechanisms and GABA$_A$ receptor-mediated inhibition in rat substantia nigra. J. Neurosci., 23: 8237–8246.

Gulledge, A.T. and Stuart, G.J. (2003) Excitatory actions of GABA in the cortex. Neuron, 37: 299–309.

Gulyas, A.I., Sik, A., Payne, J.A., Kaila, K. and Freund, T.F. (2001) The KCl cotransporter, KCC2, is highly expressed in the vicinity of excitatory synapses in the rat hippocampus. Eur. J. Neurosci., 13: 2205–2217.

Gunther, U., Benson, J., Benke, D., Fritschy, J.M., Reyes, G., Knoflach, F., Crestani, F., Aguzzi, A., Arigoni, M., Lang, Y., et al. (1995) Benzodiazepine-insensitive mice generated by targeted disruption of the γ2 subunit gene of γ-aminobutyric acid type A receptors. Proc. Natl. Acad. Sci. USA, 92: 7749–7753.

Guyon, A., Laurent, S., Paupardin-Tritsch, D., Rossier, J. and Eugene, D. (1999) Incremental conductance levels of GABA$_A$ receptors in dopaminergic neurones of the rat substantia nigra pars compacta. J. Physiol., 516: 719–737.

Haas, K.F. and Macdonald, R.L. (1999) GABA$_A$ receptor subunit γ2 and δ subtypes confer unique kinetic properties on recombinant GABA$_A$ receptor currents in mouse fibroblasts. J. Physiol. (Lond.), 514: 27–45.

Hajos, N., Nusser, Z., Rancz, E.A., Freund, T.F. and Mody, I. (2000) Cell type- and synapse-specific variability in synaptic GABA$_A$ receptor occupancy. Eur. J. Neurosci., 12: 810–818.

Hales, T.G., Dunlop, J.I., Deeb, T.Z., Carland, J.E., Kelley, S.P., Lambert, J.J. and Peters, J.A. (2006) Common determinants of single channel conductance within the large cytoplasmic loop of 5-hydroxytryptamine type 3 and α4β2 nicotinic acetylcholine receptors. J. Biol. Chem., 281: 8062–8071.

Hamann, M., Rossi, D.J. and Attwell, D. (2002) Tonic and spillover inhibition of granule cells control information flow through cerebellar cortex. Neuron, 33: 625–633.

Hedblom, E. and Kirkness, E.F. (1997) A novel class of GABA$_A$ receptor subunit in tissues of the reproductive system. J. Biol. Chem., 272: 15346–15350.

Hille, B. (2001). Ion channels of excitable membranes, Sinauer, Sunderland, Massachusetts, p. 814.

Hu, X.Q., Zhang, L., Stewart, R.R. and Weight, F.F. (2003) Arginine 222 in the pre-transmembrane domain 1 of 5-HT$_{3A}$ receptors links agonist binding to channel gating. J. Biol. Chem., 278: 46583–46589.

Huang, E.J. and Reichardt, L.F. (2003) Trk receptors: roles in neuronal signal transduction. Annu. Rev. Biochem., 72: 609–642.

Hubner, C.A., Stein, V., Hermans-Borgmeyer, I., Meyer, T., Ballanyi, K. and Jentsch, T.J. (2001) Disruption of KCC2 reveals an essential role of K-Cl cotransport already in early synaptic inhibition. Neuron, 30: 515–524.

Hull, C., Li, G.L. and von Gersdorff, H. (2006) GABA transporters regulate a standing GABA$_C$ receptor-mediated current at a retinal presynaptic terminal. J. Neurosci., 26: 6979–6984.

Inomata, N., Oomura, Y., Akaike, N. and Edwards, C. (1986) The anion selectivity of the γ-aminobutyric acid controlled chloride channel in the perfused spinal ganglion cell of frog. Neurosci. Res., 3: 371–383.

Jayaraman, V., Thiran, S. and Hess, G.P. (1999) How fast does the γ-aminobutyric acid receptor channel open? Kinetic investigations in the microsecond time region using a laser-pulse photolysis technique. Biochemistry, 38: 11372–11378.

Jefferys, J.G. (1995) Nonsynaptic modulation of neuronal activity in the brain: electric currents and extracellular ions. Physiol. Rev., 75: 689–723.

Jensen, K., Chiu, C.S., Sokolova, I., Lester, H.A. and Mody, I. (2003) GABA transporter-1 (GAT1)-deficient mice: differential tonic activation of GABA$_A$ versus GABA$_B$ receptors in the hippocampus. J. Neurophysiol., 90: 2690–2701.

Jensen, M.L., Pedersen, L.N., Timmermann, D.B., Schousboe, A. and Ahring, P.K. (2005a) Mutational studies using a cation-conducting GABA$_A$ receptor reveal the selectivity determinants of the Cys-loop family of ligand-gated ion channels. J. Neurochem., 92: 962–972.

Jensen, M.L., Schousboe, A. and Ahring, P.K. (2005b) Charge selectivity of the Cys-loop family of ligand-gated ion channels. J. Neurochem., 92: 217–225.

Jensen, M.L., Timmermann, D.B., Johansen, T.H., Schousboe, A., Varming, T. and Ahring, P.K. (2002) The β subunit determines the ion selectivity of the GABA$_A$ receptor. J. Biol. Chem., 277: 41438–41447.

Ji, F., Kanbara, N. and Obata, K. (1999) GABA and histogenesis in fetal and neonatal mouse brain lacking both the isoforms of glutamic acid decarboxylase. Neurosci. Res., 33: 187–194.

Jia, F., Pignataro, L., Schofield, C.M., Yue, M., Harrison, N.L. and Goldstein, P.A. (2005) An extrasynaptic GABA$_A$ receptor mediates tonic inhibition in thalamic VB neurons. J. Neurophysiol., 94: 4491–4501.

Johnston, G.A. (1996) GABA$_C$ receptors: relatively simple transmitter-gated ion channels? Trends Pharmacol. Sci., 17: 319–323.

Jonas, P., Bischofberger, J., Fricker, D. and Miles, R. (2004) Interneuron Diversity series: Fast in, fast out — temporal and spatial signal processing in hippocampal interneurons. Trends Neurosci., 27: 30–40.

Jones, B.L., Whiting, P.J. and Henderson, L.P. (2006). Mechanisms of anabolic androgenic steroid inhibition of Σ-containing GABA$_A$ receptors. J. Physiol.,

Jones, M.V., Sahara, Y., Dzubay, J.A. and Westbrook, G.L. (1998) Defining affinity with the GABA$_A$ receptor. J. Neurosci., 18: 8590–8604.

Jones, M.V. and Westbrook, G.L. (1995) Desensitized states prolong GABA$_A$ channel responses to brief agonist pulses. Neuron, 15: 181–191.

Kahle, K.T., Rinehart, J., de Los Heros, P., Louvi, A., Meade, P., Vazquez, N., Hebert, S.C., Gamba, G., Gimenez, I. and Lifton, R.P. (2005) WNK3 modulates transport of Cl$^-$ in and out of cells: implications for control of cell volume and neuronal excitability. Proc. Natl. Acad. Sci. USA, 102: 16783–16788.

Kaila, K. (1994) Ionic basis of GABA$_A$ receptor channel function in the nervous system. Prog. Neurobiol., 42: 489–537.

Kaila, K., Lamsa, K., Smirnov, S., Taira, T. and Voipio, J. (1997) Long-lasting GABA-mediated depolarization evoked by high-frequency stimulation in pyramidal neurons of rat hippocampal slice is attributable to a network-driven, bicarbonate-dependent K$^+$ transient. J. Neurosci., 17: 7662–7672.

Kaila, K., Voipio, J., Pasternack, M., Paalasmaa, P. and Deisz, R. (1993) The role of bicarbonate in GABA$_A$ receptor-mediated IPSPs of rat neocortical neurons. J. Physiol., 464: 273–289.

Kaila, K., Saarikoski, J. and Voipio, J. (1990) Mechanism of action of GABA on intracellular pH and on surface pH in crayfish muscle fibres. J. Physiol., 427: 241–260.

Kaila, K. and Voipio, J. (1987) Postsynaptic fall in intracellular pH induced by GABA-activated bicarbonate conductance. Nature, 330: 163–165.

Kakazu, Y., Uchida, S., Nakagawa, T., Akaike, N. and Nabekura, J. (2000) Reversibility and cation selectivity of the K$^+$-Cl$^-$ cotransport in rat central neurons. J. Neurophysiol., 84: 281–288.

Kanaka, C., Ohno, K., Okabe, A., Kuriyama, K., Itoh, T., Fukuda, A. and Sato, K. (2001) The differential expression patterns of messenger RNAs encoding K-Cl cotransporters (KCC1,2) and Na-K-2Cl cotransporter (NKCC1) in the rat nervous system. Neuroscience, 104: 933–946.

Kaneda, M., Farrant, M. and Cull-Candy, S.G. (1995) Whole-cell and single-channel currents activated by GABA and glycine in granule cells of the rat cerebellum. J. Physiol., 485: 419–435.

Kash, T.L., Dizon, M.J., Trudell, J.R. and Harrison, N.L. (2004) Charged residues in the β2 subunit involved in GABA$_A$ receptor activation. J. Biol. Chem., 279: 4887–4893.

Kash, T.L., Jenkins, A., Kelley, J.C., Trudell, J.R. and Harrison, N.L. (2003) Coupling of agonist binding to channel gating in the GABA$_A$ receptor. Nature, 421: 272–275.

Kelley, S.P., Dunlop, J.I., Kirkness, E.F., Lambert, J.J. and Peters, J.A. (2003) A cytoplasmic region determines single-channel conductance in 5-HT$_3$ receptors. Nature, 424: 321–324.

Kelsch, W., Hormuzdi, S., Straube, E., Lewen, A., Monyer, H. and Misgeld, U. (2001) Insulin-like growth factor 1 and a cytosolic tyrosine kinase activate chloride outward transport during maturation of hippocampal neurons. J. Neurosci., 21: 8339–8347.

Keramidas, A., Moorhouse, A.J., Schofield, P.R. and Barry, P.H. (2004) Ligand-gated ion channels: mechanisms underlying ion selectivity. Prog. Biophys. Mol. Biol., 86: 161–204.

Keros, S. and Hablitz, J.J. (2005) Subtype-specific GABA transporter antagonists synergistically modulate phasic and tonic GABA$_A$ conductances in rat neocortex. J. Neurophysiol., 94: 2073–2085.

Khirug, S., Huttu, K., Ludwig, A., Smirnov, S., Voipio, J., Rivera, C., Kaila, K. and Khiroug, L. (2005) Distinct properties of functional KCC2 expression in immature mouse hippocampal neurons in culture and in acute slices. Eur. J. Neurosci., 21: 899–904.

Kittler, J.T., McAinsh, K. and Moss, S.J. (2002) Mechanisms of GABA$_A$ receptor assembly and trafficking — Implications for the modulation of inhibitory neurotransmission. Mol. Neurobiol., 26: 251–268.

Korpi, E.R., Grunder, G. and Luddens, H. (2002) Drug interactions at GABA$_A$ receptors. Prog. Neurobiol., 67: 113–159.

Krnjevic, K. (1974) The chemical nature of synaptic transmission in vertebrates. Physiol. Rev., 54: 418–540.

Kuffler, S.W. and Edwards, C. (1958) Mechanism of gamma-aminobutyric acid (GABA) action and its relation to synaptic inhibition. J. Neurophysiol., 21: 589–610.

Kullmann, D.M. (2000) Spillover and synaptic cross talk mediated by glutamate and GABA in the mammalian brain. Prog. Brain Res., 125: 339–351.

Kyrozis, A., Chudomel, O., Moshe, S.L. and Galanopoulou, A.S. (2006) Sex-dependent maturation of GABA$_A$ receptor-mediated synaptic events in rat substantia nigra reticulata. Neurosci. Lett., 398: 1–5.

Lee, H., Chen, C.X., Liu, Y.J., Aizenman, E. and Kandler, K. (2005) KCC2 expression in immature rat cortical neurons is sufficient to switch the polarity of GABA responses. Eur. J. Neurosci., 21: 2593–2599.

Lee, W.Y. and Sine, S.M. (2005) Principal pathway coupling agonist binding to channel gating in nicotinic receptors. Nature, 438: 243–247.

Leinekugel, X., Medina, I., Khalilov, I., Ben-Ari, Y. and Khazipov, R. (1997) Ca^{2+} oscillations mediated by the synergistic excitatory actions of $GABA_A$ and NMDA receptors in the neonatal hippocampus. Neuron, 18: 243–255.

Lema, G.M. and Auerbach, A. (2006) Modes and models of $GABA_A$ receptor gating. J. Physiol., 572: 183–200.

Lester, H.A., Dibas, M.I., Dahan, D.S., Leite, J.F. and Dougherty, D.A. (2004) Cys-loop receptors: new twists and turns. Trends Neurosci., 27: 329–336.

Li, H., Tornberg, J., Kaila, K., Airaksinen, M.S. and Rivera, C. (2002) Patterns of cation-chloride cotransporter expression during embryonic rodent CNS development. Eur. J. Neurosci., 16: 2358–2370.

Lindquist, C.E. and Birnir, B. (2006) Graded response to GABA by native extrasynaptic GABA receptors. J. Neurochem., 97: 1349–1356.

Lindquist, C.E., Dalziel, J.E., Cromer, B.A. and Birnir, B. (2004) Penicillin blocks human α1β1 and α1β1γ2S $GABA_A$ channels that open spontaneously. Eur. J. Pharmacol., 496: 23–32.

Liu, G.S. (2003) Presynaptic control of quantal size: kinetic mechanisms for synaptic transmission and plasticity. Curr. Opin. Neurobiol., 13: 324–331.

Liu, X., Wang, Q., Haydar, T.F. and Bordey, A. (2005) Nonsynaptic GABA signaling in postnatal subventricular zone controls proliferation of GFAP-expressing progenitors. Nat. Neurosci., 8: 1179–1187.

Lorez, N., Benke, D., Luscher, B., Mohler, H. and Benson, J.A. (2000) Single-channel properties of neuronal $GABA_A$ receptors from mice lacking the γ2 subunit. J. Physiol., 527: 11–31.

LoTurco, J.J., Owens, D.F., Heath, M.J., Davis, M.B. and Kriegstein, A.R. (1995) GABA and glutamate depolarize cortical progenitor cells and inhibit DNA synthesis. Neuron, 15: 1287–1298.

Luddens, H., Pritchett, D.B., Kohler, M., Killisch, I., Keinanen, K., Monyer, H., Sprengel, R. and Seeburg, P.H. (1990) Cerebellar $GABA_A$ receptor selective for a behavioural alcohol antagonist. Nature, 346: 648–651.

Ludwig, A., Li, H., Saarma, M., Kaila, K. and Rivera, C. (2003) Developmental up-regulation of KCC2 in the absence of GABAergic and glutamatergic transmission. Eur. J. Neurosci., 18: 3199–3206.

Lummis, S.C., Beene, D.L., Lee, L.W., Lester, H.A., Broadhurst, R.W. and Dougherty, D.A. (2005) Cis-trans isomerization at a proline opens the pore of a neurotransmitter-gated ion channel. Nature, 438: 248–252.

Luscher, B. and Keller, C.A. (2004) Regulation of $GABA_A$ receptor trafficking, channel activity, and functional plasticity of inhibitory synapses. Pharmacol. Ther., 102: 195–221.

Luu, T., Cromer, B., Gage, P.W. and Tierney, M.L. (2005) A role for the 2' residue in the second transmembrane helix of the GABA A receptor γ2S subunit in channel conductance and gating. J. Membr. Biol., 205: 17–28.

Lytle, C. and Forbush, B.R. (1996) Regulatory phosphorylation of the secretory Na-K-Cl cotransporter: modulation by cytoplasmic Cl. Am. J. Physiol., 270: C437–C448.

Macdonald, R.L., Rogers, C.J. and Twyman, R.E. (1989) Kinetic properties of the $GABA_A$ receptor main conductance state of mouse spinal cord neurones in culture. J. Physiol., 410: 479–499.

Maconochie, D.J., Zempel, J.M. and Steinbach, J.H. (1994) How quickly can $GABA_A$ receptors open? Neuron, 12: 61–71.

Maksay, G., Thompson, S.A. and Wafford, K.A. (2003) The pharmacology of spontaneously open α1β3ε $GABA_A$ receptor-ionophores. Neuropharmacology, 44: 994–1002.

Martina, M., Royer, S. and Pare, D. (2001) Cell-type-specific GABA responses and chloride homeostasis in the cortex and amygdala. J. Neurophysiol., 86: 2887–2895.

Marty, A. and Llano, I. (2005) Excitatory effects of GABA in established brain networks. Trends Neurosci., 28: 284–289.

Marty, S., Wehrle, R., Alvarez-Leefmans, F.J., Gasnier, B. and Sotelo, C. (2002) Postnatal maturation of Na^+, K^+, $2Cl^-$ cotransporter expression and inhibitory synaptogenesis in the rat hippocampus: an immunocytochemical analysis. Eur. J. Neurosci., 15: 233–245.

McBain, C.J. and Fisahn, A. (2001) Interneurons unbound. Nat. Rev. Neurosci., 2: 11–23.

McCartney, M.R., Deeb, T.Z., Henderson, T.N. and Hales, T.G. (2007) Tonically active $GABA_A$ receptors in hippocampal pyramidal neurons exhibit constitutive GABA-independent gating. Mol. Pharmacol., 71: 539–548.

McClellan, A.M. and Twyman, R.E. (1999) Receptor system response kinetics reveal functional subtypes of native murine and recombinant human GABAA receptors. J. Physiol. (Lond.), 515: 711–727.

McIntire, S.L., Reimer, R.J., Schuske, K., Edwards, R.H. and Jorgensen, E.M. (1997) Identification and characterization of the vesicular GABA transporter. Nature, 389: 870–876.

McKernan, R.M. and Whiting, P.J. (1996) Which $GABA_A$-receptor subtypes really occur in the brain? Trends Neurosci., 19: 139–143.

Mercado, A., Broumand, V., Zandi-Nejad, K., Enck, A.H. and Mount, D.B. (2006) A C-terminal domain in KCC2 confers constitutive K^+-Cl^- cotransport. J. Biol. Chem., 281: 1016–1026.

Miles, R., Toth, K., Gulyas, A.I., Hajos, N. and Freund, T.F. (1996) Differences between somatic and dendritic inhibition in the hippocampus. Neuron, 16: 815–823.

Milligan, C.J., Buckley, N.J., Garret, M., Deuchars, J. and Deuchars, S.A. (2004) Evidence for inhibition mediated by coassembly of $GABA_A$ and $GABA_C$ receptor subunits in native central neurons. J. Neurosci., 24: 9241–9250.

Minichiello, L., Casagranda, F., Tatche, R.S., Stucky, C.L., Postigo, A., Lewin, G.R., Davies, A.M. and Klein, R. (1998) Point mutation in trkB causes loss of NT4-dependent neurons without major effects on diverse BDNF responses. Neuron, 21: 335–345.

Minier, F. and Sigel, E. (2004) Positioning of the α-subunit isoforms confers a functional signature to γ-aminobutyric acid type A receptors. Proc. Natl. Acad. Sci. USA, 101: 7769–7774.

Misgeld, U., Deisz, R.A., Dodt, H.U. and Lux, H.D. (1986) The role of chloride transport in postsynaptic inhibition of hippocampal neurons. Science, 232: 1413–1415.

Mitchell, S.J. and Silver, R.A. (2003) Shunting inhibition modulates neuronal gain during synaptic excitation. Neuron, 38: 433–445.

Mody, I. (2001) Distinguishing between GABA$_A$ receptors responsible for tonic and phasic conductances. Neurochem. Res., 26: 907–913.

Mody, I., De Koninck, Y., Otis, T.S. and Soltesz, I. (1994) Bridging the cleft at GABA synapses in the brain. Trends Neurosci., 17: 517–525.

Moragues, N., Ciofi, P., Lafon, P., Tramu, G. and Garret, M. (2003) GABA$_A$ receptor ε-subunit expression in identified peptidergic neurons of the rat hypothalamus. Brain Res., 967: 285–289.

Moragues, N., Ciofi, P., Tramu, G. and Garret, M. (2002) Localisation of GABA$_A$ receptor ε-subunit in cholinergic and aminergic neurones and evidence for co-distribution with the θ-subunit in rat brain. Neuroscience, 111: 657–669.

Mozrzymas, J.W. (2004) Dynamism of GABA$_A$ receptor activation shapes the "personality" of inhibitory synapses. Neuropharmacology, 47: 945–960.

Mozrzymas, J.W., Zarmowska, E.D., Pytel, M. and Mercik, K. (2003) Modulation of GABA$_A$ receptors by hydrogen ions reveals synaptic GABA transient and a crucial role of the desensitization process. J. Neurosci., 23: 7981–7992.

Mtchedlishvili, Z. and Kapur, J. (2006) High-affinity, slowly desensitizing GABA$_A$ receptors mediate tonic inhibition in hippocampal dentate granule cells. Mol. Pharmacol., 69: 564–575.

Neelands, T.R., Fisher, J.L., Bianchi, M. and Macdonald, R.L. (1999) Spontaneous and γ-aminobutyric acid (GABA)-activated GABA$_A$ receptor channels formed by ε-subunit-containing isoforms. Mol. Pharmacol., 55: 168–178.

Neelands, T.R. and Macdonald, R.L. (1999) Incorporation of the φ-subunit into functional γ-aminobutyric acid$_A$ receptors. Mol. Pharmacol., 56: 598–610.

Nguyen, L., Malgrange, B., Breuskin, I., Bettendorff, L., Moonen, G., Belachew, S. and Rigo, J.M. (2003) Autocrine/paracrine activation of the GABA$_A$ receptor inhibits the proliferation of neurogenic polysialylated neural cell adhesion molecule-positive (PSA-NCAM+) precursor cells from postnatal striatum. J. Neurosci., 23: 3278–3294.

Nielsen, T.A., DiGregorio, D.A. and Silver, R.A. (2004) Modulation of glutamate mobility reveals the mechanism underlying slow-rising AMPAR EPSCs and the diffusion coefficient in the synaptic cleft. Neuron, 42: 757–771.

Nusser, Z., Cull-Candy, S. and Farrant, M. (1997) Differences in synaptic GABA$_A$ receptor number underlie variation in GABA mini amplitude. Neuron, 19: 697–709.

Nusser, Z. and Mody, I. (2002) Selective modulation of tonic and phasic inhibitions in dentate gyrus granule cells. J. Neurophysiol., 87: 2624–2628.

Nusser, Z., Roberts, J.D., Baude, A., Richards, J.G., Sieghart, W. and Somogyi, P. (1995a) Immunocytochemical localization of the α1 and β 2/3 subunits of the GABA$_A$ receptor in relation to specific GABAergic synapses in the dentate gyrus. Eur. J. Neurosci., 7: 630–646.

Nusser, Z., Roberts, J.D., Baude, A., Richards, J.G. and Somogyi, P. (1995b) Relative densities of synaptic and extrasynaptic GABA$_A$ receptors on cerebellar granule cells as determined by a quantitative immunogold method. J. Neurosci., 15: 2948–2960.

Nusser, Z., Sieghart, W. and Mody, I. (1999) Differential regulation of synaptic GABA$_A$ receptors by cAMP — dependent protein kinase in mouse cerebellar and olfactory bulb neurones. J. Physiol., 521: 421–435.

Nusser, Z., Sieghart, W. and Somogyi, P. (1998) Segregation of different GABA$_A$ receptors to synaptic and extrasynaptic membranes of cerebellar granule cells. J. Neurosci., 18: 1693–1703.

Obata, K., Oide, M. and Tanaka, H. (1978) Excitatory and inhibitory actions of GABA and glycine on embryonic chick spinal neurons in culture. Brain Res., 144: 179–184.

Okada, M., Onodera, K., Van Renterghem, C., Sieghart, W. and Takahashi, T. (2000) Functional correlation of GABA$_A$ receptor α subunits expression with the properties of IPSCs in the developing thalamus. J. Neurosci., 20: 2202–2208.

Overstreet, L.S. and Westbrook, G.L. (2001) Paradoxical reduction of synaptic inhibition by vigabatrin. J. Neurophysiol., 86: 596–603.

Overstreet, L.S., Westbrook, G.L. and Jones, M.V. (2002). In: Quick M.W. (Ed.), Measuring and modeling the spatiotemporal profile of GABA at the synapse. Wiley-liss, Hoboken, NJ, pp 259–275.

Owens, D.F. and Kriegstein, A.R. (2002) Is there more to GABA than synaptic inhibition? Nat. Rev. Neurosci., 3: 715–727.

Owens, D.F., Liu, X. and Kriegstein, A.R. (1999) Changing properties of GABA$_A$ receptor-mediated signaling during early neocortical development. J. Neurophysiol., 82: 570–583.

Pan, Y. and Qian, H. (2005) Interactions between ρ and γ2 subunits of the GABA receptor. J. Neurochem., 94: 482–490.

Pan, Z.H., Zhang, D., Zhang, X. and Lipton, S.A. (1997) Agonist-induced closure of constitutively open γ-aminobutyric acid channels with mutated M2 domains. Proc. Natl. Acad. Sci. USA, 94: 6490–6495.

Park, J.B., Skalska, S. and Stern, J.E. (2006) Characterization of a novel tonic γ-aminobutyric acid$_A$ receptor-mediated inhibition in magnocellular neurosecretory neurons and its modulation by glia. Endocrinology, 147: 3746–3760.

Pasternack, M., Voipio, J. and Kaila, K. (1993) Intracellular carbonic anhydrase activity and its role in GABA-induced acidosis in isolated rat hippocampal pyramidal neurones. Acta Physiol. Scand., 148: 229–231.

Pastorekova, S., Parkkila, S., Pastorek, J. and Supuran, C.T. (2004) Carbonic anhydrases: current state of the art, therapeutic applications and future prospects. J. Enzyme Inhib. Med. Chem., 19: 199–229.

Payne, J.A., Rivera, C., Voipio, J. and Kaila, K. (2003) Cation-chloride co-transporters in neuronal communication, development and trauma. Trends Neurosci., 26: 199–206.

Payne, J.A., Stevenson, T.J. and Donaldson, L.F. (1996) Molecular characterization of a putative K-Cl cotransporter in rat brain. A neuronal-specific isoform. J. Biol. Chem., 271: 16245–16252.

Pearce, R.A. (1993) Physiological evidence for two distinct GABA$_A$ responses in rat hippocampus. Neuron, 10: 189–200.

Peng, Z., Hauer, B., Mihalek, R.M., Homanics, G.E., Sieghart, W., Olsen, R.W. and Houser, C.R. (2002) GABA$_A$ receptor changes in δ subunit-deficient mice: altered expression of α4 and γ2 subunits in the forebrain. J. Comp. Neurol., 446: 179–197.

Perrais, D. and Ropert, N. (1999) Effect of zolpidem on miniature IPSCs and occupancy of postsynaptic GABA$_A$ receptors in central synapses. J. Neurosci., 19: 578–588.

Peters, J.A., Hales, T.G. and Lambert, J.J. (2005) Molecular determinants of single-channel conductance and ion selectivity in the Cys-loop family: insights from the 5-HT3 receptor. Trends Pharmacol. Sci., 26: 587–594.

Pirker, S., Schwarzer, C., Wieselthaler, A., Sieghart, W. and Sperk, G. (2000) GABA$_A$ receptors: Immunocytochemical distribution of 13 subunits in the adult rat brain. Neurosci. Neurosci., 101: 815–850.

Plenz, D. (2003) When inhibition goes incognito: feedback interaction between spiny projection neurons in striatal function. Trends Neurosci., 26: 436–443.

Plotkin, M.D., Kaplan, M.R., Peterson, L.N., Gullans, S.R., Hebert, S.C. and Delpire, E. (1997) Expression of the Na$^+$-K$^+$-2Cl$^-$ cotransporter BSC2 in the nervous system. Am. J. Physiol., 272: C173–C183.

Porcello, D.M., Huntsman, M.M., Mihalek, R.M., Homanics, G.E. and Huguenard, J.R. (2003) Intact synaptic GABAergic inhibition and altered neurosteroid modulation of thalamic relay neurons in mice lacking δ subunit. J. Neurophysiol., 89: 1378–1386.

Pouille, F. and Scanziani, M. (2001) Enforcement of temporal fidelity in pyramidal cells by somatic feed-forward inhibition. Science, 293: 1159–1163.

Prenosil, G.A., Schneider Gasser, E.M., Rudolph, U., Keist, R., Fritschy, J.M. and Vogt, K.E. (2006) Specific subtypes of GABA$_A$ receptors mediate phasic and tonic forms of inhibition in hippocampal pyramidal neurons. J. Neurophysiol., 96: 846–847.

Puia, G., Santi, M.R., Vicini, S., Pritchett, D.B., Purdy, R.H., Paul, S.M., Seeburg, P.H. and Costa, E. (1990) Neurosteroids act on recombinant human GABA$_A$ receptors. Neuron, 4: 759–765.

Pytel, M., Mercik, K. and Mozrzymas, J.W. (2006) Membrane voltage modulates the GABA$_A$ receptor gating in cultured rat hippocampal neurons. Neuropharmacology, 50: 143–153.

Qian, H. and Ripps, H. (1999) Response kinetics and pharmacological properties of heteromeric receptors formed by co-assembly of GABA ρ- and γ2-subunits. Proc. R. Soc. Lond. B. Biol. Sci., 266: 2419–2425.

Race, J.E., Makhlouf, F.N., Logue, P.J., Wilson, F.H., Dunham, P.B. and Holtzman, E.J. (1999) Molecular cloning and functional characterization of KCC3, a new K-Cl cotransporter. Am. J. Physiol., 277: C1210–C1219.

Ramadan, E., Fu, Z., Losi, G., Homanics, G.E., Neale, J.H. and Vicini, S. (2003) GABA$_A$ receptor β3 subunit deletion decreases α2/3 subunits and IPSC duration. J. Neurophysiol., 89: 128–134.

Ranna, M., Sinkkonen, S.T., Moykkynen, T., Uusi-Oukari, M. and Korpi, E.R. (2006) Impact of ε and θ subunits on pharmacological properties of α3β1 GABA$_A$ receptors expressed in Xenopus oocytes. BMC Pharmacol., 6: 1.

Represa, A. and Ben-Ari, Y. (2005) Trophic actions of GABA on neuronal development. Trends Neurosci., 28: 278–283.

Richerson, G.B. and Wu, Y.M. (2003) Dynamic equilibrium of neurotransmitter transporters: Not just for reuptake anymore. .J. Neurophysiol., 90: 1363–1374.

Rivera, C., Li, H., Thomas-Crusells, J., Lahtinen, H., Viitanen, T., Nanobashvili, A., Kokaia, Z., Airaksinen, M.S., Voipio, J., Kaila, K. and Saarma, M. (2002) BDNF-induced TrkB activation down-regulates the K$^+$-Cl$^-$ cotransporter KCC2 and impairs neuronal Cl$^-$ extrusion. J. Cell. Biol., 159: 747–752.

Rivera, C., Voipio, J. and Kaila, K. (2005) Two developmental switches in GABAergic signalling: the K$^+$-Cl$^-$ cotransporter KCC2 and carbonic anhydrase CAVII. J. Physiol., 562: 27–36.

Rivera, C., Voipio, J., Payne, J.A., Ruusuvuori, E., Lahtinen, H., Lamsa, K., Pirvola, U., Saarma, M. and Kaila, K. (1999) The K$^+$/Cl$^-$ co-transporter KCC2 renders GABA hyperpolarizing during neuronal maturation. Nature, 397: 251–255.

Rivera, C., Voipio, J., Thomas-Crusells, J., Li, H., Emri, Z., Sipila, S., Payne, J.A., Minichiello, L., Saarma, M. and Kaila, K. (2004) Mechanism of activity-dependent downregulation of the neuron-specific K-Cl cotransporter KCC2. J. Neurosci., 24: 4683–4691.

Roepstorff, A. and Lambert, J.D. (1994) Factors contributing to the decay of the stimulus-evoked IPSC in rat hippocampal CA1 neurons. J. Neurophysiol., 72: 2911–2926.

Romero, M.F., Fulton, C.M. and Boron, W.F. (2004) The SLC4 family of HCO$_3^-$ transporters. Pflugers Arch., 447: 495–509.

Roos, A. and Boron, W.F. (1981) Intracellular pH. Physiol. Rev., 61: 296–434.

Rose, C.R. and Konnerth, A. (2001) NMDA receptor-mediated Na+ signals in spines and dendrites. J. Neurosci., 21: 4207–4214.

Rossi, D.J. and Hamann, M. (1998) Spillover-mediated transmission at inhibitory synapses promoted by high affinity α6 subunit GABA$_A$ receptors and glomerular geometry. Neuron, 20: 783–795.

Rossi, D.J., Hamann, M. and Attwell, D. (2003) Multiple modes of GABAergic inhibition of rat cerebellar granule cells. J. Physiol., 548: 97–110.

Rudolph, U. and Mohler, H. (2004) Analysis of GABA$_A$ receptor function and dissection of the pharmacology of benzodiazepines and general anesthetics through mouse genetics. Annu. Rev. Pharmacol. Toxicol., 44: 475–498.

Russell, J.M. (2000) Sodium-potassium-chloride cotransport. Physiol. Rev., 80: 211–276.

Ruusuvuori, E., Li, H., Huttu, K., Palva, J.M., Smirnov, S., Rivera, C., Kaila, K. and Voipio, J. (2004) Carbonic

85

anhydrase isoform VII acts as a molecular switch in the development of synchronous γ-frequency firing of hippocampal CA1 pyramidal cells. J. Neurosci., 24: 2699–2707.

Santhakumar, V., Hanchar, H.J., Wallner, M., Olsen, R.W. and Otis, T.S. (2006) Contributions of the GABA$_A$ receptor α6 subunit to phasic and tonic inhibition revealed by a naturally occurring polymorphism in the α6 gene. J. Neurosci., 26: 3357–3364.

Saxena, N.C. and Macdonald, R.L. (1996) Properties of putative cerebellar γ-aminobutyric acid A receptor isoforms. Mol. Pharmacol., 49: 567–579.

Schweizer, C., Balsiger, S., Bluethmann, H., Mansuy, I.M., Fritschy, J.M., Mohler, H. and Luscher, B. (2003) The γ2 subunit of GABA$_A$ receptors is required for maintenance of receptors at mature synapses. Mol. Cell. Neurosci., 24: 442–450.

Scimemi, A., Semyanov, A., Sperk, G., Kullmann, D.M. and Walker, M.C. (2005) Multiple and plastic receptors mediate tonic GABA$_A$ receptor currents in the hippocampus. J. Neurosci., 25: 10016–10024.

Semyanov, A., Walker, M.C. and Kullmann, D.M. (2003) GABA uptake regulates cortical excitability via cell type-specific tonic inhibition. Nat. Neurosci., 6: 484–490.

Semyanov, A., Walker, M.C., Kullmann, D.M. and Silver, R.A. (2004) Tonically active GABA$_A$ receptors: modulating gain and maintaining the tone. Trends Neurosci., 27: 262–269.

Sergeeva, O.A., Andreeva, N., Garret, M., Scherer, A. and Haas, H.L. (2005) Pharmacological properties of GABA$_A$ receptors in rat hypothalamic neurons expressing the ε-subunit. J. Neurosci., 25: 88–95.

Sergeeva, O.A., Eriksson, K.S., Sharonova, I.N., Vorobjev, V.S. and Haas, H.L. (2002) GABA$_A$ receptor heterogeneity in histaminergic neurons. Eur. J. Neurosci., 16: 1472–1482.

Shen, H., Gong, Q.H., Yuan, M. and Smith, S.S. (2005) Short-term steroid treatment increases δ GABA$_A$ receptor subunit expression in rat CA1 hippocampus: pharmacological and behavioral effects. Neuropharmacology, 49: 573–586.

Sieghart, W. and Ernst, M. (2005) Heterogeneity of GABA$_A$ receptors: revived interest in the development of subtype-selective drugs. Curr. Med. Chem. — Central Nervous System Agents, 5: 217–242.

Sieghart, W. and Sperk, G. (2002) Subunit composition, distribution and function of GABA$_A$ receptor subtypes. Curr. Top. Med. Chem., 2: 795–816.

Sigel, E., Baur, R., Malherbe, P. and Mohler, H. (1989) The rat β1-subunit of the GABA$_A$ receptor forms a picrotoxin-sensitive anion channel open in the absence of GABA. FEBS Lett., 257: 377–379.

Simon, J., Wakimoto, H., Fujita, N., Lalande, M. and Barnard, E.A. (2004) Analysis of the set of GABA$_A$ receptor genes in the human genome. J. Biol. Chem., 279: 41422–41435.

Sine, S.M. and Engel, A.G. (2006) Recent advances in Cys-loop receptor structure and function. Nature, 440: 448–455.

Sinkkonen, S.T., Hanna, M.C., Kirkness, E.F. and Korpi, E.R. (2000) GABA$_A$ receptor ε and θ subunits display unusual structural variation between species and are enriched in the rat locus ceruleus. J. Neurosci., 20: 3588–3595.

Sipila, S.T., Huttu, K., Soltesz, I., Voipio, J. and Kaila, K. (2005) Depolarizing GABA acts on intrinsically bursting pyramidal neurons to drive giant depolarizing potentials in the immature hippocampus. J. Neurosci., 25: 5280–5289.

Sipila, S., Hotto, K., Voipio, J. and Kaila, K. (2004) GABA uptake via GABA transporter-1 modulates GABAergic transmission in the immature hippocampus. J. Neurosci., 24: 5877–5880.

Sipila, S.T. and Kaila, K. (In press) GABAergic control of CA3-driven network events in the developing hippocampus. In: Darlison M.G. (Ed.), Inhibitory Regulation of Excitatory Neurotransmission. Springer, Heidelberg.

Sipila, S.T., Schuchmann, S., Voipio, J., Yamada, J. and Kaila, K. (2006) The cation-chloride cotransporter NKCC1 promotes sharp waves in the neonatal rat hippocampus. J. Physiol., 573: 765–773.

Sipila, S., Voipio, J. and Kaila, K. (2007) GAT-1 acts to limit a tonic GABA$_A$ current in rat CA3 pyramidal neurons at birth. Eur. J. Neurosci., In press.

Smirnov, S., Paalasmaa, P., Uusisaari, M., Voipio, J. and Kaila, K. (1999) Pharmacological isolation of the synaptic and nonsynaptic components of the GABA-mediated biphasic response in rat CA1 hippocampal pyramidal cells. J. Neurosci., 19: 9252–9260.

Smith, R.L., Clayton, G.H., Wilcox, C.L., Escudero, K.W. and Staley, K.J. (1995) Differential expression of an inwardly rectifying chloride conductance in rat brain neurons: a potential mechanism for cell-specific modulation of postsynaptic inhibition. J. Neurosci., 15: 4057–4067.

Somogyi, P., Fritschy, J.M., Benke, D., Roberts, J.D. and Sieghart, W. (1996) The γ2 subunit of the GABA$_A$ receptor is concentrated in synaptic junctions containing the α1 and β2/3 subunits in hippocampus, cerebellum and globus pallidus. Neuropharmacology, 35: 1425–1444.

Somogyi, P. and Klausberger, T. (2005) Defined types of cortical interneurone structure space and spike timing in the hippocampus. J. Physiol., 562: 9–26.

Spruston, N., Schiller, Y., Stuart, G. and Sakmann, B. (1995) Activity-dependent action potential invasion and calcium influx into hippocampal CA1 dendrites. Science, 268: 297–300.

Staley, K. (1994) The role of an inwardly rectifying chloride conductance in postsynaptic inhibition. J. Neurophysiol., 72: 273–284.

Stell, B.M., Brickley, S.G., Tang, C.Y., Farrant, M. and Mody, I. (2003) Neuroactive steroids reduce neuronal excitability by selectively enhancing tonic inhibition mediated by δ subunit-containing GABA$_A$ receptors. Proc. Natl. Acad. Sci. USA, 100: 14439–14444.

Stell, B.M. and Mody, I. (2002) Receptors with different affinities mediate phasic and tonic GABA$_A$ conductances in hippocampal neurons. J. Neurosci., 22: RC223.

Sur, C., Farrar, S.J., Kerby, J., Whiting, P.J., Atack, J.R. and McKernan, R.M. (1999) Preferential coassembly of α4 and δ subunits of the γ-aminobutyric acid$_A$ receptor in rat thalamus. Mol. Pharmacol., 56: 110–115.

Szabadics, J., Varga, C., Molnar, G., Olah, S., Barzo, P. and Tamas, G. (2006) Excitatory effect of GABAergic axo-axonic cells in cortical microcircuits. Science, 311: 233–235.

Takeuchi, A. and Takeuchi, N. (1971) Variations in the permeability properties of the inhibitory post-synaptic membrane of the crayfish neuromuscular junction when activated by different concentrations of GABA. J. Physiol., 217: 341–358.

Tepper, J.M. and Bolam, J.P. (2004) Functional diversity and specificity of neostriatal interneurons. Curr. Opin. Neurobiol., 14: 685–692.

Thompson, S.M. and Gahwiler, B.H. (1989) Activity-dependent disinhibition. II. Effects of extracellular potassium, furosemide, and membrane potential on ECl- in hippocampal CA3 neurons. J. Neurophysiol., 61: 512–523.

Tia, S., Wang, J.F., Kotchabhakdi, N. and Vicini, S. (1996) Developmental changes of inhibitory synaptic currents in cerebellar granule neurons: Role of GABA$_A$ receptor α6 subunit. J. Neurosci., 16: 3630–3640.

Titz, S., Hans, M., Kelsch, W., Lewen, A., Swandulla, D. and Misgeld, U. (2003) Hyperpolarizing inhibition develops without trophic support by GABA in cultured rat midbrain neurons. J. Physiol., 550: 719–730.

Tretter, V., Ehya, N., Fuchs, K. and Sieghart, W. (1997) Stoichiometry and assembly of a recombinant GABA$_A$ receptor subtype. J. Neurosci., 17: 2728–2737.

Twyman, R.E., Rogers, C.J. and Macdonald, R.L. (1990) Intraburst kinetic properties of the GABA$_A$ receptor main conductance state of mouse spinal cord neurones in culture. J. Physiol., 423: 193–220.

Unwin, N. (2005) Refined structure of the nicotinic acetylcholine receptor at 4A resolution. J. Mol. Biol., 346: 967–989.

Valeyev, A.Y., Cruciani, R.A., Lange, G.D., Smallwood, V.S. and Barker, J.L. (1993) Cl$^-$ channels are randomly activated by continuous GABA secretion in cultured embryonic rat hippocampal neurons. Neurosci. Lett., 155: 199–203.

Vardi, N., Zhang, L.L., Payne, J.A. and Sterling, P. (2000) Evidence that different cation chloride cotransporters in retinal neurons allow opposite responses to GABA. J. Neurosci., 20: 7657–7663.

Varoqueaux, F., Sigler, A., Rhee, J.S., Brose, N., Enk, C., Reim, K. and Rosenmund, C. (2002) Total arrest of spontaneous and evoked synaptic transmission but normal synaptogenesis in the absence of Munc13-mediated vesicle priming. Proc. Natl. Acad. Sci. USA, 99: 9037–9042.

Veliskova, J. and Moshe, S.L. (2006) Update on the role of substantia nigra pars reticulata in the regulation of seizures. Epilepsy Curr., 6: 83–87.

Verdoorn, T.A., Draguhn, A., Ymer, S., Seeburg, P.H. and Sakmann, B. (1990) Functional properties of recombinant rat GABA$_A$ receptors depend upon subunit composition. Neuron, 4: 919–928.

Verhage, M., Maia, A.S., Plomp, J.J., Brussaard, A.B., Heeroma, J.H., Vermeer, H., Toonen, R.F., Hammer, R.E., van den Berg, T.K., Missler, M., Geuze, H.J. and Sudhof, T.C. (2000) Synaptic assembly of the brain in the absence of neurotransmitter secretion. Science, 287: 864–869.

Vicini, S., Ferguson, C., Prybylowski, K., Kralic, J., Morrow, A.L. and Homanics, G.E. (2001) GABA$_A$ receptor α1 subunit deletion prevents developmental changes of inhibitory

synaptic currents in cerebellar neurons. J. Neurosci., 21: 3009–3016.

Vida, I., Bartos, M. and Jonas, P. (2006) Shunting inhibition improves robustness of gamma oscillations in hippocampal interneuron networks by homogenizing firing rates. Neuron, 49: 107–117.

Voipio, J. and Kaila, K. (2000) GABAergic excitation and K$^+$ -mediated volume transmission in the hippocampus. Prog. Brain Res., 125: 329–338.

Wagner, D.A., Goldschen-Ohm, M.P., Hales, T.G. and Jones, M.V. (2005) Kinetics and spontaneous open probability conferred by the ε subunit of the GABA$_A$ receptor. J. Neurosci., 25: 10462–10468.

Wall, M.J. and Usowicz, M.M. (1997) Development of action potential-dependent and independent spontaneous GABA$_A$ receptor-mediated currents in granule cells of postnatal rat cerebellum. Eur. J. Neurosci., 9: 533–548.

Wang, L., Kitai, S.T. and Xiang, Z. (2006) Activity-dependent bidirectional modification of inhibitory synaptic transmission in rat subthalamic neurons. J. Neurosci., 26: 7321–7327.

Wei, W., Zhang, N., Peng, Z., Houser, C.R. and Mody, I. (2003) Perisynaptic localization of d subunit-containing GABA$_A$ receptors and their activation by GABA spillover in the mouse dentate gyrus. J. Neurosci., 23: 10650–10661.

Weiss, D.S., Barnes, E.M.J. and Hablitz, J.J. (1988) Whole-cell and single-channel recordings of GABA-gated currents in cultured chick cerebral neurons. J. Neurophysiol., 59: 495–513.

Weiss, D.S. and Magleby, K.L. (1989) Gating scheme for single GABA-activated Cl$^-$ channels determined from stability plots, dwell-time distributions, and adjacent-interval durations. J. Neurosci., 9: 1314–1324.

Whiting, P.J. (2003) GABA-A receptor subtypes in the brain: a paradigm for CNS drug discovery? Drug Discovery Today, 8: 445–450.

Whittington, M.A. and Traub, R.D. (2003) Interneuron diversity series: inhibitory interneurons and network oscillations in vitro. Trends Neurosci., 26: 676–682.

Wisden, W., Laurie, D.J., Monyer, H. and Seeburg, P.H. (1992) The distribution of 13 GABA$_A$ receptor subunit mRNAs in the rat brain. I. Telencephalon, diencephalon, mesencephalon. J. Neurosci., 12: 1040–1062.

Wojcik, S.M., Katsurabayashi, S., Guillemin, I., Friauf, E., Rosenmund, C., Brose, N. and Rhee, J.S. (2006) A shared vesicular carrier allows synaptic corelease of GABA and glycine. Neuron, 50: 575–587.

Woodin, M.A., Ganguly, K. and Poo, M.M. (2003) Coincident pre- and postsynaptic activity modifies GABAergic synapses by postsynaptic changes in Cl$^-$ transporter activity. Neuron, 39: 807–820.

Wotring, V.E., Chang, Y.C. and Weiss, D.S. (1999) Permeability and single channel conductance of human homomeric ρ1 GABAC receptors. J. Physiol. (Lond.), 521: 327–336.

Wotring, V.E., Kaylor, T.S. and Weiss, D.S. (2001) A pair of TM2 mutations chances selectivity of the homomeric GABA P1 receptor from anionic to cationic. Biophys. J., 80: 506.

Wotring, V.E., Miller, T.S. and Weiss, D.S. (2003) Mutations at the GABA receptor selectivity filter: a possible role for effective charges. J. Physiol., 548: 527–540.

Wotring, V.E. and Weiss, D.S. (2002) Mutations at the intracellular end of M2 change ionic selectivity of the homomeric ρ1 GABA receptor. Biophys. J., 82: 1707.

Wu, Y., Wang, W. and Richerson, G.B. (2006) The transmembrane sodium gradient influences ambient GABA concentration by altering the equilibrium of GABA transporters. J. Neurophysiol., 96: 2425–2436.

Wu, Y., Wang, W. and Richerson, G.B.W. (2003) Vigabatrin induces tonic inhibition via GABA transporter reversal without increasing vesicular GABA Release. J. Neurophysiol., 89: 2021–2034.

Xiu, X., Hanek, A.P., Wang, J., Lester, H.A. and Dougherty, D.A. (2005) A unified view of the role of electrostatic interactions in modulating the gating of Cys loop receptors. J. Biol. Chem., 280: 41655–41666.

Xu, M. and Akabas, M.H. (1996) Identification of channellining residues in the M2 membrane-spanning segment of the GABA$_A$ receptor α1 subunit. J. Gen. Physiol., 107: 195–205.

Yamada, J., Okabe, A., Toyoda, H., Kilb, W., Luhmann, H.J. and Fukuda, A. (2004) Cl$^-$ uptake promoting depolarizing GABA actions in immature rat neocortical neurones is mediated by NKCC1. J. Physiol., 557: 829–841.

Yamada, J., Furukawa, T., Veno, S., Yamamoto, S. and Fukuda, A. (2006) Molecular basis for the GABA$_A$ receptormediated tonic inhibition in rat somatosensory cortex. Cereb. Cortex 2006 Sep 22 [Epub ahead of print].

Yuste, R. and Katz, L.C. (1991) Control of postsynaptic Ca^{2+} influx in developing neocortex by excitatory and inhibitory neurotransmitters. Neuron, 6: 333–344.

Zafrakas, M., Chorovicer, M., Klaman, I., Kristiansen, G., Wild, P.J., Heindrichs, U., Knuchel, R. and Dahl, E. (2006) Systematic characterisation of GABRP expression in sporadic breast cancer and normal breast tissue. Int. J. Cancer, 118: 1453–1459.

Zhang, D., Pan, Z.H., Awobuluyi, M. and Lipton, S.A. (2001) Structure and function of GABA(C) receptors: a comparison of native versus recombinant receptors. Trends Pharmacol. Sci., 22: 121–132.

Zheng, J., Lee, S. and Zhou, Z.J. (2006) A transient network of intrinsically bursting starburst cells underlies the generation of retinal waves. Nat. Neurosci., 9: 363–371.

GABAergic Microcircuits of the Basal Ganglia

Tepper, Abercrombie & Bolam (Eds.)
Progress in Brain Research, Vol. 160
ISSN 0079-6123

CHAPTER 6

GABAergic inhibition in the neostriatum

Charles J. Wilson[*]

Department of Biology, University of Texas at San Antonio, 6900 N. Loop 1604 W, San Antonio, TX 78249, USA

Abstract: In the neostriatum, GABAergic inhibition arises from the action of at least two classes of inhibitory interneurons, and from recurrent collaterals of the principal cells. Interneurons receive excitatory input only from extrinsic sources, and so act in a purely feedforward capacity. Feedback inhibition arises from the recurrent collaterals of the principal cells. These two kinds of inhibition have functionally distinct effects on the principal cells. Inputs from interneurons are not very convergent. There are few inhibitory neurons, and so each principal cell receives inhibitory synaptic input from very few interneurons. But, they are individually powerful, and a single interneuron can substantially delay action potentials in a group of nearby principal cells. Recurrent inhibition is highly convergent, with each principal cell receiving inhibitory input from several hundred other such cells. Feedback inhibitory synaptic inputs individually have very weak effects, as seen from the soma. The differences in synaptic strength are not caused by differences in the release of transmitter or in sensitivity of the postsynaptic membrane. Rather, they arise from differences in the number of synaptic contacts formed on individual principal cells by feedforward or feedback axons, and from differences in synaptic location. Interneurons form their powerful synapses near the somata of principal cells, while most feedback synapses are more distal, where they interact with the two-state nonlinear properties of the principal cells' dendrite. This arrangement suggests that feedforward inhibition may serve in the traditional role for inhibition, adjusting the excitability of the principle neuron near the site of action potential generation. Feedback inhibitory synapses may interact with voltage-sensitive conductances in the dendrite to alter the electrotonic structure of the spiny cell.

Keywords: striatal spiny cell; parvalbumin; convergence; feedforward inhibition; feedback inhibition; competitive network

Most striatal neurons are GABAergic, but most striatal synapses are not

The striatum apparently possesses no native species of excitatory neuron. With the exception of the cholinergic interneuron, all the cell types so far identified in the striatum are GABAergic, and presumably inhibitory in function. This includes both interneurons and the principal cells. The abundant principal neurons of the striatum, the spiny (Sp) cells, which constitute the vast majority of the neurons, are of two types. The two subclasses are present in approximately equal proportions and are usually called the direct pathway cells and the indirect pathway cells. Both of these are GABAergic, but differ in their extrastriatal axonal projections and cotransmitters (e.g., Kawaguchi et al., 1990; Wang et al., 2006). The direct pathway Sp cells have axonal projections to the basal ganglia output nuclei (GP_i/entopeduncular nucleus and SNr), as well as to the external pallidum, and

[*]Corresponding author. Tel.: (210) 458-5654;
E-mail: Charles.Wilson@utsa.edu

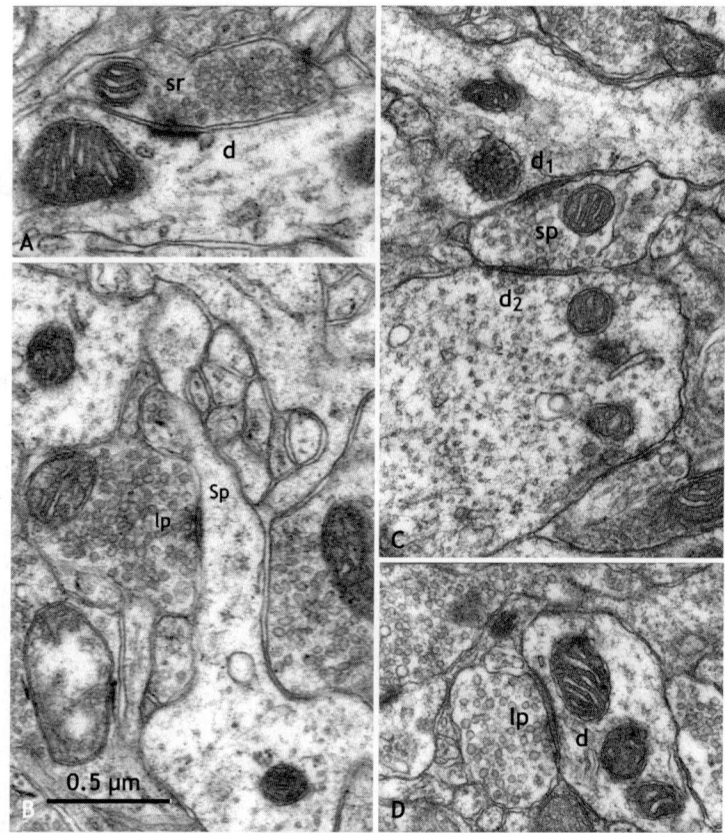

Fig. 1. Synapse types common in the neostriatum. (A) An sr-type synapse typical of those formed by excitatory striatal afferents. This example is on the shaft of an aspiny dendrite (d), probably a GABAergic interneuron. The characteristic features are small round synaptic vesicles (sr) and an asymmetrical synapse. This type of synapse, especially those made onto dendritic spines of spiny neurons, dominates the striatal neuropil. (B) A large presynaptic terminal containing large pleomorphic vesicles (lp) forms a symmetric synapse with the shaft of a dendritic spine (sp). (C) A small axon terminal containing small pleomorphic vesicles (sp) makes two symmetric synapses, one on a small dendritic shaft (d1), and one on a larger primary dendrite (d2). (D) A terminal containing large pleomorphic vesicles (lp) forming a symmetrical synapse on a small dendrite (d).

contain substance P as a cotransmitter. The indirect pathway Sp cells project to the external pallidum only, and contain the cotransmitters enkephalin and dynorphin. In addition to the Sp cells, two other GABAergic cell types have been identified, the parvalbumin-containing fast-spiking (FS) GABAergic interneurons, and the somatostatin (SOM) and nitric oxide synthetase (NOS)-containing GABAergic interneurons. There may be additional varieties of interneurons, or subclasses of these two types, which have not yet been identified, but if so they are most likely GABAergic in nature (Kawaguchi, 1993).

Despite the dominance of GABAergic cell types in the striatum, the vast majority of synapses are asymmetrical, glutamatergic synapses formed by striatal afferent axons, mainly from the cerebral cortex, and the thalamus (Fig. 1). In the electron microscope, the asymmetrical synapses formed by glutamatergic synapses are primarily on spines and, to a lesser extent, on the dendrites of Sp cells and other neurons. Together, synapses of this kind constitute about 80% of all the synapses in the striatum (Kemp and Powell, 1971a; Ingham, et al., 1998). The synapses formed by cortical axons end almost exclusively (about 95%) on dendritic spines

(Kemp and Powell, 1971b; Somogyi et al., 1981; Xu et al., 1989). Thalamic inputs from the parafascicular/centromedian complex are primarily made onto dendritic shafts while thalamic inputs from the intralaminar nuclei resemble those from the cortex, and are almost completely axospinous (Dube et al., 1988; Xu et al., 1991; Smith et al., 1994). The total number of axospinous synapses, and the total number of asymmetrical synapses per Sp neuron can be estimated from existing data. Each dendritic spine receives one asymmetric synapse, and the average number of dendritic spines per Sp neuron has been estimated to be about 10,000 (Wilson, 1986). Most asymmetric synapses on Sp neurons are axospinous, so this is close to the number of asymmetric synapses per Sp cell. Another estimate of the total number of asymmetric synapses per neuron can be obtained from the estimate of the volume density of asymmetric synapses from Ingham et al. (1998) (0.92/μm^3) and the volume density of neurons from Oorschot (1996) (84,000/mm^3). Assuming that practically all the asymmetric synapses are on Sp cells, gives a value of slightly less than 11,000 asymmetric synapses per Sp cell. That approximate agreement between these two estimates implies that the assumptions that all asymmetric synapses are on Sp cells and that all asymmetric synapses are on spines do not introduce a large error. Of course, those assumptions are strictly wrong. Inputs from parafascicular nucleus are made mostly on dendritic shafts, and there are asymmetric synapses on aspiny neurons, including the cholinergic cell and the FS neurons, but they do not introduce a large error, presumably because the axodendritic asymmetric synapses from the parafascicular/centromedian complex, and the other axodendritic asymmetric synapses are not a large proportion of the total. In any case, this provides a rough estimate of about 10,000 asymmetrical synapses on each Sp cell.

Spatial distribution of GABAergic synapses on spiny cells

It would be useful to know how many GABAergic synapses there are on each Sp projection neuron,

and what proportion of these originates from interneurons and what from other projection cells. GABAergic axons arising from the Sp cells and interneurons form symmetrical synapses. Symmetrical contacts account for about 20% of all synapses in the striatum (Ingham et al., 1998). Assuming that most of these are on Sp cells, it would suggest that there are about 2500 symmetrical synapses per Sp cell. However, not all symmetrical synapses are on Sp cells, and not all of them are GABAergic. Prominent contributors to this group of synapses are the dopaminergic synapses formed by afferent axons from the substantia nigra, pars compacta (e.g., Smith et al., 1994), the serotoninergic projection from the dorsal raphe (e.g., Soghomonian et al., 1989), and the axonal arborizations of cholinergic interneurons (e.g., Izzo and Bolam, 1988). Kemp and Powell (1971a) and subsequent authors have generally recognized two different kinds of symmetrical synapses: one characterized by relatively large synaptic vesicles and another with relatively small vesicles. Synaptic vesicles in each of these may exhibit various degrees of flattening, and so are referred to as pleomorphic. The synapses with larger vesicles are the more numerous. Included in this category are synapses formed by the collaterals of the GABAergic Sp projection neurons (Wilson and Groves, 1980; Somogyi et al., 1981), and also dopaminergic axons (e.g., Smith et al., 1994). Axons forming the presynaptic elements with smaller vesicles include the parvalbumin-containing FS interneurons (Kita et al., 1990; Bennett and Bolam, 1994) and the axons of cholinergic interneurons (e.g., Izzo and Bolam, 1988). Thus, neither category represents a pure sample of GABAergic synapses. This makes it difficult to estimate the proportion of symmetric synapses that are GABAergic, and also to know the proportion of GABAergic synapses that arise from Sp cells and from interneurons. The density of dopaminergic synapses is known (Roberts et al., 2002) to be 0.03/μm^3, or about 13% of the total density of symmetric synapses (which is 0.235/μm^3 according to Ingham et al., 1998). Thus, there should be about 325 dopaminergic synapses on each Sp neuron. The cholinergic synaptic density has not been similarly measured, but if it is about the same as the dopaminergic innervation, then

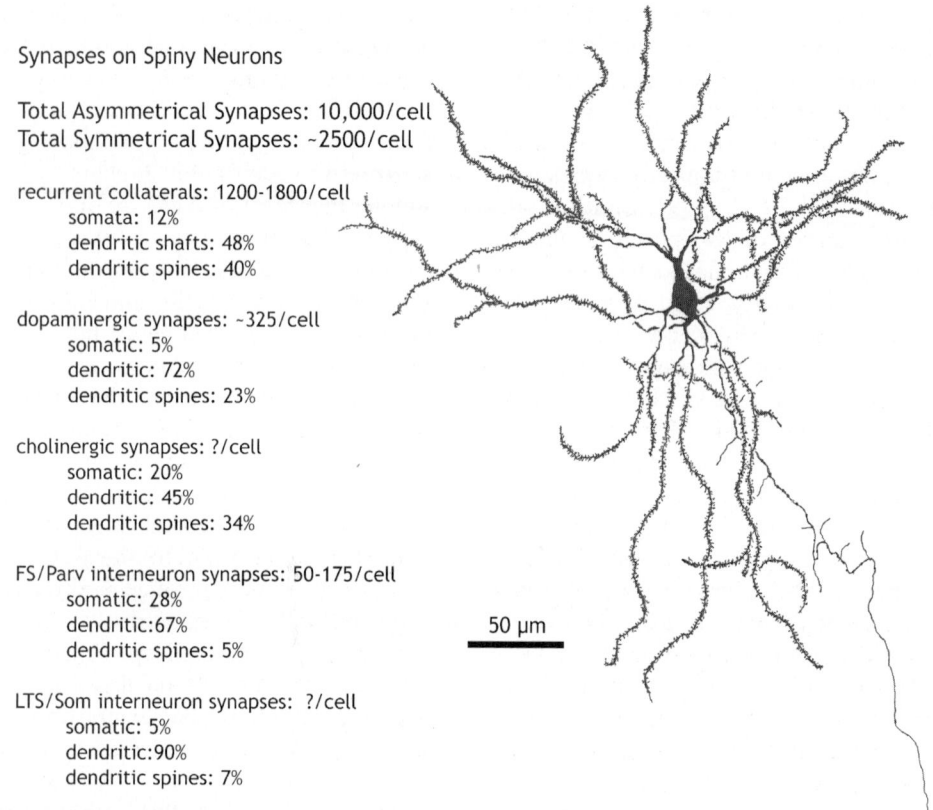

Synapses on Spiny Neurons

Total Asymmetrical Synapses: 10,000/cell
Total Symmetrical Synapses: ~2500/cell

recurrent collaterals: 1200-1800/cell
    somata: 12%
    dendritic shafts: 48%
    dendritic spines: 40%

dopaminergic synapses: ~325/cell
    somatic: 5%
    dendritic: 72%
    dendritic spines: 23%

cholinergic synapses: ?/cell
    somatic: 20%
    dendritic: 45%
    dendritic spines: 34%

FS/Parv interneuron synapses: 50-175/cell
    somatic: 28%
    dendritic:67%
    dendritic spines: 5%

LTS/Som interneuron synapses: ?/cell
    somatic: 5%
    dendritic:90%
    dendritic spines: 7%

50 µm

Fig. 2. Distribution of synapses from all sources on the spiny neuron. The proportions of synapses of each type have been measured a number of times by various authors, and are in reasonable agreement. The particular numbers shown here are as follows: (a) recurrent collaterals, Wilson and Groves (1980), (b) dopaminergic synapses, Smith et al. (1994), (c) cholinergic synapses, Izzo and Bolam (1988), and (d) GABAergic interneurons, Kubota and Kawaguchi (2000). The number of dopaminergic synapses is from Roberts et al. (2002), the number of recurrent collateral cell synapses and FS cell synapses are based on results of Koós et al. (2004).

these two non-GABAergic inputs would account for about 25% of the total number of symmetrical synapses, or about 625 of the 2500 symmetrical synapses per Sp neuron. The remaining (1875) symmetrical synapses would be the combined GABAergic innervation from FS, SOM/NOS, and other Sp cells (Fig. 2).

How can inhibition be effective when so outnumbered?

If inhibition is to be effective in controlling the effects of the massive excitatory innervation of the Sp cell, it must be either because few excitatory inputs are active at any one time, or because inhibitory synapses are given some advantage over excitation, for example a larger or more long-lived conductance change, a higher probability of transmitter release, or a more advantageous location on the neuron. The effectiveness of synapses formed by GABAergic interneurons may rely in part on their location on the Sp neuron, but probably this is not so for the synapses formed between Sp cells. Sp→Sp synapses have been examined using electron microscopy of the axonal arborizations of intracellularly stained Sp cells (Wilson and Groves, 1980; Somogyi et al., 1981), and also by electron microscopy of axons positive for substance P (for the direct pathway neurons) and enkephalin (for the indirect pathway cells) (Somogyi et al., 1982; Aronin et al., 1986; Bolam and Izzo, 1988). These studies are in reasonable agreement, and suggest that the Sp→Sp synapses are primarily formed on the spiny region of the dendrites and on the shafts of dendritic spines. In one typical study (Wilson and Groves, 1980), intracellularly stained Sp cell axons made 48% of their synapses on the spiny region of dendritic shafts, and 40% on dendritic spines. The latter group of synapses always shared the spine with an asymmetrical synapse, and formed its contact on a more proximal part of the spine, either on the side of the spine head, or on the shaft of the spine. Only 12% of Sp cell synapses were formed with somata or the initial segments of axons. It was not possible to determine how many of these were formed on Sp cell somata and axons, and how many are formed by some other cell type.

In contrast, the synapses formed by the FS cell (identified by the presence of parvalbumin immunoreactivity) are reported to be mainly on the somata of Sp cells (Kita et al., 1990; Bennett and Bolam, 1994). The SOM/NOS neuron can be effectively identified by the presence of Somatostatin or NOS immunoreactivity. Kubota and Kawaguchi (2000) compared the synapses made by FS cells and SOM/NOS cells in the striatum. In their sample, 28% of FS cell synapses were axosomatic, 67% were made onto dendrites and 5% were made onto dendritic spines. In contrast, identified LTS (SOM/NOS) cells formed only 5% of their synapses onto somata, 90% were formed with dendrites and 7% were formed onto dendritic spines. This kind of arrangement, in which some GABAergic cells preferentially contact somata and some the dendrites, is common in telencephalic structures (e.g., Kawaguchi and Kubota, 1998; Cope et al., 2002).

One major difference between both of the GABAergic interneurons and the Sp cell is the near absence of interneuron synapses on spine necks. It is often speculated that the placement of dopaminergic, or GABAergic, synapses on the shafts of dendritic spines of Sp cells gives those synapses special properties, for example a specific veto authority for the asymmetric synapse on that spine or control over synaptic plasticity at that specific location. An early estimate suggested that this kind of dual innervation of spines was common, at least for dopaminergic synapses (Freund et al., 1984).

However, more recent estimates using more accurate methods produced a much different result. The stereological estimate of the number of spines receiving a second symmetrical synapse by Ingham et al. (1998) is a reliable one, being based on a large sample and using unbiased measures. Those authors showed that only 9% of dendritic spines receive a second, symmetrical, synapse. This agrees with an earlier estimate of 8%, obtained using reconstructions from serial sections (Wilson et al., 1983), and is actually a little less than the number that would be expected based on the overall innervation of the Sp cell membrane. If there are about 2500 symmetrical synapses in the dendritic tree of each Sp cell, and about half of the dendritic membrane is in the spines (Wilson, 1986), then about 1250 of 10,000 spines (12.5%) should have a symmetrical synapse, even if this were happening entirely by chance. It should also be remembered that of the 8–10% of symmetrical second inputs to spines, only a small proportion are dopaminergic. Almost half of all recurrent collaterals synapses made by Sp cells are formed on spine shafts, and these are much more numerous than dopaminergic axons. The proportion of dopaminergic axons that make synapses on spine necks is actually somewhat smaller than that of recurrent collaterals, according to Smith et al. (1994), who reported that 72% of all dopaminergic synapses are formed on dendritic shafts, and only 23% on spines (compared to about 40% for recurrent collaterals). Of course, when looking at dopaminergic synapses in tissue stained for a dopaminergic cell marker like tyrosine hydroxylase, this is a relatively high proportion of all labelled synapses, so it appears to be a large number. This kind of experiment can be misleading, unless it is tempered with knowledge of the total proportion of synapses represented by the axon type being studied. In the case of GABAergic synapses, there is little reason for expecting that those formed on spine shafts are especially directed at the control of that one particular spine. The same axon that forms the inhibitory input on a spine shaft often forms a similar axodendritic synapse a nearby part of the same dendrite (e.g., Wilson and Groves, 1980).

Feedforward and feedback inhibition

Feedback and feedforward inhibition are separated in the striatum by the inhibitory nature of the Sp cell. Although collaterals of the Sp cells may make synaptic contacts with FS and SOM interneurons and so may influence the feedforward pathways, they do not excite those cells and so cannot evoke lateral inhibition through that pathway. The Sp cells, via their connections among each other, exclusively control feedback inhibition. To be faithful to the original meaning of these words, feedforward inhibition should reflect the overall level of input to the Sp neurons, and feedback inhibition should reflect the average output. However, these theoretical ideas about feedback and feedforward inhibition are mostly derived from control theory or artificial neural networks. To insure that these conditions are met in the nervous system, the feedforward inhibitory neurons must sample the same input as the principal cells. We certainly cannot say that in the striatum the feedforward interneurons are getting a faithful copy of the input to the Sp cells. They do receive corticostriatal synapses, as do the Sp cells (e.g., Lapper et al., 1992; Ramanathan et al., 2002). But it is unlikely that their corticostriatal (or other) inputs are the same as those of the Sp neurons. For example, Sp neurons fall into two categories, direct pathway cells receiving input from one set of cortical neurons, and indirect pathway cells receiving synapses from a largely different set (Lei et al., 2004). Are there two corresponding different sets of FS cells or SOM cells? Or do FS and SOM cells get input from both corticostriatal pathways? Moreover, neurons of the globus pallidus project to the FS and SOM/NOS cells, but apparently not to the Sp neurons (Bevan et al., 1998). In addition to these possible qualitative differences, there are certainly important quantitative ones. The Sp cells receive many more synapses overall than the interneurons do, and they are much more diverse. Cortical (and probably thalamic) axons spread out to innervate a large volume of striatum, and each Sp cell samples a very small proportion of the axons in the vicinity (Zheng and Wilson, 2001). As a result, the cortical input to each Sp cell is substantially different from the next. The inhibitory interneurons sample a smaller, not a larger subset of the cortical inputs to the same part of the striatum. Thus, they cannot really have a copy of the input to any one Sp cell, much less a reliable average of the diverse inputs to a group of nearby Sp cells. There could be a variety of different feedforward pathways, carrying different signals, coexisting in the same space in the striatum.

If there were many feedforward interneurons that converge onto each Sp cell, convergence of those diverse inhibitory inputs to the Sp cell could perform the averaging function necessary to synthesize a reliable average of feedforward inhibitory input, but there are not. There are a small number of these interneurons compared to the Sp cells (each Sp cell receives input from 4-27 FS cells, according to Koós and Tepper, 1999), and their inhibitory function does not rely much on convergence. Individual FS and SOM interneurons responsible for feedforward inhibition exert powerful IPSPs that can prevent or delay action potentials in Sp neurons (Koós and Tepper, 1999; Gustafson et al., 2005). These pathways do not require synchronous activation in large numbers of cells to alter the timing of action potentials in striatal projection neurons. The strength of this synaptic connection, and its rapid time course, gives FS cells the potential to control the firing of striatal projection neurons on a millisecond time scale. FS cells also respond rapidly to neostriatal afferents, raising the possibility that there may be some excitatory neostriatal inputs that could excite interneurons and inhibit Sp neurons before they were able to respond to the same excitation (Gustafson et al., 2005). The identity of an excitatory input that would trigger such an inhibition is not known, and likewise the function of this inhibition in the neostriatal network is unclear. In networks that have strong recurrent excitation among the principal cells (like the cerebral cortex), fast inhibition is often proposed to be a mechanism for preventing runaway positive feedback (e.g., Lawrence and McBain, 2003; Sun et al., 2006), or as a mechanism for spatial or temporal focusing of excitation in networks consisting of competing lateral excitation and inhibition (e.g., Bruno and Simons, 2002; Miller, 2003; Swadlow, 2003). These mechanisms are less applicable to the neostriatum, where there is no comparable source of positive feedback. A temporal role for feedforward inhibition in the neostriatum is a possibility. Because it is more rapid, feedforward inhibition could contribute to the delay in response of Sp cells to synaptic excitation, making the cell less sensitive to brief excitatory inputs. Short-term synaptic depression of inhibitory synapses (Koós et al., 2004; Gustafson et al., 2005) could also contribute to a role in temporal filtering. Another common proposal for the function of feedforward inhibition is gain control. If FS interneurons did sample the same set of inputs as the Sp neurons, then feedforward inhibition could reduce the excitability of Sp cells in proportion to the level of input. This would have a normalizing function, maintaining contrast between the most strongly activated Sp cells and the rest (e.g., Mallet et al., 2005).

Winner-take-all inhibition in the striatum?

Feedback inhibition has often been proposed to play an important functional role in the striatum (e.g., Wickens, 1993; Rolls and Treves, 1998; Plenz and Kitai, 2000; Bar-Gad and Bergman, 2001). Feedback inhibition can offer a powerful computational advantage in a network of neurons receiving a common set of afferents, and having a mechanism of use-dependent synaptic plasticity in the input pathway. If the inhibition is strong enough to limit the number of neurons that can respond to any one input pattern, it can limit the number of synapses changed by the experience of this pattern. In so doing, it increases the memory capacity of the network (e.g., Hertz et al., 1991). Over time, each input activity pattern comes to specifically invoke a particular output pattern in the network. Could collateral inhibition serve this function in the striatum?

Despite the predominance of excitation overall in the striatum, every new student of striatal structure and function is struck by the dominance of inhibition in striatal internal organization. To the first approximation, the striatal network appears to be a single layer of mutually inhibitory neurons, and so it appears to operate like one of the classic archetypes of neural network theory, the competitive network (Hertz et al., 1991). This impression

is reinforced by the presence of long-term depression and potentiation at corticostriatal synapses, and the dependence of these forms of synaptic plasticity on postsynaptic depolarization (e.g., Lovinger and Tyler, 1996; Kerr and Wickens, 2001; Centonze et al., 2003). Competitive networks can learn to classify patterns of their inputs, and to encode categories of patterns in an efficient manner. The requirements for a network of this kind are: (a) a high degree of input sharing (so many neurons receive about the same set of patterns that need to be categorized), (b) synaptic plasticity of the type that requires simultaneous presynaptic activation and postsynaptic depolarization (so that inhibited neurons do not learn), and (c) strong mutual inhibition that severely limits the number of neurons that can be active simultaneously. In a network of this kind, a small group of the first neurons to respond to any particular input pattern will suppress firing in all the others. The active synapses on the firing neurons will be strengthened by the synaptic plasticity mechanism, while the corresponding inputs to the other neurons will not (they may undergo depression). In future presentations of that same input pattern those neurons that responded first previously will be even more likely to respond first and most strongly again, and again suppress the others. Different groups of neurons would likewise come to respond to other input patterns. Over time, the separation between the groups of neurons responding to different input patterns will grow, and new input patterns that are similar to familiar ones will evoke activity in the same output cells, and thus can be said to have been identified as belonging to the same category.

Certain aspects of this kind of network resonate with well-known features of the striatum. For example, phasically active (mostly the spiny) striatal cells have a low background firing rate, and a relatively small proportion of neurons responds to any one stimulus or to any one phase of movement during performance of a task. Thus, movements and stimuli are represented in the striatum by the activation of a relatively small proportion of Sp neurons, as expected for a competitive network. Some variants of the competitive network idea seem especially appropriate for application to the striatum. Early theoretical studies of competitive networks emphasized that stability of responses to learned patterns could be achieved by a separate input that could enable or disable synaptic plasticity. The discovery that dopamine can control synaptic plasticity suggests that a reward (or reward prediction error) signal from the substantia nigra might adjust the rate of learning in the striatal network, turning it off to gain stability during readout (when rewards match expectations) and on during training (when they do not). Networks of this type would have the advantage of learning to categorize input patterns; not only on the basis of how often they have occurred (as in ordinary competitive networks) but also on the basis of whether they occurred in the context of unexpected success. Input patterns that occur when there are no surprises would be classified according to categories established in the past, whereas the establishment of new categories or rearrangement of old ones would be enabled when rewards did not match expectations (e.g., Bar-Gad and Bergman, 2001). Models of this kind are very appealing, especially to those studying the basal ganglia who have long endured the accusation that we do not have a clear proposal for the function of these structures. For that reason, we may not wish to abandon them on the basis of one or two pieces of contrary information. The fundamental requirement for winner-take-all inhibition among Sp neurons makes this a hinge issue for the competitive network idea, however, and so this has taken on a greater importance than it might otherwise have.

The first direct study of Sp cell interactions (Jaeger et al., 1994) failed to detect any synaptic interconnections at all. This study, which employed sharp electrode recordings and did not alter the chloride equilibrium potential, was designed to detect the powerful winner-take-all inhibition expected at the time. Competitive networks require that a small number of winners can suppress the activity of a large number of losers. Thus, this idea suggested that inhibition among Sp neurons would be strong enough that the inhibitory effect of one Sp cell might be readily detected with no special provisions for enhancing the detection of very weak effects. Pairs of nearby Sp neurons were both made to fire repetitively by passing small depolarizing currents through the recording electrodes, and the effects of action potentials in one cell on the timing of action potentials in the other were directly monitored, as well as looking for changes in membrane potential. The authors concluded that mutual inhibition among Sp neurons is weak, or perhaps even (functionally) non-existent. The presence or absence of GABAergic synaptic connections among the cells was of course not in doubt, as the existence of synapses among Sp neurons had been known for 15 years at that time (Wilson and Groves, 1980; Somogyi et al., 1981). The first demonstration of the IPSP generated by the synapses among Sp cells was obtained by Tunstall, et al. (2002), who averaged hundreds of trials to reveal the tiny IPSPs evoked in Sp cells by action potentials in nearby cells of the same kind (Fig. 3). Subsequently, these IPSPs were demonstrated without averaging using whole cell recording, and especially by manipulation of the reversal potential to dramatically increase the driving force for chloride current (Czubayko and Plenz, 2002; Guzman et al., 2003; Koós et al., 2004; Taverna et al., 2004; Venance et al., 2004). But individually, action potentials in Sp neurons are too weak to have a perceptible effect on the firing of nearby cells, in contrast with the effects exerted by FS and LTS cells.

Could groups of spiny cells compete with each other?

Starting with the Tunstall et al. (2002) paper, an effort was made to measure the connectivity among Sp cells. The results consistently showed a connectivity of about 0.16, that is, any particular Sp neuron is found to make synapses with about 1/6 of its nearby neighbours. This high connectivity indicates that every Sp neuron receives many synapses from other Sp neurons. Assuming an axonal field of about 400 μm diameter, and a connectivity of 1/6, each Sp neuron would receive synapses from about 450 other Sp cells (out of about 2800 within the connection volume). If a large number of those cells were active at once, it is possible that this innervation would produce a powerful inhibition, even if the synapses were individually weak. Is it

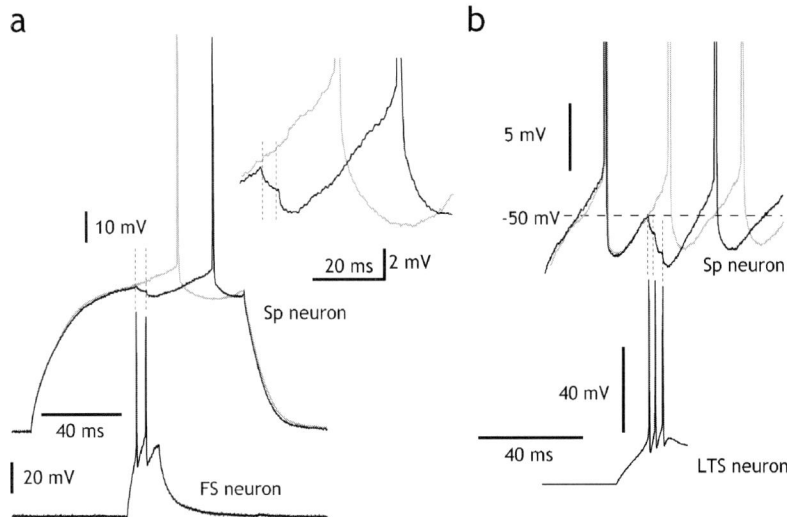

Fig. 3. Unitary IPSPs and delay of firing evoked in spiny neurons by firing of a single nearby FS cell (a) or LTS cell (b). In both cases the Sp cell is depolarized to firing threshold by passage of a depolarizing current pulse. The response, and firing of the Sp neuron in response to the current pulse alone is shown by the grey lines. The time of firing of the interneurons is indicated by dotted vertical lines. With repetitive firing, the IPSPs evoked by single interneurons sum and can substantially delay spontaneous firing. [Redrawn and simplified from Kóos and Tepper (1999)]. Adapted by permission from Macmillan Publishers Ltd. Nature Neuroscience, Copyright 1999.

possible that this large convergent input could still serve a competitive inhibition role? Of course the answer is no in the most general sense, because competition requires that a small number of cells be able to inhibit a large number, and not the other way around. However, it might still be possible for one group of neurons to collectively inhibit the other cells, and still achieve the same effect if the connectivity were properly sculpted. One possibility is that cells are preassigned to groups that work together. Members of the same group must not inhibit each other, but should inhibit cells in the other groups. This kind of preassigned group membership is consistent with the observed connectivity, which, while high, is considerably less than one. Receiving synapses from the same (or nearly the same) cortical and thalamic inputs, and not inhibiting each other, cells within a group would tend to become selective for the same patterns of synaptic input. When cells of one group fired they might collectively produce sufficient inhibition on the cells of other groups to prevent their responses to the same inputs. Cell groups would compete, rather than cells. Of course, all of

this should occur on a spatial scale comparable to the axonal field of the Sp neuron. It would be possible to detect the existence of such connectivity groups using the existing connectivity measurements, and Tunstall et al. (2002) recognized this. Because cells of a group would not be connected to each other, the observed connectivity of Sp cells would really be the superposition of two connectivity probabilities, one for cells of the same group (whose connectivity is zero) and one for cells of different groups, whose connectivity is higher than the average. For example, if there were six equal size connectivity groups represented among striatal Sp neurons within synaptic distance of each other, and cells in each group were not connected, cells of different groups would have a connectivity of about 1/5, and the average connectivity would still be 1/6. Thus, if one found that cells were connected in one direction (and hence are in different groups), there would be a higher than average chance that they are connected in the other direction as well. Bidirectional connectivity would occur according to the joint probability of connections among cells of different groups. In the above case, the proportion of cells that are connected in both directions would be 1 in 25; whereas if connections were random across all cells (no groups), the proportion of bidirectionally connected cells would be 1 in 36. Experiments with suitably large samples and which tested for connectivity in both directions are few. Tunstall et al. (2002) examined 45 pairs for connections in either direction, and observed nine one-way connections. This corresponds to a connectivity of 1/10, and they saw no reciprocally connected pairs. Venance et al. (2004) reported connections among 17% of the 72 pairs they analyzed, and they saw no reciprocally connected cells. Taverna et al. (2004) studied 38 pairs, and found 13 that were connected in one direction, yielding a connectivity of about 1/6. They saw a single reciprocally connected pair, (1/38). It is hard to be confident in a ratio generated from a single observation, but this is close to the expectation for randomly connected neurons. Slightly higher connectivity was seen in explant cultures of the striatum by Czubayko and Plenz (2002), who reported 26/69 cell pairs examined were connected in one direction (connectivity is about 1/5) but these authors reported a much higher proportion (8/69) of reciprocally connected neurons than those seen in slices (1/8.6 vs. 1/25 expected by chance). These results provide no support for the presence of connectivity groups in slices of striatum, but suggest that such groups may occur in explant cultures.

It should be noted that competing internally unconnected groups could never achieve a substantial across-group connectivity anyway, unless there were only 2 or 3 groups. For an average connectivity of 1/6, if there were only two competing groups, the connectivity across groups could only reach 1/3. Increasing the number of groups will reduce the connectivity across groups, with an asymptotic connectivity of 1/6 as the number of groups increases (Fig. 4). Of course, if the groups are competing, the number of different categories that could be learned by the network is dependent on the number of groups, not the number of neurons. For the striatum to be an effective competitive network, the number of competing groups within axon-collateral range of each other should be substantially greater than 2.

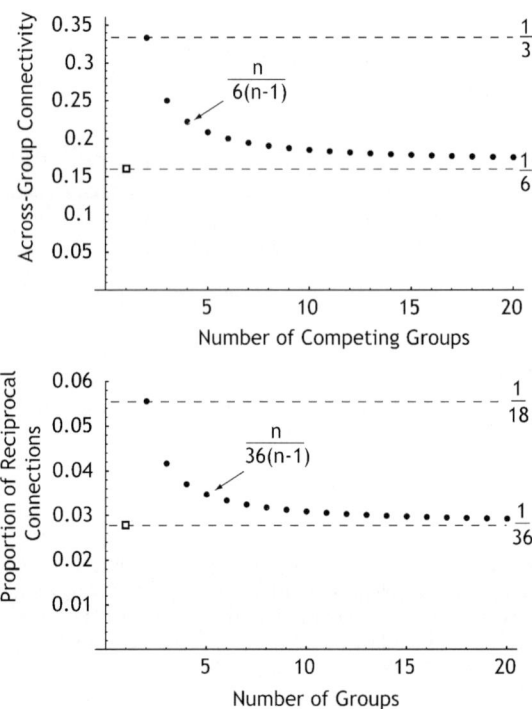

Fig. 4. Connectivity groups cannot rescue the mutual inhibition idea. Imagine all the Sp cells are divided into groups, with zero connectivity within groups. The connectivity across groups will go up, but to get a substantial increase in across-group connectivity required to get competitive inhibition, the number of groups must be very small. (A) The across-group connectivity versus number of groups, given an average connectivity of 1/6, as measured for striatal Sp neurons in rats. The square corresponds to the connectivity in the absence of any groups (1/6). When there are only two groups, connectivity across groups doubles, but decays rapidly. As the number of groups increases, the across-group connectivity again approaches 1/6. (B) The proportion of reciprocal connections expected for connectivity groups. The square corresponds to random connections among equivalent spiny cells (i.e., there is only one group of cells). There, only one of 36 pairs of cells should be reciprocally connected. Reciprocal connections double with two groups, but rapidly decrease again as the number of groups increases.

Inhibition and the mechanism of up and down states in vivo

In vivo, the membrane potentials of Sp neurons transition between the Up and Down states under the influence of synaptic inputs. In unanesthetized animals, the transitions are irregular, and cells can spend minutes in the Down state (Wilson and

Groves, 1981). In animals anesthetized with urethane, or with ketamine, highly organized slow rhythmic changes in cortical activity imposes similar slow changes in Sp neurons, making it relatively easy to study these transitions (Wilson and Kawaguchi, 1996). These oscillations are a good opportunity to study synaptic activation of the Sp cells, because large numbers of corticostriatal (Stern et al., 1997), and presumably also thalamostriatal cells, fire during the depolarized phase of the oscillation, and then cease during the hyperpolarized phase. Thus, the excitatory inputs to Sp cells, and the Sp cells themselves, are active during the depolarizing phase of the oscillation. FS cells also fire during the Up state. This has been clearly demonstrated in explant cocultures of the cortex and striatum (Plenz and Kitai, 1998), and in vivo, by Mallet et al. (2005). This pattern of correlated excitation and inhibition is of course expected, given that the Sp cells and FS cells receive very similar excitatory inputs.

The relative proportion of excitatory and inhibitory synaptic conductances experienced by the Sp cell during the Up state can be estimated by measuring the reversal potential of the membrane potential during the Up and Down states, as reported by Wilson and Kawaguchi (1996). To remove the contribution of voltage-sensitive conductances in vivo, it is necessary to apply ion channel antagonists that act from the cytoplasmic side, and that do not cross the cell membrane. The quartenary lidocaine QX-314 was used to block sodium channels, the quartenary verapimil D-890 was used to block calcium channels, and potassium channels were partially antagonized by loading the cells with cesium, all through the recording micropipette. The blockade of potassium channels was only partial, as indicated by the continued presence of the membrane potential, although the cells were generally depolarized. Applied current was used to hold the average membrane potential at different levels, and the occurrence and amplitude of the membrane potential oscillations were measured. If the excitation and inhibition occurred out of phase, the oscillations might change qualitatively (e.g., change frequency) as the membrane potential changed. This did not happen, in keeping with the evidence that excitation and inhibition are both maximal during the Up state. The membrane potential in the Up and Down states both changed with applied current, but showed different slopes (Fig. 5), as expected because the membrane conductance is lower in the Down state (there is less synaptic conductance) and higher in the Up state (when synaptic conductance is high). The point at which the Up and Down states reverse is the point at which the current from excitatory synapses and the current from inhibitory synapses are equal. If this point occurred near the reversal point for the glutamatergic synapses (near 0 mV), it would indicate that nearly all of the Up state synaptic conductance was excitatory in nature, while a reversal potential near that of the GABA synapses (near −60 mV, see below) would indicate a preponderance of inhibitory conductance responsible for the Up and Down states. The reversal potential for Sp neurons in vivo was between −10 and −20 mV, suggesting that the excitatory conductance was roughly 2–5 times as great as the inhibitory conductance. This is probably an underestimate, because it assumes that the location of inhibitory synapses and excitatory synapses share the same spatial distribution on the cell. This is approximately true for Sp→Sp synapses, but not for FS→Sp synapses, whose conductance would be overestimated in this measurement due to their proximal location (cf. Kotaleski et al., 2006). This dominance of excitation in the Up state in the striatum stands in contrast with the situation for layer V pyramidal cells in the somatosensory cortex under the same conditions, whose Up state equilibrium potential is about −40 mV (Sachdev et al., 2004), indicating something close to equality in synaptic conductances from excitatory and inhibitory inputs.

It should also be noted that the Up and Down states in an individual Sp cell reversed at the same potential as the cortically evoked EPSP. This is not expected if a large proportion of the inhibition came from Sp cells. Spiny cells are late to fire in response to cortical excitation (e.g., Wilson, 1986; Kawaguchi et al., 1989), so Sp→Sp cell inhibition should be slow to exert its influence on the response to cortical input. The reversal potential for the early part of the response should be more positive than that of the later part. This is true in the cerebral cortex, for example, in which there is an

102

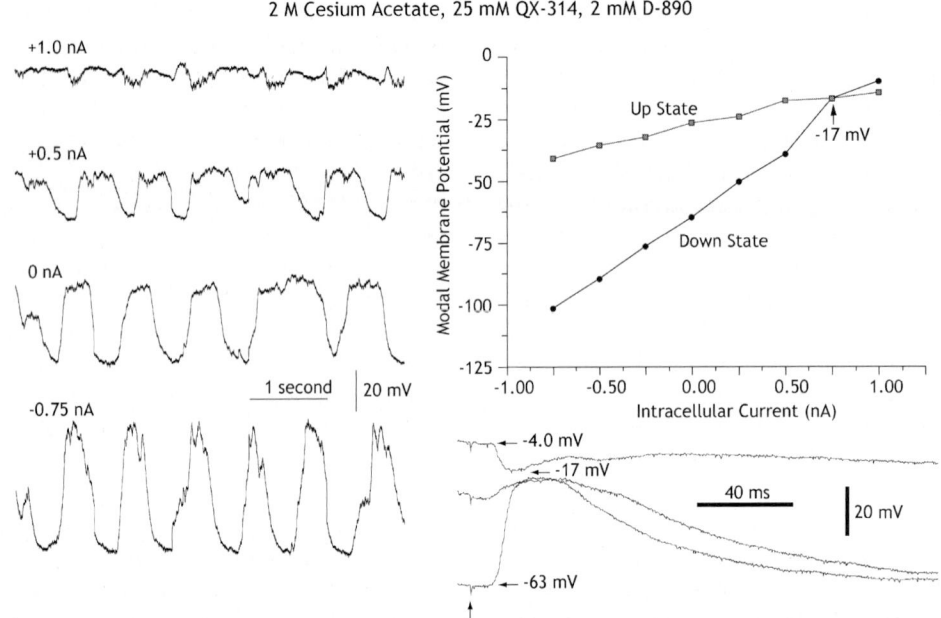

2 M Cesium Acetate, 25 mM QX-314, 2 mM D-890

Fig. 5. The proportion of excitation and inhibition during the spontaneous Up and Down states of Sp neurons in anesthetized rats. The measurements are taken in the presence of cesium, QX-314, and D-890 to block voltage-dependent currents in the Sp cell. The difference between the Up and Down membrane potentials increases with hyperpolarization well below the equilibrium potential for chloride, indicating that chloride currents are not responsible for the Down state. The input resistance (slope of the curve) in the Down state is greater than that of the Up state, indicating that synaptic conductance is greater in the Up state. The difference between Up and Down states is reduced with depolarization, vanishes at about −17 mV, and reverses at more positive voltages. The reversal potential for the Up state corresponds to the reversal potential for the synaptic potential evoked by cortical stimulation. This reversal potential is slightly negative to that expected for glutamatergic excitation, indicating a small contribution of inhibition in the Up state.

early part of evoked EPSPs that reverses at more positive potentials than the later part (e.g., Cowan and Wilson, 1994). That this does not happen in Sp neurons indicates that the inhibition responsible for shifting the reversal potential negative to 0 mV is already present before Sp cells respond to the input. A likely source of such fast inhibitory input to the Sp neuron is the FS cell. The speed of FS→Sp inhibition is a result of the rapid response of the FS cell to cortical input, which results from its short time constant and high excitability. The inhibition exerted by the FS cell relies on the magnitude of the input to the striatum, not on the magnitude of the Sp cell response (Fig. 6).

Why is Sp→Sp inhibition so weak?

Although the functional implications of fast, powerful and non-convergent feedforward inhibition

and weak convergent feedback are not known, we do know the cellular reasons why the strengths of FS→Sp and Sp→Sp synapses are so different. Kóos et al. (2004) studied the reasons for the strength of FS→Sp cell synapses, using quantal analysis to compare the synapses on Sp cells derived from FS cells to the weaker synapses formed by Sp cells. These experiments showed that the conductance change evoked by a single quantum at the two synapses were identical, both in amplitude and in time course. The probability of quantal release was also relatively high (over 0.5) in both cases. The difference between FS→Sp and Sp→Sp connections was attributable to two causes, the number of release sites per axon, and the difference in synaptic location. The number of transmitter release sites per axon was higher for FS→Sp synapses, sometimes as high as 18 per neuron (average about 6). Sp→Sp synapses were

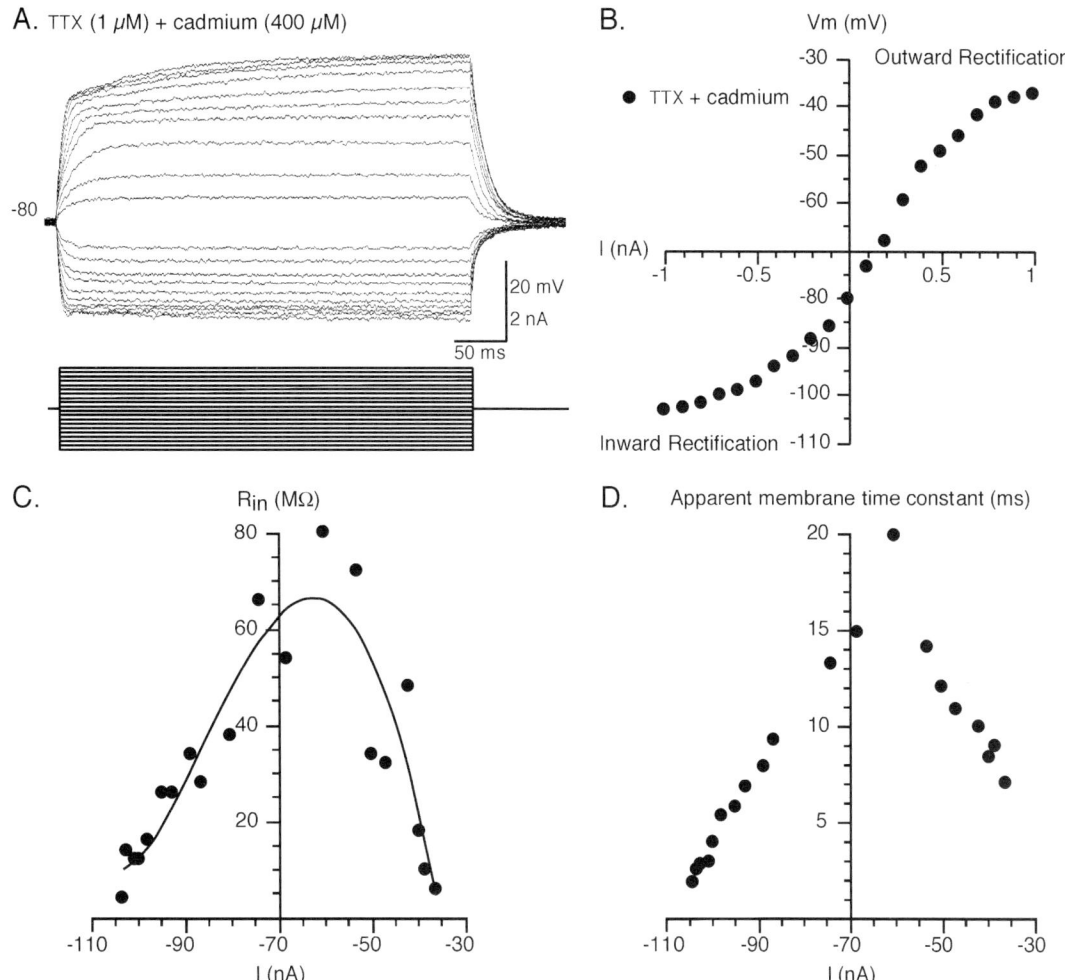

Fig. 6. The contribution of potassium currents to the input resistance and the membrane time constant of Sp cells. (A) The input resistance and time constant of Sp neurons are measured by passage of current pulses in the presence of TTX and cadmium to block sodium and calcium currents (C), respectively. (B) The whole cell steady state I–V curve calculated from the traces in A. Note the sigmoidal nature of the I–V curve. The whole cell input resistance of the cell, calculated from the responses shown in (A) and (B). The points are the input resistance calculated from the individual points in B. The smooth curve is the derivative of a polynomial fit to the points in B. Input resistance increases by a factor of about 6 as the cell is depolarized over the subthreshold range, reaching a peak near −60 mV, and decreasing again after that. (D) The corresponding change in membrane time constant, measured from the onset of the curves in (A). The membrane time constant also varies greatly with membrane potential, reaching a peak near −60 mV.

characterized by small numbers of release sites, (average was about 3). This difference, which on average is responsible for a factor of about 2 in the strength of feedforward versus feedback pathway probably reflects a difference in the number of synaptic contacts formed by the axons of FS versus Sp neurons. FS cell axons often make clusters of synaptic contacts on the somata of Sp cells (Kita et al., 1990), whereas Sp→Sp synapses are made mostly one at a time, en passant as axons cross the dendrites of the Sp cells (Wilson and Groves, 1980). In slices from immature animals, in mature cells whose potassium channels are compromised by intracellular cesium, and in explant cultures of the striatum, FS cell synapses are about 2–3 times more effective than Sp synapses, reflecting this difference in the number of release sites alone (Koós et al., 2004; Gustafson et al., 2005). This is probably because the high input resistance of the Sp cells in these preparations reduces the electrotonic distance between the soma and dendrites, and obviates the importance of synaptic location. In adult neostriatal Sp cells, synaptic location apparently plays a critical role in determining the strength of feedforward and feedback inhibition.

The importance of synapse location is amplified by the voltage sensitivity of cable properties of Sp cell dendrites. The input resistance and time constant of Sp neurons vary with voltage in the subthreshold range because of the action of two potassium currents, one activated at hyperpolarized potentials and one activated when the cell is more depolarized but still below the spike thres-hold (e.g., Nisenbaum and Wilson, 1995; Wilson and Kawaguchi, 1996; Shen et al., 2004). The hyper-polarization-activated K^+ conductance (KIR_2) is most influential near the resting membrane potential of the Sp cell, and as a result of this conductance, the input resistance of the Sp cell near rest is low, ranging from about 20 to 60 Megohms, and the resting membrane potential is very hyper-polarized, between −80 and −95 mV (Kawaguchi et al., 1989; Nisenbaum and Wilson, 1995; Reyes et al., 1998). This conductance deactivates as the cell is depolarized, and the Sp cell input resistance and time constant are maximum at about –60 mV. The hyperpolarization-activated conductance develops relatively late in Sp cells, not being fully mature un-til postnatal day 25–28 (Tepper et al., 1998). This conductance must be present on the dendritic mem-brane, because otherwise it could not account for the additional 2- to 10-fold reduction in the size of Sp→Sp synapses seen in adult animals, which can be removed by blockade of the channels with intracellular cesium (Koós et al., 2004). Computer modelling shows that the reduced input resistance per se is not responsible for this effect of K^+ con-ductance on axodendritic inhibition, but rather the associated change in the effective length con-stant of the dendrite, and so the electrotonic dis-tance between the synapse and the soma. The effect of this voltage-sensitivity of the electrotonic struc-ture of the Sp neuron is to make the effectiveness of both dendritic excitation and inhibition dependent on the membrane potential. This is important because it means that the sizes of local synaptic po-tentials out in the dendrite are not nearly as much affected by membrane potential as are the somatic potentials seen by experimenters using somatic intracellular recordings (Fig. 7).

At membrane potentials between −60 mV and spike threshold, the input resistance and time con-stant of Sp neurons again decrease, reflecting the activation of subthreshold depolarization-activated K^+ currents (Kawaguchi et al., 1989; Nisenbaum and Wilson, 1995; Shen et al., 2004). At membrane potentials near the action potential threshold, the responsiveness of the Sp cell is determined by the interaction between a host of voltage-sensitive conductances, including persistent sodium con-ductance and L-type calcium conductance (e.g., Hernandez-Lopez et al., 2000) as well as the sub-threshold potassium current, and the excitatory and inhibitory synaptic conductances along the dendrite. The voltage-sensitive conductances like-wise alter the electrotonic length of the dendrites, and in turn modulate the effectiveness of dendritic excitation and inhibition.

The reversal potential of GABA$_A$

The reversal potential of GABA$_A$ inhibition in Sp cells has repeatedly been shown to be positive to the resting potential, and negative to the action potential threshold, resulting in depolarizing IPSPs when measured in slices at the resting potential (Misgeld et al., 1982; Plenz, 2003; Bracci and Panzeri, 2006). Measurements of the reversal potential of GABA$_A$ IPSCs in gramicidin perforated patch recordings in slices have con-firmed that the chloride equilibrium potential in Sp neurons lies approximately at the point of highest input resistance for these neurons (Koós et al., 2004; Gustafson et al., 2005). At this voltage, the dendritic tree is the most electrically compact, and small differences in excitatory synaptic input can make the difference between a transition to the Up state or a return to the Down state. Perhaps this offers some clue as to the functional importance of synapses specifically located on the dendrites of Sp

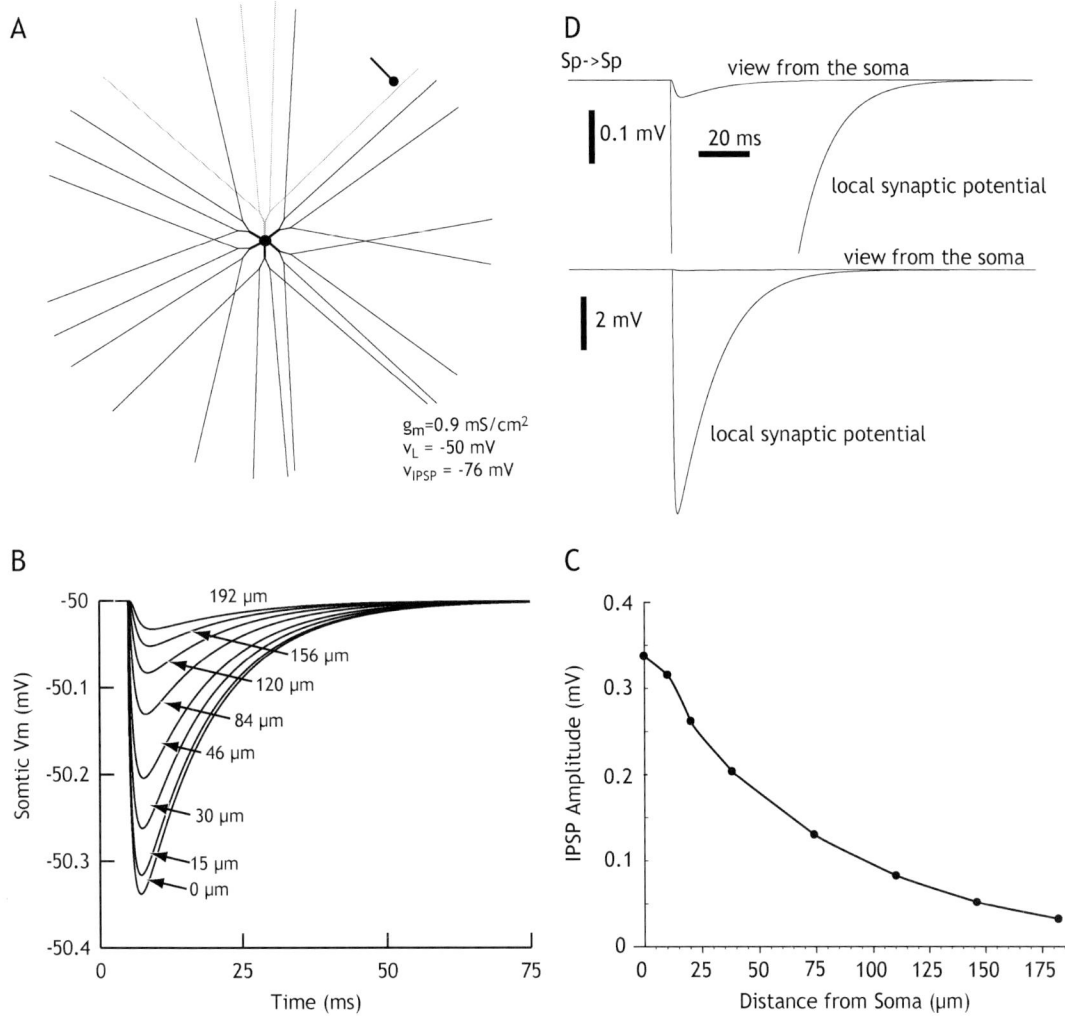

Fig. 7. Computer simulation of single synaptic potentials generated at various parts of the dendritic tree of a spiny neuron, from a starting membrane potential of -50 mV (the Up state), and with a synaptic reversal potential of -76 mV. (A) The spiny neuron is represented by six primary dendrites that bifurcate twice near the soma to give rise to 24 spine-laden tertiary dendrites (spines are not shown). The resting membrane conductance is set at 0.9 mS/cm^2, which gives an input resistance of about 15 Megohm, corresponding to the spiny neuron in the Up state. (B) IPSPs as recorded at the soma, computed for synapses located at various distances from the soma. (C) Peak amplitudes of somatically recorded synaptic potentials as a function of distance from the soma. (D) Synaptic potentials for an input located on the distal dendrite, 192 μm from the soma, as seen from the soma, and as it would appear locally at the position of the synapse, illustrating the large IPSP evoked in the dendrite. IPSPs are shown at two different scales, to allow a better comparison of the IPSP as seen from the two locations.

neurons. Dendritic Sp→Sp inhibition will depolarize Sp dendrites when the cells are most deeply hyperpolarized, and may therefore promote escape of the membrane potential from the hyperpolarization-activated K$^+$ current active at these potentials (Plenz, 2003; Bracci and Panzeri, 2006). If active at more depolarized membrane potentials near action potential threshold, GABA$_A$ inhibition will act to hyperpolarize the cell membrane, and will be inhibitory. Because its reversal potential is centered on this transition point for the voltage-sensitive conductances, distributed inhibition along the surface of the dendrite would act to counteract the nonlinearity introduced by the voltage-sensitive potassium currents. This effect of $GABA_A$ conductance on the I–V curve of the Sp neuron dendrites, and hence on the electrotonic length of the dendrites, is illustrated in Fig. 8. This figure shows that a 220 μm Sp cell dendrite varies in effective electrotonic length over a large range, being electrically compact (<0.5 length constants) between −60 and −70 mV, and very extended (up to 3–4 length constants) when hyperpolarized or depolarized relative to that range. Adding a small $GABA_A$ conductance preferentially alters the electrotonic structure of the dendrite in its compact range. In the presence of dendritic $GABA_A$ inhibition, the dendrite remains electrotonically

long over the entire voltage range (>2 length constants). It should be noted that the conductance required to achieve this is not extremely large. The middle line in Fig. 8B–D, 0.3 mS/cm², is the peak conductance produced by only about 9 Sp→Sp $GABA_A$ quanta distributed along a 220 μm Sp cell dendrite, assuming a peak quantal conductance of 630 pS (Koós et al., 2004). If one synapse can release only one quantum at a time, and if each Sp cell receives about 1200 Sp→Sp cell synapses arising from 400 different nearby Sp cells distributed on about 26 dendritic branches, as in the Koós et al. (2004) model, each dendrite would receive about 46 independent Sp→Sp synapses. Thus, to

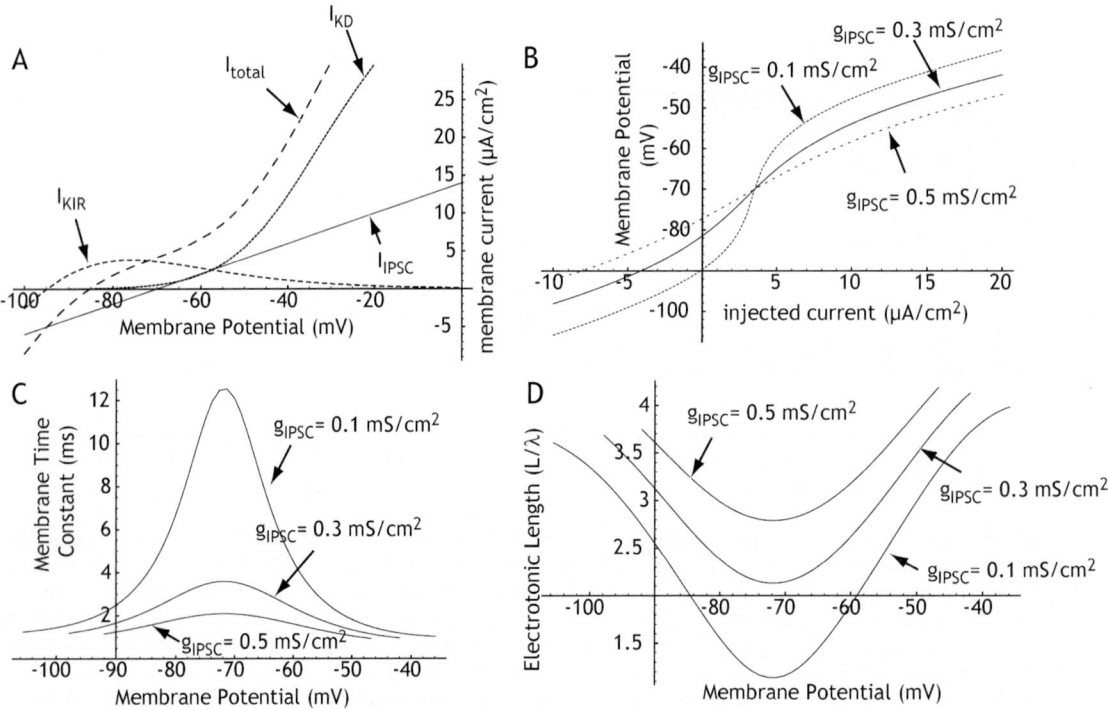

Fig. 8. The reversal potential of the dendritic Sp→Sp synapse corresponds to the point of maximum membrane resistivity of the Sp cell. This will cause it to reduce the voltage-dependence of the Sp cell dendritic cable properties. (A) The voltage-dependence of the two subthreshold potassium currents, the hyperpolarization-activated potassium current (I_{KIR}), the depolarization-activated potassium current (I_{KD}), and the IPSP current (I_{IPSC}). For these curves $g_{KIR} = 0.75$ mS/cm², $g_{KD} = 0.45$ mS/cm², and $g_{IPSP} = 0.2$ mS/cm². (B) The whole cell I–V curve (as in Fig. 6), for an isopotential model cell with currents as in (A) and various amounts of superimposed inhibition. Note how the nonlinearity responsible for the Up and Down states is removed by inhibition acting at the same location as the voltage-sensitive currents. (C) The effect of varying amounts of inhibition on the time constant of an isopotential neuron, with membrane capacitance of 1 μF/cm². Compare with Fig. 6. The time constant would grow even higher if g_{IPSC} were zero (not shown). (D) The effect of varying amounts of local inhibition on the electrotonic length of a Sp cell dendrite, with geometry as described in Fig. 7. Note that for this to occur, the inhibition would have to be dendritic, so it could interact with the dendritic voltage-dependent conductances.

achieve $0.3\,mS/cm^2$ would require about 20% of the local Sp→Sp cell synapses to be active at any one time.

At more hyperpolarized membrane potentials corresponding to the Down state, or when the cell is depolarized closer to action potential threshold, the inhibitory synaptic conductance will be dwarfed by the much larger voltage-sensitive channels activated in those membrane potential ranges, and by excitatory synaptic conductances, which outnumber them many times over. It would not be a very effective kind of inhibition, in the sense of counteracting synaptic excitation and preventing action potentials.

Dendritic inhibition probably acts in the dendrites

Although the inhibition exerted by Sp cell synapses is small when viewed from the Sp cell soma, it is very large at locations in the dendrite near the synapse. Possibly, the large local IPSPs generated by these synapses may exert a function not fully appreciated from the soma, or from the site of action potential generation. One possible dendritic effect of inhibition is suggested by the nonlinear membrane characteristics of the cell in the subthreshold range. As described above, the membrane resistance of the Sp cell dendrites changes radically over the subthreshold voltage range, which alters the electrotonic structure of the cell. Locally, the effectiveness of a synaptic input to the dendrite depends very much on the membrane potential of the dendrite, especially near the synapse. Excitatory synapses activated in isolation, so that the dendrite is in the Down state at the time of synaptic activation, ought to be relatively ineffective, because of the low-input resistance of the dendrite and the large electrotonic distance between dendritic locations and the soma. Excitatory synapses along the dendrite may summate supralinearly, because the depolarization caused by one input can deactivate the hyperpolarization-activated potassium current locally, raising the input resistance and shortening the electrotonic path to the soma. This form of nonlinear synaptic interaction in the Sp cell probably contributes to its peculiar firing pattern in vivo,

which is characterized by a low level of background firing, and discrete episodes of firing. It is also likely to be a contributor to the sparse nature of network activity in the striatum during movements. The spatial arrangement of the Sp→Sp synapse and the nature of intrastriatal connectivity suggests that mutual inhibition does not produce strong competitive interactions that could be responsible either for the low level of background activity, the phasic nature of Sp cell firing, or the sparseness of striatal activity during movement. On the contrary, high levels of convergent mutual inhibition among Sp neurons are predicted to linearize the response of the Sp cell to its input. Under conditions that evoke activity in a large proportion of nearby Sp neurons, the resulting feedback inhibition should cause Sp cells to respond in a more graded, continuous fashion to changes in the level of excitatory input. Possibly, this allows the striatal network to possess two different styles of synaptic integration; one characterized by very sparse activity containing discrete episodes of activity superimposed on almost complete silence, and one in which many neurons are engaged, and firing of individual neurons is continuously distributed over a wide range.

References

Aronin, N., Chase, K. and DiFiglia, M. (1986) Glutamic acid decarboxylase and enkephalin immunoreactive axon terminals in the rat neostriatum synapse with striatonigral neurons. Brain Res., 365: 151–158.

Bar-Gad, I. and Bergman, H. (2001) Stepping out of the box: information processing in the neural networks of the basal ganglia. Curr. Opin. Neurobiol., 11: 689–695.

Bracci, E. and Panzeri, S. (2006) Excitatory GABAergic effects in striatal projection neurons. J. Neurophysiol., 95: 1285–1290.

Bennett, B.D. and Bolam, J.P. (1994) Synaptic input and output of parvalbumin-immunoreactive neurons in the neostriatum of the rat. Neuroscience, 62: 707–719.

Bevan, M.D., Booth, P.A.C., Eaton, S.A. and Bolam, J.P. (1998) Selective innervation of neostriatal interneurons by a subclass of neuron in the globus pallidus of the rat. J. Neurosci., 18: 9438–9452.

Bolam, J.P. and Izzo, P.N. (1988) The postsynaptic targets of substance P-immunoreactive terminals in the rat neostriatum with particular reference to identified spiny striatonigral neurons. Exp. Brain Res., 70: 361–377.

108

Bruno, R.M. and Simons, D.J. (2002) Feedforward mechanisms of excitatory and inhibitory cortical receptive fields. J. Neurosci., 22: 10966–10975.

Centonze, D., Gubellini, P., Pisani, A., Bernardi, G. and Calabresi, P. (2003) Dopamine, acetylcholine and nitric oxide systems interact to induce corticostriatal synaptic plasticity. Rev. Neurosci., 14: 207–216.

Cope, D.W., Maccaferri, G., Marton, L.F., Roberts, J.D., Cobden, P.M. and Somogyi, P. (2002) Cholecystokinin-immunopositive basket and Schaffer collateral-associated interneurones target different domains of pyramidal cells in the CA1 area of the rat hippocampus. Neuroscience, 109: 63–80.

Cowan, R.L. and Wilson, C.J. (1994) Spontaneous firing patterns and axonal projections of single corticostriatal neurons in the rat medial agranular cortex. J. Neurophysiol., 71: 17–32.

Czubayko, U. and Plenz, D. (2002) Fast synaptic transmission between striatal spiny projection neurons. Proc. Natl. Acad. Sci. USA, 99: 15764–15769.

Dube, L., Smith, A.D. and Bolam, J.P. (1988) Identification of synaptic terminals of thalamic or cortical origin in contact with distinct medium-size spiny neurons in the rat neostriatum. J. Comp. Neurol., 267: 455–471.

Freund, T.F., Powell, J.F. and Smith, A.D. (1984) Tyrosine hydroxylase-immunoreactive boutons in synaptic contact with identified striataonigral neurons, with particular reference to dendritic spines. Neuroscience, 13: 1189–1215.

Guzman, J.N., Hernandez, A., Galarraga, E., Tapia, D., Laville, A., Vergara, R., Aceves, J. and Bargas, J. (2003) Dopaminergic modulation of axon collaterals interconnecting spiny neurons of the rat striatum. J. Neurosci., 23: 8931–8940.

Gustafson, N., Gireesh-Dharmaraj, E., Czubayko, U., Blackwell, K.T. and Plenz, D. (2005) A comparative voltage and current-clamp analysis of feedback and feedforward synaptic transmission in the striatal microcircuit in vitro. J. Neurophysiol., 95: 737–752.

Hertz, J.A., Krogh, A.S. and Palmer, R.G. (1991) Introduction to the Theory of Neural Computation. Perseus Books, Cambridge, MA 327 p.

Hernandez-Lopez, S., Tkatch, T., Perez-Garci, E., Galarraga, E., Bargas, J., Hamm, H. and Surmeier, D.J. (2000) D2 dopamine receptors in striatal medium spiny neurons reduce L-type Ca^{2+} currents and excitability via a novel PLC[beta]1-IP3-calcineurin-signaling cascade. J. Neurosci., 20: 8987–8995.

Ingham, C.A., Hood, S.H., Taggart, P. and Arbuthnott, G.W. (1998) Plasticity of synapses in the rat neostriatum after unilateral lesion of the nigrostriatal dopaminergic pathway. J. Neurosci., 18: 4732–4743.

Izzo, P.N. and Bolam, J.P. (1988) Cholinergic synaptic input to different parts of spiny striatonigral neurons in the rat. J. Comp. Neurol., 269: 219–234.

Jaeger, D., Kita, H. and Wilson, C.J. (1994) Surround inhibition among projection neurons is weak or nonexistent in the rat neostriatum. J. Neurophysiol., 72: 2555–2558.

Kawaguchi, Y. (1993) Physiological, morphological and histochemical characterization of three classes of interneurons in rat neostriatum. J. Neurosci., 13: 4908–4923.

Kawaguchi, Y. and Kubota, Y. (1998) Neurochemical features and synaptic connections of large physiologically-identified GABAergic cells in the frontal cortex. Neuroscience, 85: 677–701.

Kawaguchi, Y., Wilson, C.J. and Emson, P.C. (1989) Intracellular recording of identified neostriatal patch and matrix spiny cells in a slice preparation preserving cortical inputs. J. Neurophysiol., 62: 1052–1068.

Kawaguchi, Y., Wilson, C.J. and Emson, P.C. (1990) Projection subtypes of rat neostriatal matrix cells revealed by intracellular injection of biocytin. J. Neurosci., 10: 3421–3438.

Kemp, J.M. and Powell, T.P.S. (1971a) The synaptic organization of the caudate nucleus. Phil. Trans. R. Soc. Lond. B., 262: 403–412.

Kemp, J.M. and Powell, T.P.S. (1971b) The site of termination of afferent fibres in the caudate nucleus. Phil. Trans. R. Soc. Lond. B., 2262: 413–427.

Kerr, J.N. and Wickens, J.R. (2001) Dopamine D-1/D-5 receptor activation is required for long-term potentiation in the rat neostriatum in vitro. J. Neurophysiol., 85: 117–124.

Kita, H., Kosaka, T. and Heizmann, C.W. (1990) Parvalbumin-immunoreactive neurons in the rat neostriatum: a light and electron microscopic study. Brain Res., 536: 1–15.

Koós, T. and Tepper, J.M. (1999) Inhibitory control of neostriatal projection neurons by GABAergic interneurons. Nature Neurosci., 2: 467–472.

Koós, T., Tepper, J.M. and Wilson, C.J. (2004) Comparison of IPSCs evoked in spiny and fast-spiking neurons in the neostriatum. J. Neurosci., 24: 7916–7922.

Kotaleski, J.H., Plenz, D. and Blackwell, K.T. (2006) Using potassium currents to solve signal-to-noise problems in inhibitory feedforward neworks of the striatum. J. Neurophysiol., 95: 331–341.

Kubota, Y. and Kawaguchi, Y. (2000) Dependence of GABAergic synaptic areas on the interneuron type and target size. J. Neurosci., 20: 375–386.

Lapper, S.R., Smith, Y., Sadikot, A.F., Parent, A. and Bolam, J.P. (1992) Cortical input to parvalbumin-immunoreactive neurons in the putamen of the squirrel monkey. Brain Res., 580: 215–224.

Lawrence, J.J. and McBain, C.J. (2003) Interneuron diversity series: containing the detonation—feedforward inhibition in the CA3 hippocampus. Trends Neurosci., 26: 631–640.

Lei, W., Jiao, Y., Del Mar, N. and Reiner, A. (2004) Evidence for differential cortical input to direct pathway versus indirect pathway striatal projection neurons. J. Neurosci., 24: 82989–88299.

Lovinger, D.M. and Tyler, E. (1996) Synaptic transmission and modulation in the neostriatum. Int. Rev. Neurobiol., 39: 77–111.

Mallet, N., Le Moine, C., Charpier, S. and Gonon, F. (2005) Feedforward inhibition of projection neurons by fast-spiking

GABA interneurons in the rat striatum in vivo. J. Neurosci., 25: 3857–3869.

Miller, K.D. (2003) Understanding layer 4 of the cortical circuit: a model based on cat V1. Cereb. Cortex, 13: 73–82.

Misgeld, U., Wagner, A. and Ohno, T. (1982) Depolarizing IPSPs and depolarization by GABA of rat neostriatum cells in vitro. Exp. Brain Res., 45: 108–114.

Nisenbaum, E.S. and Wilson, C.J. (1995) Potassium currents responsible for inward and outward rectification in rat neostriatal spiny projection neurons. J. Neurosci., 15: 4449–4463.

Oorschot, D.E. (1996) Total number of neurons in the neostriatal, pallidal, subthalamic and substantia nigral nuclei of the rat basal ganglia: a stereological study using the cavalieri and optical dissector methods. J. Comp. Neurol., 366: 580–599.

Plenz, D. (2003) When inhibition goes incognito: feedback interaction between spiny projection neurons in striatal function. Trends Neurosci., 26: 436–443.

Plenz, D. and Kitai, S.T. (1998) Up and down states in striatal medium spiny neurons simultaneously recorded with spontaneous fast-spiking interneurons studied in cortex-striatum-substantia nigra organotypic cultures. J. Neurosci., 18: 266–283.

Plenz, D. and Kitai, S.T. (2000) Adaptive classification of cortical input to the striatum by competitive learning. In: Miller R. and Wickens J.R. (Eds.), Brain Dynamics and the Striatal Complex. Academic Publishers, Harwood, Australia, pp. 165–178.

Ramanathan, S., Hanley, J.J., Deniau, J.M. and Bolam, J.P. (2002) Synaptic convergence of motor and somatosensory cortical afferents onto GABAergic interneurons in the rat striatum. J. Neurosci., 22: 8158–8169.

Reyes, A., Galarraga, E., Flores-Hernandez, J., Tapia, D. and Bargas, J. (1998) Passive properties of neostriatal neurons during potassium conductance blockade. Exp. Brain Res., 120: 70–84.

Roberts, R.C., Force, M. and Kung, L. (2002) Dopaminergic synapses in the matrix of the ventrolateral striatum after chronic haloperidol treatment. Synapse, 45: 78–85.

Rolls, E.T. and Treves, A. (1998) Neural Networks and Brain Function. Oxford University Press, Oxford 418 p.

Sachdev, R.N., Ebner, F.F. and Wilson, C.J. (2004) Effect of subthreshold up and down states on the whisker-evoked response in somatosensory cortex. J. Neurophysiol., 92: 3511–3521.

Shen, W., Hernandez-Lopez, S., Tkatch, T., Held, J.E. and Surmeier, D.J. (2004) Kv.2-containing K^+ channels regulate subthreshold excitability of striatal medium spiny neurons. J. Neurophysiol., 91: 1337–1349.

Smith, Y., Bennett, B.D., Bolam, J.P., Parent, A. and Sadikot, A.F. (1994) Synaptic relationships between dopaminergic afferents and cortical or thalamic input in the sensorimotor territory of the striatum in monkey. J. Comp. Neurol., 344: 1–19.

Soghomonian, J.J., Descarries, L. and Watkins, K.C. (1989) Serotonin innervation in adult rat neostriatum. II. Ultrastructural features: a radioautographic and immunocytochemical study. Brain Res., 481: 67–86.

Somogyi, P., Bolam, J.P. and Smith, A.D. (1981) Monosynaptic cortical input and local axon collaterals of identified striatonigral neurons. A light and electron microscopic study using the Golgi-peroxidase transport-degeneration procedure. J. Comp. Neurol., 195: 567–584.

Somogyi, P., Priestley, J.V., Cuello, A.C., Smith, A.D. and Takagi, H. (1982) Synaptic connections of enkephalin-immunoreactive nerve terminals in the neostriatum: a correlated light and electron microscopic study. J. Neurocytol., 11: 779–807.

Stern, E.A., Kincaid, A.E. and Wilson, C.J. (1997) Spontaneous subthreshold membrane potential fluctuations and action potential variability of rat corticostriatal and striatal neurons. J. Neurophysiol., 77: 1697–1715.

Sun, Q.Q., Huguenard, J.R. and Prince, D.A. (2006) Barrel cortex microcircuits: thalamocortical feedforward inhibition in spiny stellate cells is mediated by a small number of fast-spiking interneurons. J. Neurosci., 26: 1219–1230.

Swadlow, H.A. (2003) Fast-spike interneurons and feedforward inhibition in awake sensory neocortex. Cereb. Cortex, 13: 25–32.

Taverna, S., Van Dongen, Y.C., Groenewegen, H.J. and Pennartz, C.M. (2004) Direct physiological evidence for synaptic connectivity between medium-sized spiny neurons in rat nucleus accumbens in situ. J. Neurophysiol., 91: 1111–1121.

Tepper, J.M., Sharpe, N.A., Koós, T.Z. and Trent, F. (1998) Postnatal development of the rat neostriatum: electrophysiological, light- and electron-microscopic studies. Developmental Neurosci., 20: 125–145.

Tunstall, M.J., Oorschot, D.E., Kean, A. and Wickens, J.R. (2002) Inhibitory interactions between spiny projection neurons in the rat striatum. J. Neurophysiol., 88: 1263–1269.

Venance, L., Glowinski, J. and Giaume, C. (2004) Electrical and chemical transmission between striatal GABAergic output neurons in rat brain slices. J. Physiol., 559: 215–230.

Wang, H.B., Laverghetta, A.V., Foehring, R., Deng, Y.P., Sun, Z., Yamamoto, K., Lei, W.L., Jiao, Y. and Reiner, A. (2006) Single-cell RT-PCR, in situ hybridization histochemical, and immunohistochemical studies of substance P and enkephalin co-occurrence in striatal projection neurons in rats. J. Chem. Neuroanat., 3: 178–199.

Wickens, J. (1993) A Theory of the Striatum. Pergamon Press, Oxford 182 p.

Wilson, C.J. (1986) Postsynaptic potentials evoked in spiny neostriatal neurons by stimulation of ipsilateral and contralateral cortex. Brain Res., 367: 201–213.

Wilson, C.J. (1986) Three-dimensional analysis of dendritic spines by means of HVEM. J. Electron Microsc., 35: 1151–1155.

Wilson, C.J. and Groves, P.M. (1980) Fine structure and synaptic connections of the common spiny neuron of the rat neostriatum: a study employing intracellular injection of horseradish peroxidase. J. Comp. Neurol., 194: 599–615.

110

Wilson, C.J. and Groves, P.M. (1981) Spontaneous firing patterns of identified spiny neurons in the rat neostriatum. Brain Res., 220: 67–80.

Wilson, C.J., Groves, P.M., Kitai, S.T. and Linder, J.C. (1983) Three-dimensional structure of dendritic spines in the rat neostriatum. J. Neurosci., 3: 383–398.

Wilson, C.J. and Kawaguchi, Y. (1996) The origins of two-state spontaneous membrane potential fluctuations of neostriatal spiny neurons. J. Neurosci., 16: 2397–2410.

Xu, Z.C., Wilson, C.J. and Emson, P.C. (1989) Restoration of the corticostriatal projection in rat neostriatal grafts: electron microscopic analysis. Neuroscience, 29: 539–550.

Xu, Z.C., Wilson, C.J. and Emson, P.C. (1991) Restoration of the thalamostriatal projection in rat neostriatal grafts: an electron microscopic analysis. J. Comp. Neurol., 303: 22–34.

Zheng, T. and Wilson, C.J. (2001) Corticostriatal combinatorics: the implications of corticostriatal axonal arborizations. J. Neurophysiol., 87: 1007–1017.

Tepper, Abercrombie & Bolam (Eds.)
Progress in Brain Research, Vol. 160
ISSN 0079-6123

CHAPTER 7

Globus pallidus external segment

Hitoshi Kita*

*Department of Anatomy and Neurobiology, The University of Tennessee Memphis, 855 Monroe Avenue,
Memphis, TN 38163, USA*

Abstract: The external segment of the pallidum (GP_e) is a relatively large nucleus located caudomedial to the neostriatum (Str). The GP_e receives major inputs from two major basal ganglia input nuclei, the Str and the subthalamic nucleus (STN), and sends its output to many basal ganglia nuclei including the STN, the Str, the internal pallidal segment (GP_i), and the substantia nigra (SN). Thus, the GPe can be placed at the center of the basal ganglia connection diagram (Fig. 1(A)). From the viewpoint that emphasizes the direct and indirect pathways of the basal ganglia, the GP_e is a component of the indirect pathway that relays Str inputs to the STN. The indirect pathway can be traced in Fig. 1(A), although it comprises only a part of multiple indirect pathways. This chapter begins with a brief description of the anatomical organization of the GP_e followed by physiological and pharmacological characterizations of GABAergic responses in the GP_e.

Keywords: globus pallidus; GABAergic response; neostriatum; subthalamic nucleus; basal ganglia

Anatomical organization of the GP_e

Morphologically distinct types of GP_e neurons

Figure 1 shows the major connections of the basal ganglia. A majority of GP_e neurons are projection neurons (Kita, 1994; Kita and Kitai, 1994) that contain glutamate decarboxylase (Mugnaini and Oertel, 1985; Smith et al., 1987; Kita, 1994). These GABAergic projection neurons have been subtyped by various morphological and physiological criteria. Single axon and retrograde tracing studies in rodents revealed that approximately 1/3 of GP_e neurons project to the Str (Kita and Kitai, 1994; Kita et al., 1999; Kita and Kita, 2001). Both the striatal and non-striatal projecting neurons also project to the GP_i and STN (Staines and Fibiger, 1984; Kita

and Kitai, 1994; Bevan et al., 1998) (Fig. 2). These striatal and non-striatal projecting neurons have other differences. The former tend to have sparsely spiny dendrites forming a radiating dendritic field and are often immunonegative for the Ca^{2+}-binding protein parvalbumin (PV), but often express preproenkephalin mRNA (Hoover and Marshall, 2002). The latter neuron type has large aspiny primary dendrites, varicose secondary and tertiary dendrites with occasional complex endings (i.e., thick apparati terminally located on dendrites having many appendages of various types) forming a discoidal dendritic field in a flat plane parallel to the border between the GP_e and the Str, and are PV-positive (Kita and Kitai, 1994; Kita, 1996; Kita and Kita, 2001). Approximately 30% of GP_e neurons in monkeys also project to the Str (Kita et al., 1999). However, in monkeys, each GP_e neuron has a more specific synaptic target such that neurons projecting to the Str do not project to other nuclei, and other GP_e neurons either project to the GP_i and

*Corresponding author. Tel.: +901-448-5234;
Fax: +901-448-7193; E-mail: hkita@utmem.edu

DOI: 10.1016/S0079-6123(06)60007-1

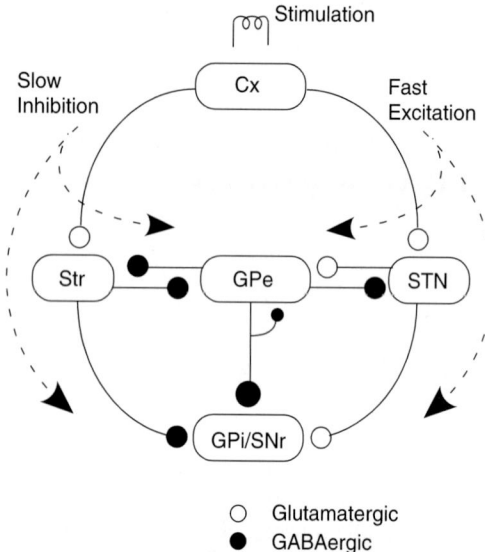

Fig. 1. A diagram of the major connections of the basal ganglia. The external segment of the pallidum (GP$_e$) receives major inputs from two input nuclei of the basal ganglia, the neostriatum (Str) and the subthalamic nucleus (STN). The internal segment of the pallidum (GP$_i$) and the substantia nigra pars reticulata (SNr) are the main output nuclei of the basal ganglia. The cortico (Cx)-Str-GP$_i$/SNr pathway is a slowly conducting inhibitory connection and the Cx-STN-GP$_i$/SN, pathway is a fast excitatory connection.

STN or to the substantia nigra pars reticulata (SNr) and STN (Sato et al., 2000).

The existence of a small number of interneurons in the GP$_e$ was suggested in early Golgi studies (DiFiglia et al., 1982; Falls et al., 1983). A recent study suggested that less than 1% of small calretinin-containing neurons in rats were interneurons (Cooper and Stanford, 2002). In monkeys, calretinin-containing neurons are either large or small and their projection sites have not been identified (Fortin and Parent, 1994).

Physiologically distinct types of GP$_e$ neurons

In vivo unit recordings from monkey GP$_e$ distinguish two types of neurons. The most numerous type of neuron exhibits high-frequency firing interspersed with spontaneous pauses, while the other type exhibits low-frequency firing and bursts (DeLong, 1971; Gardiner and Kitai, 1992; Magill

et al., 2000; Ni et al., 2000). Unit recordings from rats also distinguish two neuronal types with distinct extracellular waveforms and different responses to systemic administration of the dopamine agonist apomorphine (Kelland et al., 1995). Results of in vivo and in vitro intracellular recordings suggest that GP$_e$ neurons can be classified by differences in membrane properties such as spike accommodation, existence of a prominent hyperpolarization activated inward current (I_h), low-threshold Ca^{2+} spikes, and an early K$^+$ current (Kita and Kitai, 1991; Nambu and Llinás, 1994; Cooper and Stanford, 2000). The neurons that fire with high frequency and pauses in vivo may correspond to the neurons that show weak I_h, no rebound Ca^{2+} spikes, and no prominent spike accommodation in vitro. The neurons that fire with low frequency and bursts may correspond to the type with prominent I_h, prominent rebound Ca^{2+} spikes, and no prominent spike accommodation in vitro. Although some GP$_e$ neurons have clearly identifiable membrane properties that can be used to classify them, in other neurons, the differences are often not distinct enough to unequivocally classify them into different subclasses. A small number of neurons have membrane properties similar to Str projection neurons such as prominent spike accommodation and long-duration spikes and are readily distinguishable from the rest with no prominent spike accommodation (Kita, 1996). Also, caution is needed in that in vivo studies may not distinguish cholinergic neurons that are mostly located ventrocaudal to the GP$_e$.

Afferent projections of the GP$_e$

The GP$_e$ receives massive GABAergic afferent fibers from the Str and glutamatergic afferent fibers from the STN. The GP$_e$ also receives sparse afferents from the cerebral cortex, intralaminar thalamic nuclei, GP$_i$, substantia nigra pars compacta (SNc), raphe, and pedunclopontine tegmentum (Hazrati et al., 1990; Fink-Jensen and Mikkelsen, 1991; Kita, 1994; Naito and Kita, 1994; Parent and Hazrati, 1995; Deschenes et al., 1996; Yasukawa et al., 2004). In both rats and monkeys, most, if not all, of the Str projection neurons project to

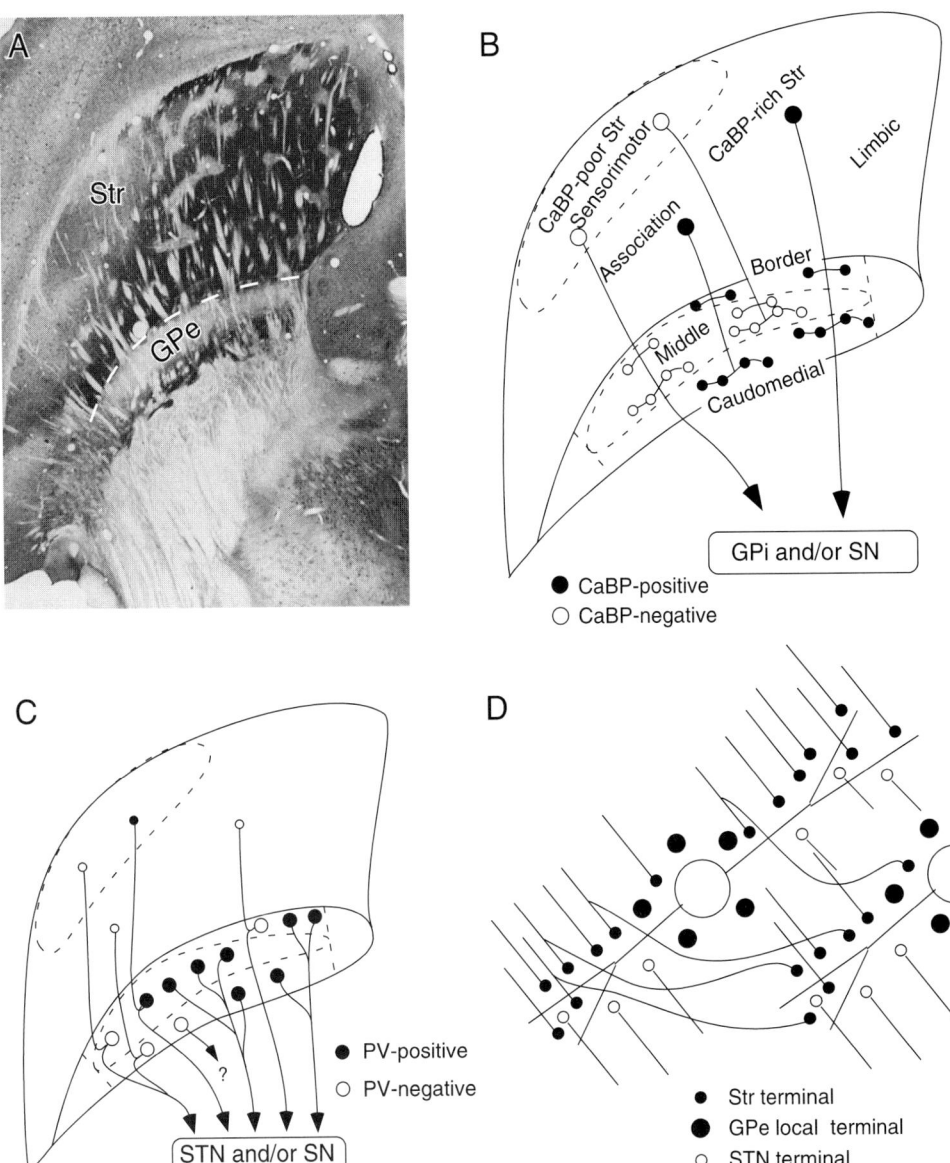

Fig. 2. (A) Photomicrograph of a CaBP-immunostained horizontal section of a rat brain. CaBP staining is poor in the patch compartment and the matrix of the dorsolateral sensorimotor region of the Str, while staining is heavy in the matrix of the middle associate to the medial limbic regions of the Str. The dashed lines mark the border between the Str and the GP$_e$. In the GP$_e$, CaBP heavily stained the border and the caudomedial regions. The border region is approximately 50–150 μm thick facing to the Str. The area between the border and the caudomedial regions is the CaBP-poor middle region of the GP$_e$. (B) A diagram for Str-GP$_e$ projections. Str projection neurons either project only to the GP$_e$ or to the GP$_e$, the internal segment of pallidum (GP$_i$) and/or the substantia nigra (SN). Most Str axons form two main terminal fields in the GP$_e$. Axons originating from the CaBP-poor and -rich parts of the Str form terminal fields in the CaBP-poor and -rich regions of the GP$_e$, respectively. (C) The GP$_e$-Str projections are reciprocal to the Str-GP$_e$ projections in that the CaBP-poor and -rich regions of the GP$_e$ project to the CaBP-poor and -rich parts of the Str, respectively. The majority of the GP$_e$-Str neurons are PV-negative, although some PV-positive neurons in the middle GP$_e$ region also project to the GP$_e$. (D) A diagram for the synaptic arrangement of the GP$_e$. Str boutons are most abundant and form synapses mainly on dendrites of GP$_e$ neurons. Each Str axon evokes very small unitary IPSCs suggesting that boutons from a large number of Str neurons terminate on a GP$_e$ neuron. GP$_e$ local collateral axons form synapses on the somata and proximal dendrites and evoke large unitary IPSCs. Glutamatergic axons from the STN are less numerous and terminate mostly on dendrites.

the GP$_e$, with approximately half of the neurons projecting to the GP$_e$ only and the other half emitting collaterals to the GP$_e$ on the way to the GP$_i$ and the substantia nigra pars reticulata (SNr) (Kawaguchi et al., 1990; Parent et al., 1995; Wu et al., 2000). The number of boutons belonging to the latter collateral axons is approximately half of that belonging to the striatal axons projecting only to the GP$_e$ (Kawaguchi et al., 1990).

The Str can be divided into functional territories based on the cortical and thalamic afferent localization (Gerfen et al., 1985; Donoghue and Herkenham, 1986; McGeorge and Faull, 1989; Berendse et al., 1992). Anterograde neurotracing and immunohistochemistry for calcium-binding protein (CaBP; CaBP is contained in striatal axons originating from the matrix compartment of the associative and limbic territories but not in those from the sensorimotor territory) revealed that the GP$_e$ can also be divided into functional territories (Fig. 2) (Hontanilla et al., 1994; Rajakumar et al., 1994; Kita and Kita, 2001; Karachi et al., 2002). Afferent fibers from each territory of the Str form two narrow band-shaped terminal fields in the corresponding territory of the GP$_e$. The rostrocaudal levels of the bands depend upon the rostrocaudal levels of the originating location of the afferent fibers in the Str (Chang et al., 1981; Wilson and Phelan, 1982; Kawaguchi et al., 1990; Kita and Kita, 2001). Although the striatal-GPe projection is massive (the ratio of the number of neurons in Str:GP$_e$ is approximately 60:1 in the rat, Oorshot, 1996), the number of boutons belonging to each axon is relatively small, between 100 and 250, and the boutons are sparsely distributed (Kawaguchi et al., 1990; Yelnik et al., 1996). STN afferent fibers also form multiple terminal fields in the GP$_e$ (Kita and Kitai, 1987; Smith et al., 1990).

All GP$_e$ projection neurons that were examined with single cell staining had local collateral axons, some having dense axonal arborizations with numerous boutons and others having short axons with less numerous boutons (Kita and Kitai, 1994). Most of the terminal fields of collateral axons could be found in the same medio-lateral zone of the GP$_e$, where the parent cells were located (Kita, 1996).

Electron microscopy revealed that large areas of the somata and dendrites of GP$_e$ projection neurons were covered with synaptic boutons. Over 80% of terminals formed symmetrical synapses of which approximately 80% belong to striatal axons and the remaining to local collateral axons. Striatal boutons are small or medium in size (less than 1 μm), contain large pleomorphic vesicles, and form synapses mainly on the dendrites of GP$_e$ neurons. Boutons of local collateral axons are larger (over 1 μm) than striatal boutons, contain small round or elongated vesicles, and synapse on the somata and proximal dendrites (DiFiglia et al., 1982; Falls et al., 1983; Okoyama et al., 1987; Kita, 1994; Shink and Smith, 1995). The boutons forming asymmetric synapses (i.e., less than 20% of total synapses in the GP$_e$) belong mainly to the fibers from the STN, while others are from the frontal cortex and the intralaminar thalamic nuclei. The boutons of the STN are medium in size, contain small, pleomorphic vesicles, and form asymmetrical synapses mainly on the dendritic shafts of both PV-positive and -negative neurons. The details of other aminergic afferents are unknown.

Efferent projections of the GP$_e$

The main projection sites to the GP$_e$ are the GP$_i$, STN, and Str as mentioned above. A small number of GP$_e$ neurons also innervate the dorsal thalamus, inferior colliculus, and the pedunculopontine tegmentum. GP$_e$ axons form large symmetric boutons that contain small round or elongated vesicles and multiple mitochondria, and form symmetric synapses on the somata and proximal dendrites of GP$_i$ and STN neurons, similar to the local collateral axons (Chang et al., 1983; Shink and Smith, 1995). In the Str, GP$_e$ axons terminate on aspiny GABAergic interneurons (Bevan et al., 1998) and the dendritic shafts of spiny projection neurons. The topographic arrangements of the GP$_e$-STN and GP$_e$-striatal projections are in register with that of the STN-GP$_e$ and striatal-GP$_e$ projections, suggesting the existence of precise reciprocal loops.

Postsynaptic GABAergic responses

GABAergic responses evoked by striatal and local axons in vitro

Electrical stimulation of the Str through a bipolar electrode in slice preparations evoked IPSPs/IPSCs

with at least two different latencies in GP_e neurons (Ogura and Kita, 2000). In the example shown in Fig. 3, Str stimulation at threshold intensity evoked very small (less than 3 mV or 10 pA) IPSPs/IPSCs with a latency of 7–10 ms in a GP_e neuron dialyzed with a high (50 mM) Cl^- containing electrolyte. When the stimulus intensity

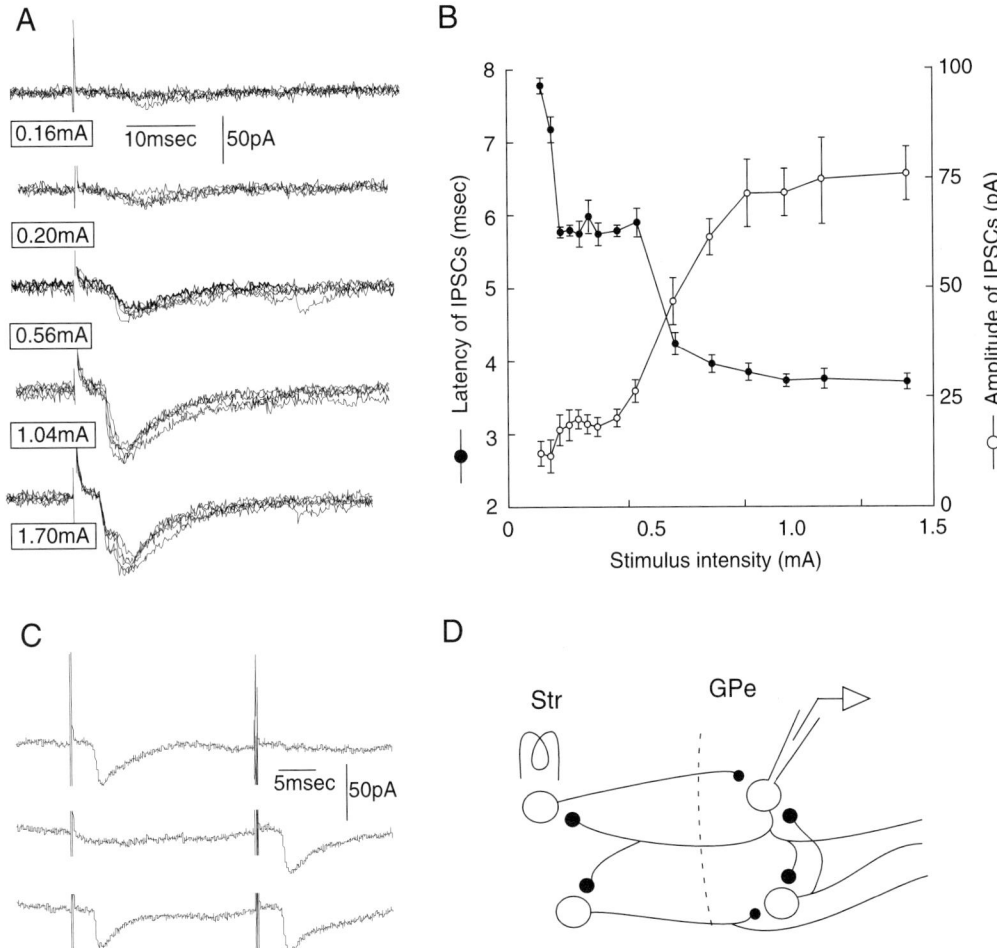

Fig. 3. Stimulation of the Str induced short- and long-latency IPSCs in slice preparations from rats. Recordings were made from GP_e neurons with a pipette containing a high Cl^--containing electrolyte. Artificial cerebrospinal fluid (ACSF) contained CPP (10 µM) and DBQX (50 µM) to block glutamatergic responses. (A) Sample traces show that Str stimulation-evoked IPSCs in a GP_e neuron have various intensities. Amplitudes and latencies of IPSCs changed with the change in stimulus intensity. (B) Plots of the amplitude and the latency of IPSCs against the stimulus intensity indicate that Str stimulation-induced IPSCs consist of multiple components. (C) IPSCs recorded from another GP_e neuron. Sample traces show large unitary IPSCs to paired Str stimulation. The IPSCs had relatively constant amplitude and a constant latency (3.1 ms) with stimulus intensities from threshold (35 µA) to 90 µA. These large, short latency IPSCs were considered to be evoked by GP_e local collateral axons. (D) A schematic drawing shows the experimental setup and neuronal circuits that are involved.

was gradually increased, the latency and the response amplitude showed two relatively stable regions, one with about a 6 ms latency and a 20 pA amplitude and the other with a 4 ms latency and a 75 pA amplitude (Fig. 3(B)). The half-width of these IPSCs were approximately 10 ms. When strong stimulation was applied, the long latency components were blended into the short latency component and could not be isolated in most of the neurons. These results suggested that electrical stimulation of the Str activated IPSPs/IPSCs of multiple origins. It is expected that Str stimulation activates both striatal projection neurons and axons of GP$_e$ neurons projecting to the Str that also have local collateral axons in the GP$_e$ (Kita and Kitai, 1994; Nambu and Llinás, 1997; Bevan et al., 1998; Smith et al., 1998; Kita et al., 1999). Based on known conduction velocities and/or conduction times of striatal-GP$_e$ and GP$_e$-striatal axons (Park et al., 1982; Walker et al., 1989; Kita and Kitai, 1991; Nambu and Llinás, 1994), we assume that the IPSPs/IPSCs with 6 and 4 ms latencies were evoked by the Str and GP$_e$ axons, respectively. The origins of the lowest threshold 7–10 ms responses are unknown.

Our in vitro study also suggested that unitary IPSPs/IPSCs evoked by striatal-GP$_e$ axons were very small because a gradual increase in the stimulus intensity resulted in a gradual increase in the response amplitude from zero (Ogura and Kita, 2000). We have reexamined this by evoking disynaptic cortico-striatal-GP$_e$ responses in slice preparations. Direct stimulation of the GP$_e$-striatal axons was avoided by stimulating the cortex. Figure 4 shows that stimulation of cortical white matter by three repetitive pulses induced small IPSCs, less than 10 pA on average, with 18–22 ms latencies in a GP$_e$ neuron dialyzed with Cs$^+$ ions and TEA to block potassium currents (Watanabe and Kita, unpublished observation). These data confirmed that the size of striatal axon-induced unitary IPSCs recorded at the somata of GP$_e$ neurons is very small. The precise size is not yet known because paired Str and GP$_e$ cell recording has not been performed, and the input is too small to be tested with the minimal stimulation

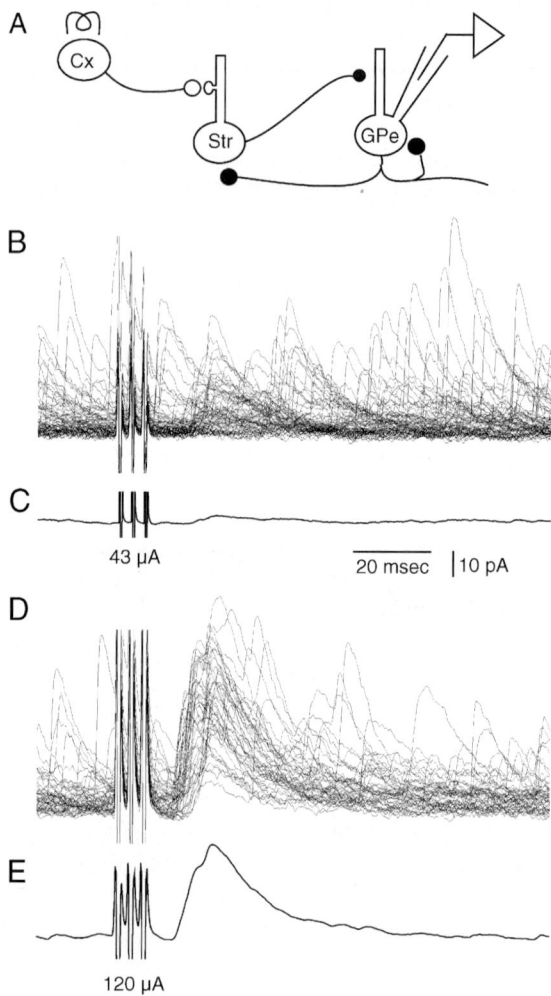

Fig. 4. (A) Responses of a GP$_e$ neuron dialyzed with CsSO$_4$ and TEA to stimulation of cortical white matter. CsSO$_4$ and TEA were contained in the whole cell recording pipette to reduce potassium currents and to increase the membrane resistance for easier detection of small IPSCs. Repetitive stimulation of the cortical white matter was used to induce Str mediated disynaptic IPSCs without evoking stimulation of GP$_e$ axon mediated IPSCs. The neuron was voltage clamped at 0 mV. (B) Overlapped traces show responses to 50 stimulus pulses with intensities near the threshold, which was determined by the averaged trace shown in (C). (D) and (E) Individual and average responses to approximately three times higher intensity stimuli than the threshold. This neuron had large spontaneous IPSCs that were considered to be evoked by spontaneously active GP$_e$ neurons.

technique, although the same technique can be used for the short latency IPSCs presumably evoked by GP_e axons (Fig. 3(C)). In the Str, paired cell recordings showed that the axon collaterals of striatal projection neurons evoke very small IPSCs in postsynaptic striatal neurons (Tunstall et al., 2002; Koos et al., 2004). These physiological observations are consistent with anatomical data that show that each striatal axon has a relatively small number of terminal boutons in a relatively large terminal area and that the terminals form synapses mainly on the dendrites of GP_e neurons. Because each striatal-GP_e axon evokes only a small response, the simultaneous activation of a group of striatal neurons may be required to produce strong inhibition in the GP_e.

These data also suggest that the electrical stimulation of the Str has to be used carefully to excite striatal-GP_e inputs without also exciting GP_e axon collateral inputs. Striatal neurons in slice preparations have resting membrane potentials of approximately $-80 \, mV$ and require relatively large currents to depolarize the membrane to spike threshold.

GP_e neurons recorded in slice preparations show a relatively high rate of spontaneous IPSPs/IPSCs. The amplitude of the IPSCs ranged from 10 to over 100 pA with a Cl^- driving force of approximately 50 mV (Fig. 4) (Ogura and Kita, 2000; Matsui and Kita, 2003; Watanabe and Kita, unpublished observation). These spontaneous IPSCs originate from the GP_e local axon collaterals. The evidence for this is that GP_e neurons, but not striatal neurons, are spontaneously active in slice preparations at relatively high rates (Ogura and Kita, 2000; Matsui and Kita, 2003; Kaneda and Kita, 2005) and that GP_e axons form relatively large boutons with large synapses that terminate mainly on the somata and proximal dendrites (Falls et al., 1983; Okoyama et al., 1987; Kita, 1994; Shink and Smith, 1995), which would tend to produce large amplitude IPSCs recorded at the soma. Conversely, the spontaneous IPSCs cannot be attributed to striatal axons because striatal neurons are totally silent in slice preparations and because striatal axons evoke only very small IPSCs in GP_e neurons, as discussed above.

The large intra-GP_e collateral and GP_e-GP_i inputs play a significant role in controlling the firing rate and pattern of GP_e/GP_i neurons in vivo. Figure 5(A) shows that when a small spontaneous IPSP occurred in the recovery phase of the spike after hyperpolarization of a GP_i neuron, the next spike was very effectively delayed. Figure 5(B) shows that GP_e stimulation evoked IPSPs very effectively reset the firing of the GP_i neuron. These observations suggest that the inhibition induced by the GP_e neuron can have a strong effect in controlling the timing of firing of GP_e/GP_i neurons that are located in the same functional territory. If the IPSPs were strong enough, firing of neurons postsynaptic to an axon can be synchronized. One of the most remarkable observations made in experimental parkinsonian monkeys is the synchronization of firing of a large number of GP_e/GP_i neurons (Raz et al., 2000). It is possible that the synchronization was due to an increase in the efficacy of the GP_e axons after a loss of dopaminergic innervation. Supporting evidence for this speculation is the increase in the level of ambient GABA in the GP_e, as discussed later.

GABAergic responses evoked by Str and Cx stimulation in vivo

Several interesting observations have been made in unit recordings from awake monkeys. Electrical stimulation of the motor cortex evoked a sequence of responses including an early excitation, an inhibition, and a late excitation (Fig. 6(A)). The origin of both the early and the late excitations was the STN, because injection of the GABA agonist, muscimol, into the STN diminished these responses. Muscimol injections into the STN have been used to block the activity of the STN and the pathways that include this nucleus. The pathway for the early excitation is cortico-STN-GP_e. The pathway for the late excitation was considered to be cortico-striatal-GP_e-STN-GP_e connections because injection of the GABA antagonist, gabazine, into the GP_e diminished the response (Fig. 6(C)). After blockade of the STN, single pulse stimulation

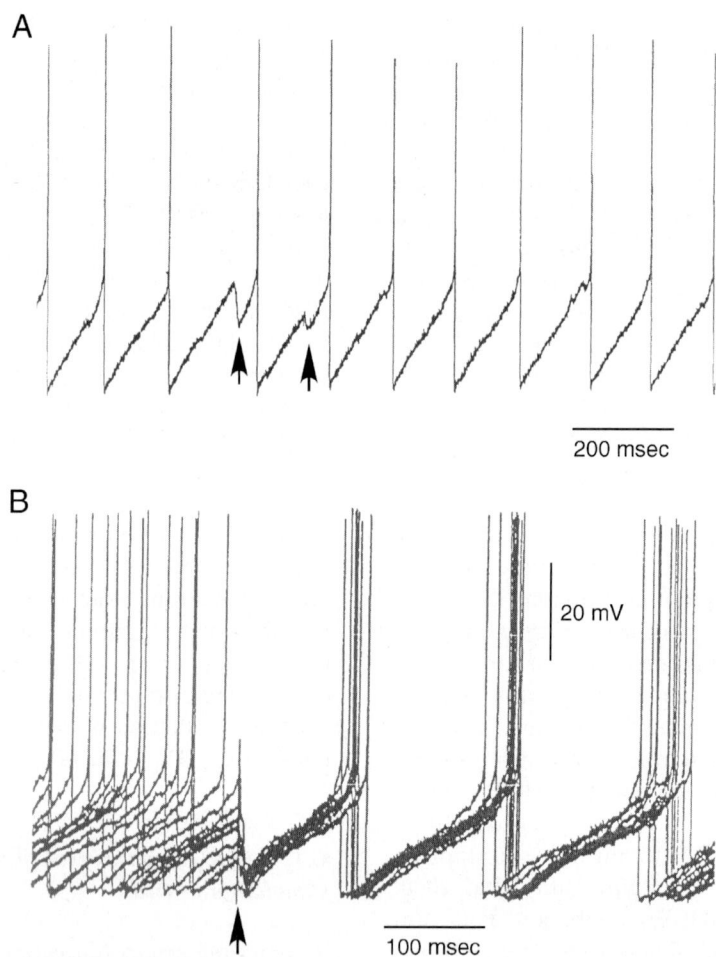

Fig. 5. (A) Recording of a continuously firing GP_i neuron in a slice preparation shows spontaneous IPSPs (some are marked by arrows). The spontaneous IPSPs were very effective in delaying the next spike generation. (B) IPSPs recorded from a GP_i neuron to GP_e stimulation. The amplitude of the IPSPs was small when it was evoked at the peak of the spike after hyperpolarization. In contrast, the amplitude was large when it was evoked in the recovery phase of the after-hyperpolarization; thus, the IPSPs effectively delayed the next spike generation. This observation suggests that synchronized activation of the GP_e reset the firing of a large number of the GP_i neurons.

of the cortex or Str evoked a very long lasting inhibition, with duration of up to 800 ms in GP_e neurons (Fig. 7) (Nambu et al., 2000). The inhibition was completely blocked by local application of gabazine. A similar long inhibition was also recorded in normal monkeys by adjusting the intensity of striatal stimulation (Tremblay and Filion, 1989) or by local application of ionotropic glutamate receptor blockers (Kita et al., 2004). In vitro experiments showed that single pulse stimulation of the Str evoked IPSPs with much

shorter duration in the GP_e than the inhibition recorded in vivo. Under normal conditions in slices, the duration of the IPSP evoked in GP_e after striatal stimulation was less than 100 ms (Nambu and Llinás, 1990) and was similar to single pulse electrical stimulation-induced IPSPs in many other brain areas. The reasons for the long inhibition in monkey GP_e are not known at this point. However, the ability to evoke long inhibition in vivo has important functional implications, as discussed later.

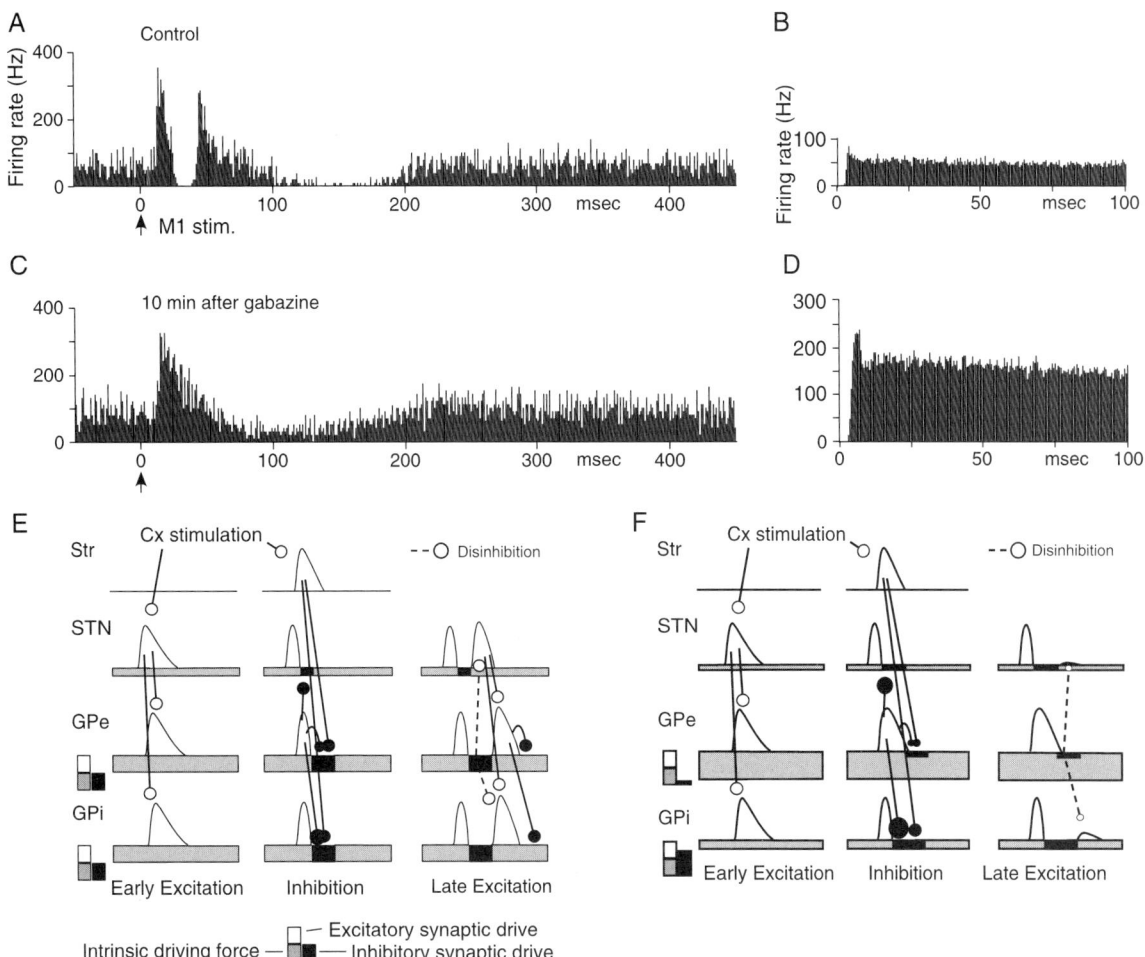

Fig. 6. Effects of local injection of gabazine on a GP$_e$ neuron in a monkey. (A) and (B) A peri-stimulus time histogram (PSTH) shows responses to primary motor cortex (M1) stimulation, and an autocorrelogram shows firing pattern in the control. (C) and (D) Responses obtained 10 min after local gabazine injection (1 mM in saline, 0.2 μl). Note that gabazine greatly increased the spontaneous firing rate and also diminished the inhibition and late excitation. (E) and (F) The diagrams show the time sequence and the major driving forces that evoke a sequence of early excitation, inhibition, and late excitation in pallidal neurons to Cx stimulation in control (E) and after gabazine (F). The early excitation of the GP$_e$ and GP$_i$ is due to activation of the fast STN-mediated pathway. The inhibition is due to activations of the GP$_e$ and the slow Str-mediated pathway. The late excitation is due to the late activation of the STN (i.e., the indirect pathway) and the disinhibition of the GP$_e$ inputs. The thickness of base bars represents the level of spontaneous firing that is set by the sum of the strengths of the glutamatergic excitatory component (white column), the GABAergic inhibitory component (black column), and others such as intrinsic driving forces (gray column).

Local application of gabazine (1 mM in saline up to 0.2 μl) abolished single cortical or putamen stimulation-induced inhibition of GP$_e$ neurons recorded in monkeys. Gabazine also increased the rate and the regularity of firing of most GP$_e$ neurons (Fig. 6) (Kita et al., 2004). However, after gabazine application, the spontaneous pauses disappeared in some neurons while they became more apparent in others, thus suggesting that at least in some cells, the pauses were not due to GABAergic inhibition. It is likely that the increase in the level and regularity of the spontaneous firing by local gabazine is due, at least in part, to blockade of relatively high-frequency spontaneous

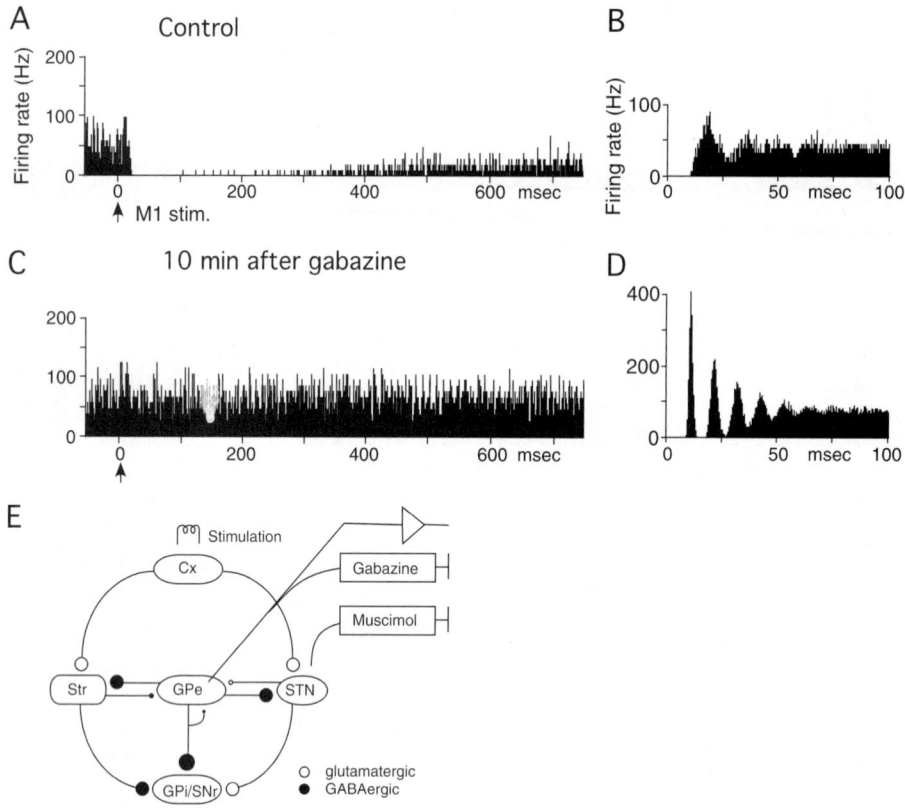

Fig. 7. (A) A PSTH shows a long inhibitory response of a GP$_e$ neuron to M1 stimulation observed after muscimol (0.5 μg/μl, 1.0 μl) injection in the STN. (C) Local injection of gabazine (1 mM, 0.2 μl) abolished the inhibition to M1 stimulation. (B) and (D) Auto-correlograms show that gabazine greatly regularized the firing. (E) A diagram shows the experiment setup.

postsynaptic GABA$_A$ receptor (GABA$_A$-R) mediated inhibition that stemmed from local collaterals of GP$_e$ projection neurons. However, it is also likely that extrasynaptic GABA$_A$-Rs that were activated continuously by ambient GABA were blocked by the high concentration of gabazine we used. Microdialysis studies reported that the ambient level of GABA in the GP$_e$ is relatively high, approximately 0.5 μM (Robertson et al., 1991; Galvan et al., 2005), and may be sufficient to stimulate extrasynaptic GABA$_A$-Rs that exist on the dendrites of monkey GP$_e$ neurons (Charara et al., 2005). It has been shown in other brain areas that some extrasynaptic GABA$_A$-Rs have different subunit compositions to postsynaptic receptors, have a high affinity for GABA, and have greatly reduced desensitization (Saxena and Macdonald, 1994), although the existence of

such receptors has not yet been demonstrated in the GP$_e$.

Muscimol blockade of the STN greatly decreased the firing rate of GP$_e$ neurons, in some cases completely silencing them. However, 5–10 min after the muscimol injection, the activity began to increase with repeated occurrences of short, grouped spike discharges. As time progressed, the activity further increased and developed into repeated occurrences of 2–12 s of a very high frequency active phase followed by a completely silent period with a similar duration (Fig. 8) (Nambu et al., 2000; Kita et al., 2004). In most neurons, the intervals between active phases were fairly regular, as shown in the auto-correlograms, with clearly identifiable multiple peaks, although the duration of each active and silent phase was somewhat variable. In the GP$_e$ neurons with alternately occurring active and silent

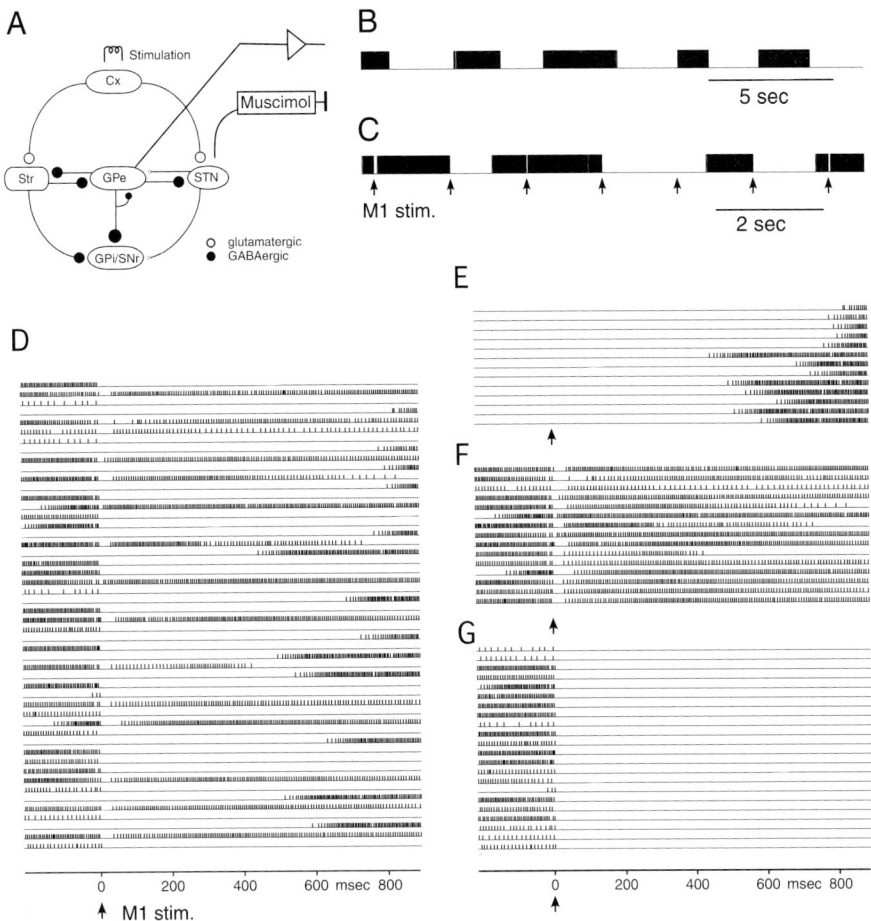

Fig. 8. The responses of a GP$_e$ neuron to stimulation of the M1 recorded after muscimol (0.5 μg/μl, 1.0 μl) injection in the STN. (A) A diagram shows the experimental setup. (B) Digitized spike traces in a slow-sweep speed show the repeated occurrence of active and silent phases in approximately 3 s intervals. (C) A slow-sweep recording shows that M1 stimulation (marked by arrows) induced different responses depending on the timing of stimulus in the slow oscillation. (D) Fast-sweep digitized unit responses to 50 consecutive stimulations. (E–G) The responses shown in (D) can be rearranged to be grouped into three basic types: ones applied during the silent phase that evoked spiking after a 400–800 ms delay (E), ones having a short-duration (less than 100 ms) inhibition (F), and ones terminating the spiking activity (G).

phases, cortical stimulation induced different responses depending on the phase of the firing at the time of stimulation (Fig. 8). Stimulation applied during a silent phase triggered an active phase approximately 400–800 ms after the stimulation. Stimulation applied during an active phase induced different responses depending on the intensity and timing of the stimulation. Low-intensity stimulation and stimulation applied during an early part of the active phase often induced a short inhibition. Conversely, strong stimulation and stimulation

applied during a late part of the active phase often terminated the active phase.

After muscimol blockade of the STN, local application of gabazine greatly increased the firing rate and the regularity of the interspike intervals (Fig. 7(D)). Gabazine also totally abolished single cortical or striatal stimulation-induced inhibition (Fig. 7(C)). In addition, gabazine eliminated the silent phases in most, but not all, GP$_e$ neurons. Our recent in vitro study suggested that long active and silent phases could be induced in GP$_e$ neurons by depolarizing the dendrites by activating Na^+ currents and, at the same time, hyperpolarizing the somata (Hashimoto and Kita, 2006). If this is true, a loss of tonic glutamatergic inputs and blockade of GABAergic inputs shifts the membrane in and out of the voltage range that is suitable for generation of the active and silent phases, and strong GABAergic inputs may also function as a trigger of the silent phases.

GABA$_B$-R mediated responses

Immunohistochemical and in situ hybridization studies showed expression of GABA$_B$-R1 and GABA$_B$-R2 subunits in the GP$_e$ (Charara et al., 2000, 2004; Smith et al., 2000; Chen et al., 2004; Waldvogel et al., 2004). Electron microscopic studies localized GABA$_B$-Rs mainly on extrasynaptic membranes and some on presynaptic membranes of GABAergic and glutamatergic boutons (Charara et al., 2000, 2005; Chen et al., 2004). We have studied properties of synaptically-induced GABA$_B$-R mediated responses in the GP$_e$ (Kaneda and Kita, 2005). Repetitive local stimulation (20 pulses at 50 Hz) induced fast EPSPs with or without accompanying action potentials followed by a slow IPSP in GP$_e$ neurons in slice preparations (Fig. 9). Application of gabazine (10 μM) significantly increased the amplitude of the fast EPSPs and increased the number of the action potentials triggered by the EPSPs,

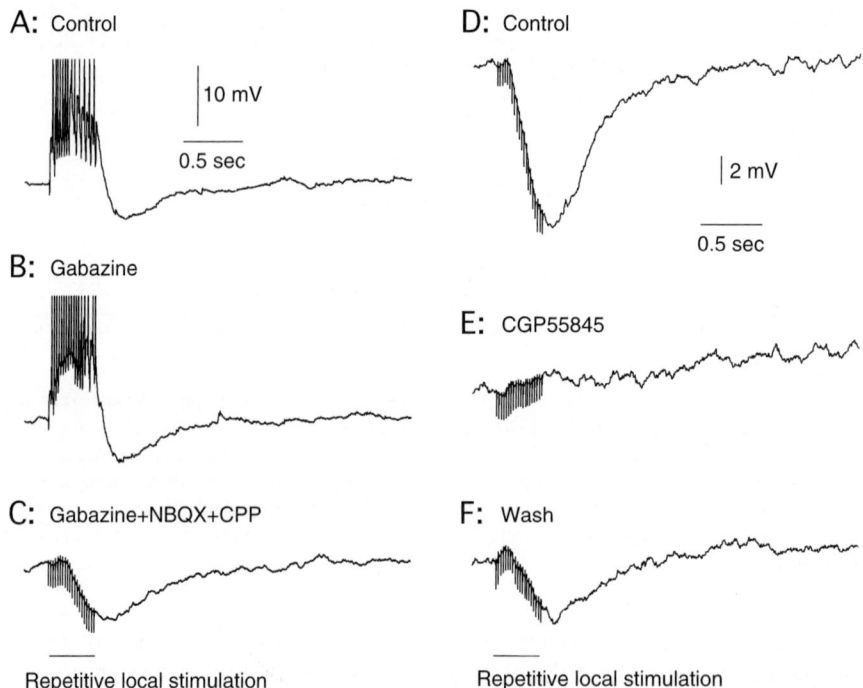

Fig. 9. (A–C) Pharmacological isolation of slow IPSPs evoked in a GP$_e$ neuron in a slice preparation. (A) Repetitive local stimulation (20 pulses at 50 Hz, thick bars) evoked fast EPSPs with several action potentials followed by a slow IPSP in a GP$_e$ neuron current-clamped at approximately −70 mV in standard ACSF. (B) Bath application of gabazine (10 μM) increased the amplitude of the EPSPs and the number of action potentials triggered by the EPSPs. Gabazine did not change the amplitude of the slow IPSP in this neuron. (C) Addition of NBQX (10 μM) and CPP (30 μM) completely suppressed the EPSPs and, at the same time, shortened the onset and increased the duration of the slow IPSP. (D—F) Recording from another GP$_e$ neuron. The ACSF contained NBQX (10 μM), CPP (30 μM), gabazine (10 μM), and the mGluR1 antagonist LY367385 (50 μM). The neuron was current-clamped at approximately −70 mV. (D) Repetitive local stimulation induced a slow IPSP in control conditions. (E) Bath application of CGP55845 (0.3 μM) blocked the slow IPSP. (F) A partial recovery from the CGP55845 effect by washing. The action potentials and stimulus artifacts were truncated.

123

suggesting that the EPSPs were overlapped by GABAergic IPSPs. Gabazine did not change the amplitude of the slow IPSPs. Addition of the ionotropic glutamate receptor blockers NBQX (10 μM) and CPP (30 μM) abolished the fast EPSPs, resulting in a faster onset and an enhancement of the amplitude of the slow IPSPs (Fig. 9(C)). These results suggest that either the fast EPSPs partially overlapped with the slow IPSPs and decreased their amplitude or that activation of ionotropic glutamate receptors at the synaptic terminals was suppressing GABA release (Rodriguez-Moreno et al., 1997). Thus, the slow IPSP induction was attributable to a direct activation of GABAergic axons that did not require glutamatergic activation of other intermediate neurons. The slow IPSPs were considered to be postsynaptic GABA_B-R mediated responses because bath application of the GABA_B-R antagonists, CGP55845 or CGP52432, blocked the responses (Fig. 9(E)) and because the reversal potential of the slow IPSPs was close to the potassium

equilibrium potential, similar to GABA_B-R mediated responses in other brain areas (Kaneda and Kita, 2005).

We tested whether striatal-GP_e inputs can evoke GABA_B-R mediated responses in monkeys. Repetitive electrical stimulation of the putamen through a concentric bipolar electrode induced repeated occurrences of GABA_A-R mediated inhibition and glutamatergic excitation during stimulation that were followed by another excitation in most of GP_e neurons in control conditions (Fig. 10(A-a)). GABA_B-like slow inhibitions were observed only in a small number of neurons. However, when gabazine was applied to block GABA_A responses, most of the neurons exhibited a slow inhibition after the termination of the repetitive stimulation (Fig. 10(A-b)). This may be because GABA_A responses may have effectively shunted the slow inhibition. Gabazine increased the firing rate (i.e., depolarized neurons) and made the slow inhibition easy to detect in unit recordings. Another

A-a: GPe neuron Control
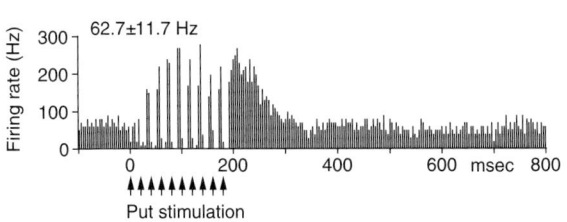

B-a: A GPe neuron after local gabazine
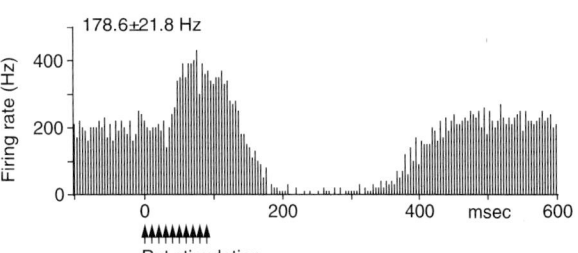

A-b: After local gabazine
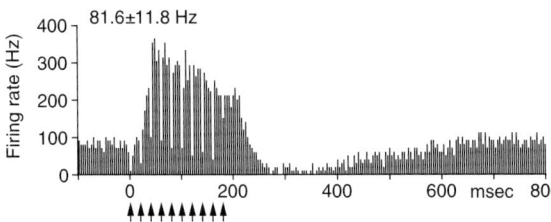

B-b: After local gabazine and CGP55845
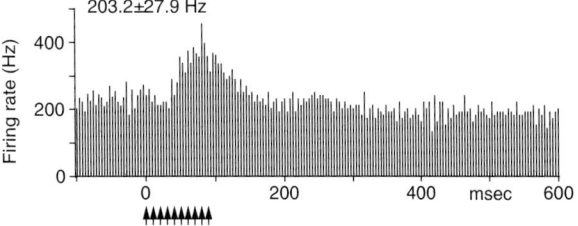

Fig. 10. (A) Responses of a GP_e neuron to repetitive putamen stimulation recorded before and after local application of gabazine (0.2 mM, 0.2 μl) recorded in awake monkey. (A-a) Repetitive stimulation induced repeated occurrences of inhibitions and excitations during stimulation that were followed by another excitation. (A-b) Application of gabazine disclosed slow inhibition. (B) Effects of CGP55845 on a GP_e neuron recorded after local application of gabazine (0.2 mM, 0.2 μl). (B-a) Repetitive putamen stimulation induced excitation during stimulation that was followed by a strong slow inhibition. (B-b) Local injection of CGP55845 abolished the slow inhibition. CGP55845 also slightly increased the spontaneous firing rate of this neuron. PSTHs were constructed with 4 ms bins, and the mean ± S.D. of the prestimulus firing frequency are indicated at the top left of each PSTH.

possibility applicable to the GP_e is that gabazine increased the firing during repetitive stimulation, thus evoking stronger slow inhibition by local axon collaterals. The latter possibility is based on the assumption that GP_e activation is the source of the slow inhibition. Direct postsynaptic interactions between $GABA_A$ and $GABA_B$ are unlikely because our previous in vitro study showed no significant direct gabazine effects on the amplitude of local repetitive stimulation-induced $GABA_B$ responses in rat GP_e (Kaneda and Kita, 2005). In approximately half of the GP_e neurons, the $GABA_B$-R blocker, CGP55845, eliminated the slow inhibition, suggesting that it was mediated by $GABA_B$-R and other, yet undetermined receptors (Fig. 10(B)).

We also found that excitation during repetitive stimulation was required to evoke the slow inhibition. For instance, the blockade of the STN eliminated not only excitation to repetitive putamen stimulation but also the slow CGP55845-sensitive inhibition, while the gabazine-sensitive responses remained. Thus, the GP_e neurons that form GABAergic synapses in the GP_e effectively evoke $GABA_B$-R mediated responses but synapses from putamen axons do not. Receptor localization studies suggest both striatal and local collateral synapses may evoke $GABA_B$ responses (Chen et al., 2004). However, at this point, there are no reports of striatal axon-induced $GABA_B$-R mediated postsynaptic inhibition in the GP_e. Some possibilities for the failure of putamen axons to evoke an appreciable $GABA_B$ response are that the putamen axon terminals may have a small GABA release probability or, conversely, that the GABA released from putamen terminals is taken up before reaching the receptors. Another factor may be that the distal dendritic location of the putamen axon synapses, as opposed to the proximal dendritic location of GP_e axon synapses, greatly dampens the response recorded at the soma.

Application of the $GABA_B$-R antagonist, CGP55845, to slice preparations does not change the firing activity of GP_e neurons (Kaneda and Kita, 2005). However, in monkey unit recordings, local application of CGP55845 did increase the firing rate of GP_e neurons (Galvan et al., 2005; Kita et al., 2006). CGP55845 did not alter firing patterns such as pauses and bursting of GP_e

neurons, although the duration of pauses was decreased as the firing frequency increased (Galvan et al., 2005; Kita et al., submitted). The long pauses in the GP_e observed after muscimol blockade of the STN were also not abolished by CGP55845. The anatomical and physiological observations suggest that GABA released mainly from local collateral axon terminals maintain a relatively high ambient level of GABA, approximately $0.5\,\mu M$, that is enough to activate extrasynaptic $GABA_B$-Rs in GP_e in vivo.

From this and previous in vivo and in vitro findings, it can be concluded that $GABA_B$-Rs at GP_e innervation sites, including the GP_e, GP_i, and the STN, are activated by GABA released from GP_e axons and that they contribute significantly to the feedback and feedforward control of the neurons in the target structures (Hallworth and Bevan, 2005; Galvan et al., 2005; Kaneda and Kita, 2005). It has been reported that the expression of $GABA_B$-Rs in the pallidum and STN was altered in patients and in animal models of Parkinson's disease (Calon et al., 2003; Johnston and Duty, 2003).

Presynaptic modulation

The synaptic release of GABA in the GP_e is controlled by several types of presynaptic receptor, including $GABA_B$-R, mGluR, opioids, adenosine, and dopamine. The following is a short summary of recent findings.

Presynaptic $GABA_B$ autoreceptors

Our recent in vitro study revealed a presynaptic $GABA_B$ action. Presynaptic $GABA_B$-R mediated effects were studied under conditions that minimized the induction of postsynaptic $GABA_B$ responses. The neurons recorded were filled with an electrolyte containing high Cl^- and QX-314 and superfused with artificial cerebrospinal fluid (ACSF) containing the ionotropic glutamate receptor blockers, NBQX and CPP. QX-314 is known to inhibit the postsynaptic $GABA_B$ response as well as the sodium current (Nathan et al., 1990; Andrade, 1991; McLean et al., 1996). Under these conditions, repetitive local stimulation (20 pulses at $50\,Hz$)

125

evoked gabazine-sensitive IPSCs (Fig. 11). As shown in Fig. 11, the repetitive stimulation-induced IPSCs were summated and the onset point of each stimulus pulse-coupled IPSC deviated from the base line, especially during the latter part of the repetitive stimulation. Bath application of CGP55845 significantly increased the amplitude of IPSCs from the fourth to the last stimulation (Fig. 11). This result suggested that GABA$_B$ auto-receptors inhibit GABA release from GABAergic

terminals in GP$_e$. To confirm this, we examined the effects of the GABA$_B$-R agonist, baclofen, on TTX-insensitive mIPSCs. mIPSCs were recorded in the presence of NBQX, CPP, and TTX (1 μM) using pipettes containing the high Cl$^-$ and QX-314. Bath application of baclofen significantly reduced the mIPSCs frequency without altering the mean amplitude or the amplitude distribution of the mIPSCs. Baclofen also decreased the amplitude of stimulation-induced IPSCs in the GP$_e$, an effect

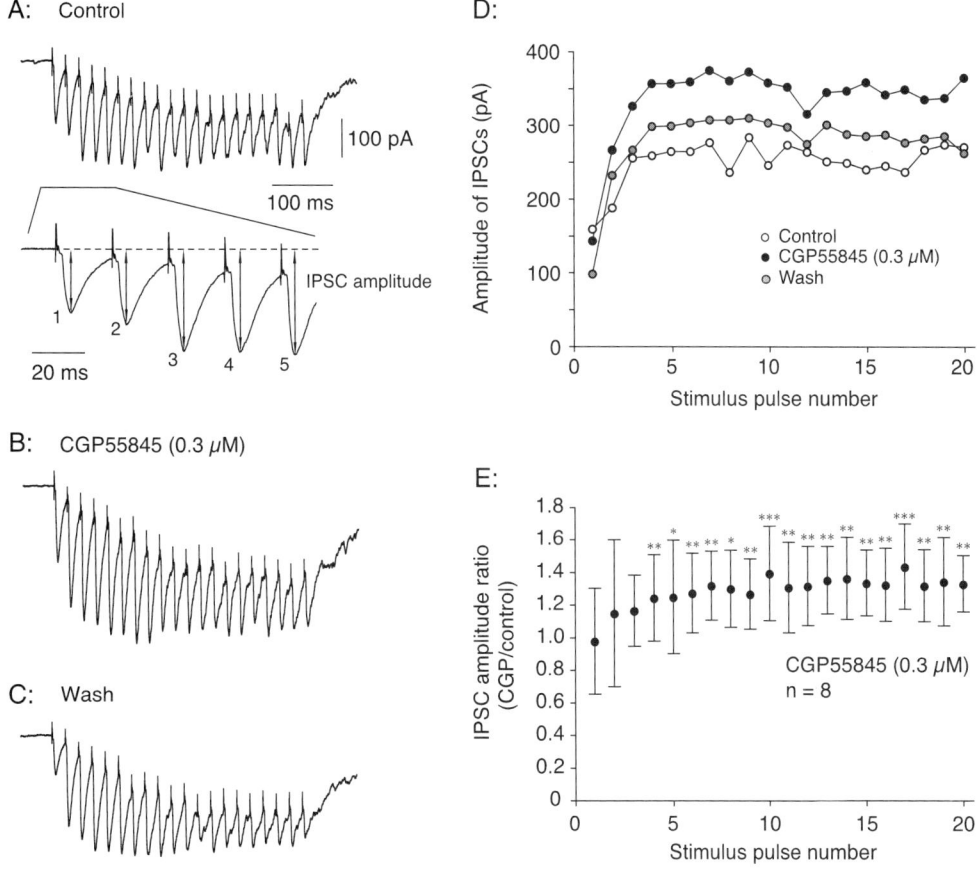

Fig. 11. Blockade of presynaptic GABA$_B$-Rs increased the amplitudes of GABAergic IPSCs. In a GP$_e$ neuron in a slice preparation. Pipettes contained high Cl$^-$ and QX-314 (3 mM). QX-314 was included to block postsynaptic GABA$_B$ responses and sodium-dependent action potentials. Neurons were voltage-clamped at −80 mV. The ACSF contained NBQX (10 μM) and CPP (30 μM) in control conditions. (A-c) Repetitive local stimulation (20 pulses at 50 Hz) evoked IPSCs before, during, and after bath application of 0.3 μM CGP55845. The traces shown are averages of two successive trials separated by 3 min. An expanded trace of the control record in (A) shows the method of IPSC amplitude measurements. (D) Plots of IPSC amplitude versus the stimulus pulse sequence for the neuron shown in (A-c). CGP55845 reversibly increased the amplitudes of the IPSCs. (E) Plots of IPSC amplitude ratios for eight GP$_e$ neurons tested with 0.3 μM CGP55845. The ratio was determined by dividing the IPSC amplitude in the presence of CGP55845 by the amplitude of the corresponding control. The IPSCs evoked by the fourth to the last stimulation were significantly enhanced by CGP55845 ($^*P<0.05$, $^{**}P<0.01$, $^{***}P<0.001$; paired t-test).

blocked by CGP55845 (Chen and Yung., 2003; Charara et al. 1999; Kaneda and Kita, 2005). Thus, these results are consistent with the suggestion that activation of the presynaptic GABA$_B$ receptors reduces the release of GABA. In the GP$_e$ in vivo, the regulation of GABA release may be achieved not only by autoregulation of individual terminals but also by relatively high levels of ambient GABA.

Metabotropic glutamate receptors

Glutamatergic inputs to the GP$_e$ modulate GABAergic synaptic transmissions in the GP$_e$. Both striatal-pallidal and local axon collateral terminals have group III mGluRs (Bradley et al., 1999; Corti et al., 2002). Activation of these receptors by exogenous agonists reduced the frequency of mIPSCs and stimulation-induced GABA$_A$-R mediated IPSCs (Marino et al., 2003; Matsui and Kita, 2003). Thus, an increase in the subthalamo-GP$_e$ inputs not only postsynaptically excites GP$_e$ neurons but also presynaptically inhibits GABAergic inputs. Electron microscopy showed that GABAergic axon terminals that are probably striatal in origin express kainate receptors, suggesting possible presynaptic modulation of GABA release by kainate receptors (Charara et al., 1999; Kane-Jackson and Smith, 2003) although their physiological nature remains to be shown.

Opioid receptors

Opioids contained in the striatal-GP$_e$ axons might also presynaptically inhibit GABA release not only from the striatal axon terminals but also from local collateral axon terminals. GABAergic terminals in the GP$_e$ express mu, delta, and kappa opioid receptors (Morris and Herz, 1986; Mansour et al., 1988; Sharif and Hughes, 1989). Application of kappa- or mu-opioid receptor agonists greatly reduced the amplitude of postsynaptic GABAergic responses evoked by both striatal-GP$_e$ and local collateral terminals (Ogura and Kita, 2000). Application of mu- or delta-opioid receptor agonists exerted similar effects in the GP$_e$ (Stanford and Cooper, 1999).

Adenosine and D2 receptors

Another interesting presynaptic modulator is adenosine. mRNA for the adenosine A2A receptor is expressed in striatal-GP$_e$ neurons (Augood and Emson, 1994; Shindou et al., 2003). The increase in extracellular GABA levels in the Str and GP$_e$ observed after a lesion of nigrostriatal dopamine projections was blocked by systemic injection of adenosine A2A antagonist (Ochi et al., 2000). Activation of adenosine A2A receptors increased the electrically stimulated release of GABA (Mayfield et al., 1993) and the striatal stimulation-induced IPSCs in the GP$_e$ via a cyclic AMP-dependent mechanism (Shindou et al., 2003). In vitro studies showed that dopamine D2 receptor activation suppresses IPSCs of GP$_e$ neurons to striatal stimulation (Cooper and Stanford, 2001) and electrical or high K$^+$ stimulation-induced GABA release in the GP$_e$ (Dayne Mayfield et al., 1996; Floran et al., 2005). Recent studies suggested functional interactions between A2A and D2 receptors such that the GABA release caused by activation of A2A receptors was blocked when D2 receptors were blocked. In turn, activation of A2A receptors decreased the D2 suppression of GABA release (Floran et al., 2005). In other studies in animals and patients of Parkinson's disease, adenosine A2A antagonists significantly enhanced behavioral or neuropharmacological effects of L-dopa or D2 agonists (Kanda et al., 2000; Kase et al., 2003; Tanganelli et al., 2004; Floran et al., 2005). The mechanisms for the interactions are unclear but an interesting possibility is that A2A and D2 receptors form heteromers (Ferre et al., 2004). Activation of adenosine A1 receptors on GP$_e$ axons may suppress GABAergic transmission, an effect opposite to A2A action, because IPSCs evoked in STN neurons were greatly suppressed by activation of A1 receptors (Shen and Johnson, 2003).

Cannabinoid receptors

The striatal-GP$_e$ axons express abundant cannabinoid CB1 receptors (Sanudo-Pena et al., 1998). A recent ultrastructural study showed that preterminal portions of axons close to terminal boutons exhibited more intense staining than the terminal boutons (Matyas et al., 2006). In several brain areas, CB1 receptors are involved in the depolarization-induced suppression of inhibition (or excitation), in which strong depolarization of the postsynaptic neuron is followed by endocannabinoid-mediated activation of presynaptic CB1 receptors, which suppresses GABA or glutamate release. Recent in vitro studies reported the depolarization-induced suppression of both inhibition and excitation in the GP$_e$ (Engler et al., 2006; Freiman and Szabo, 2005). A surprising observation reported recently was that the CB1 agonist, WIN55,212-2, as well as its antagonist, AM251, could block action potential generation through a CB1-independent mechanism (Matyas et al., 2006). Thus, the use of these drugs needs to be undertaken with some caution.

It has been suggested that a deficiency in endocannabinoid transmission may contribute to levodopa-induced dyskinesias. In animal models of Parkinson's disease, the level of CB1 receptor mRNA was decreased in the Str and GP$_e$ (Hurley et al., 2003) while the level of endocannabinoids in the GP$_e$ was elevated (Di Marzo et al., 2000). Stimulation of cannabinoid receptors reduces L-dopa induced dyskinesia in the MPTP-lesioned primate model of Parkinson's disease (Fox et al., 2002; Ferrer et al., 2003). CB1 receptors are also considered to be involved in modulation of GABA uptake (Maneuf et al., 1994; Di Marzo et al., 1998; Venderova et al., 2005), although details are not known.

Postsynaptic modulations

There are not many postsynaptic modulations that are specific to GABAergic transmission. Activation of postsynaptic dopamine D4 receptors reduced GABAergic currents through the suppression of PKA activity (Shin et al., 2003). Similar to other brain areas, activation of benzodiazepine sites of GABA$_A$-Rs significantly prolonged the half-decay time of both miniature and spontaneous IPSCs (Chen et al., 2004). The GP$_e$ has abundant benzodiazepine binding sites

(Somogyi et al., 1996) and their density is reduced in animal models and patients of Parkinson's disease (Griffiths et al., 1990; Chadha et al., 2000).

Possible modulations by ambient GABA

The ambient GABA level in the GP$_e$ is relatively high as mentioned above. The level of GABA, of course, is controlled by its release and uptake. The release is most likely to be synaptic. Thus, the increase in the GABA level in the GP$_e$ that has been reported in animal models of Parkinson's disease is considered to be an indication of increased striatal-GP$_e$ activity (Robertson et al., 1991; Ochi et al., 2000; Schroeder and Schneider, 2002). However, non-synaptic reversal of transporters may also occur under some conditions such as elevation of intracellular Na^{2+} concentration during very high frequency activation of GP$_e$ neurons (Richerson and Wu, 2003).

GP$_e$ activity in patients and animal models of Parkinson's disease

According to the theory that unbalanced activity of the direct and indirect pathways underlies the development of Parkinson's disease, the activity of the striatal-GP$_e$ GABAergic input, the initial stage of the indirect pathway, is increased after the loss of nigrostriatal dopaminergic projections. A number of physiological observations made in patients undergoing surgical treatments and in animal models of Parkinson's disease were consistent with this theory. Bursting activity in the GP$_e$ increased with or without a decrease in the mean firing rate (Filion et al., 1991; Hutchison et al., 1994; Rothblat and Schneider, 1995; Boraud et al., 1998; Magill et al., 2001). Responses of GP$_e$ neurons to joint stimulation became less selective, responding to single joint movement in normal conditions but responding to multiple joints after being rendered parkinsonian (Filion et al., 1988; Rothblat and Schneider, 1995). Responses of GP$_e$ neurons to striatal stimulation increased not only the response amplitude but also the number of responsive areas such that many GP$_e$ neurons showed convergent responses to both the caudate nucleus and the putamen (Tremblay et al., 1989). The ambient GABA level in the GP_e is increased (Robertson et al., 1991; Ochi et al., 2000; Schroeder and Schneider, 2002). Drugs that reduce striatal-GP_e GABAergic IPSPs, such as A2A antagonists (Mori and Shindou, 2003) and bicuculline (Maneuf et al., 1994), have anti-parkinsonian effects.

The other change observed in the GP_e of parkinsonian subjects was the occurrence of synchronized or correlated oscillatory neuronal firing, which may be attributable to an increase in direct synaptic connections or common collateral afferent inputs as mentioned above. In MPTP treated monkeys, correlated firing was observed not only between nearby pallidal neurons and between tonically active pallidal and striatal interneurons, but also between neurons in different functional territories, suggesting a breakdown of territory-independent processing (Nini et al., 1995; Bergman et al., 1998; Raz et al., 2001). However, in recordings performed in parkinsonian patients, oscillatory synchronization between pairs of GP_i or GP_e neurons was found only in patients with limb tremor, suggesting that overt neuronal synchronization was not present in the basal ganglia of patients with Parkinson's disease (Levy et al., 2002).

Recent observations from parkinsonian patients showed that local field potentials recorded from the pallidum by deep brain stimulation electrodes exhibited stronger power in the 10–30 Hz band that could be decreased by dopaminomimetics (Brown et al., 2001; Silberstein et al., 2003, 2005; Brown and Williams, 2005). Although it is not totally clear whether the local field potential depicts synchronized firing of neurons or other such synaptic activities, the observations suggest that some significant physiological changes take place in the GP_e of parkinsonian subjects.

Many of the observations are consistent with the suggestion that the activity of the indirect pathway is increased in parkinsonian subjects. However, the locus of the changes is uncertain at this time. The changes can occur in the cortex, Str, or in the GP_e. Indeed, increasing evidence suggests significant alterations occur in the STN and the cerebral cortex in patients and in animal models of Parkinson's disease (Vila et al., 1996; Magill et al., 2001; Bevan et al., 2002; Goldberg et al., 2003).

Conclusions

The GP_e is rich in GABAergic synapses that exert significant effects in controlling the firing rate and pattern of GP_e neurons. The GP_e projects to many nuclei in the basal ganglia. Thus, GABAergic controls in the GP_e, in turn, have a major role in the signal processing function in the basal ganglia. Most of the observations reviewed in this chapter were made during the last 10 years. It has become evident that GABAergic transmission in the GP_e is controlled by GABAergic and non-GABAergic inputs and also by non-synaptic factors. The understanding the role of the GP_e in basal ganglia functions, including motor control, requires further study.

Acknowledgments

I thank Drs. Satomi Chiken, Kenji Hashimoto, Katsuyuki Kaneda, Takako Kita, Toshihiro Matsui, Atsushi Nambu, Mitsuhiro Ogura, Yoshihisa Tachibana, Masahiko Takada, and Katsushige Watanabe for their hard work to obtain data presented in this chapter and Ms. Dawn Merrick for editing the manuscript. This chapter was prepared under the support of NIH grants NS-42762 and NS-47085.

References

Andrade, R. (1991) Blockade of neurotransmitter-activated K^+ conductance by QX-314 in the rat hippocampus. Eur. J. Pharmacol., 199: 259–262.

Augood, S.J. and Emson, P.C. (1994) Adenosine A2a receptor mRNA is expressed by enkephalin cells but not by somatostatin cells in rat striatum: a co-expression study. Brain Res. Mol. Brain Res., 22: 204–210.

Berendse, H.W., Graff, Y.G.d. and Groenewegen, H.J. (1992) Topographical organization and relationship with ventral striatal compartments of prefrontal corticostriatal projections in the rat. J. Comp. Neurol., 316: 314–347.

Bergman, H., Feingold, A., Nini, A., Raz, A., Slovin, H., Abeles, M. and Vaadia, E. (1998) Physiological aspects of information processing in the basal ganglia of normal and parkinsonian primates. Trends Neurosci., 21: 32–38.

Bevan, M.D., Booth, P.A., Eaton, S.A. and Bolam, J.P. (1998) Selective innervation of neostriatal interneurons by a subclass of neuron in the globus pallidus of the rat. J. Neurosci., 18: 9438–9452.

Bevan, M.D., Magill, P.J., Terman, D., Bolam, J.P. and Wilson, C.J. (2002) Move to the rhythm: oscillations in the subthalamic nucleus-external globus pallidus network. Trends Neurosci., 25: 525–531.

Boraud, T., Bezard, E., Guehl, D., Bioulac, B. and Gross, C. (1998) Effects of L-DOPA on neuronal activity of the globus pallidus externalis (GPe) and globus pallidus internalis (GPi) in the MPTP-treated monkey. Brain Res., 787: 157–160.

Bradley, S.R., Standaert, D.G., Levey, A.I. and Conn, P.J. (1999) Distribution of group III mGluRs in rat basal ganglia with subtype-specific antibodies. Ann. N Y Acad. Sci., 868: 531–534.

Brown, P., Oliviero, A., Mazzone, P., Insola, A., Tonali, P. and Di Lazzaro, V. (2001) Dopamine dependency of oscillations between subthalamic nucleus and pallidum in Parkinson's disease. J. Neurosci., 21: 1033–1038.

Brown, P. and Williams, D. (2005) Basal ganglia local field potential activity: character and functional significance in the human. Clin. Neurophysiol., 116: 2510–2519.

Calon, F., Morissette, M., Rajput, A.H., Hornykiewicz, O., Bedard, P.J. and Di Paolo, T. (2003) Changes of GABA receptors and dopamine turnover in the postmortem brains of parkinsonians with levodopa-induced motor complications. Mov. Disord., 18: 241–253.

Chadha, A., Howell, O., Atack, J.R., Sur, C. and Duty, S. (2000) Changes in [^{3}H]zolpidem and [^{3}H]Ro 15-1788 binding in rat globus pallidus and substantia nigra pars reticulata following a nigrostriatal tract lesion. Brain Res., 862: 280–283.

Chang, H.T., Kita, H. and Kitai, S.T. (1983) The fine structure of the rat subthalamic nucleus: an electron microscopic study. J. Comp. Neurol., 221: 113–123.

Chang, H.T., Wilson, C.J. and Kitai, S.T. (1981) Single neostriatal efferent axons in the globus pallidus: a light and electron microscopic study. Science, 213: 915–918.

Charara, A., Blankstein, E. and Smith, Y. (1999) Presynaptic kainate receptors in the monkey striatum. Neuroscience, 91: 1195–1200.

Charara, A., Galvan, A., Kuwajima, M., Hall, R.A. and Smith, Y. (2004) An electron microscope immunocytochemical study of GABA(B) R2 receptors in the monkey basal ganglia: a comparative analysis with GABA(B) R1 receptor distribution. J. Comp. Neurol., 476: 65–79.

Charara, A., Heilman, T.C., Levey, A.I. and Smith, Y. (2000) Pre- and postsynaptic localization of GABA(B) receptors in the basal ganglia in monkeys. Neuroscience, 95: 127–140.

Charara, A., Pare, J.F., Levey, A.I. and Smith, Y. (2005) Synaptic and extrasynaptic GABA-A and GABA-B receptors in the globus pallidus: an electron microscopic immunogold analysis in monkeys. Neuroscience, 131: 917–933.

Chen, L., Savio Chan, C. and Yung, W.H. (2004) Electrophysiological and behavioral effects of zolpidem in rat globus pallidus. Exp. Neurol., 186: 212–220.

Chen, L. and Yung, W.H. (2003) Effects of the GABA-uptake inhibitor tiagabine in rat globus pallidus. Exp. Brain Res., 152: 263–269.

Cooper, A.J. and Stanford, I.M. (2000) Electrophysiological and morphological characteristics of three subtypes of rat globus pallidus neurone in vitro. J. Physiol., 527(Pt 2): 291–304.

Cooper, A.J. and Stanford, I.M. (2001) Dopamine D2 receptor mediated presynaptic inhibition of striatopallidal GABA(A) IPSCs in vitro. Neuropharmacology, 41: 62–71.

Cooper, A.J. and Stanford, I.M. (2002) Calbindin D-28k positive projection neurones and calretinin positive interneurones of the rat globus pallidus. Brain Res., 929: 243–251.

Corti, C., Aldegheri, L., Somogyi, P. and Ferraguti, F. (2002) Distribution and synaptic localisation of the metabotropic glutamate receptor 4 (mGluR4) in the rodent CNS. Neuroscience, 110: 403–420.

Dayne Mayfield, R., Larson, G., Orona, R.A. and Zahniser, N.R. (1996) Opposing actions of adenosine A2a and dopamine D2 receptor activation on GABA release in the basal ganglia: evidence for an A2a/D2 receptor interaction in globus pallidus. Synapse, 22: 132–138.

DeLong, M.R. (1971) Activity of pallidal neurons during movement. J. Neurophysiol., 34: 414–427.

Deschenes, M., Bourassa, J., Doan, V.D. and Parent, A. (1996) A single-cell study of the axonal projections arising from the posterior intralaminar thalamic nuclei in the rat. Eur. J. Neurosci., 8: 329–343.

Di Marzo, V., Hill, M.P., Bisogno, T., Crossman, A.R. and Brotchie, J.M. (2000) Enhanced levels of endogenous cannabinoids in the globus pallidus are associated with a reduction in movement in an animal model of Parkinson's disease. FASEB. J., 14: 1432–1438.

Di Marzo, V., Melck, D., Bisogno, T. and De Petrocellis, L. (1998) Endocannabinoids: endogenous cannabinoid receptor ligands with neuromodulatory action. Trends Neurosci., 21: 521–528.

DiFiglia, M., Pasik, P. and Pasik, T. (1982) A Golgi and ultrastructural study of the monkey globus pallidus. J. Comp. Neurol., 212: 53–75.

Donoghue, J.P. and Herkenham, M. (1986) Neostriatal projections from individual cortical fields conform to histochemically distinct striatal compartments in the rat. BRES, 365: 397–403.

Engler, B., Freiman, I., Urbanski, M. and Szabo, B. (2006) Effects of exogenous and endogenous cannabinoids on GABAergic neurotransmission between the caudate-putamen and the globus pallidus in the mouse. J. Pharmacol. Exp. Ther., 316: 608–617.

Falls, W.M., Park, M.R. and Kitai, S.T. (1983) An intracellular HRP study of the rat globus pallidus. II. Fine structural characteristics and synaptic connnections of medially located large GP neurons. J. Comp. Neurol., 220: 229–245.

Ferre, S., Ciruela, F., Canals, M., Marcellino, D., Burgueno, J., Casado, V., Hillion, J., Torvinen, M., Fanelli, F., Benedetti Pd, P., Goldberg, S.R., Bouvier, M., Fuxe, K., Agnati, L.F., Lluis, C., Franco, R. and Woods, A. (2004) Adenosine A2A-dopamine D2 receptor-receptor heteromers. Targets for neuro-psychiatric disorders. Parkinsonism Relat. Disord., 10: 265–271.

Ferrer, B., Asbrock, N., Kathuria, S., Piomelli, D. and Giuffrida, A. (2003) Effects of levodopa on endocannabinoid levels in rat basal ganglia: implications for the treatment of levodopa-induced dyskinesias. Eur. J. Neurosci., 18: 1607–1614.

Filion, M., Tremblay, L. and Bedard, P.J. (1988) Abnormal influences of passive limb movement on the activity of globus pallidus neurons in parkinsonian monkeys. Brain Res., 444: 165–176.

Filion, M., Tremblay, L. and Bedard, P.J. (1991) Effects of dopamine agonists on the spontaneous activity of globus pallidus neurons in monkeys with MPTP-induced parkinsonism. Brain Res., 547: 152–161.

Fink-Jensen, A. and Mikkelsen, J.D. (1991) A direct neuronal projection from the entopeduncular nucleus to the globus pallidus. A PHA-L anterograde tracing study in the rat. Brain Res., 542: 175–179.

Floran, B., Gonzalez, B., Floran, L., Erlij, D. and Aceves, J. (2005) Interactions between adenosine A(2a) and dopamine D2 receptors in the control of [(3)H]GABA release in the globus pallidus of the rat. Eur. J. Pharmacol., 520: 43–50.

Fortin, M. and Parent, A. (1994) Calretinin labels a specific neuronal subpopulation in primate globus pallidus. Neuroreport, 5: 2097–2100.

Fox, S.H., Henry, B., Hill, M., Crossman, A. and Brotchie, J. (2002) Stimulation of cannabinoid receptors reduces levodopa-induced dyskinesia in the MPTP-lesioned nonhuman primate model of Parkinson's disease. Mov. Disord., 17: 1180–1187.

Freiman, I. and Szabo, B. (2005) Cannabinoids depress excitatory neurotransmission between the subthalamic nucleus and the globus pallidus. Neuroscience, 133: 305–313.

Galvan, A., Villalba, R.M., West, S.M., Maidment, N.T., Ackerson, L.C., Smith, Y. and Wichmann, T. (2005) GABAergic modulation of the activity of globus pallidus neurons in primates: in vivo analysis of the functions of GABA receptors and GABA transporters. J. Neurophysiol., 94: 990–1000.

Gardiner, T.W. and Kitai, S.T. (1992) Single-unit activity in the globus pallidus and neostriatum of the rat during performance of a trained head movement. Exp. Brain Res., 88: 517–530.

Gerfen, C.R., Baimbridge, K.G. and Miller, J.J. (1985) The neostriatal mosaic: Compartmental distribution of calcium-binding protein and parvalbumin in the basal ganglia of the rat and monkey. Proc. Natl. Acad. Sci. USA, 82: 8780–8784.

Goldberg, J.A., Kats, S.S. and Jaeger, D. (2003) Globus pallidus discharge is coincident with striatal activity during global slow wave activity in the rat. J. Neurosci., 23: 10058–10063.

Griffiths, P.D., Sambrook, M.A., Perry, R. and Crossman, A.R. (1990) Changes in benzodiazepine and acetylcholine receptors in the globus pallidus in Parkinson's disease. J. Neurol. Sci., 100: 131–136.

Hallworth, N.E. and Bevan, M.D. (2005) Globus pallidus neurons dynamically regulate the activity pattern of subthalamic nucleus neurons through the frequency-dependent activation of postsynaptic GABAA and GABAB receptors. J. Neurosci., 25: 6304–6315.

Hashimoto, K. and Kita, H. (2006) Slow oscillatory activity of rat globus pallidus neurons in vitro. Eur. J. Neurosci., 23: 443–453.

Hazrati, L.N., Parent, A., Mitchell, S. and Haber, S.N. (1990) Evidence for interconnections between the two segments of the globus pallidus in primates: a PHA-L anterograde tracing study. Brain Res., 533: 171–175.

Hontanilla, B., Parent, A. and Gimenez-Amaya, J.M. (1994) Compartmental distribution of parvalbumin and calbindin D-28k in rat globus pallidus. Neuroreport, 5: 2269–2272.

Hoover, B.R. and Marshall, J.F. (2002) Further characterization of preproenkephalin mRNA-containing cells in the rodent globus pallidus. Neuroscience, 111: 111–125.

Hurley, M.J., Mash, D.C. and Jenner, P. (2003) Expression of cannabinoid CB1 receptor mRNA in basal ganglia of normal and parkinsonian human brain. J. Neural. Transm., 110: 1279–1288.

Hutchison, W.D., Lozano, A.M., Davis, K.D., Saint-Cyr, J.A., Lang, A.E. and Dostrovsky, J.O. (1994) Differential neuronal activity in segments of globus pallidus in Parkinson's disease patients. Neuroreport, 5: 1533–1537.

Johnston, T. and Duty, S. (2003) GABA(B) receptor agonists reverse akinesia following intranigral or intracerebroventricular injection in the reserpine-treated rat. Br. J. Pharmacol., 139: 1480–1486.

Kanda, T., Jackson, M.J., Smith, L.A., Pearce, R.K., Nakamura, J., Kase, H., Kuwana, Y. and Jenner, P. (2000) Combined use of the adenosine A(2A) antagonist KW-6002 with L-DOPA or with selective D1 or D2 dopamine agonists increases antiparkinsonian activity but not dyskinesia in MPTP-treated monkeys. Exp. Neurol., 162: 321–327.

Kane-Jackson, R. and Smith, Y. (2003) Pre-synaptic kainate receptors in GABAergic and glutamatergic axon terminals in the monkey globus pallidus. Neuroscience, 122: 285–289.

Kaneda, K. and Kita, H. (2005) Synaptically released GABA activates both pre- and postsynaptic GABA(B) receptors in the rat globus pallidus. J. Neurophysiol., 94: 1104–1114.

Karachi, C., Francois, C., Parain, K., Bardinet, E., Tande, D., Hirsch, E. and Yelnik, J. (2002) Three-dimensional cartography of functional territories in the human striatopallidal complex by using calbindin immunoreactivity. J. Comp. Neurol., 450: 122–134.

Kase, H., Aoyama, S., Ichimura, M., Ikeda, K., Ishii, A., Kanda, T., Koga, K., Koike, N., Kurokawa, M., Kuwana, Y., Mori, A., Nakamura, J., Nonaka, H., Ochi, M., Saki, M., Shimada, J., Shindou, T., Shiozaki, S., Suzuki, F., Takeda, M., Yanagawa, K., Richardson, P.J., Jenner, P., Bedard, P., Borrelli, E., Hauser, R.A. and Chase, T.N. (2003) Progress in pursuit of therapeutic A2A antagonists: the adenosine A2A receptor selective antagonist KW6002: research and development toward a novel nondopaminergic therapy for Parkinson's disease. Neurology, 61: S97–S100.

Kawaguchi, Y., Wilson, C.J. and Emson, P.C. (1990) Projection subtypes of rat neostriatal matrix cells revealed by intracellular injection of biocytin. J. Neurosci., 10: 3421–3438.

Kelland, M.D., Soltis, R.P., Anderson, L.A., Bergstrom, D.A. and Walters, J.R. (1995) In vivo characterization of two cell types in the rat globus pallidus which have opposite responses to dopamine receptor stimulation: comparison of electrophysiological properties and responses to apomorphine, dizocilpine, and ketamine anesthesia. Synapse, 20: 338–350.

Kita, H. (1994) Parvalbumin-immunopositive neurons in rat globus pallidus: a light and electron microscopic study. Brain Res., 657: 31–41.

Kita, H. (1996) Two pathways between the cortex and the basal ganglia output nuclei and the globus pallidus. In: Ohye C., Kimura M. and McKenzie J.S. (Eds.), The basal ganglia V. Plenum Press, New York, pp. 77–94.

Kita, H., Chiken, S., Tachibana, Y. and Nambu, A. (2006) Origins of GABA(A) and GABA(B) receptor mediated responses of globus pallidus induced after stimulation of the putamen in the monkey. J. Neurosci., 26: 6554–6562.

Kita, H. and Kita, T. (2001) Number, origins, and chemical types of rat pallidostriatal projection neurons. J. Comp. Neurol., 437: 438–448.

Kita, H. and Kitai, S.T. (1987) Efferent projections of the subthalamic nucleus in the rat: Light and electron microscopic analysis with the PHA-L Method. J. Comp. Neurol., 260: 435–452.

Kita, H. and Kitai, S.T. (1991) Intracellular study of rat globus pallidus neurons: membrane properties and responses to neostriatal, subthalamic and nigral stimulation. Brain Res., 564: 296–305.

Kita, H. and Kitai, S.T. (1994) The morphology of globus pallidus projection neurons in the rat: an intracellular staining study. Brain Res., 636: 308–319.

Kita, H., Nambu, A., Kaneda, K., Tachibana, Y. and Takada, M. (2004) Role of ionotropic glutamatergic and GABAergic inputs on the firing activity of neurons in the external pallidum in awake monkeys. J. Neurophysiol., 92: 3069–3084.

Kita, H., Tokuno, H. and Nambu, A. (1999) Monkey globus pallidus external segment neurons projecting to the neostriatum. Neuroreport, 10: 1467–1472.

Koos, T., Tepper, J.M. and Wilson, C.J. (2004) Comparison of IPSCs evoked by spiny and fast-spiking neurons in the neostriatum. J. Neurosci., 24: 7916–7922.

Levy, R., Hutchison, W.D., Lozano, A.M. and Dostrovsky, J.O. (2002) Synchronized neuronal discharge in the basal ganglia of parkinsonian patients is limited to oscillatory activity. J. Neurosci., 22: 2855–2861.

Magill, P.J., Bolam, J.P. and Bevan, M.D. (2000) Relationship of activity in the subthalamic nucleus-globus pallidus network to cortical electroencephalogram. J. Neurosci., 20: 820–833.

Magill, P.J., Bolam, J.P. and Bevan, M.D. (2001) Dopamine regulates the impact of the cerebral cortex on the subthalamic nucleus-globus pallidus network. Neuroscience, 106: 313–330.

Maneuf, Y.P., Mitchell, I.J., Crossman, A.R. and Brotchie, J.M. (1994) On the role of enkephalin cotransmission in the GABAergic striatal efferents to the globus pallidus. Exp. Neurol., 125: 65–71.

Mansour, A., Khachaturian, H., Lewis, M.E., Akil, H. and Watson, S.J. (1988) Anatomy of CNS opioid receptors. Trends Neurosci., 11: 308–314.

Marino, M.J., Valenti, O., O'Brien, J.A., Williams Jr., D.L. and Conn, P.J. (2003) Modulation of inhibitory transmission in the rat globus pallidus by activation of mGluR4. Ann. N Y Acad. Sci., 1003: 435–437.

Matsui, T. and Kita, H. (2003) Activation of group III metabotropic glutamate receptors presynaptically reduces both GABAergic and glutamatergic transmission in the rat globus pallidus. Neuroscience, 122: 727–737.

Matyas, F., Yanovsky, Y., Mackie, K., Kelsch, W., Misgeld, U. and Freund, T.F. (2006) Subcellular localization of type 1 cannabinoid receptors in the rat basal ganglia. Neuroscience, 137: 337–361.

Mayfield, R.D., Suzuki, F. and Zahniser, N.R. (1993) Adenosine A2a receptor modulation of electrically evoked endogenous GABA release from slices of rat globus pallidus. J. Neurochem., 60: 2334–2337.

McGeorge, A.J. and Faull, R.L.M. (1989) The organization of the projection from the cerebral cortex to the striatum in the rat. Neuroscience, 29: 503–537.

McLean, H.A., Caillard, O., Khazipov, R., Ben-Ari, Y. and Gaiarsa, J.L. (1996) Spontaneous release of GABA activates GABAB receptors and controls network activity in the neonatal rat hippocampus. J. Neurophysiol., 76: 1036–1046.

Mori, A. and Shindou, T. (2003) Modulation of GABAergic transmission in the striatopallidal system by adenosine A2A receptors: a potential mechanism for the antiparkinsonian effects of A2A antagonists. Neurology, 61: S44–S48.

Morris, B.J. and Herz, A. (1986) Autoradiographic localization in rat brain of k opiate binding sites labelled by [3 H]bremazocine. Neuroscience, 19: 839–846.

Mugnaini, E. and Oertel, W.H. (1985) An atlas of the distribution of GABAergic neurons and terminals in the rat CNS as revealed by GAD immunohistochemistry. Elsevier, Amsterdam.

Naito, A. and Kita, H. (1994) The cortico-pallidal projection in the rat: an anterograde tracing study with biotinylated dextran amine. Brain Res., 653: 251–257.

Nambu, A. and Llinás, R. (1990) Electrophysiology of the globus pallidus neurons: an in vitro study in guinea pig brain slices. Soc. Neurosci. Abstr., 16: 428.

Nambu, A. and Llinás, R. (1994) Electrophysiology of globus pallidus neurons in vitro. J. Neurophysiol., 72: 1127–1139.

Nambu, A. and Llinás, R. (1997) Morphology of globus pallidus neurons: its correlation with electrophysiology in guinea pig brain slices. J. Comp. Neurol., 377: 85–94.

Nambu, A., Tokuno, H., Hamada, I., Kita, H., Imanishi, M., Akazawa, T., Ikeuchi, Y. and Hasegawa, N. (2000) Excitatory cortical inputs to pallidal neurons via the subthalamic nucleus in the monkey. J. Neurophysiol., 84: 289–300.

Nathan, T., Jensen, M.S. and Lambert, J.D. (1990) The slow inhibitory postsynaptic potential in rat hippocampal CA1

132

neurones is blocked by intracellular injection of QX-314. Neurosci. Lett., 110: 309–313.

Ni, Z., Bouali-Benazzouz, R., Gao, D., Benabid, A.L. and Benazzouz, A. (2000) Changes in the firing pattern of globus pallidus neurons after the degeneration of nigrostriatal pathway are mediated by the subthalamic nucleus in the rat. Eur. J. Neurosci., 12: 4338–4344.

Nini, A., Feingold, A., Slovin, H. and Bergman, H. (1995) Neurons in the globus pallidus do not show correlated activity in the normal monkey, but phase-locked oscillations appear in the MPTP model of parkinsonism. J. Neurophysiol., 74: 1800–1805.

Ochi, M., Koga, K., Kurokawa, M., Kase, H., Nakamura, J. and Kuwana, Y. (2000) Systemic administration of adenosine A(2A) receptor antagonist reverses increased GABA release in the globus pallidus of unilateral 6-hydroxydopamine-lesioned rats: a microdialysis study. Neuroscience, 100: 53–62.

Ogura, M. and Kita, H. (2000) Dynorphin exerts both postsynaptic and presynaptic effects in the Globus pallidus of the rat. J. Neurophysiol., 83: 3366–3376.

Okoyama, S., Nakamura, Y., Moriizumi, T. and Kitao, Y. (1987) Electron microscopic analysis of the synaptic organization of the globus pallidus in the cat. J. Comp. Neurol., 265: 323–331.

Oorshot, D.E. (1996) Total number of neurons in the neostriatal, pallidal, subthalamic, and substantia nigral nuclei of the rat basal ganglia: a stereological study using the cavalieri and optical disector methods. J. Comp. Neurol., 366: 580–599.

Parent, A., Charara, A. and Pinault, D. (1995) Single striatofugal axons arborizing in both pallidal segments and in the substantia nigra in primates. Brain Res., 698: 280–284.

Parent, A. and Hazrati, L.N. (1995) Functional anatomy of the basal ganglia. II. The place of subthalamic nucleus and external pallidum in basal ganglia circuitry. Brain Res. Brain Res. Rev., 20: 128–154.

Park, M.R., Falls, W.M. and Kitai, S.T. (1982) An intracellular HRP study of the rat globus pallidus. I. Responses and light microscopic analysis. J. Comp. Neurol., 211: 284–294.

Rajakumar, N., Rushlow, W., Naus, C.C., Elisevich, K. and Flumerfelt, B.A. (1994) Neurochemical compartmentalization of the globus pallidus in the rat: an immunocytochemical study of calcium-binding proteins. J. Comp. Neurol., 346: 337–348.

Raz, A., Frechter-Mazar, V., Feingold, A., Abeles, M., Vaadia, E. and Bergman, H. (2001) Activity of pallidal and striatal tonically active neurons is correlated in MPTP-treated monkeys but not in normal monkeys. J. Neurosci., 21: RC128.

Richerson, G.B. and Wu, Y. (2003) Dynamic equilibrium of neurotransmitter transporters: not just for reuptake anymore. J. Neurophysiol., 90: 1363–1374.

Robertson, R.G., Graham, W.C., Sambrook, M.A. and Crossman, A.R. (1991) Further investigations into the pathophysiology of MPTP-induced parkinsonism in the primate: an intracerebral microdialysis study of gamma-aminobutyric acid in the lateral segment of the globus pallidus. Brain Res., 563: 278–280.

Rodriguez-Moreno, A., Herreras, O. and Lerma, J. (1997) Kainate receptors presynaptically downregulate GABAergic inhibition in the rat hippocampus. Neuron, 19: 893–901.

Rothblat, D.S. and Schneider, J.S. (1995) Alterations in pallidal neuronal responses to peripheral sensory and striatal stimulation in symptomatic and recovered parkinsonian cats. Brain Res., 705: 1–14.

Sanudo-Pena, M.C., Patrick, S.L., Khen, S., Patrick, R.L., Tsou, K. and Walker, J.M. (1998) Cannabinoid effects in basal ganglia in a rat model of Parkinson's disease. Neurosci. Lett., 248: 171–174.

Sato, F., Lavallee, P., Levesque, M. and Parent, A. (2000) Single-axon tracing study of neurons of the external segment of the globus pallidus in primate. J. Comp. Neurol., 417: 17–31.

Saxena, N.C. and Macdonald, R.L. (1994) Assembly of GABAA receptor subunits: role of the delta subunit. J. Neurosci., 14: 7077–7086.

Schroeder, J.A. and Schneider, J.S. (2002) GABA-opioid interactions in the globus pallidus: [D-Ala2]-Met-enkephalinamide attenuates potassium-evoked GABA release after nigrostriatal lesion. J. Neurochem., 82: 666–673.

Sharif, N.A. and Hughes, J. (1989) Discrete mapping of brain Mu and delta opioid receptors using selective peptides: quantitative autoradiography, species differences and comparison with kappa receptors. Peptides, 10: 499–522.

Shen, K.Z. and Johnson, S.W. (2003) Presynaptic inhibition of synaptic transmission by adenosine in rat subthalamic nucleus in vitro. Neuroscience, 116: 99–106.

Shin, R.M., Masuda, M., Miura, M., Sano, H., Shirasawa, T., Song, W.J., Kobayashi, K. and Aosaki, T. (2003) Dopamine D4 receptor-induced postsynaptic inhibition of GABAergic currents in mouse globus pallidus neurons. J. Neurosci., 23: 11662–11672.

Shindou, T., Richardson, P.J., Mori, A., Kase, H. and Ichimura, M. (2003) Adenosine modulates the striatal GABAergic inputs to the globus pallidus via adenosine A2A receptors in rats. Neurosci. Lett., 352: 167–170.

Shink, E. and Smith, Y. (1995) Differential synaptic innervation of neurons in the internal and external segments of the globus pallidus by the GABA- and glutamate-containing terminals in the squirrel monkey. J. Comp. Neurol., 358: 119–141.

Silberstein, P., Kuhn, A.A., Kupsch, A., Trottenberg, T., Krauss, J.K., Wohrle, J.C., Mazzone, P., Insola, A., Di Lazzaro, V., Oliviero, A., Aziz, T. and Brown, P. (2003) Patterning of globus pallidus local field potentials differs between Parkinson's disease and dystonia. Brain, 126: 2597–2608.

Silberstein, P., Oliviero, A., Di Lazzaro, V., Insola, A., Mazzone, P. and Brown, P. (2005) Oscillatory pallidal local field potential activity inversely correlates with limb dyskinesias in Parkinson's disease. Exp. Neurol., 194: 523–529.

Smith, Y., Bevan, M.D., Shink, E. and Bolam, J.P. (1998) Microcircuitry of the direct and indirect pathways of the basal ganglia. Neuroscience, 86: 353–387.

Smith, Y., Charara, A., Hanson, J.E., Paquet, M. and Levey, A.I. (2000) GABA(B) and group I metabotropic glutamate receptors in the striatopallidal complex in primates. J. Anat., 196(Pt 4): 555–576.

Smith, Y., Hazrati, L.N. and Parent, A. (1990) Efferent projections of the subthalamic nucleus in the squirrel monkey as studied by the PHA-L anterograde tracing method. J. Comp. Neurol., 294: 306–323.

Smith, Y., Parent, A., Seguela, P. and Descarries, L. (1987) Distibution of GABA-immunoreactive neurons in the basal ganglia of the squirrel monkey (Saimiri sciureus). J. Comp. Neurol., 259: 50–64.

Somogyi, P., Fritschy, J.M., Benke, D., Roberts, J.D. and Sieghart, W. (1996) The gamma 2 subunit of the GABAA receptor is concentrated in synaptic junctions containing the alpha 1 and beta 2/3 subunits in hippocampus, cerebellum and globus pallidus. Neuropharmacology, 35: 1425–1444.

Staines, W.A. and Fibiger, H.C. (1984) Collateral projections of neurons of the rat globus pallidus to the striatum and substantia nigra. Exp. Brain Res., 56: 217–220.

Stanford, I.M. and Cooper, A.J. (1999) Presynaptic mu and delta opioid receptor modulation of GABAA IPSCs in the rat globus pallidus in vitro. J. Neurosci., 19: 4796–4803.

Tanganelli, S., Sandager Nielsen, K., Ferraro, L., Antonelli, T., Kehr, J., Franco, R., Ferre, S., Agnati, L.F., Fuxe, K. and Scheel-Kruger, J. (2004) Striatal plasticity at the network level. Focus on adenosine A2A and D2 interactions in models of Parkinson's Disease. Parkinsonism Relat. Disord., 10: 273–280.

Tremblay, L. and Filion, M. (1989) Responses of pallidal neurons to striatal stimulation in intact waking monkeys. Brain Res., 498: 1–16.

Tremblay, L., Filion, M. and Bedard, P.J. (1989) Responses of pallidal neurons to striatal stimulation in monkeys with MPTP-induced parkinsonism. Brain Res., 498: 17–33.

Tunstall, M.J., Oorschot, D.E., Kean, A. and Wickens, J.R. (2002) Inhibitory interactions between spiny projection neurons in the rat striatum. J. Neurophysiol., 88: 1263–1269.

Venderova, K., Brown, T.M. and Brotchie, J.M. (2005) Differential effects of endocannabinoids on [(3)H]-GABA uptake in the rat globus pallidus. Exp. Neurol., 194: 284–287.

Vila, M., Levy, R., Herrero, M.T., Faucheux, B., Obeso, J.A., Agid, Y. and Hirsch, E.C. (1996) Metabolic activity of the basal ganglia in parkinsonian syndromes in human and non-human primates: a cytochrome oxidase histochemistry study. Neuroscience, 71: 903–912.

Waldvogel, H.J., Billinton, A., White, J.H., Emson, P.C. and Faull, R.L. (2004) Comparative cellular distribution of GABAA and GABAB receptors in the human basal ganglia: immunohistochemical colocalization of the alpha 1 subunit of the GABAA receptor, and the GABABR1 and GABABR2 receptor subunits. J. Comp. Neurol., 470: 339–356.

Walker, R.H., Arbuthnott, G.W. and Wright, A.K. (1989) Electrophysiological and anatomical observations concerning the pallidostriatal pathway in the rat. Exp. Brain Res., 74: 303–310.

Wilson, C.J. and Phelan, K.D. (1982) Dual topographic representation of neostriatum in the globus pallidus. Brain Res., 243: 354–359.

Wu, Y., Richard, S. and Parent, A. (2000) The organization of the striatal output system: a single-cell juxtacellular labeling study in the rat. Neurosci. Res., 38: 49–62.

Yasukawa, T., Kita, T., Xue, Y. and Kita, H. (2004) Rat intralaminar thalamic nuclei projections to the globus pallidus: a biotinylated dextran amine anterograde tracing study. J. Comp. Neurol., 471: 153–167.

Yelnik, J., Francois, C., Percheron, G. and Tande, D. (1996) A spatial and quantitative study of the striatopallidal connection in the monkey. Neuroreport, 7: 985–988.

Tepper, Abercrombie & Bolam (Eds.)
Progress in Brain Research, Vol. 160
ISSN 0079-6123

CHAPTER 8

Globus pallidus internal segment

Atsushi Nambu*

Division of System Neurophysiology, National Institute for Physiological Sciences, 38 Nishigo-naka, Myodaiji, Okazaki 444-8585, Japan

Abstract: The internal segment of the globus pallidus (GP$_i$) gathers many bits of information including movement-related activity from the striatum, external segment of the globus pallidus (GP$_e$), and subthalamic nucleus (STN), and integrates them. The GP$_i$ receives rich GABAergic inputs from the striatum and GP$_e$, and γ-aminobutyric acid (GABA) receptors are distributed in the GP$_i$ in a specific manner. Thus, inputs from the striatum and GP$_e$ may control GP$_i$ activity in a different way. The GP$_i$ finally conveys processed information outside the basal ganglia. Changes in GABAergic neurotransmission have been reported in movement disorders and suggested to play an important role in the pathophysiology of the symptoms.

Keywords: GABA receptors; hyperdirect; direct and indirect pathways; motor control; movement disorders

Introduction

The globus pallidus is divided by the internal medullary lamina into the external (GP$_e$) and internal (GP$_i$) segments in primates. In nonprimate mammals, the GP$_i$ is small and buried in the fibers of the internal capsule to form the entopeduncular nucleus, and the GP$_e$ is referred to simply as the globus pallidus. The GP$_i$ in primates and the entopeduncular nucleus in rodents are homologous structures. In this chapter, they are both dealt with without any distinction and called simply as the GP$_i$.[1] Although the GP$_i$ and GP$_e$ share similar morphology and a common neurotransmitter,

GABA, they are distinct. The GP$_e$ has strong interconnections with the STN, while the GP$_i$ projects toward outside the basal ganglia.

GP$_i$ and substantia nigra pars reticulata (SNr) are rather considered to be similar: They both receive inputs from the striatum, GP$_e$ and STN, and project to the cerebral cortex through the thalamus. These observations suggest a view of the GP$_i$ and SNr as a continuum, accidentally separated by dense fiber bundles of the internal capsule, and in dolphins they actually form one structure. The GP$_i$ and SNr have similar functional roles in principle, and their differences merely reflect their topographical organization. The GP$_i$ has strong interactions with the motor cortices and controls somato-motor behaviors. On the other hand, the SNr has interactions with the prefrontal cortex and is considered to control associative functions. The GP$_i$ and SNr are the output nuclei of the basal ganglia and relay the final signals outside the basal ganglia. Thus, investigating the GP$_i$ is crucial for understanding the basal ganglia functions.

---

*Corresponding author. Tel.: +81-564-55-7771;
Fax: +81-564-52-7913; E-mail: nambu@nips.ac.jp

[1]The GP$_e$ and GP$_i$ in primates may not be totally identical to the globus pallidus and entopeduncular nucleus in rodents, respectively. In addition to the entopeduncular nucleus, the SNr in rodents has motor functions.

DOI: 10.1016/S0079-6123(06)60008-3

Cellular types of GP_i neurons

The cellular morphology of GP_i neurons is similar to that of GP_e neurons. GP_i neurons have relatively large (20–60 μm) triangular or polygonal soma with thick, sparsely-spined, poorly branching dendrites (Yelnik et al., 1984). The dendrites appear as a disk-like territory oriented perpendicular to incoming striatal fibers (Percheron et al., 1984). This topographic organization of dendrites and fibers enables convergence of incoming information in the GP_i as in the GP_e. Unlike GP_e neurons, GP_i neurons do not have extensive local axon collaterals (Nakanishi et al., 1991; Parent et al., 1999).

In vivo unit recordings from monkey GP_i recognize a single neuronal group. Unlike GP_e neurons that show pauses of activity, GP_i neurons fire spontaneously at high frequency (10–110 Hz) without pauses (DeLong, 1971; DeLong et al., 1985). All GP_i neurons use GABA as their neurotransmitter (Smith et al., 1987; Ilinsky et al., 1997). In addition to the main group of GP_i neurons, there is another group of neurons named "border cells." They are generally located near the perimeter of pallidal segments, or adjacent to or in the external or internal medullary laminae. Border cells exhibit a medium frequency (20–50 Hz), regular discharge pattern, and are considered to be aberrant cholinergic neurons of the nucleus basalis of Meynert (DeLong, 1971). In vitro intracellular recordings also recognize two types of neurons (Nakanishi et al., 1990, 1991; Kita, 2001). Type I neurons show a strong, time-dependent anomalous rectification sensitive to Cs^+, the ability to generate high frequency, repetitive firing with little spike adaptation, a strong rebound excitation with a low threshold Ca^{2+}-spike and a burst of fast spikes, and spontaneous repetitive firing. Type II neurons show a weak anomalous rectification, a strong spike adaptation, and a ramp-shaped repolarization probably an A-current. Type I and Type II neurons seem to correspond to principal GP_i neurons and border cells recorded in monkeys, respectively.

Basic circuitry of the GP_i

In the current model of basal ganglia organization, the basal ganglia receive cortical inputs, process the information, and send it back to the cerebral cortex via the thalamus. Thus, the cerebral cortex and the basal ganglia form a cortico-basal ganglia loop circuit (Fig. 1). The GP_i and SNr are the output nuclei of the basal ganglia. On the other hand, the striatum is one of the input stations and receives direct excitatory cortical inputs. The striatum projects to the output nuclei via two major projection systems, the *direct* and *indirect* pathways (Alexander and Crutcher, 1990). The direct pathway arises from GABAergic striatal neurons containing substance P and projects monosynaptically to the GP_i/SNr. The indirect pathway arises from GABAergic striatal neurons containing enkephalin and projects polysynaptically to the GP_i/SNr by way of a sequence of connections involving the GP_e and STN.

In addition to the classical indirect pathway through the GP_e and STN, the GP_e gives rise to GABAergic projections to the GP_i/SNr, the reticular nucleus of the thalamus, and the striatum

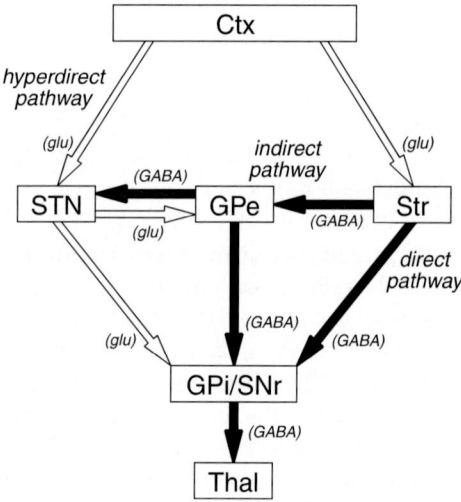

Fig. 1. Basic circuitry of the basal ganglia with respect to the GP_i. Circuits through the basal ganglia comprise the cortico–STN–GP_i/SNr "hyperdirect," cortico–striato–GP_i/SNr "direct" and cortico–striato–GP_e–STN–GP_i/SNr "indirect" pathways. Open and filled arrows represent excitatory glutamatergic (glu) and inhibitory GABAergic (GABA) projections, respectively. Ctx, cerebral cortex; GP_e, external segment of the globus pallidus; GP_i, internal segment of the globus pallidus; SNr, substantia nigra pars reticulata; STN, subthalamic nucleus; Str, striatum; Thal, thalamus. (Modified from Nambu et al., 2002 with permission from Elsevier.)

(Parent and Hazrati 1995a, b; Smith et al., 1998). In the indirect pathway, the STN and GP_e have intimate interconnections. STN neurons possess axon collaterals that innervate both the GP_i and GP_e. Interconnected groups of neurons in the GP_e and STN innervate the same population of neurons in the GP_i (Shink et al., 1996; Smith et al., 1998). The cerebral cortex projects topographically to the STN as well as to the stratum. Thus, recent studies propose the idea that the STN is another input station of the basal ganglia that forms the cortico-STN-GP_i/SNr *hyperdirect* pathway, which is in addition to the direct and indirect pathways (Fig. 1) (Nambu et al., 2000, 2002). Thus, the GP_i and SNr receive inputs from other basal ganglia nuclei, such as the striatum, STN and GP_e. Other inputs to the GP_i include glutamatergic inputs from the intralaminar nuclei, serotonergic inputs from the dorsal raphe nucleus, glutamatergic and cholinergic inputs from the pedunculopontine tegmental nucleus (PPN), and dopaminergic inputs from the substantia nigra pars compacta (SNc) (Lavoie et al., 1989; Lavoie and Parent, 1990, 1994).

The GP_i projects to the ventral anterior/ventral lateral thalamic complex, i.e., oral part of the ventral lateral nucleus (VLo) and parvicellular part of the ventral anterior nucleus (VApc), and gives off axon collaterals to the centromedian nucleus (CM) in primates (Kuo and Carpenter, 1973; Kim et al., 1976). Pallidal fibers entering the thalamus give off several collaterals and terminate primarily onto the soma and proximal dendrites of thalamic projection neurons. In addition, pallidal fibers contact GABAergic inhibitory interneurons in the thalamus. This terminal organization suggests that pallidal inputs not only directly inhibit thalamic projection neurons but also disinhibit projection neurons via thalamic interneurons (Ilinsky et al., 1997). The terminal field from the GP_i is distinct from that of the SNr or that from the deep cerebellar nuclei. In addition, GP_i neurons project to the habenular nucleus and PPN (Parent et al., 1981; Parent and De Bellefeuille, 1983). Thalamic neurons in the VApc, VLO, and CM project to the striatum as well as to the motor cortices, such as the primary motor cortex, supplementary motor area, and premotor cortex. Thus the basal ganglia

form a loop circuit with or without the cerebral cortex (McFarland and Haber, 2000).

Every nucleus of the basal ganglia can be divided, at least, into motor, associative, and limbic territories based on their fiber connections and their functions (Alexander et al., 1986; Parent, 1990). One territory of one nucleus projects basically to the corresponding territory of another nucleus, although an overlap of projections exists. Thus, cortico–basal ganglia loops are considered to be composed of several segregated and parallel loops, such as motor, associative, and limbic loops. However, these territories should be viewed more as a continuum rather than subdivisions with strict boundaries. The GP_i can be subdivided into the motor (ventral two-thirds of the GP_i), associative (dorsal one-third), and limbic (medial tip) territories.

Origin of GABA in the GP_i

The major source of GABAergic afferents to the GP_i arises from medium-sized GABAergic neurons in the striatum. About 70% of the total number of synaptic terminals in contact with individual GP_i neurons originate from striatal spiny neurons (Shink and Smith, 1995). Two other major inputs to the GP_i are glutamatergic afferents from the STN (about 10%) and GABAergic afferents from the GP_e (about 15%). In contrast to GP_e neurons, which give off intranuclear collaterals terminating on neighboring neurons, GP_i neurons have no intranuclear connections (Nakanishi et al., 1991; Parent et al., 1999). Synaptic terminals of these inputs differ in their localization on the GP_i neurons, as well as in their density (Smith et al., 1994; Shink and Smith, 1995) (Fig. 2). Glutamatergic afferents from the STN are homogeneously distributed on the soma and dendrites of GP_i neurons. In contrast, GABAergic terminals are unevenly distributed. GABAergic terminals from the striatum terminate more heavily on the distal dendrites than on the soma and proximal dendrites. GABAergic terminals from the GP_e terminate almost exclusively on the soma and proximal dendrites. Striatal projection neurons are silent at rest and active only when they receive inputs, while GP_e neurons

138

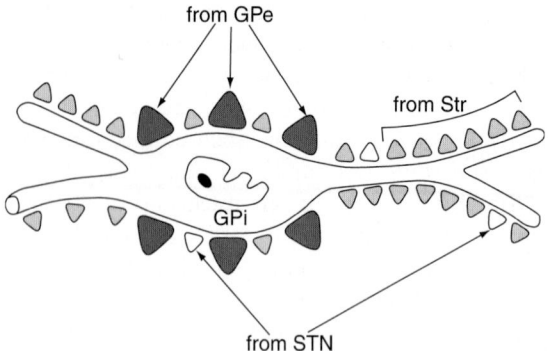

from GPe

from Str

GPi

from STN

Fig. 2. Distribution and relative densities of GABAergic and glutamatergic inputs to GP$_i$ neurons. (Modified from Shink and Smith, 1995, with permission of John Wiley & sons, Inc.)

spontaneously fire at high frequency (10–100 Hz). These differences of terminal distributions and spontaneous activity suggest that GABAergic inputs from the striatum and GP$_e$ modulate GP$_i$ activity in a different way. In particular, GABA released from GP$_e$ terminals is expected to accumulate in the extracellular space in the GP$_i$.

Localization of GABA receptors and transporters in the GP$_i$

GABAergic transmission is mediated by ionotropic GABA$_A$ and metabotropic GABA$_B$ receptors. Both types of receptors are expressed in the GP$_i$ in primates and rodents, but their functions and distribution patterns are different (Fig. 3). The action of GABA is terminated by its diffusion from the synaptic cleft and its uptake by GABA transporters (GATs). Therefore, the effects of GABA on GP$_i$ activity depend on the distribution and activity of GABA receptors and GATs. The following findings indicate that ionotropic and metabotropic GABA receptors are located to subserve different functions in the GP$_i$. GABA$_A$ receptors located at symmetric synapses are consistent with their role in fast inhibitory synaptic transmission, and the extrasynaptic distribution of both GABA$_A$ and GABA$_B$ receptors provides a substrate for complex modulatory functions that rely predominantly on the spillover of GABA. Under conditions of relatively low GABA release, the activation of

synaptic GABA$_A$ receptors could be responsible for most of the neuronal responses to GABA, whereas increased GABA release may result in a more substantial spillover, resulting in the activation of extrasynaptic receptors, particularly of the GABA$_B$ type.

GABA$_A$ receptors

The overall expression patterns of GABA$_A$ receptor subunit mRNAs are similar in GP$_i$ and GP$_e$ neurons in monkeys (Kultas-Ilinsky et al., 1998). In both nuclei, $\alpha1$, $\beta2$, and $\gamma2$ subunit mRNAs are expressed at a high level in all neurons. The $\beta3$ mRNA expression level is low and $\beta1$ subunit expression is almost absent. The $\gamma1$ mRNA expression level is low in the GP$_e$ and absent in the GP$_i$. All other subunit mRNAs ($\alpha2$, $\alpha3$, $\alpha4$, and δ subunits) are expressed at a moderate level. In GP$_i$ and GP$_e$ neurons in rats, $\alpha1$, $\beta2$, and $\gamma2$ subunit mRNAs are expressed at a high to low level (Wisden et al., 1992). Other subunit mRNAs are almost absent in GP$_i$ neurons of rats. Thus, the predominant subunits expressed are $\alpha1$, $\beta2$, and $\gamma2$ in the GP$_i$ and GP$_e$ of both monkeys and rats. An electron microscopic immunocytochemical study demonstrated a colocalization of these three subunits at the same synapse in the rat GP$_e$ (Somogyi et al., 1996). Thus, the subunit combination of $\alpha1\beta2\gamma2$ is most probable in the GP$_i$ and GP$_e$. Actually this combination of receptor subtype is most abundant (almost 50%) in all GABA$_A$ receptors of the mammalian brain (Benke et al., 1991). It corresponds to the BZ I subtype of benzodiazepine receptors (Pritchett et al., 1989). It is also highly probable that this subunit combination is associated with the striato-GP$_e$/GP$_i$ synapses, which are the most numerous GABAergic synapses densely covering GP$_i$ and GP$_e$ neurons (see previous section). The expression of additional subunit mRNAs, such as $\alpha2$, $\alpha3$, $\alpha4$, and δ, in the GP$_i$ suggests that some other receptor subtypes may exist in the GP$_i$ as well (Kultas-Ilinsky et al., 1998). These receptors may be associated with GABAergic synapses from other sources such as from the GP$_e$.

Immunoreactivity for GABA$_A$ receptor subunits is found on dendritic shafts and perikarya

(Charara et al., 2005). GABA$_A$ receptors are mostly bound to the plasma membrane, while only a small portion is expressed intracellularly (associated with a variety of organelles) in the GP$_i$ and GP$_e$. GABA$_A$ receptors are aggregated in the active zone of symmetric synapses, presumably of GABAergic striatal- and pallidal-terminals (Fig. 3). GABA$_A$ receptors are exclusively postsynaptic. These synaptic GABA$_A$ receptors are most likely responsible for the GABA$_A$ mediated fast monosynaptic inhibitory postsynaptic potentials (IPSPs) observed in rat GP$_i$ slices after stimulation of the striatum or GP$_e$ (Nakanishi et al., 1990; Kita, 2001).

In addition to strong synaptic expression, GABA$_A$ receptors also display a significant degree of extrasynaptic labeling in the GP$_i$ and GP$_e$ (Charara et al., 2005) (Fig. 3). Their physiology and pharmacology are different from those of synaptic receptors, possibly because of the differences in subunit composition between extrasynaptic and synaptic receptors (Semyanov et al., 2004). In the hippocampus and cerebellum, extrasynaptic GABA$_A$ receptors contain a benzodiazepine-insensitive δ subunit, although the mRNA for the δ subunit is present at moderate level in the monkey GP$_i$ (Kultas-Ilinsky et al., 1998). The activation of these extrasynaptic receptors requires a spillover of GABA outside synaptic cleft or nonsynaptic GABA release (Semyanov et al., 2004).

GABA$_B$ receptors

GABA$_B$ receptors are heterodimers made up of GABA$_B$ R1 and GABA$_B$ R2 subunits. In both GP$_i$ and GP$_e$ neurons of monkeys, virtually all cell bodies exhibit a moderate GABA$_B$ R1 immunoreactivity (Charara et al., 2000). GABA$_B$ R2 immunoreactivity is also found in the GP$_i$ and GP$_e$ (Charara et al., 2004). Similar distribution is also observed in the GP$_i$ and GP$_e$ of rats (Durkin et al., 1999; Margeta-Mitrovic et al., 1999). The relative distribution of GABA$_B$ R2 immunoreactivity is largely consistent with that of GABA$_B$ R1, but a significantly larger proportion of presynaptic elements is labeled for GABA$_B$ R1 than GABA$_B$ R2 in the GP$_e$ (Charara et al., 2004). These findings suggest the existence of additional GABA$_B$ subunits other than GABA$_B$ R1/R2 and specific combinations depending on various sites.

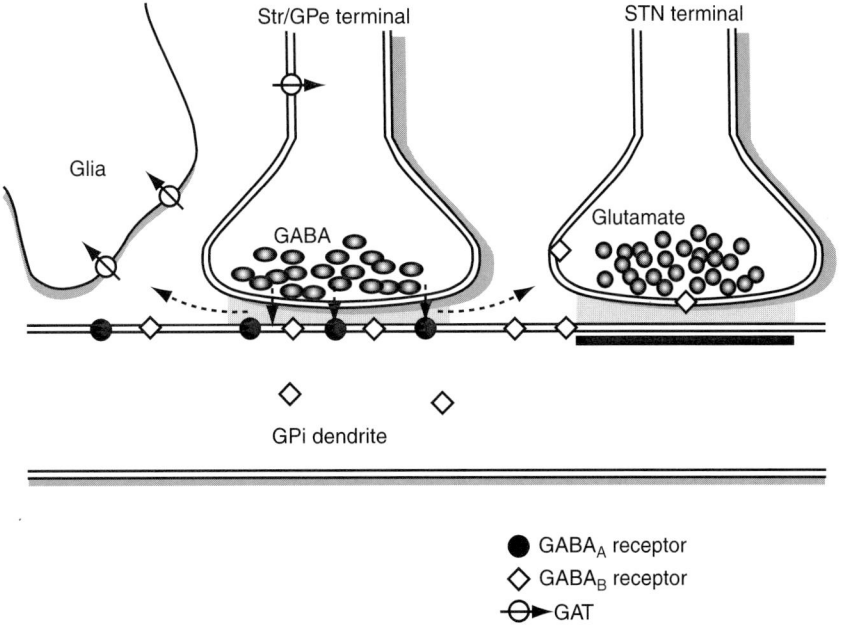

Fig. 3. Subsynaptic localization of GABA$_A$ receptors, GABA$_B$ receptors and GABA transporters (GATs) in the GP$_i$.

As described above for $GABA_A$ receptors, $GABA_B$ R1 labeling is also mainly found in dendritic shafts and perikarya in the GP_i. However, the pattern of subcellular distribution of $GABA_B$ R1 immunoreactivity is strikingly different from that of $GABA_A$ subunits (Charara et al., 2005). In contrast to $GABA_A$ receptor subunits, which are mainly associated with the plasma membrane, a high proportion of $GABA_B$ R1 labeling is located at intracellular sites (the endoplasmic reticulum or vesicular organelles in large dendrites and cell bodies) in the GP_i and GP_e (Fig. 3). Plasma membrane-associated $GABA_B$ R1 immunoreactivity occurs mostly (around 70%) at extrasynaptic sites, although a small portion is expressed postsynaptically at the active zone of symmetric synapses, presumably of GABAergic striatal- and pallidal-terminals.

Immunoreactive preterminal axonal segments are also frequently encountered in the GP_i and GP_e (Charara et al., 2004) (Fig. 3). Moreover, $GABA_B$ R1 immunoreactivity is found presynaptically in asymmetric synapses likely formed by glutamatergic subthalamic terminals (Charara et al., 2005). These observations suggest that $GABA_B$ receptors in the presynaptic terminals may modulate transmitter release in the GP_i and GP_e. The modulation of glutamatergic and GABAergic transmission by $GABA_B$ receptors has been shown in the rat GP_e (Chen et al., 2002; Kaneda and Kita, 2005). The application of baclofen, a $GABA_B$ receptor agonist, reduces the frequency of excitatory postsynaptic potentials (EPSPs) in rat GP_e slices, and the unilateral injection of baclofen into the GP_e induces glutamate receptor-dependent ipsilateral rotation in rats, suggesting a presynaptic regulation of glutamatergic transmission. Synaptically released GABA activates presynaptic $GABA_B$ autoreceptors, decreases GABA release and reduces the amplitude of IPSPs in rat GP_e slices.

GATs

Given the prominence of GABA receptors at extrasynaptic sites, GABA receptor activation may be mediated by the spillover of synaptic GABA, as has been reported in other brain areas. The concentration of GABA at extrasynaptic sites is considered to be primarily determined by the action of plasma membrane-bound GATs. Among GAT families of at least four distinct members, namely GAT1–3 and B-GAT, only GAT-1 and GAT-3 mRNAs and proteins have been reported to express in the rodent and primate GP_i and GP_e. At a light microscopic level, strong labeling for GAT-1 and GAT-3 is apparent in the GP_i and GP_e (Wang and Ong, 1999; Ng et al., 2000; Galvan et al., 2005). GAT-1 staining is observed in axon terminals, and GAT-1 positive axon terminals form symmetrical synapses, presumably of GABAergic terminals (Wang and Ong, 1999). On the other hand, GAT-3 staining is present in astrocytes (Ng et al., 2000). Densely labeled glial processes are observed to surround unlabeled axon terminals and dendrites. Based on a recent report on the GP_e (Galvan et al., 2005), GAT-1 is primarily found in preterminal axonal segments and glial processes, while GAT-3 labeling is found almost exclusively in glial processes. Thus, these observations suggest that GATs are usually located distant from GABAergic synapses, and that GABA released from terminals may diffuse to extrasynaptic locations and then be captured by GATs (Fig. 3). A complex of dendrites and terminals (a rosette) is often surrounded by GAT-3-labeled glial processes (Galvan et al., 2005). This spatial arrangement suggests that synaptically released GABA may diffuse only inside a GAT-containing glial barrier.

Interactions with glutamate receptors

Metabotropic glutamatergic receptors (group I mGluRs) are found not only at postsynaptic sites of asymmetric synapses, but also at postsynaptic sites of symmetric synapses presumably formed by striatal GABAergic terminals in the GP_i and GP_e (Hanson and Smith, 1999). This observation suggests a possibility that these mGluRs may modulate GABAergic striato-GP_i neurotransmission.

GP_i activity controlled by GABA

The major sources of GABAergic inputs to GP_i are the striatum and GP_e. Striatal axons form synapses at distal portions of GP_i neurons, while GP_e axons form synapses on the somata and proximal dendrites. In a rat slice experiment (Kita, 2001), striatal stimulation induced monosynaptic IPSPs with long latency (7.2 ± 2.7 ms) in GP_i neurons, and the unitary IPSPs induced by striato–GP_i fibers were very small, as expected from the slow conduction velocity of striatal efferent axons and the location of synapses at distal portions of GP_i neurons. These results suggest that a large number of synchronized striatal outputs may be required to induce large IPSPs in GP_i neurons. On the other hand, GP_e stimulation induced monosynaptic IPSPs with shorter latency (2.8 ± 1.4 ms) in GP_i neurons, and the unitary IPSPs induced by GP_e-GP_i fibers were large, as expected from the fast conduction velocity of GP_e axons and the location of synapses on the somata and proximal dendrites. $GABA_A$ receptors, not $GABA_B$ receptors, are involved in IPSPs evoked by striatal or GP_e stimulation, because the response is completely blocked by bicuculline, a $GABA_A$ receptor antagonist, but not altered by CGP-55845, a $GABA_B$ receptor antagonist. Repetitive stimulation of striatum and GP_e fuses individual IPSPs together and induces long lasting hyperpolarization, which is induced by $GABA_A$ receptors, but not $GABA_B$ receptors. It is likely that the high-frequency firing of GP_e neurons produces tonic IPSPs mediated by $GABA_A$ receptors that are not time locked to the individual spikes. On the other hand, the low-frequency firing of striatal neurons may induce discrete IPSPs mediated by $GABA_A$ receptors.

In awake monkeys, striatal stimulation induces an inhibition followed by an excitation in GP_i and GP_e neurons (Yoshida et al., 1993; Kita et al., 2006). Burst striatal stimulation induces repeated occurrences of short inhibition and excitation. The inhibition induced by single and repetitive stimulation is mediated by striato-GP_i GABAergic inputs and $GABA_A$ receptors because local application of gabazine, a $GABA_A$ receptor antagonist, but not CGP-55845, a $GABA_B$ receptor antagonist, abolishes the inhibition (Kita et al., 2006). Burst striatal stimulation also induces a late excitation and a late inhibition. The late inhibition is greatly decreased by CGP-55845 application or

blockade of the STN by muscimol, a $GABA_A$ receptor agonist, injection[2] (Kita et al., 2006). Thus, the late inhibition is considered to be mediated by the GP_e–GP_i projections and $GABA_B$ receptors. These results suggest that $GABA_A$ responses are effectively induced by single or repetitive activation of striatal terminals and $GABA_B$ responses are induced by repetitive activation of GP_e terminals. While GP_e stimulation induces an inhibition in GP_i neurons in awake monkeys (Tachibana et al., in press), the precise mechanism of such inhibition remains to be clarified.

The stimulation of motor-related cortical areas induces triphasic responses composed of an early, short-latency excitation, followed by an inhibition and a late excitation in monkey GP_i neurons (Nambu et al., 2000; Kita et al., 2004; Tachibana et al., in press). The latter two components are very similar to the GP_i responses evoked by striatal stimulation. The inhibition is mediated by the cortico–striato–GP_i direct pathway and mainly $GABA_A$ receptors because injecting gabazine in the vicinity of recording neurons abolished the inhibition.[3]

GP_i neurons maintain high-frequency discharge. In contrast to GP_e neurons that show intervening pauses in their high-frequency discharge, GP_i neurons fire without pause. They fire randomly in the normal state and autocorrelograms calculated from GP_i activity are flat (Nambu et al., 2000; Tachibana et al., in press). The activity of GP_i neurons is controlled mainly by GABAergic inputs from the striatum and GP_e, and glutamatergic inputs from the STN. Local application of muscimol or baclofen strongly reduces the firing rate of GP_i neurons (Galvan et al., 2005). Consistent with this result, local application of gabazine or CGP-55845 increases the firing rate and induces an increase in the tendency of burst firing or oscillatory firing of

---

[2]The excitation evoked by single or burst striatal stimulation was mediated by glutamatergic inputs from the STN (through the striato–GP_e–STN indirect pathway or through the continuous excitatory inputs from the STN), because STN blockade by muscimol injection increased the duration of inhibition and decreased the amplitude of excitations greatly.

[3]Other responses, the early excitation and late excitation are derived from the cortico–STN–GPi hyperdirect and cortico–striato–GP_e–STN–GPi indirect pathways, respectively.

142

GP$_i$ neurons (Galvan et al., 2005; Kita et al., 2006; Tachibana et al., in press). Application of SKF-89976A, a selective GAT-1 inhibitor, or SNAP-5114, a semiselective GAT-3 inhibitor, decreases GP$_i$ activity, which suggests that a substantial proportion of released GABA diffuses away from the synapse to activate extrasynaptic GABA$_A$ and GABA$_B$ receptors in the GP$_i$ (Galvan et al., 2005). Blockade of GP$_e$ activity by application of muscimol increases the firing rate of GP$_i$ neurons, but blockade of striatal activity does not (Tachibana et al., in press; Nambu et al., unpublished observations). These results suggest that synaptic and extrasynaptic GABA$_A$ and GABA$_B$ receptors in GP$_i$ neurons are tonically activated by GABA in the GP$_i$, probably released from GP$_e$ axon terminals. They also support the hypothesis that ambient GABA levels in GP$_e$ projection sites play a significant role in the control of the rate and pattern generations in neuronal activity (Galvan et al., 2005; Hallworth and Bevan, 2005; Kaneda and Kita, 2005).

In addition to GABA$_A$ receptor blockade, application of ($\pm$)-3-(2-carboxypiperazin-4-yl)-propyl-1-phosphonic acid (CPP), an NMDA receptor antagonist and 1,2,3,4-tetrahydro-6-nitro-2,3-dioxo-benzo[f]quinoxaline-7-sulfonamide disodium (NBQX), an AMPA/kainate receptor antagonist, induces high-frequency regular oscillatory firing in GP$_i$ neurons (Tachibana et al., in press). These data suggest that both GABAergic inputs mainly from the GP$_e$ and glutamatergic inputs from the STN account for the irregular nature of GP$_i$ firing.

Together with the above-mentioned physiological experiments and the distribution of GABA receptors, the following can be suggested (Fig. 4):

(1) Striatal inputs with very low activity induce phasic inhibitory effects on GP$_i$ neurons mainly through GABA$_A$ receptors.
(2) GP$_e$ neurons firing at high frequency induce tonic inhibitory effects on GP$_i$ neurons through phasic but repetitive activation of synaptic GABA$_A$ receptors and tonic activation of extrasynaptic GABA$_A$ and GABA$_B$ receptors. These GP$_e$ inputs together with glutamatergic STN inputs control spontaneous GP$_i$ activity.

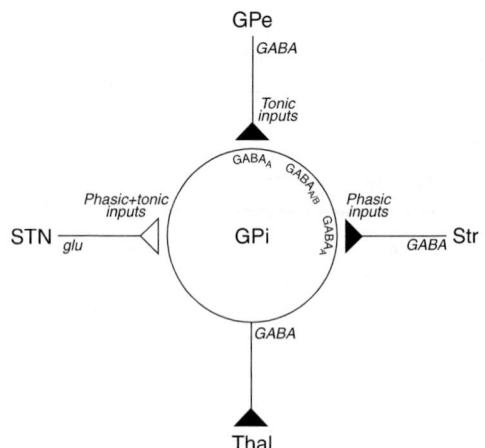

Fig. 4. GABAergic inputs to GP$_i$ neurons and their hypothesized functions.

GP$_i$ activity during movements

The GP$_i$ receives topographical inputs from the cerebral cortex through the hyperdirect, direct, and indirect pathways. In the motor territories of the GP$_i$ (ventral two-thirds of caudal GP$_i$), the hindlimb, forelimb, and oro-facial regions are somatotopically represented from the dorsal part to the ventral part (DeLong, 1971; DeLong et al., 1985; Yoshida et al., 1993; Strick et al., 1995). The most dorsal part of the SNr represents oro-facial area and is considered to be a continuation of oro-facial area of the GP$_i$. GP$_i$ neurons in one somatotopical area show activity change, either inhibition or excitation, in relation to the movements of the corresponding body part (Georgopoulos et al., 1983; Anderson and Horak, 1985; DeLong et al., 1985; Mitchell et al., 1987a; Hamada et al., 1990; Nambu et al., 1990; Mink and Thach, 1991a; Turner and Anderson, 1997). For example, GP$_i$ neurons in the forelimb region of the motor territory change their activity in relation to forelimb movements. GP$_i$ neurons in more rostral and dorsal parts to the motor territory receive inputs from the association cortex, such as prefrontal cortex, and show more complex activity changes (Nambu et al., 1990). Border cells also show similar response pattern as GP$_i$ and GP$_e$ neurons, suggesting that they share similar inputs (Mitchell et al., 1987b).

These changes are presumably the result of a combination of inhibitory GABAergic inputs from the striatum and GP$_e$, and excitatory glutamatergic inputs from the STN, in other words, information through the direct, indirect, and hyperdirect pathways. We do not know much about the contribution of each input to GP$_i$ activity during movements. However, the interesting finding is that GP$_i$ activity during voluntary limb movements always displays an increase rather than a decrease in discharge (Georgopoulos et al., 1983; Anderson and Horak, 1985; Mitchell et al., 1987a; Hamada et al., 1990; Nambu et al., 1990; Mink and Thach, 1991a; Turner and Anderson, 1997), suggesting a large contribution of STN inputs.

GP$_i$/SNr neurons are GABAergic neurons and fire at high frequency, thus neurons in their target structures, such as the thalamus, are inhibited continuously (Fig. 5). When striatal neurons are activated by cortical inputs, the striatal neurons inhibit GP$_i$/SNr activity through the striato-GP$_i$/SNr direct pathway. The continuous inhibition from the output nuclei to the target structures is transiently removed (disinhibition) and neurons in the target structures, including thalamic and cortical neurons, are activated resulting in the release of a selected motor program. These mechanisms have been investigated in saccadic eye movements (Hikosaka and Wurtz, 1983a, b; Hikosaka et al., 2000). On the other hand, signals through the hyperdirect and indirect pathways have excitatory effects on the GP$_i$/SNr and thus have inhibitory effects on thalamic neurons. Considering the respective axonal conduction velocities, signals through the hyperdirect pathway first actively inhibit thalamic neurons, then those through the direct pathway disinhibit them, and finally those through the indirect pathway inhibit them again. Thus, signals through the hyperdirect and indirect pathways make clear initiation and termination of the selected motor program (Nambu et al., 2002).

In addition to such a temporal aspect, the enhancement by differential inputs through the hyperdirect, direct, and indirect pathways may work in spatial domain as well. Anatomical studies

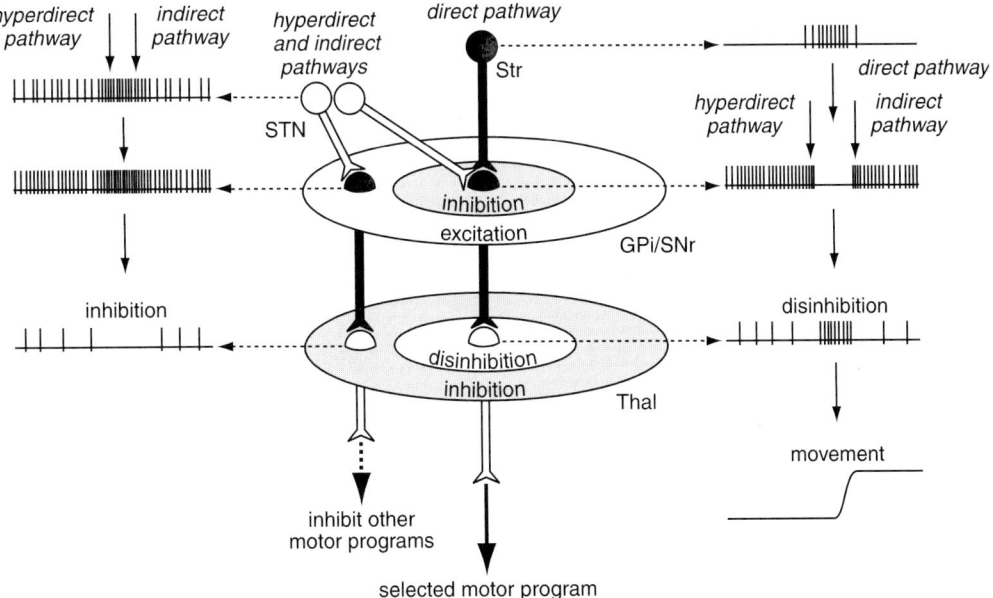

Fig. 5. Spatial and temporal distributions of basal ganglia activity during voluntary movement based on the center-surround model. Signals through the direct pathway inhibit GP$_i$ neurons in the center area, activate thalamic neurons by disinhibition and finally release a selected motor program. On the other hand, signals through the hyperdirect and indirect pathways have broad excitatory effects on GP$_i$ neurons in temporal and spatial domains, make clear initiation and termination of the selected motor program, and inhibit other unnecessary competing motor programs. Open neurons = glutamatergic projections, filled neurons = GABAergic projections.

144

have reported that STN-GP$_i$ fibers arborize more widely and terminate on more proximal neuronal elements than striato-GP$_i$ fibers (Hazrati and Parent, 1992a, b). These anatomical observations suggest that a "center-surround model" of basal ganglia functions proposes focused selection of an appropriate motor program and inhibition of competing motor programs (Mink and Thach, 1993; Mink 1996; Hikosaka et al., 2000; Nambu et al., 2002) (Fig. 5). Thalamic neurons, the target structure of the GP$_i$, are usually inhibited by continuous GABAergic inputs from the GP$_i$. Signals through the direct pathway inhibit a specific group of GP$_i$ neurons in the center area and thalamic neurons are released from inhibition and become active by disinhibition. The increased activity of the thalamus is transmitted to the motor cortex and finally evokes the selected motor program. On the other hand, signals through the hyperdirect and indirect pathways excite other groups of GP$_i$ neurons in the surrounding area and increases inhibition on thalamic neurons mediating other competing motor programs. Inhibition in the center area of the GP$_i$ by striatal GABAergic inputs and excitation in the surrounding area by glutamatergic STN inputs help the execution of the selected motor program and the inhibition of other competing motor programs.

Based on the temporal and spatial inputs to the target structures through the hyperdirect, direct, and indirect pathways, only the selected motor program is executed at the selected timing and other competing motor programs are canceled. In line with this hypothesis, blocking GP$_i$ activity resulted in inaccurate movements probably due to unnecessary muscular contraction (Mink and Thach, 1991b; Kato and Kimura, 1992; Inase et al., 1996). In addition to the motor loop, the oculomotor, prefrontal, and limbic loops control the activity of corresponding cortical areas in a similar manner.

Thalamic neurons receiving GP$_i$ inputs show movement-related activity in a somatotopical manner (Nambu et al., 1991). There are some arguments that thalamic activity change is induced by cortical activity, but not by GP$_i$ activity (Inase et al., 1996). However, it is reasonable to assume that the activity change in thalamic neurons is

induced by activity change in the GP$_i$. The contribution of GP$_i$ activity to thalamic activity changes remains to be investigated.

GABA in the GP$_i$ and movement disorders

The current hypothesis of basal ganglia functions suggests that normal motor behaviors depend on a critical balance in the information processing through the direct and indirect pathways from the striatum to the pallidum, and an imbalance between the two pathways induces movement disorders (Albin et al., 1989; DeLong 1990). Such an imbalance must evoke abnormal activity in the output nuclei, i.e., the GP$_i$ and SNr, which may be accompanied by changes in GABA contents and GABA receptors.

Parkinson's disease

Parkinson's disease is characterized by a loss of dopaminergic neurons in the SNc, and symptoms such as tremor (at rest), muscle rigidity, and akinesia (poverty or slowness of movement). A dramatic decrease in dopamine concentration in the striatum results in increased activity through the indirect pathway and reduced activity through the direct pathway. Both effects together lead to increased excitation of GP$_i$ and SNr neurons, inhibition of thalamocortical neurons and reduced excitation of the cortex, and finally the manifestation of the symptoms of Parkinson's disease (Albin et al., 1989; DeLong, 1990). It has been shown in 1-methyl-4-phenyl-1,2,3,6-tetrahydropyridine (MPTP)-treated primates or in parkinsonian patients that neurons in the STN and GP$_i$ have high discharge rates and show changes in their discharge patterns including burst discharge, synchronized discharge between neighboring neurons, and an increased proportion of neurons with responses to somatosensory inputs (Filion et al., 1988; Miller and DeLong, 1987; Filion and Tremblay, 1991; Nini et al., 1995; Bergman et al., 1998).

In contrast to predictions from the above-mentioned theory, no significant change in GABA contents has been observed in the parkinsonian brain (Calon et al., 1999). The mechanism of changes in their discharge patterns remains to be studied. However, an upregulation and a down-regulation of GABA$_A$ receptors in the GP$_i$ and GP$_e$, respectively, were reported in parkinsonian rats (Gnanalingham and Robertson, 1993), monkeys (Robertson et al., 1990; Calon et al., 1995, 1999), and patients (Griffiths et al., 1990; Calon et al., 2003). These changes in GABA$_A$ receptors can be attributed mainly to the changes in α1 and β2 GABA$_A$ receptor subunit gene expression (Chadha et al., 2000). These receptor changes could be reversed by treatment with a long-acting D2 receptor agonist, but not by a D1 receptor agonist (Robertson et al., 1990; Calon et al., 1995, 1999). The receptor changes in the GP$_i$ and GP$_e$ can be considered to be a compensatory response following the hypoactive striato-GP$_i$ and hyper-active striato-GP$_e$ pathways. On the other hand, GABA$_A$/benzodiazepine binding is higher in the GP$_i$ of monkeys and patients exhibiting dyskinesia with chronic dopaminergic treatments in comparison to those that do not (Calon et al., 1995, 1999, 2003). This suggests that GABA$_A$ receptor upregulation in the GP$_i$ might lead to a supersensitive state of the GP$_i$ to GABAergic inputs. GABAergic inputs to the GP$_i$ could generate an intensified inhibitory drive and produce a decrease in firing rate of GP$_i$ neurons toward the thalamus. Consequently exaggerated thalamo-cortical discharge might impair the generation of normal movements and even promote involuntary movements such as dyskinesia.

GABA$_B$ receptors are also upregulated in the GP$_i$ of parkinsonian monkeys (Calon et al., 2000) and rats (Johnston and Duty, 2003). Changes in GABA$_B$ receptors may be attributed mainly to changes in GABA$_B$ R1 mRNA expression. While this elevation was maintained in MPTP monkeys treated chronically with a D1 agonist, it was partly reversed in D2 agonist-treated MPTP monkeys. Animals with dopaminomimetic-induced dyskinesia have a tendency to show higher levels of GABA$_B$ receptors in the GP$_i$. These observations suggest that upregulation of GABA$_B$ receptors may be a compensatory response and play a role in pathophysiology of dopaminomimetic-induced dyskinesia.

Huntington's disease

In Huntington's disease, a marked upregulation of GABA$_A$ receptors is observed within the GP$_e$ in the very early stages of the disease and sustained in the more advanced stage. Upregulation of GABA$_A$ receptors within the GP$_i$ is not observed in the early stages, becomes evident in the middle stages, and is further increased in the advanced stages (Glass et al., 2000). These data are consistent with the previous findings showing that striato-GP$_e$ terminals degenerate early in the disease and the degeneration of striato-GP$_i$ terminals occurs in the later stages. The increased GABA$_A$-binding sites might reflect a compensatory mechanism for a decrease in GABA release due to striatal cell death.

Dystonia

Reduced rate and synchronized firing of GP$_i$ neurons has been reported in human patients and animal models of dystonia (Vitek et al., 1999; Gernert et al., 2002; Starr et al., 2005) suggesting increased activity through both the direct and indirect pathways. In focal dystonia, GABA content has been reported to be decreased in the lentiform nuclei (Levy and Hallett, 2002). Oral or intrathecal administration of baclofen may be effective in the treatment of dystonia, but its site of action remains to be elucidated (Jankovic, 2004).

Mechanism of deep brain stimulation (DBS)

Recent development of stereotaxic surgery showed that the lesion or high-frequency stimulation of basal ganglia ameliorates motor disabilities of movement disorders. Abnormal high-frequency discharges or abnormal oscillatory firings in the basal ganglia are considered to cause motor symptoms, thus the nuclei showing such abnormal firing are the targets for surgery. In Parkinson's disease, the GP$_i$ and STN show increased firing rate and the thalamus, particularly the ventrointermediate nucleus (Vim), shows tremor-related oscillatory firing. For technical reasons, thalamotomy, pallidotomy, GP$_i$-DBS or STN-DBS are usually employed. Although the high-frequency stimulation

146

shows clinical effects similar to those of lesions, the mechanism(s) for its effectiveness is still under debate and may include depolarization block, activation of inhibitory pathways, "jamming" of abnormal activity through output pathways, and normalization in the pattern of neuronal activity. GP_i stimulation induces inhibitory responses in neighboring GP_i neurons in human patients (Dostrovsky et al., 2000). This inhibition is considered to be generated by stimulation of GABAergic inhibitory afferent fibers from the striatum and/or the GP_e. Consistent with this result, GP_i stimulation increases GABA release in human patients (Ogura et al., 2004). Repetitive stimulation of the STN produces inhibition in the GP_i, while single pulse stimulation produces excitation (Kita et al., 2005). Single pulse stimulation of the STN produces excitation in the GP_e and GP_i through the glutamatergic excitatory STN-GP_e/GP_i pathway. Repetitive stimulation of the STN summates excitations in the GP_e, which inhibit GP_i activity through the GABAergic inhibitory GP_e–GP_i pathway and $GABA_A$ receptors, overcoming the excitatory STN-GP_i pathway. This is considered to be one mechanism for the effectiveness of STN-DBS.

Stereotaxic surgery involving the GP_i is also employed in other movement disorders such as dystonia, chorea, and tardive dyskinesia. Bilateral GP_i-DBS has a dramatic effect on idiopathic torsion dystonia including DYT1. Benefits take several weeks and sometimes months to occur. Thus, some plastic changes may occur. The effects of GP_i-DBS on GABA release, GABA uptake, and GABA receptors remain to be studied.

Conclusions

The GP_i is one of the output nuclei of the basal ganglia and relay the final signals to targets outside the basal ganglia. Thus, investigating the GP_i is crucial for understanding basal ganglia functions. The GP_i receives rich GABAergic inputs from the striatum and GP_e, and GABA receptors are distributed in the GP_i in a specific manner, suggesting that inputs from the striatum and GP_e may control GP_i activity in different ways. Changes in GABAergic neurotransmission have

been reported in movement disorders. Although much information on various aspects of GP_i functional organization, such as receptor distribution, activity patterns of GP_i neurons, and anatomical connections has been obtained, it remains to be elucidated how these elements work as one integrated system in behaving animals in normal and pathological conditions.

Abbreviations

CM	centromedian nucleus of the thalamus
CPP	($\pm$)-3-(2-carboxypiperazin-4-yl)-propyl-1-phosphonic acid
DBS	deep brain stimulation
GAT	GABA transporter
GP_e	external segment of the globus pallidus
GP_i	internal segment of the globus pallidus
mGluR	metabotropic glutamatergic receptor
MPTP	1-methyl-4-phenyl-1,2,3,6-tetrahydro-pyridine
NBQX	1,2,3,4-tetrahydro-6-nitro-2,3-dioxo-benzo[f]quinoxaline-7-sulfonamide disodium
PPN	pedunculopontine tegmental nucleus
SNc	substantia nigra pars compacta
SNr	substantia nigra pars reticulata
STN	subthalamic nucleus
VApc	parvicellular part of the ventral anterior nucleus of the thalamus
Vim	ventrointermediate nucleus of the thalamus
VLo	oral part of the ventral lateral nucleus of the thalamus

Acknowledgments

Supported by a Grant-in-Aid for Scientific Research from the Ministry of Education, Culture, Sports, Science and Technology of Japan (18300135) and the Uehara Memorial Foundation.

References

Albin, R.L., Young, A.B. and Penney, J.B. (1989) The functional anatomy of basal ganglia disorders. Trends Neurosci., 12: 366–375.

Alexander, G.E. and Crutcher, M.D. (1990) Functional architecture of basal ganglia circuits: neural substrates of parallel processing. Trends Neurosci., 13: 266–271.

Alexander, G.E., DeLong, M.R. and Strick, P.L. (1986) Parallel organization of functionally segregated circuits linking basal ganglia and cortex. Annu. Rev. Neurosci., 9: 357–381.

Anderson, M.E. and Horak, F.B. (1985) Influence of the globus pallidus on arm movements in monkeys. III. Timing of movement-related information. J. Neurophysiol., 54: 433–448.

Benke, D., Mertens, S., Trzeciak, A., Gillessen, D. and Mohler, H. (1991) GABA$_A$ receptors display association of gamma 2-subunit with alpha 1- and beta 2/3-subunits. J. Biol. Chem., 266: 4478–4483.

Bergman, H., Feingold, A., Nini, A., Raz, A., Slovin, H., Abeles, M. and Vaadia, E. (1998) Physiological aspects of information processing in the basal ganglia of normal and parkinsonian primates. Trends Neurosci., 21: 32–38.

Calon, F., Goulet, M., Blanchet, P.J., Martel, J.C., Piercey, M.F., Bedard, P.J. and Di Paolo, T. (1995) Levodopa or D2 agonist induced dyskinesia in MPTP monkeys: correlation with changes in dopamine and GABA$_A$ receptors in the striatopallidal complex. Brain Res., 680: 43–52.

Calon, F., Morissette, M., Goulet, M., Grondin, R., Blanchet, P.J., Bedard, P.J. and Di Paolo, T. (1999) Chronic D1 and D2 dopaminomimetic treatment of MPTP-denervated monkeys: effects on basal ganglia GABA(A)/benzodiazepine receptor complex and GABA content. Neurochem. Int., 35: 81–91.

Calon, F., Morissette, M., Goulet, M., Grondin, R., Blanchet, P.J., Bedard, P.J. and Di Paolo, T. (2000) 125I-CGP 64213 binding to GABA(B) receptors in the brain of monkeys: effect of MPTP and dopaminomimetic treatments. Exp. Neurol., 163: 191–199.

Calon, F., Morissette, M., Rajput, A.H., Hornykiewicz, O., Bedard, P.J. and Di Paolo, T. (2003) Changes of GABA receptors and dopamine turnover in the postmortem brains of parkinsonians with levodopa-induced motor complications. Mov. Disord., 18: 241–253.

Chadha, A., Dawson, L.G., Jenner, P.G. and Duty, S. (2000) Effect of unilateral 6-hydroxydopamine lesions of the nigrostriatal pathway on GABA(A) receptor subunit gene expression in the rodent basal ganglia and thalamus. Neuroscience, 95: 119–126.

Charara, A., Galvan, A., Kuwajima, M., Hall, R.A. and Smith, Y. (2004) An electron microscope immunocytochemical study of GABA(B) R2 receptors in the monkey basal ganglia: a comparative analysis with GABA(B) R1 receptor distribution. J. Comp. Neurol., 476: 65–79.

Charara, A., Heilman, T.C., Levey, A.I. and Smith, Y. (2000) Pre- and postsynaptic localization of GABA(B) receptors in the basal ganglia in monkeys. Neuroscience, 95: 127–140.

Charara, A., Pare, J.F., Levey, A.I. and Smith, Y. (2005) Synaptic and extrasynaptic GABA-A and GABA-B receptors in the globus pallidus: an electron microscopic immunogold analysis in monkeys. Neuroscience, 131: 917–933.

Chen, L., Chan, S.C. and Yung, W.H. (2002) Rotational behavior and electrophysiological effects induced by GABA(B) receptor activation in rat globus pallidus. Neuroscience, 114: 417–425.

DeLong, M.R. (1971) Activity of pallidal neurons during movement. J. Neurophysiol., 34: 414–427.

DeLong, M.R. (1990) Primate models of movement disorders of basal ganglia origin. Trends Neurosci., 13: 281–285.

DeLong, M.R., Crutcher, M.D. and Georgopoulos, A.P. (1985) Primate globus pallidus and subthalamic nucleus: functional organization. J. Neurophysiol., 53: 530–543.

Dostrovsky, J.O., Levy, R., Wu, J.P., Hutchison, W.D., Tasker, R.R. and Lozano, A.M. (2000) Microstimulation-induced inhibition of neuronal firing in human globus pallidus. J. Neurophysiol., 84: 570–574.

Durkin, M.M., Gunwaldsen, C.A., Borowsky, B., Jones, K.A. and Branchek, T.A. (1999) An in situ hybridization study of the distribution of the GABA(B2) protein mRNA in the rat CNS. Brain Res. Mol. Brain Res., 71: 185–200.

Filion, M. and Tremblay, L. (1991) Abnormal spontaneous activity of globus pallidus neurons in monkeys with MPTP-induced parkinsonism. Brain Res., 547: 142–151.

Filion, M., Tremblay, L. and Bedard, P.J. (1988) Abnormal influences of passive limb movement on the activity of globus pallidus neurons in parkinsonian monkeys. Brain Res., 444: 165–176.

Galvan, A., Villalba, R.M., West, S.M., Maidment, N.T., Ackerson, L.C., Smith, Y. and Wichmann, T. (2005) GABA-ergic modulation of the activity of globus pallidus neurons in primates: in vivo analysis of the functions of GABA receptors and GABA transporters. J. Neurophysiol., 94: 990–1000.

Georgopoulos, A.P., DeLong, M.R. and Crutcher, M.D. (1983) Relations between parameters of step-tracking movements and single cell discharge in the globus pallidus and subthalamic nucleus of the behaving monkey. J. Neurosci., 3: 1586–1598.

Gernert, M., Bennay, M., Fedrowitz, M., Rehders, J.H. and Richter, A. (2002) Altered discharge pattern of basal ganglia output neurons in an animal model of idiopathic dystonia. J. Neurosci., 22: 7244–7253.

Glass, M., Dragunow, M. and Faull, R.L. (2000) The pattern of neurodegeneration in Huntington's disease: a comparative study of cannabinoid, dopamine, adenosine and GABA(A) receptor alterations in the human basal ganglia in Huntington's disease. Neuroscience, 97: 505–519.

Gnanalingham, K.K. and Robertson, R.G. (1993) Chronic continuous and intermittent L-3,4-dihydroxyphenylalanine treatments differentially affect basal ganglia function in 6-hydroxydopamine lesioned rats—an autoradiographic study using [^{3}H]flunitrazepam. Neuroscience, 57: 673–681.

Griffiths, P.D., Sambrook, M.A., Perry, R. and Crossman, A.R. (1990) Changes in benzodiazepine and acetylcholine receptors in the globus pallidus in Parkinson's disease. J. Neurol. Sci., 100: 131–136.

148

Hallworth, N.E. and Bevan, M.D. (2005) Globus pallidus neurons dynamically regulate the activity pattern of subthalamic nucleus neurons through the frequency-dependent activation of postsynaptic GABA$_A$ and GABA$_B$ receptors. J. Neurosci., 25: 6304–6315.

Hamada, I., DeLong, M.R. and Mano, N. (1990) Activity of identified wrist-related pallidal neurons during step and ramp wrist movements in the monkey. J. Neurophysiol., 64: 1892–1906.

Hanson, J.E. and Smith, Y. (1999) Group I metabotropic glutamate receptors at GABAergic synapses in monkeys. J. Neurosci., 19: 6488–6496.

Hazrati, L.N. and Parent, A. (1992a) Differential patterns of arborization of striatal and subthalamic fibers in the two pallidal segments in primates. Brain Res., 598: 311–315.

Hazrati, L.N. and Parent, A. (1992b) Convergence of subthalamic and striatal efferents at pallidal level in primates: an anterograde double-labeling study with biocytin and PHA-L. Brain Res., 569: 336–340.

Hikosaka, O. and Wurtz, R.H. (1983a) Visual and oculomotor functions of monkey substantia nigra pars reticulata. I. Relation of visual and auditory responses to saccades. J. Neurophysiol., 49: 1230–1253.

Hikosaka, O. and Wurtz, R.H. (1983b) Visual and oculomotor functions of monkey substantia nigra pars reticulata. IV. Relation of substantia nigra to superior colliculus. J. Neurophysiol., 49: 1285–1301.

Hikosaka, O., Takikawa, Y. and Kawagoe, R. (2000) Role of the basal ganglia in the control of purposive saccadic eye movements. Physiol. Rev., 80: 953–978.

Ilinsky, I.A., Yi, H. and Kultas-Ilinsky, K. (1997) Mode of termination of pallidal afferents to the thalamus: a light and electron microscopic study with anterograde tracers and immunocytochemistry in Macaca mulatta. J. Comp. Neurol., 386: 601–612.

Inase, M., Buford, J.A. and Anderson, M.E. (1996) Changes in the control of arm position, movement, and thalamic discharge during local inactivation in the globus pallidus of the monkey. J. Neurophysiol., 75: 1087–1104.

Jankovic, J. (2004) Dystonia: medical therapy and botulinum toxin. In: Fahn S., Hallett M. and DeLong M.R. (Eds.), Dystonia 4, Advances in Neurology, Vol. 94. Lippincott Williams & Wilkins, Philadelphia, pp. 275–286.

Johnston, T. and Duty, S. (2003) Changes in GABA(B) receptor mRNA expression in the rodent basal ganglia and thalamus following lesion of the nigrostriatal pathway. Neuroscience, 120: 1027–1035.

Kaneda, K. and Kita, H. (2005) Synaptically released GABA activates both pre- and postsynaptic GABA(B) receptors in the rat globus pallidus. J. Neurophysiol., 94: 1104–1114.

Kato, M. and Kimura, M. (1992) Effects of reversible blockade of basal ganglia on a voluntary arm movement. J. Neurophysiol., 68: 1516–1534.

Kim, R., Nakano, K., Jayaraman, A. and Carpenter, M.B. (1976) Projections of the globus pallidus and adjacent structures: an autoradiographic study in the monkey. J. Comp. Neurol., 169: 263–290.

Kita, H. (2001) Neostriatal and globus pallidus stimulation induced inhibitory postsynaptic potentials in entopeduncular neurons in rat brain slice preparations. Neuroscience, 105: 871–879.

Kita, H., Chiken, S., Tachibana, Y. and Nambu, A. (2006) Origins of GABA(A) and GABA(B) receptor-mediated responses of globus pallidus induced after stimulation of the putamen in the monkey. J. Neurosci., 26: 6554–6562.

Kita, H., Nambu, A., Kaneda, K., Tachibana, Y. and Takada, M. (2004) Role of ionotropic glutamatergic and GABAergic inputs on the firing activity of neurons in the external pallidum in awake monkeys. J. Neurophysiol., 92: 3069–3084.

Kita, H., Tachibana, Y., Nambu, A. and Chiken, S. (2005) Balance of monosynaptic excitatory and disynaptic inhibitory responses of the globus pallidus induced after stimulation of the subthalamic nucleus in the monkey. J. Neurosci., 25: 8611–8619.

Kultas-Ilinsky, K., Leontiev, V. and Whiting, P.J. (1998) Expression of 10 GABA(A) receptor subunit messenger RNAs in the motor-related thalamic nuclei and basal ganglia of Macaca mulatta studied with in situ hybridization histochemistry. Neuroscience, 85: 179–204.

Kuo, J.S. and Carpenter, M.B. (1973) Organization of pallidothalamic projections in the rhesus monkey. J. Comp. Neurol., 151: 201–236.

Lavoie, B. and Parent, A. (1990) Immunohistochemical study of the serotoninergic innervation of the basal ganglia in the squirrel monkey. J. Comp. Neurol., 299: 1–16.

Lavoie, B. and Parent, A. (1994) Pedunculopontine nucleus in the squirrel monkey: projections to the basal ganglia as revealed by anterograde tract-tracing methods. J. Comp. Neurol., 344: 210–231.

Lavoie, B., Smith, Y. and Parent, A. (1989) Dopaminergic innervation of the basal ganglia in the squirrel monkey as revealed by tyrosine hydroxylase immunohistochemistry. J. Comp. Neurol., 289: 36–52.

Levy, L.M. and Hallett, M. (2002) Impaired brain GABA in focal dystonia. Ann. Neurol., 51: 93–101.

Margeta-Mitrovic, M., Mitrovic, I., Riley, R.C., Jan, L.Y. and Basbaum, A.I. (1999) Immunohistochemical localization of GABA(B) receptors in the rat central nervous system. J. Comp. Neurol., 405: 299–321.

McFarland, N.R. and Haber, S.N. (2000) Convergent inputs from thalamic motor nuclei and frontal cortical areas to the dorsal striatum in the primate. J. Neurosci., 20: 3798–3813.

Miller, W.C. and DeLong, M.R. (1987) Altered tonic activity of neurons in the globus pallidus and subthalamic nucleus in the primate MPTP model of parkinsonism. In: Carpenter M.B. and Jayaraman A. (Eds.), The Basal Ganglia II, Structure and Function-Current Concepts. Plenum Press, New York, pp. 415–427.

Mink, J.W. (1996) The basal ganglia: focused selection and inhibition of competing motor programs. Prog. Neurobiol., 50: 381–425.

Mink, J.W. and Thach, W.T. (1991a) Basal ganglia motor control. II. Late pallidal timing relative to movement onset and inconsistent pallidal coding of movement parameters. J. Neurophysiol., 65: 301–329.

Mink, J.W. and Thach, W.T. (1991b) Basal ganglia motor control. III. Pallidal ablation: normal reaction time, muscle cocontraction, and slow movement. J. Neurophysiol., 65: 330–351.

Mink, J.W. and Thach, W.T. (1993) Basal ganglia intrinsic circuits and their role in behavior. Curr. Opin. Neurobiol., 3: 950–957.

Mitchell, S.J., Richardson, R.T., Baker, F.H. and DeLong, M.R. (1987a) The primate globus pallidus: neuronal activity related to direction of movement. Exp. Brain Res., 68: 491–505.

Mitchell, S.J., Richardson, R.T., Baker, F.H. and DeLong, M.R. (1987b) The primate nucleus basalis of Meynert: neuronal activity related to a visuomotor tracking task. Exp. Brain Res., 68: 506–515.

Nakanishi, H., Kita, H. and Kitai, S.T. (1990) Intracellular study of rat entopeduncular nucleus neurons in an in vitro slice preparation: electrical membrane properties. Brain Res., 527: 81–88.

Nakanishi, H., Kita, H. and Kitai, S.T. (1991) Intracellular study of rat entopeduncular nucleus neurons in an in vitro slice preparation: response to subthalamic stimulation. Brain Res., 549: 285–291.

Nambu, A., Tokuno, H. and Takada, M. (2002) Functional significance of the cortico-subthalamo-pallidal 'hyperdirect' pathway. Neurosci. Res., 43: 111–117.

Nambu, A., Tokuno, H., Hamada, I., Kita, H., Imanishi, M., Akazawa, T., Ikeuchi, Y. and Hasegawa, N. (2000) Excitatory cortical inputs to pallidal neurons via the subthalamic nucleus in the monkey. J. Neurophysiol., 84: 289–300.

Nambu, A., Yoshida, S. and Jinnai, K. (1990) Discharge patterns of pallidal neurons with input from various cortical areas during movement in the monkey. Brain Res., 519: 183–191.

Nambu, A., Yoshida, S. and Jinnai, K. (1991) Movement-related activity of thalamic neurons with input from the globus pallidus and projection to the motor cortex in the monkey. Exp. Brain Res., 84: 279–284.

Ng, C.H., Wang, X.S. and Ong, W.Y. (2000) A light and electron microscopic study of the GABA transporter GAT-3 in the monkey basal ganglia and brainstem. J. Neurocytol., 29: 595–603.

Nini, A., Feingold, A., Slovin, H. and Bergman, H. (1995) Neurons in the globus pallidus do not show correlated activity in the normal monkey, but phase-locked oscillations appear in the MPTP model of parkinsonism. J. Neurophysiol., 74: 1800–1805.

Ogura, M., Nakao, N., Nakai, E., Uematsu, Y. and Itakura, T. (2004) The mechanism and effect of chronic electrical stimulation of the globus pallidus for treatment of Parkinson disease. J. Neurosurg., 100: 997–1001.

Parent, A. (1990) Extrinsic connections of the basal ganglia. Trends Neurosci., 13: 254–258.

Parent, A. and De Bellefeuille, L. (1983) The pallidointralaminar and pallidonigral projections in primate as studied by retrograde double-labeling method. Brain Res., 278: 11–27.

Parent, A. and Hazrati, L.N. (1995a) Functional anatomy of the basal ganglia. I. The cortico-basal ganglia-thalamo-cortical loop. Brain Res. Brain Res. Rev., 20: 91–127.

Parent, A. and Hazrati, L.N. (1995b) Functional anatomy of the basal ganglia. II. The place of subthalamic nucleus and external pallidum in basal ganglia circuitry. Brain Res. Brain Res. Rev., 20: 128–154.

Parent, A., Gravel, S. and Boucher, R. (1981) The origin of forebrain afferents to the habenula in rat, cat and monkey. Brain Res. Bull., 6: 23–38.

Parent, M., Levesque, M. and Parent, A. (1999) The pallidofugal projection system in primates: evidence for neurons branching ipsilaterally and contralaterally to the thalamus and brainstem. J. Chem. Neuroanat., 16: 153–165.

Percheron, G., Yelnik, J. and Francois, C. (1984) A Golgi analysis of the primate globus pallidus. III. Spatial organization of the striato-pallidal complex. J. Comp. Neurol., 227: 214–227.

Pritchett, D.B., Luddens, H. and Seeburg, P.H. (1989) Type I and type II GABA$_A$-benzodiazepine receptors produced in transfected cells. Science, 245: 1389–1392.

Robertson, R.G., Clarke, C.A., Boyce, S., Sambrook, M.A. and Crossman, A.R. (1990) The role of striatopallidal neurones utilizing gamma-aminobutyric acid in the pathophysiology of MPTP-induced parkinsonism in the primate: evidence from [3 H]flunitrazepam autoradiography. Brain Res., 531: 95–104.

Semyanov, A., Walker, M.C., Kullmann, D.M. and Silver, R.A. (2004) Tonically active GABA A receptors: modulating gain and maintaining the tone. Trends Neurosci., 27: 262–269.

Shink, E. and Smith, Y. (1995) Differential synaptic innervation of neurons in the internal and external segments of the globus pallidus by the GABA- and glutamate-containing terminals in the squirrel monkey. J. Comp. Neurol., 358: 119–141.

Shink, E., Bevan, M.D., Bolam, J.P. and Smith, Y. (1996) The subthalamic nucleus and the external pallidum: two tightly interconnected structures that control the output of the basal ganglia in the monkey. Neuroscience, 73: 335–357.

Smith, Y., Bevan, M.D., Shink, E. and Bolam, J.P. (1998) Microcircuitry of the direct and indirect pathways of the basal ganglia. Neuroscience, 86: 353–387.

Smith, Y., Parent, A., Seguela, P. and Descarries, L. (1987) Distribution of GABA-immunoreactive neurons in the basal ganglia of the squirrel monkey (Saimiri sciureus). J. Comp. Neurol., 259: 50–64.

Smith, Y., Wichmann, T. and DeLong, M.R. (1994) Synaptic innervation of neurones in the internal pallidal segment by the subthalamic nucleus and the external pallidum in monkeys. J. Comp. Neurol., 343: 297–318.

Somogyi, P., Fritschy, J.M., Benke, D., Roberts, J.D. and Sieghart, W. (1996) The gamma 2 subunit of the GABA$_A$ receptor is concentrated in synaptic junctions containing the alpha 1 and beta 2/3 subunits in hippocampus, cerebellum and globus pallidus. Neuropharmacology, 35: 1425–1444.

Starr, P.A., Rau, G.M., Davis, V., Marks Jr., W.J., Ostrem, J.L., Simmons, D., Lindsey, N. and Turner, R.S. (2005) Spontaneous pallidal neuronal activity in human dystonia:

comparison with Parkinson's disease and normal macaque. J. Neurophysiol., 93: 3165–3176.

Strick, P.L., Dum, R.P. and Picard, N. (1995) Macro-organization of the circuits connecting the basal ganglia with the cortical motor areas. In: Houk J.C., Davis J.L. and Beiser D.G. (Eds.), Models of Information Processing in the Basal Ganglia. MIT, Cambridge, pp. 117–130.

Turner, R.S. and Anderson, M.E. (1997) Pallidal discharge related to the kinematics of reaching movements in two dimensions. J. Neurophysiol., 77: 1051–1074.

Vitek, J.L., Chockkan, V., Zhang, J.Y., Kaneoke, Y., Evatt, M., DeLong, M.R., Triche, S., Mewes, K., Hashimoto, T. and Bakay, R.A. (1999) Neuronal activity in the basal ganglia in patients with generalized dystonia and hemiballismus. Ann. Neurol., 46: 22–35.

Wang, X.S. and Ong, W.Y. (1999) A light and electron microscopic study of GAT-1 in the monkey basal ganglia. J. Neurocytol., 28: 1053–1061.

Wisden, W., Laurie, D.J., Monyer, H. and Seeburg, P.H. (1992) The distribution of 13 $GABA_A$ receptor subunit mRNAs in the rat brain. I. Telencephalon, diencephalon, mesencephalon. J. Neurosci., 12: 1040–1062.

Yelnik, J., Percheron, G. and Francois, C. (1984) A Golgi analysis of the primate globus pallidus. II. Quantitative morphology and spatial orientation of dendritic arborizations. J. Comp. Neurol., 227: 200–213.

Yoshida, S., Nambu, A. and Jinnai, K. (1993) The distribution of the globus pallidus neurons with input from various cortical areas in the monkeys. Brain Res., 611: 170–174.

Tepper, Abercrombie & Bolam (Eds.)
Progress in Brain Research, Vol. 160
ISSN 0079-6123

CHAPTER 9

The pars reticulata of the substantia nigra: a window to basal ganglia output

J.M. Deniau[1,*], P. Mailly[2], N. Maurice[1] and S. Charpier[1]

[1]*Dynamique et Physiopathologie des Réseaux Neuronaux, INSERM U667, UPMC, Collège de France, 11 Place Marcelin Berthelot, 75231 Paris, Cedex 05, France*
[2]*Neurobiologie des Signaux Intercellulaires, CNRS UMR 7101, UPMC, 4 Place Jussieu, 75005 Paris, France*

Abstract: Together with the internal segment of the globus pallidus (GP$_i$), the pars reticulata of the substantia nigra (SNr) provides a main output nucleus of the basal ganglia (BG) where the final stage of information processing within this system takes place. In the last decade, progress on the anatomical organization and functional properties of BG output neurons have shed some light on the mechanisms of integration taking place in these nuclei and leading to normal and pathological BG outflow. In this review focused on the SNr, after describing how the anatomical arrangement of nigral cells and their afferents determines specific input–output registers, we examine how the basic electrophysiological properties of the cells and their interaction with synaptic inputs contribute to the spatio-temporal shaping of BG output. The reported data show that the intrinsic membrane properties of the neurons subserves a tonic discharge allowing BG to gate the transmission of information to motor and cognitive systems thereby contributing to appropriate selection of behavior.

Keywords: electrophysiology; three-dimensional reconstructions; dendrites; membrane properties; selection of behavior

The pars reticulata of the substantia nigra (SNr) and the internal segment of the globus pallidus (GP$_i$) provide major output nuclei of the basal ganglia (BG) where the final stage of information processing within this system takes place (Fig. 1). These cell groups are mainly composed of GABAergic neurons, integrate inputs from all other component nuclei of the BG (striatum, globus pallidus [GP$_e$], subthalamic nucleus, [STN]) and elaborate the message sent by the BG to extrinsic structures. From rodents to primates, the SNr and GP$_i$ innervate thalamic and brain stem nuclei connected to motor,

prefrontal, parietal and temporal associative cortical areas offering an access for BG to control motor, cognitive, as well as emotional–motivational processes (Rinvik et al., 1976; Beckstead et al., 1979, 1981; Graybiel and Ragsdale, 1979; Gerfen et al., 1982; François et al., 1984, 2002; Alexander et al., 1986; Olazabal and Moore, 1989; Huerta et al., 1991; Kuroda and Price, 1991; Deniau and Chevalier, 1992; Sakai and Patton, 1993; Tokuno et al., 1993; Deniau et al., 1994; Joel and Weiner, 1994; Miyamoto and Jinnai, 1994; Parent and Hazrati., 1995a; Middleton and Strick., 1996, 2002; Montaron et al., 1996; Percheron et al., 1996; Groenewegen et al., 1997; Grofova and Zhou, 1998; Sakai et al., 1998; Hoover and Strick,

*Corresponding author. E-mail: jean-michel.deniau@college-de-france.fr

DOI: 10.1016/S0079-6123(06)60009-5

152

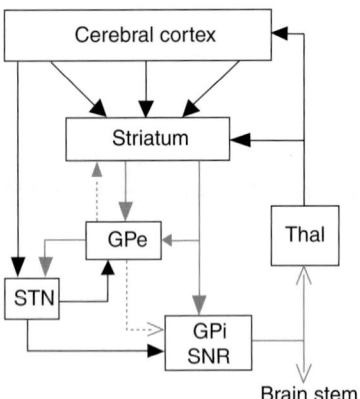

Fig. 1. Functional position of the substantia nigra pars reticulata (SNr) in the basal ganglia circuitry. GABAergic pathways are represented in gray and glutamatergic pathways in black. Abbreviations: GP_e, external segment of globus pallidus; GP_i, internal segment of the globus pallidus; STN, subthalamic nucleus; Thal, thalamus.

1999; Harting et al., 2001; Bar-Gad et al., 2003; Groenewegen, 2003; Haber, 2003; Takakusaki et al., 2003; Clower et al., 2005). In rodents, specie in which the GP_i is less developed than in primates, it is essentially through the SNr that BG influence motor centers. In rats, the SNr innervates the ventral medial and parafascicular thalamic nuclei from which projections to the motor cortex arise (Beckstead et al., 1979; Di Chiara et al., 1979; Graybiel and Ragsdale, 1979; Herkenham, 1979; Gerfen et al., 1982; Deniau and Chevalier, 1992; Deniau et al., 1994; Kha et al., 2001; Mitchell and Cauller, 2001; Tsumori et al., 2003). The SNr also projects to the superior colliculus which is implicated in orienting behavior and to regions of the pontine tegmentum controlling postural tone and locomotion (Clavier et al., 1976; Rinvik et al., 1976; Beckstead et al., 1979, 1981; Di Chiara et al., 1979; Graybiel and Ragsdale, 1979; Gerfen et al., 1982; Deniau and Chevalier, 1984; Rye et al., 1987; Williams and Faull, 1988; Appell and Behan, 1990; Bickford and Hall, 1992; Redgrave et al., 1992; Grofova and Zhou, 1998; Takakusaki et al., 2004). The functional role of BG output nuclei in motor and cognitive functions is particularly exemplified in diseases such as Parkinsonism, Huntington's chorea, hemiballism, Tourette yndrome, obsessive–compulsive disorder and

schizophrenia (Chesselet and Delfs, 1996; Levy et al., 1997; Wichmann and DeLong, 1998, 2003; Graybiel and Rauch, 2000; Graybiel and Canales, 2001; Boraud et al., 2002; Hutchison et al., 2004). In these diseases, the functional impairment of BG circuitry and/or of their afferents is thought to generate an abnormal activity in the BG output nuclei which, transmitted to their thalamic and brain stem targets, would contribute to the observed motor and cognitive disorders. In addition, as a major component of the cortico-BG-thalamo-cortical loop circuits, the BG output nuclei are considered to influence cortical electrogenesis, a mechanism through which BG output nuclei and the SNr in particular would control and/or contribute to the propagation of cortical paroxysmal activity during epileptic seizures (Deransart and Depaulis, 2002; Slaght et al., 2002; Vercueil and Hirsch, 2002; Paz et al., 2005).

In the last decade, progress on the anatomical organization and functional properties of BG output neurons have shed some light on the mechanisms of integration taking place in these nuclei that lead to normal or abnormal BG outflow in health and pathological states. In this review focused on the SNr, after presenting the cellular organization and the functional compartmentalization of this nucleus, the basic electrophysiological properties of SNr cells and the contribution of their synaptic inputs to the spatio-temporal shaping of SNr outflow will be examined.

Neuronal types

Compared to the overlying cell rich pars compacta, the SNr is characterized by a low neuronal density, the cells being interspaced within a dense neuropil of radiating dendrites issued from both the SNr and SNc neurons. Laterally, the SNr merges within the substantia nigra pars lateralis (SNl), a cluster of cells considered as a lateral extension of the SN where SNr and SNc neurons intermingle. The SNr is essentially composed of projection neurons. These cells have been characterized in Golgi-stained tissue from rats (Gulley and Wood, 1971; Juraska et al., 1977; Danner and Pfister, 1982), cats (Rinvik and Grofova, 1970), monkeys (Yelnik et al., 1987) and humans

(Braak and Braak, 1986; Yelnik et al., 1987) and their morphology has been further detailed using single cell labeling techniques in rats (Grofova et al., 1982; Mailly et al., 2001) and cats (Karabelas and Purpura, 1980). In agreement with early observations from Nissl-stained sections (Hanaway et al., 1970; Rinvik and Grofova, 1970; Hajdu et al., 1973; Domesick et al., 1983; Poirier et al., 1983), the soma of SNr neurons range from medium to large size and their pleomorphic shape varies from triangular, fusiform, ovoid to polygonal. The SNr neurons whose isodendritic character is maintained from rat to humans possess sparsely branched dendrites extending for distances as long as 1.5 mm in their longest axis. Dendrites are smooth in their initial portion and bear an increasing number of pleomorphic appendages in their distal parts. Although most SNr cells synthesize GABA (Oertel and Mugnaini, 1984; Smith et al., 1987), SNr also contains dopaminergic (DA) (Deutch et al., 1986; Decavel et al., 1987; Campbell and Takada, 1989; German and Manaye, 1993; Lynd-Balta and Haber, 1994; Nelson et al., 1996; Richards et al., 1997; Gonzalez-Hernandez and Rodriguez, 2000) and cholinergic neurons (Gould and Butcher, 1986; Martinez-Murillo et al., 1989). In rats, DA neurons are more numerous in the caudal and ventral parts of the SNr where they aggregate into distinct clusters. Interestingly, some of these DA neurons are also immunoreactive for the GABA synthetic enzyme, glutamate decarboxylase (Hattori, 1993). Nigral cholinergic neurons are mainly located caudally and might represent ectopically located neurons of the ponto-mesencephalo-tegmental cholinergic system (Gould and Butcher, 1986).

In addition to projection neurons, putative local circuit neurons have been described in Golgi impregnated material of the rat (Gulley and Wood, 1971; Juraska et al., 1977) and monkey SNr (Schwyn and Fox, 1974). In this latter species a subpopulation of neurons characterized by a small soma size, a lack of axon and short dendrites giving rise to thin filiform processes has been observed (François et al., 1979). However, these neurons appear to be very few in number. The functional role of these neurons is still unknown. Former physiological studies had hypothesized

the existence of nigral interneurons based on observations of local synaptic interactions between the SNr and SNc (Grace and Bunney, 1985) but these interactions are likely supported by the intranuclear axon collateral network of projection neurons, the SNr cells exerting the dual function of projection neurons and interneurons (Deniau et al., 1982; Grofova et al., 1982; Tepper et al., 1995; Mailly et al., 2003).

Cytoarchitecture and functional compartmentalization

Information from virtually the entire cerebral qcortex is transmitted to the SNr via the striato-nigral projections (McGeorge and Faull, 1989; Deniau et al., 1996; and see for review: Alexander et al., 1986; Groenewegen and Berendse, 1994; Parent and Hazrati, 1995a; Deniau and Thierry, 1997; Smith et al., 1998; Haber, 2003). Evidence that the topographic arrangement of the cortico-striatal projection imposes to the striatum a rigorous functional compartmentalization led to investigations of whether or not the intrinsic organization of the SNr maintains the segregation of its cortico-striatal inputs in its outflow. To that goal, the topographic arrangement and geometry of the dendritic arborizations of nigral efferent neurons have been analyzed in relation to the topography of cortico-striato-nigral projections. Available data in rats (Deniau and Chevalier, 1992; Deniau et al., 1996, Deniau and Thierry, 1997; Mailly et al., 2001, 2003), cats (Tokuno et al., 1990) and primates (Alexander et al., 1986; Tokuno et al., 1993; Kitano et al., 1998) support the concept that the spatial arrangement of SNr cells provides a channeling mechanism largely preserving the segregation of corticostriatal inputs through SNr.

In rats where the three-dimensional organization of the SNr is best known at cellular and circuit levels, neurons and their striatal afferents are ordered along longitudinal and curved laminae arranged in an onion-like manner (Gerfen, 1985; Deniau and Chevalier, 1992; Deniau et al., 1996; Mailly et al., 2001, 2003). They envelop a central core located dorsolaterally and extending throughout the rostrocaudal extent of the SNr (Fig. 2). Thus, in the lateral half of the SNr innervated by striatal subterritories processing inputs from sensory and motor cortical areas, the neurons located in the core of the nigral 'onion' occupy spherical or longitudinally oriented cylindrical domains that conform to the geometry of striatal projections. By contrast, in more peripheral regions of SNr (including pars lateralis), neurons have flat dendritic fields curving around the central core (Fig. 3). Similar rules apply in the medial part of SNr innervated by striatal subterritories related to prefrontal and limbic cortical areas. In this region neuronal dendritic fields are also arranged in layers. Whereas, the dorsal neurons extend longitudinally along the plane of the overlying pars compacta, the ventral cells orient parallel to the crus cerebri to extend within a ventral lamina. Such a lamellar cytoarchitecture of the rat SNr is consistent with the layered organization first proposed by Juraska et al. (1977) based on Golgi impregnation.

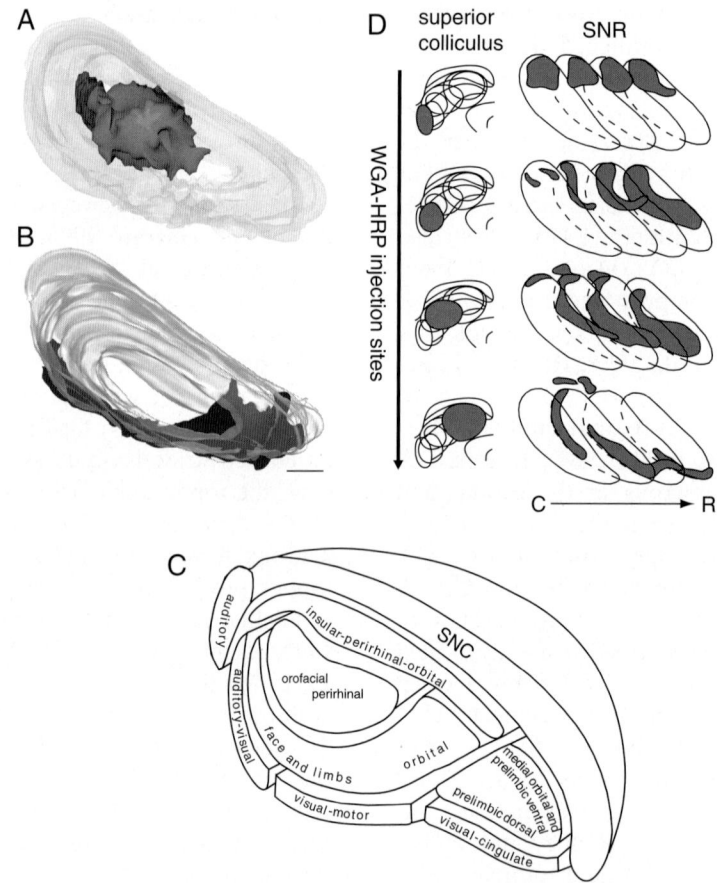

Fig. 2. An 'onion'-like lamellar architecture rules the topographic arrangement of striato-nigral projections and nigral output neurons. (A, B) Three-dimensional reconstructions of striato-nigral projections from striatal subterritories related to the orofacial and visual cortical areas respectively, visualized by anterograde tracing following injection of WGA-HRP in the striatum. The limits of the SNr are indicated in gray and the striatal projections in black. (C) Summary diagram of the ordered representation of the cortico-striatal functional mosaic within the SNr. Note that such a representation applies for the mid-rostrocaudal portion of the SNr. (D) Lamellar topographical distribution of the nigral cells projecting to the superior colliculus as revealed by retrograde axonal transport of WGA-HRP. Left column: injection sites of retrograde axonal tracer in the superior colliculus. Right column: distribution of the labeled cells in the SNr. For each injection site, the distribution of the labeled neurons is shown in four coronal sections arranged from rostral (right) to caudal (left).

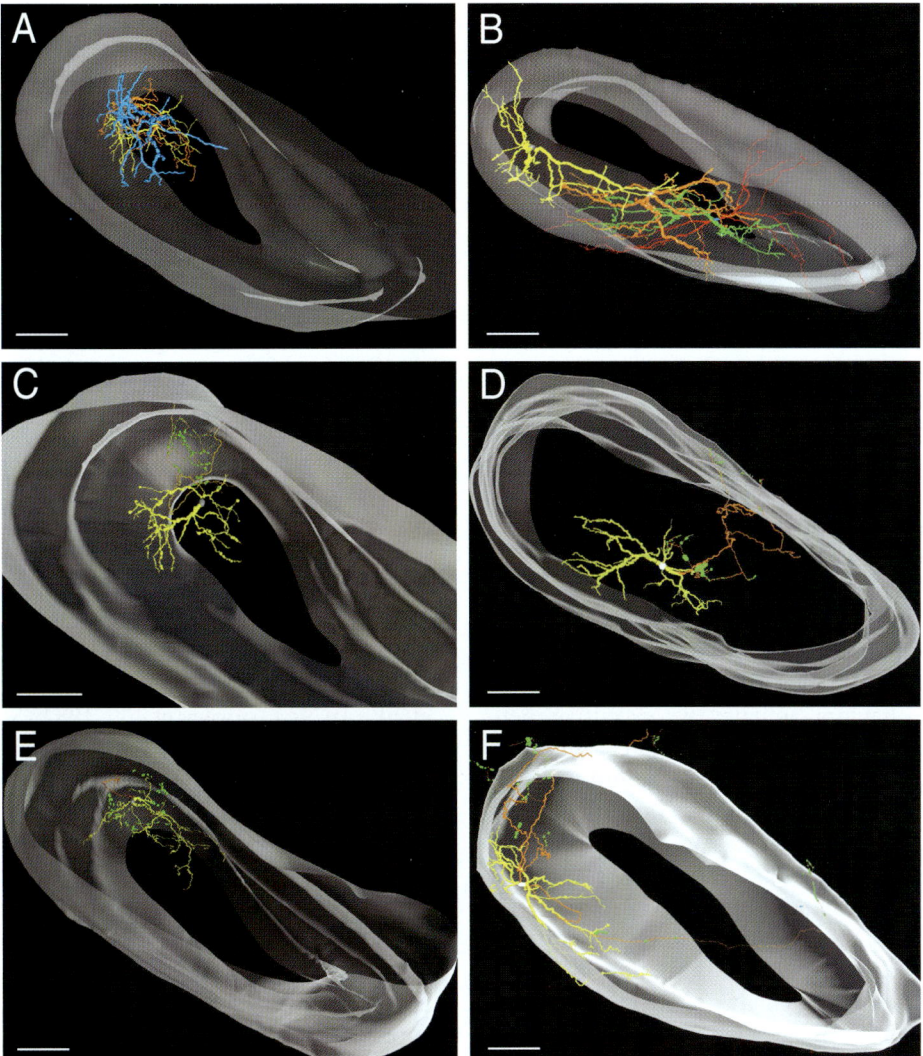

Fig. 3. Spatial arrangement of the dendrites and axons of SNr cells as visualized by single cell labeling. (A) Superimposition of the dendritic arborizations of three neurons located in the orofacial central core of SNr (A) and of four neurons located in a more peripheral region of SNr. Note that the geometry of the cells conforms to the lamellar arrangement of the striato-nigral projections. (C), (D), (E), (F) illustrate the dendrites (yellow), axon (orange) and local axon terminal boutons (green) from individual neurons labeled in different regions of the SNr. Note that the axons originating from ventral neurons take curve paths enveloping laterally and medially the dorsolateral core of the nigral onion-like structure. Scale bar: 350 μm.

Certainly, a remarkable feature of the rat SNr organization is that the cells giving rise to the three main nigral efferent systems (i.e., nigrothalamic, nigrocollicular and nigrotegmental) are topographically ordered along the same onion-like architectural plan. In the case of the nigrocollicular neurons (Fig. 2), those located ventrolaterally in the peripeduncular lamina (innervated by striatal

subterritories receiving inputs from visual and auditory cortical areas) project to the intermediate gray layer of the superior colliculus with a mediolateral topography (Rhoades et al., 1982; Deniau and Chevalier, 1992; Redgrave et al., 1992; Mana and Chevalier, 2001). By contrast, nigral neurons located in more dorsal laminae and in the central core (regions receiving inputs from the striatal subterritories related to somatic sensorimotor cortical areas) innervate the intermediate white layers of the superior colliculus in a series of patches organized with a precise topography. Nigrocollicular neurons project to more lateral patches in the superior colliculus as they occupy a more dorsal position in the SNr. Similar topographical rules conforming to the lamellar organization of striatonigral projections also apply to the nigrothalamic and nigrotegmental projections.

Altogether these data suggest that the lamination of the rat SNr underlies the formation of specific input–output registers connecting defined components of the corticostriatal functional mosaic to particular sets of SNr projection neurons. Based on the present knowledge of input–output registers, the SNr can be subdivided in two main parts implicated in sensorimotor and associative functions, respectively. In the sensorimotor subdivision, the lamellar nigral architecture would promote the formation of sensorimotor links through which information related to the completion of particular behaviors actuate the executive networks supporting these behaviors. To illustrate this notion, the neurons lying in the dorsolateral core of the SNr are in a position to integrate inputs originating from the orofacial sensorimotor and gustatory cortical areas. In turn, they project to thalamic regions connected to the orofacial motor cortical area and to a region of the superior colliculus processing orofacial information and supporting head-orienting behavior. Therefore during feeding behavior, it is expected that neurons in the dorsolateral core of SNr integrates the various sensory and motor information normally associated with this behavior and through their thalamic and collicular projections promote the selection of the motor programs supporting the same behavior. In the associative subdivision of the SNr, the input–output registers established by the nigral lamination would contribute to different cognitive or affective/motivational aspects of behavior.

Early studies on the functional compartmentalization of the rat SNr proposed the subdivision of the SNr into two subnuclei based on a segregation of the cells innervating the thalamus and the superior colliculus (Faull and Mehler, 1978). In this scheme the nigrothalamic subnucleus was considered to provide the efferent limb of a BG motor channel connecting the sensorimotor striatum to the motor thalamus whereas the nigrocollicular subnucleus provided the efferent limb of a visuo-motor circuit channeling visual striatal inputs to the superior colliculus. Although this scheme provided significant conceptual advance since it was the first proposal in favor of a modular architecture of the BG circuits, more recent data on the distribution of nigrothalamic, nigro and nigrotegmental neurons indicated that the proposed segregation of nigrothalamic and nigrocollicular neurons resulted from partial visualization of the overall populations of projection neurons (Deniau and Chevalier, 1992). In fact, nigrothalamic neurons distribute throughout the SNr and overlap the populations of nigrocollicular and nigrotegmental neurons, these two latter cell populations being partially segregated. Because SNr cells possess highly collateralized axons directed to both thalamus, superior colliculus and/or tegmentum (Deniau et al., 1978b; Bentivoglio et al., 1979; Grofova et al., 1982; Steindler and Deniau, 1980; Yasui et al., 1995; Nishimura et al., 1997), we favor a model of SNr organization based on branched neurons whose pattern of axonal branching is specific to each region of the SNr.

Is this organizational principle applicable to other species? In the cat, evidence in favor of a laminar architecture of the SNr has also been provided. As visualized in semihorizontal sections planes, nigral neurons projecting to the thalamus, superior colliculus and pontine tegmentum and their striatal afferents are organized into bands oriented in the same anteromedial-posterolateral direction (Tokuno et al., 1990). Moreover, as observed in rats, striatonigral projections and nigral efferent neurons are topographically organized (Kemel et al., 1988; Hontanilla et al., 1994; Desban et al., 1995) and axons of nigral neurons are highly collateralized, the projections to thalamus, superior colliculus and pontine reticular formation originating from same population of branched SNr cells (Anderson and Yoshida, 1977, 1980; Karabelas and Purpura, 1980; Niijima and Yoshida, 1982). Therefore it is likely that in cats as in rats, a laminar architecture of SNr provides a mechanism allowing defined corticostriatal inputs to be directed to specific and functionally associated sites in the thalamus, superior colliculus and tegmentum.

In primates, different aspects of the functional architecture of the SNr remain to be clarified. Based on a collection of data on the topography of cortico-striato-nigral projections and the distribution of nigrothalamic neurons, a model of segregated cortico-BG-thalamo-cortical circuits has been elaborated (Alexander et al., 1986). In this model, the SNr is subdivided into dorsoventral, mediolateral and anteroposterior regions involved in distinct motor, oculomotor, 'prefrontal' and 'limbic' functions. However, this model relies on the distribution of the somata of nigrothalamic neurons and does not take into account their main integrative component: the dendritic arborizations. The spatial organization of the dendritic fields of nigral neurons has been examined but in these studies, the dendritic orientations has been compared to the global direction of the axons of the striatonigral bundle but not to the specific projection fields from defined functional subterritories of striatum (François et al., 1987). Although the precise input–output relationships of SNr neurons remain to be clarified, some of the characteristic features of the rodent and cat SNr appear to be preserved in primates. As observed in rats and cats, SNr neurons display a high degree of axonal branching (Parent et al., 1983) and based on the intranigral distribution of nigrothalamic, nigrocollicular and nigrotegmental neurons (Beckstead and Frankfurter, 1982; François et al., 1984) the axonal branching pattern is also expected to be highly specific. Moreover, the striatonigral and nigrocollicular neurons form parallel bands in the anteromedial-posterolateral direction suggesting that a laminar organization similar to that described in cat might rule the input–output relationships of the monkey SNr (Tokuno et al., 1993). Interestingly, the orientation of these bands is reminiscent of the anteromedial-posterolateral distribution of nigrocollicular neurons in the ventral part of the rat SNr (see Fig. 2). Therefore, the spatial distribution of neurons within the SNr and the topographic organization of cortico-striato-nigral projections (Alexander et al., 1986; François et al., 1994; Parent and Hazrati, 1994,

1995a; Kitano et al., 1998; Haber, 2003) suggest that in primates as in rats and cats, the neuronal architecture of SNr provides a mechanism allowing defined corticostriatal inputs to be directed to specific and functionally associated sites in the thalamus, superior colliculus and tegmentum. However, in the current state of knowledge, a precise model of the intrinsic organization of SNr such as established in the rat cannot be elaborated for the monkey brain.

Functional properties of SNr cells

The GABAergic output neurons of the SNr exhibit electrophysiological properties clearly distinct from those of the DA nigral neurons. Contrasting with the large action potentials, low-frequency firing rate and strong spike accommodation of the DA nigrostriatal neurons, the GABAergic nigrothalamic, nigrocollicular and nigrotegmental neurons are characterized by short-duration action potentials, a spontaneous repetitive firing reaching 40–80 Hz in vivo and an ability to generate high-frequency discharges with restricted spike accommodation (Wilson et al., 1977; Deniau et al., 1978b; Guyenet and Aghajanian, 1978; Nakanishi et al., 1987; Atherton and Bevan, 2005). The specific electrical membrane properties underlying these firing properties have now been disclosed (Fig. 4). SNr cells display a strong voltage-dependent K^+ conductance responsible for a delayed rectification and short-duration action potential. This membrane conductance contributes to the sustained high-frequency spike discharge of SNr cells since it prevents the membrane potential from reaching the Na^+ inactivation level even when these neurons are strongly depolarized (Nakanishi et al., 1987). In brain slices, the repetitive firing of SNr cells is preserved and moderately affected by a blockade of fast synaptic transmission indicating that it is essentially autonomous in nature (Nakanishi et al., 1987; Yung et al., 1991; Richards et al., 1997; Atherton and Bevan, 2005). This pacemaker activity is essentially driven by two types of subthreshold inward currents that contribute to depolarize SNr cells up to the action potential threshold: a slowly

158

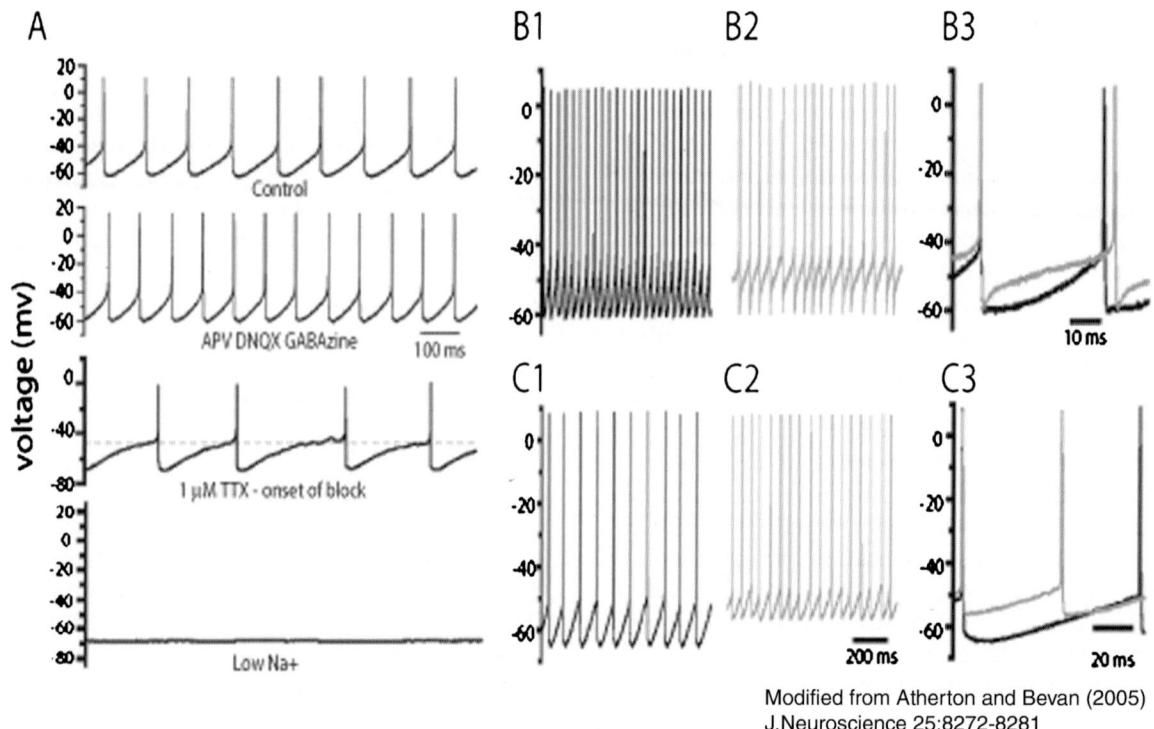

Modified from Atherton and Bevan (2005)
J.Neuroscience 25:8272-8281

Fig. 4. Electrophysiological properties of SNr cells as determined in brain slices using patch clamp recordings. (A) From top to bottom, traces show successively the spontaneous firing of SNr cells recorded in control condition, after blockade of fast glutamatergic and GABAergic synaptic transmission, immediately following 1 μM tetrodotoxin (TTX) and finally after the bath solution was changed to a cerebrospinal fluid (CSF) in which N-methyl-D-glucosamine-Cl (NMDG-Cl) was substituted for NaCl. These observations indicate that the spontaneous discharge of SNr cells is autonomous in nature and partly involve a TTX-sensitive Na^+ current. (B) illustrates the effect of blockade of SK channels with apamine. Compared to control condition (B1), apamine application (B2 and B3) reduces the precision and frequency of firing. The expanded plot in B3 (taken from B1 and B2) illustrates the reduction of the single spike after hyper polarization induced by apamin. (C) Effect of blockade of class 2.2 voltage-dependent Ca^{2+} channels by application of 1 μMω-conotoxin-GIVA. Expanded traces from C1 (control) and C2 (under conotoxin) illustrate the reduction in single spike after hyperpolarization and the elevated action potential threshold.

inactivating voltage-dependent tetrodotoxin (TTX)-sensitive Na^+ current and a TTX-insensitive current mediated in part by Na + (Atherton and Bevan, 2005). In addition, the entry of Ca^{2+} during the action potential discharge activates the small conductance calcium-activated potassium channels (SK) responsible for a post-spike hyperpolarization that controls the precision of the autonomous activity. Finally, SNr cells also possess the cationic conductance, I_h, and are able to generate slow calcium spikes when depolarized from a resting membrane potential more negative than -60 mv. These conductances encountered in many pacemaker neuronal systems allow the triggering of post hyperpolarization spike discharges

thereby contributing to the maintenance of the spontaneous firing of the SNr cells (Nakanishi et al., 1987; Atherton and Bevan, 2005). These electrophysiological features are essential to the functions of the BG. Thanks to their regular and high level of spontaneous activity, SNr cells exert a tonic inhibitory influence on their target structures. It is by modulating this tonic inhibitory influence that BG controls the excitability of thalamocortical and brain stem premotor circuits and shape behavioral outcome. This mechanism requires precise regulation in the balance of inhibitory and excitatory influence that SNr cells receive from striatum, GPe, STN and local circuits. This balance is submitted to the regulatory influence of aminergic systems including the DA neurons of the SNc, the serotonergic neurons of the raphe, the cholinergic neurons of the pedunculopontine nucleus and the histaminergic neurons of the tuberomammillary nucleus.

The striatonigral input

Medium-sized spiny neurons, the main neuronal population of striatum, provide a major source of GABAergic inputs to SNr neurons (see for review Dray, 1979; Graybiel and Ragsdale, 1979; Graybiel, 1984; McGeer et al., 1984; Smith and Bolam, 1990a, b; Langer et al., 1991; Parent and Hazrati, 1995a; Smith et al., 1998; Bolam et al., 2000). The dendrites of nigrothalamic and nigrotegmental neurons are ensheathed by the terminal boutons of striatal fibers that form symmetrical synaptic contacts. The precise identity of the striatal neurons innervating the SNr is still a matter of debate. Retrograde tracing experiments have led to the classical view that striatonigral projections originate from the subpopulation of spiny neurons located in the striatal matrix compartment and expressing the peptides substance P, dynorphin, neurokinin A and neurotensin (Castel et al., 1993 and see for review Graybiel, 1990; Berretta et al., 1993; Parent et al., 1995; Steiner and Gerfen, 1998; Beaujouan et al., 2004). These neurons are supposed to form a subpopulation distinct from the striatal neurons innervating the GP$_e$ and expressing enkephalin and neurokinin B (Gerfen, 1985, 1992a, b; Gerfen and Young, 1988). However, single cell labeling studies in rats and monkey have clearly shown that an important proportion of striatonigral neurons, in particular those expressing substance P, also innervates the GP$_e$ (Kawaguchi et al., 1990; Wu et al., 2000; Levesque et al., 2003). In addition, recent studies using neurotoxic molecules targeting the subpopulation of striosomal neurons revealed that besides the striatal matrix, striosomes also contribute to the innervation of SNr (Tokuno et al., 2002).

The synaptic influence exerted by striatal neurons on SNr cells is well characterized. Electrical stimulation of the striatum elicits short duration inhibitory synaptic events (Goswell and Sedgwick, 1971; McNair et al., 1972; Feger and Ohye, 1975;

Kitai et al., 1975; Deniau et al., 1976; Dray et al., 1976; Collingridge and Davies, 1981; Yoshida et al., 1981; Hikosaka et al., 1993; Wallmichrath and Szabo, 2002). The reversal potential as well as the complete blockade of the striatal-evoked synaptic events by bicuculline or picrotoxin indicate that neurotransmission between striatonigral fibers and SNr cells is due to activation of GABA$_A$ receptors with chloride ions as charge carriers (Precht and Yoshida, 1971; Dray et al., 1976; Collingridge and Davies, 1981; Yoshida et al., 1981; Wallmichrath and Szabo, 2002). The presence of GABA$_B$ receptor subtypes BR1 and BR2 on the presynaptic and postsynaptic membranes of GABAergic striatonigral synapses and glutamatergic STN-like terminals indicates that besides GABA$_A$, GABA$_B$ receptors are also implicated in GABAergic striatonigral transmission (Chan et al., 1998; Charara et al., 2000; Smith et al., 2001; Boyes and Bolam, 2003). Physiological, electrophysiological and neurochemical studies suggest that GABA$_B$ receptors would mainly participate in the control of the release of GABA, the role of postsynaptic GABA$_B$ receptors in SNr cells being minimal (Floran et al., 1988; Rick and Lacey, 1994; Chan and Yung, 1999). In vivo, in addition to inhibitory responses, electrical, or chemical striatal stimulation induces excitatory events in SNr cells (Feger and Ohye, 1975; Dray et al., 1976; Collingridge and Davies, 1981). These excitatory responses have been thought to result from a release of substance P from striatonigral neurons (Kanazawa and Yoshida, 1980; Cuello et al., 1982). However, likely sources for striatonigral evoked excitations include: (1) a spurious axon reflex in the glutamatergic subthalamonigral projections elicited by antidromic activation of the subthalamostriatal neurons (Kita and Kitai, 1987; Smith et al., 1990); (2) a disinhibition of the subthalamonigral neurons via the so-called indirect striato-nigro-subthalamic circuit (Fujimoto and Kita, 1992; Ryan and Sanders, 1994; Smith et al., 1998; Maurice et al., 1999) and (3) a disinhibition of SNr neurons via local synaptic interactions between SNr cells (Deniau et al., 1982; Deniau and Chevalier, 1985). Indeed, because SNr cells are tonically active and exert mutual inhibitory synaptic interactions via their local axon collateral network (Deniau et al., 1982), inhibition of a subset of SNr neurons is expected to disinhibit their partners. A function of the striato-nigral pathway is to induce a transient interruption of the repetitive firing of SNr cells. Whereas under resting conditions SNr cells are tonically discharging, striatonigral neurons are maintained in a quasi-silent state by their specific intrinsic membrane properties (Nisenbaum and Wilson, 1995; Wilson and Kawaguchi, 1996). To discharge, striatal neurons require a minimal level of synchronization in their glutamatergic cortical afferents (Wilson and Kawaguchi, 1996; Mahon et al., 2001). When this condition is reached, the silencing of SNr cells generates a potent increased excitability in BG target nuclei via a mechanism of disinhibition (Chevalier and Deniau, 1990). This process is regarded as the basic mechanism by which BG actuate associative and motor networks for promoting action (Joseph and Boussaoud, 1985; Wurtz and Hikosaka, 1986; Chevalier and Deniau, 1990; Hauber, 1998; Hikosaka et al., 2000). In accordance with a detailed topographic arrangement of connections in the cortico-striato-nigral circuits, this disinhibitory process is specifically organized. Stimulation of defined cortical areas evokes an inhibition in restricted subpopulations of SNr cells and the spatial distribution of the inhibited neurons largely conforms to the representation of the corticostriatal functional mosaic in the SNr (Kitano et al., 1998; Kolomiets et al., 2003).

The pallidonigral input

Besides the striatum, the GP_e provides an additional source of GABAergic inputs to the SNr (McBride and Larsen, 1980; Parent and De Bellefeuille, 1983; Smith and Bolam, 1989; Von Krosigk et al., 1992; Bevan et al., 1996; Shink et al., 1996). As shown in rodent, this projection arises from the subpopulations of pallidal neurons belonging to the categories of aspiny and spiny neurons located mainly in the lateral part of the GPe (Totterdell et al., 1984; Smith and Bolam, 1989; Kita and Kitai, 1994). Pallido-nigral terminals display specific ultrastructural features

(Totterdell et al., 1984; Smith and Bolam, 1989, 1990 a, b; Von Krosigk et al., 1992). They have a large size, contain pleomorphic vesicles, numerous mitochondria and form symmetric synaptic contacts preferentially with the perikarya and proximal dendrites of the SNr projection neurons. These features contrast those of the striatonigral terminals that are of a smaller size, possess few mitochondria and contact predominantly more distal portions of the dendrites of SNr cells (Von Krosigk et al., 1992). The proximal localization of pallidonigral terminals suggests that GP_e may have a significant impact on the discharge of SNr cells. Accordingly, in rat brain slices preparation, comparison of the synaptic effects induced by striatal and GP_e stimulations in BG output nuclei revealed that GP_e stimulation evokes IPSPs of larger amplitude than those evoked from striatum and this inhibitory synaptic effect is strong enough to reset the ongoing rhythmic firing of the neurons (Kita, 2001).

Pallidonigral neurons receive a major input from the striatum (Totterdell et al., 1984). Thus, in addition to the direct striatonigral pathway, the striatum can influence the SNr via an indirect striato-pallido-nigral circuit. Functionally the direct and trans-pallidal circuits are expected to exert opposite effects on SNr cells. Indirect striato-pallidonigral pathway is composed of two successive GABAergic inhibitory links and a majority of pallidonigral neurons are spontaneously active (Kita and Kitai, 1991). Therefore, contrary to the inhibitory function of the direct striatonigral pathway, the indirect striato-pallido-nigral circuit should exert an excitatory influence on SNr cells via a disinhibitory mechanism. The way inputs from the striatum and GP_e combine to shape the SNr outflow is not yet understood. SNr and GP_e are innervated by virtually the entire striatum suggesting that each subdivision of the striatal functional mosaic influences SNr directly and indirectly via the GP_e (Deniau et al., 1996; Smith et al., 1998). Moreover, the striatonigral and pallidonigral pathways are topographically organized (Smith et al., 1998; François et al., 2004) and converge onto the same SNr output neurons (Smith and Bolam, 1990 a, b; Von Krosigk et al., 1992). As will be further detailed below, besides the prominent role of SNr inhibition in eliciting motor responses, excitations may be essential to shape behavioral output by precisely organizing the spatio-temporal patterns of disinhibitory signals issued from the direct striatonigral pathway. It remains now to determine (1) how the topographic arrangement in the striato-pallido-nigral pathway relates to the ordered representation of the striatal functional mosaic in SNr; (2) the relative timing and discharge characteristics of striatonigral and striatopallidal neurons in behavioral situation.

The subthalamonigral input

The STN is the only glutamatergic component of the BG and provides the main source of glutamatergic inputs to BG output nuclei (see for review Parent and Hazrati, 1995b; Smith et al., 1998). In rodent, the STN projection to the SNr originates from the whole nucleus, with most STN neurons sending branched axons to GPe, entopeduncular nucleus (EN) and SNr (Deniau et al., 1978a, b; Van Der Kooy and Hattori, 1980). In primates, retrograde labeling studies have suggested that the subthalamonigral projections originate mainly from the ventromedial part of the STN (Parent and Smith, 1987). However, this picture might result from a partial labeling since in anterograde tracing experiments the lateral STN was also found to project to the SNr (Smith et al., 1990). Within the SNr, subthalamonigral fibers arborize profusely and form asymmetrical synaptic contacts on the proximal and distal dendrites of projection neurons and more rarely on their cell bodies (Rinvik and Ottersen, 1993; Bevan et al., 1994). As determined using anatomical and electrophysiological approaches, STN projections converge with striatonigral and the pallidonigral terminals on the same SNr neurons (Fujimoto and Kita, 1992; Bevan et al., 1994; Ryan and Sanders, 1994; Kitano et al., 1998; Maurice et al., 1999; Kolomiets et al., 2003; Nambu, 2004). The synaptic influence exerted by STN on SNr output neurons is well characterized. In accordance with the glutamatergic nature of the STN-SNr transmission, STN stimulation elicits monosynaptic excitatory postsynaptic potentials leading to action potential

discharge (Hammond et al., 1978; Nakanishi et al., 1987; Robledo and Feger, 1990). In vivo, the excitatory influence of STN contributes to the tonic discharge of SNr cells as shown by the marked reduction in SNr firing rate induced by a pharmacological blockade of STN activity (Feger and Robledo, 1991). The effects of glutamate on SNr cells is mediated by the three principal types of glutamate receptors: α-amino-3hydroxy-5methyl-4-isoxaline propionic acid/kainate (AMPA), N-methyl-D-aspartate (NMDA) and metabotropic receptors (Martin et al., 1993; Standaert et al., 1994; Zhang et al., 1994; Paquet and Smith, 1996; Gotz et al., 1997; Yung, 1998; Schmitt et al., 1999; Chatha et al., 2000; Tse and Yung, 2000; Hubert et al., 2001; Smith et al., 2001; Marino et al., 2002; Hedberg et al., 2003; Hubert and Smith, 2004). Briefly, SNr neurons display the subunits of the ionotropic glutamate receptors GluR1, GluR2, GluR2/3, GluR4, NMDAR1 and NMDAR2A/B and the subunits of the group I metabotropic glutamatergic receptors mGluR1 and mGluR5. Group II and group III mGluRs are also present in SNr but at presynaptic level where they exert an inhibitory control on the glutamatergic STN-SNr and GABAergic striato-nigral transmission (Bradley et al., 2000; Wittmann et al., 2001). Microiontophoretic application of selective agonists and antagonists of excitatory amino acid receptors have confirmed the sensitivity of SNr cells to activation of AMPA, NMDA and metabotropic receptors and stressed the prominent role of NMDA transmission in the spontaneous firing rate of these neurons. The specific contribution of the various ionotropic and metabotropic receptors to the synaptic influence exerted by STN on SNr cells remains to be determined.

Together with the striatum, the STN provides additional routes through which cortical information is transmitted to BG output nuclei. The STN receives a direct innervation from frontal and prefrontal cortical areas and constitutes a relay station of the so-called indirect striato-pallido-subthalamonigral circuit (see for review Joel and Weiner, 1997; Smith et al., 1998). In this indirect circuit, which originates from the entire cerebral cortex, a striatal discharge inhibits the GABAergic pallido-STN neurons leading to a disinhibition of STN cells.

162

In order to better understand how the trans-striatal and trans-subthalamic circuits interact to shape the SNr outflow, stimulation was performed in various cortical areas and the effects evoked in SNr were analyzed. As documented in rats and monkeys, cortical stimulation evokes various patterns of electrophysiological events composed of an early excitation followed or not by an inhibition and a late excitation (Fujimoto and Kita, 1992; Kita and Kitai, 1994; Ryan and Sanders, 1994; Maurice et al., 1999; Kolomiets et al., 2003; Nambu, 2004). The early excitation results from the cortico-subthalamo-nigral circuit ('hyperdirect circuit'), the inhibition is due to the direct striato-nigral circuit and the late excitation to the indirect striato-pallido-subthalamo-nigral circuit (Fig. 5). The pallidonigral pathway might also contribute to the late excitation by desinhibiting directly SNr cells conjointly to the STN. However, contribution of this pathway to the cortical evoked responses in SNr remains to be determined.

In the majority of cases, the inhibitory component of the cortically evoked response is associated with the excitatory components. In addition to confirming the existence of a convergence between striatal and STN projections onto single SNr cells, this observation led to the proposal that through this convergence, the STN contributes to the calibration of the duration of the disinhibitory signals conveyed by the striato-nigral pathway (Mink and Thach, 1993; Kolomiets et al., 2003; Nambu, 2004). It should be well understood that the pattern of excitatory-inhibitory sequence evoked by an electrical cortical stimulation does not necessarily reproduce the sequence of events occurring in a physiological situation. Electrical stimulation induces an artificial synchronization in the different subpopulations of cortical cells innervating the STN, the striatonigral and striatopallidal neurons whereas these subpopulations might receive distinct inputs from distinct cortical neurons acting with a different time sequence during behavior (Lei et al., 2004; Paz et al., 2005). Nevertheless this does not invalidate the proposed concept. The excitations evoked from a given cortical area are more widely distributed within the SNr than the inhibitions. Consequently, SNr cells inhibited from a cortical area can respond with pure

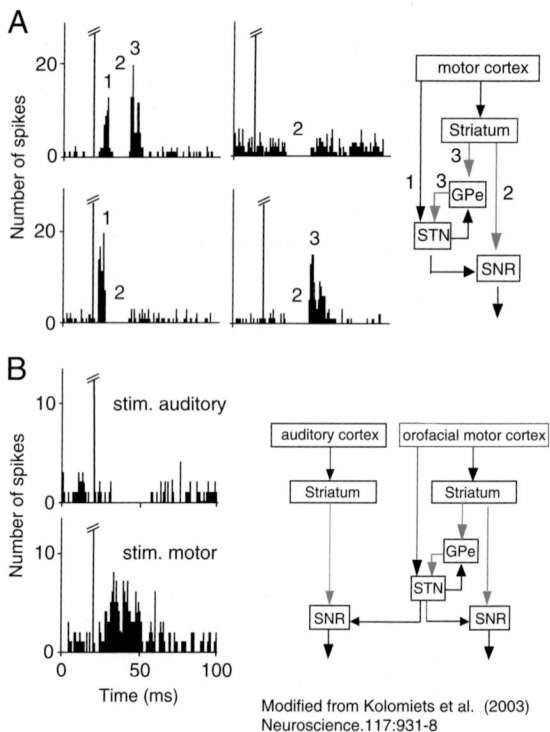

Modified from Kolomiets et al. (2003)
Neuroscience.117:931-8

Fig. 5. (A) Patterns of inhibitory-excitatory sequences induced in SNr cells following stimulation of the motor cortex. The inhibition that is due to the activation of the direct trans-striatal circuit (2) can be preceded by an early excitation resulting from activation of the direct trans-subthalamic circuit (1) and succeeded by a late excitation resulting from the activation of the indirect striato-pallido-subthalamo-nigral circuit (3). (B) Example of an SNr cell receiving an inhibitory influence from the direct striato-nigral circuit originating from the auditory cortex and an excitatory influence from the trans-subthalamic circuits originating from the orofacial motor cortex. This example illustrates the integration of striatal and subthalamic inputs by SNr cells in the process of action selection. In response to activation of a given cortical area (orofacial motor cortex in this example), a subpopulation of SNr cells is inhibited thereby and promotes the expression of a given behavior by disinhibition of its specific thalamic and brain stem targets. Simultaneously, other subpopulations of SNr cells are activated thereby preventing the expression of concurrent behaviors.

excitations to stimulation of other cortical regions (Fig. 5). For example, the neurons located in the lateral SNr within the projection field of the striatal subterritory related to the auditory cortex, display an inhibitory excitatory sequence to stimulation of the auditory cortex but respond with a strong excitation to the orofacial motor cortex

(Kolomiets et al., 2003). This surround excitatory influence provided by the STN likely deserves a dual functional role: (1) it focuses the striatal disinhibitory signals on a precise subpopulation of SNr output neurons and (2) it prevents the activation of concurrent motor programs or cognitive processes by inhibiting the corresponding thalamic and brain stem cell groups.

Local interactions between SNr cells

SNr projection neurons possess a recurrent axon collateral network through which these cells are engaged in mutual inhibitory interactions. The spatial arrangement of this local axonal network has been precisely determined in the rat using single cell labeling and 3D reconstructions (Mailly et al., 2003). Despite important variability, some general rules could be drawn from the axonal population studied. The trajectories of axons largely conform to the lamellar architecture of the SNr. They follow curved paths enveloping a dorsolateral core. The terminal fields of axons distribute both within and outside the dendritic field of the parent neurons with a ventral to dorsal polarity; the terminal boutons being mostly distributed in regions of the SNr more dorsal than the parent cell body. Within the dendritic field of the parent cells, the terminal boutons remain mostly at some distance from the dendrites of the parent neurons. Beyond the dendritic field of the parent cells, the terminal boutons remain in the same or immediately adjacent functional subdivision of the parent neurons in regions where the dendritic arborizations of the parent and target neurons overlap. This spatial arrangement suggests that the local axonal network of SNr cells supports mainly mutual interactions between neurons sharing partly the same inputs. When operating between adjacent neurons, these interactions may contribute to desynchronize the discharge of SNr cells innervating the same targets, thereby reinforcing the tonic nature of their synaptic influence. When operating between distant neurons only partially sharing the same inputs, these interactions likely participate in surrounding mechanisms of contrast acting in synergy with the indirect striato-pallido-subthalamic circuit.

Regulation by neuromodulators

Various peptides (tachykinins, dynorphin, enkephalin, neurotensin, secretoneurin, orexin, cholecystokinin, neuropeptides K and Y, Nociceptin/orphanin), endocannabinoids and neuromodulators can influence SNr cells discharge either directly or indirectly via a presynaptic control of GABAergic and glutamatergic transmissions (Graybiel, 1986; Lavin and Garcia-Munoz, 1986; Valentino et al., 1986; Robertson et al., 1987; Hokfelt et al., 1991; Martin et al., 1991; Castel et al., 1993; Zhang and Freeman, 1994; Parent et al., 1995, Pickel et al., 1995; You et al., 1996; Reiner et al., 1999; Di Giovanni et al., 2001; Ferraro et al., 2001; Korotkova et al., 2002; Marti et al., 2002; Wallmichrath and Szabo, 2002; Beaujouan et al., 2004; Misgeld, 2004; Threlfell et al., 2004). These molecules contribute with different time and space scales to the regulation of BG outflow but in most of the cases, their exact role in the overall function of the BG still remains to be clarified. Among these modulatory agents, a particular interest has been paid to the DA released in SNr by the dendrites of SNc neurons (Cheramy et al., 1981). Early in vivo studies (Ruffieux and Schultz, 1980; Waszczak and Walters, 1984, 1986) revealed that DA applied in SNr by microiontophoresis or endogenously released by application of amphetamine generates an increased firing in SNr cells and attenuates the ability of GABA to inhibit the firing of SNr cells. However, a more complicated picture emerged from subsequent studies reporting a decrease of SNr activity induced by endogenous release of DA (Timmerman and Abercrombie, 1996). In fact, the net effect of DA appears to depend on the types of receptors involved. Activation of D1 receptors generates a concurrent increased firing in SNr cells, similar to that induced by DA application, together with an increased release of GABA from striato-nigral terminals (Martin and Waszczak, 1994; Radnikow and Misgeld, 1998; Trevitt et al., 2002). A component of the GABAergic synaptic input impinging onto distal regions of SNr cells is also concurrently submitted to presynaptic inhibition via the D1 receptors (Miyazaki and Lacey, 1998). Finally, the D2 receptors appear to contribute to the attenuation of SNr cell sensitivity to GABA via a direct postsynaptic action (Martin and Waszczak, 1996). The molecular mechanisms underlying these effects are not fully elucidated. In addition to the implication of a pertussis toxin-sensitive G protein (Martin and Waszczak, 1994), a permissive role for the NMDA receptor activation in the ability of D1 receptors stimulation to both enhance and reduce SNr activity has been reported (Huang et al., 1998). An additional complexity is added by observations that DA facilitates the inhibitory influence exerted by mgluR type II and III on the release of GABA from striatonigral terminals as well as the inhibitory control exerted by the mgluR type II receptors on the release of glutamate by the subthalamonigral fibers (Wittmann et al., 2002).

Conclusion and perspectives

As an output nucleus of the BG, the SNr exerts major functions. Thanks to their ability to generate repetitive discharges, SNr neurons gate the transmission of information to motor and cognitive systems thereby contributing to the selection of behavior, and through the arrest of their discharge favors behavioral output (Redgrave et al., 1999; Gurney et al., 2001). The role of the SNr however cannot be restricted to a simple steering mechanism based upon simple inhibitory-disinhibitory processes. Oscillatory mechanisms are considered to favor the binding of distributed cortical neurons into functional assemblies subserving cognitive and motor functions (Buzsaki and Draguhn, 2004). The BG circuits that are characterized by extensive mutual interactions between their elements possess specific oscillatory properties that have only begun to be explored (Bevan et al., 2002; Deransart and Deapulis, 2002; Slaght et al., 2002; Deransart et al., 2003; Mahon et al., 2003; Ruskin et al., 2003; Paz et al., 2005; Williams et al., 2005). Understanding the way oscillations in BG output nuclei interact dynamically with the thalamo-cortical circuits to shape cortical activity constitutes a major goal for understanding BG functions in health and disease.

References

Alexander, G.E., DeLong, M.R. and Strick, P.L. (1986) Parallel organization of functionally segregated circuits linking basal ganglia and cortex. Annu. Rev. Neurosci., 9: 357–381.

Anderson, M. and Yoshida, M. (1977) Electrophysiological evidence for branching nigral projections to the thalamus and the superior colliculus. Brain Res., 137: 361–364.

Anderson, M.E. and Yoshida, M. (1980) Axonal branching patterns and location of nigrothalamic and nigrocollicular neurons in the cat. J. Neurophysiol., 43: 883–895.

Appell, P.P. and Behan, M. (1990) Sources of subcortical GABAergic projections to the superior colliculus in the cat. J. Comp. Neurol., 302: 143–158.

Atherton, J.F. and Bevan, M.D. (2005) Ionic mechanisms underlying autonomous action potential generation in the somata and dendrites of GABAergic susbtantia nigra pars reticulata neurons in vitro. J. Neurosci., 25: 8272–8281.

Bar-Gad, I., Morris, G. and Bergman, H. (2003) Information processing, dimensionality reduction and reinforcement learning in the basal ganglia. Prog. Neurobiol., 71: 439–473.

Beaujouan, J.C., Torrens, Y., Saffroy, M., Kemel, M.L. and Glowinski, J. (2004) A 25 year adventure in the field of tachykinins. Peptides, 25: 339–357.

Beckstead, R.M., Domesick, V.B. and Nauta, W.J. (1979) Efferent connections of the substantia nigra and ventral tegmental area in the rat. Brain Res., 175: 191–217.

Beckstead, R.M., Edwards, S.B. and Frankfurter, A. (1981) A comparison of the intranigral distribution of nigrotectal neurons labeled with horseradish peroxidase in the monkey, cat, and rat. J. Neurosci., 1: 121–125.

Beckstead, R.M. and Frankfurter, A. (1982) The distribution and some morphological features of substantia nigra neurons that project to the thalamus, superior colliculus and pedunculopontine nucleus in the monkey. Neuroscience, 7: 2377–2388.

Bentivoglio, M., van der Kooy, D. and Kuypers, H.G. (1979) The organization of the efferent projections of the substantia nigra in the rat. A retrograde fluorescent double labeling study. Brain Res., 174: 1–17.

Berretta, S., Robertson, H.A. and Graybiel, A.M. (1993) Neurochemically specialized projection neurons of the striatum respond differentially to psychomotor stimulants. Prog. Brain Res., 99: 201–205.

Bevan, M.D., Bolam, J.P. and Crossman, A.R. (1994) Convergent synaptic input from the neostriatum and the subthalamus onto identified nigrothalamic neurons in the rat. Eur. J. Neurosci., 6: 320–334.

Bevan, M.D., Magill, P.J., Terman, D., Bolam, J.P. and Wilson, C.J. (2002) Move to the rhythm: oscillations in the subthalamic nucleus-external globus pallidus network. Trends Neurosci., 25: 525–531.

Bevan, M.D., Smith, A.D. and Bolam, J.P. (1996) The substantia nigra as a site of synaptic integration of functionally diverse information arising from the ventral pallidum and the globus pallidus in the rat. Neuroscience, 75: 5–12.

Bickford, M.E. and Hall, W.C. (1992) The nigral projection to predorsal bundle cells in the superior colliculus of the rat. J. Comp. Neurol., 319: 11–33.

Bolam, J.P., Hanley, J.J., Booth, P.A. and Bevan, M.D. (2000) j Synaptic organisation of the basal ganglia. J. Anat., 196: 527–542.

Boraud, T., Bezard, E., Bioulac, B. and Gross, C.E. (2002) From single extracellular unit recording in experimental and human Parkinsonism to the development of a functional concept of the role played by the basal ganglia in motor control. Prog. Neurobiol., 66: 265–283.

Boyes, J. and Bolam, J.P. (2003) The subcellular localization of GABA(B) receptor subunits in the rat substantia nigra. Eur. J. Neurosci., 18: 3279–3293.

Braak, H. and Braak, E. (1986) Nuclear configuration and neuronal types of the nucleus niger in the brain of the human adult. Hum. Neurobiol., 5: 71–82.

Bradley, S.R., Marino, M.J., Wittmann, M., Rouse, S.T., Awad, H., Levey, A.I. and Conn, P.J. (2000) Activation of group II metabotropic glutamate receptors inhibits synaptic excitation of the substantia nigra pars reticulata. J. Neurosci., 20: 3085–3094.

Buzsaki, G. and Draguhn, A. (2004) Neuronal oscillations in cortical networks. Science, 304: 1926–1929.

Campbell, K.J. and Takada, M. (1989) Bilateral tectal projection of single nigrostriatal dopamine cells in the rat. Neuroscience, 33: 311–321.

Castel, M.N., Morino, P., Frey, P., Terenius, L. and Hokfelt, T. (1993) Immunohistochemical evidence for a neurotensin striatonigral pathway in the rat brain. Neuroscience, 55: 833–847.

Chan, P.K., Leung, C.K. and Yung, W.H. (1998) Differential expression of pre- and postsynaptic GABA(B) receptors in rat substantia nigra pars reticulata neurones. Eur. J. Pharmacol., 349: 187–197.

Chan, P.K. and Yung, W.H. (1999) Inhibitory postsynaptic currents of rat substantia nigra pars reticulata neurons: role of GABA receptors and GABA uptake. Brain Res., 838: 18–26.

Charara, A., Heilman, T.C., Levey, A.I. and Smith, Y. (2000) Pre- and postsynaptic localization of GABA(B) receptors in the basal ganglia in monkeys. Neuroscience, 95: 127–140.

Chatha, B.T., Bernard, V., Streit, P. and Bolam, J.P. (2000) Synaptic localization of ionotropic glutamate receptors in the rat substantia nigra. Neuroscience, 101: 1037–1051.

Cheramy, A., Leviel, V. and Glowinski, J. (1981) Dendritic release of dopamine in the substantia nigra. Nature, 289: 537–542.

Chesselet, M.F. and Delfs, J.M. (1996) Basal ganglia and movement disorders: an update. Trends Neurosci., 19: 417–422.

Chevalier, G. and Deniau, J.M. (1990) Disinhibition as a basic process in the expression of striatal functions. Trends Neurosci., 13: 277–280.

Clavier, R.M., Atmadja, S. and Fibiger, H.C. (1976) Nigrothalamic projections in the rat as demonstrated by orthograde and retrograde tracing techniques. Brain Res. Bull., 1: 379–384.

Clower, D.M., Dum, R.P. and Strick, P.L. (2005) Basal ganglia and cerebellar inputs to 'AIP'. Cereb. Cortex, 15(7): 913–920.

Collingridge, G.L. and Davies, J. (1981) The influence of striatal stimulation and putative neurotransmitters on identified neurones in the rat substantia nigra. Brain Res., 212: 345–359.

Cuello, A.C., Priestley, J.V. and Matthews, M.R. (1982) Localization of substance P in neuronal pathways. Ciba. Found Symp., 91: 55–83.

Danner, H. and Pfister, C. (1982) 7 types of neurons in the substantia nigra of the rat. Golgi rapid-impregnation study. J. Hirnforsch., 23: 553–556.

Decavel, C., Lescaudron, L., Mons, N. and Calas, A. (1987) First visualization of dopaminergic neurons with a monoclonal antibody to dopamine: a light and electron microscopic study. J. Histochem. Cytochem., 35: 1245–1251.

Deniau, J.M. and Chevalier, G. (1984) Synaptic organization of the basal ganglia: an electroanatomical approach in the rat. Ciba. Found Symp., 107: 48–63.

Deniau, J.M. and Chevalier, G. (1985) Disinhibition as a basic process in the expression of striatal functions. II. The striatonigral influence on thalamocortical cells of the ventromedial thalamic nucleus. Brain Res., 334: 227–233.

Deniau, J.M. and Chevalier, G. (1992) The lamellar organization of the rat substantia nigra pars reticulata: distribution of projection neurons. Neuroscience, 46: 361–377.

Deniau, J.M., Feger, J. and Le Guyader, C. (1976) Striatal evoked inhibition of identified nigro-thalamic neurons. Brain Res., 104: 152–156.

Deniau, J.M., Hammond, C., Chevalier, G. and Feger, J. (1978a) Evidence for branched subthalamic nucleus projections to susbtantia nigra, entopeduncular nucleus and globus pallidus. Neurosci. Lett., 9: 117–121.

Deniau, J.M., Hammond, C., Riszk, A. and Feger, J. (1978b) Electrophysiological properties of identified output neurons of the rat substantia nigra (pars compacta and pars reticulata): evidences for the existence of branched neurons. Exp. Brain Res., 32: 409–422.

Deniau, J.M., Kitai, S.T., Donoghue, J.P. and Grofova, I. (1982) Neuronal interactions in the substantia nigra pars reticulata through axon collaterals of the projection neurons. An electrophysiological and morphological study. Exp. Brain Res., 47: 105–113.

Deniau, J.M., Menetrey, A. and Charpier, S. (1996) The lamellar organization of the rat substantia nigra pars reticulata: segregated patterns of striatal afferents and relationship to the topography of corticostriatal projections. Neuroscience, 73: 761–781.

Deniau, J.M., Menetrey, A. and Thierry, A.M. (1994) Indirect nucleus accumbens input to the prefrontal cortex via the substantia nigra pars reticulata: a combined anatomical and electrophysiological study in the rat. Neuroscience, 61: 533–545.

Deniau, J.M. and Thierry, A.M. (1997) Anatomical segregation of information processing in the rat substantia nigra pars reticulata. Adv. Neurol., 74: 83–96.

166

Deransart, C. and Depaulis, A. (2002) The control of seizures by the basal ganglia? A review of experimental data. Epileptic Disord., 4(Suppl 3): S61–S72.

Deransart, C., Hellwig, B., Heupel-Reuter, M., Leger, J.F., Heck, D. and Lucking, C.H. (2003) Single-unit analysis of substantia nigra pars reticulata neurons in freely behaving rats with genetic absence epilepsy. Epilepsia, 44: 1513–1520.

Desban, M., Gauchy, C., Glowinski, J. and Kemel, M.L. (1995) Heterogeneous topographical distribution of the striatonigral and striatopallidal neurons in the matrix compartment of the cat caudate nucleus. J. Comp. Neurol., 352: 117–133.

Deutch, A.Y., Goldstein, M. and Roth, R.H. (1986) The ascending projections of the dopaminergic neurons of the substantia nigra, zona reticulata: a combined retrograde tracer-immunohistochemical study. Neurosci. Lett., 71: 257–263.

Di Chiara, G., Porceddu, M.L., Morelli, M., Mulas, M.L. and Gessa, G.L. (1979) Evidence for a GABAergic projection from the substantia nigra to the ventromedial thalamus and to the superior colliculus of the rat. Brain Res., 176: 273–284.

Di Giovanni, G., Di Matteo, V., La Grutta, V. and Esposito, E. (2001) m-Chlorophenylpiperazine excites non-dopaminergic neurons in the rat substantia nigra and ventral tegmental area by activating serotonin-2C receptors. Neuroscience, 103: 111–116.

Domesick, V.B., Stinus, L. and Paskevich, P.A. (1983) The cytology of dopaminergic and nondopaminergic neurons in the substantia nigra and ventral tegmental area of the rat: a light- and electron-microscopic study. Neuroscience, 8: 743–765.

Dray, A. (1979) The striatum and substantia nigra: a commentary on their relationships. Neuroscience, 4: 1407–1439.

Dray, A., Gonye, T.J. and Oakley, N.R. (1976) Caudate stimulation and substantia nigra activity in the rat. J. Physiol., 259: 825–849.

Faull, R.L. and Mehler, W.R. (1978) The cells of origin of nigrotectal, nigrothalamic and nigrostriatal projections in the rat. Neuroscience, 3: 989–1002.

Feger, J. and Ohye, C. (1975) The unitary activity of the substantia nigra following stimulation of the striatum in the awake monkey. Brain Res., 89: 155–159.

Feger, J. and Robledo, P. (1991) the effects of activation or inhibition of the subthalamic nucleus on the metabolic and electrophysiological activities within the pallidal complex and substantia nigra in the rat. Eur. J. Neurosci., 3: 947–952.

Ferraro, L., Tomasini, M.C., Fernandez, M., Bebe, B.W., O'Connor, W.T., Fuxe, K., Glennon, J.C., Tanganelli, S. and Antonelli, T. (2001) Nigral neurotensin receptor regulation of nigral glutamate and nigroventral thalamic GABA transmission: a dual-probe microdialysis study in intact conscious rat brain. Neuroscience, 102: 113–120.

Floran, B., Silva, I., Nava, C. and Aceves, J. (1988) Presynaptic modulation of the release of GABA by GABA_A receptors in pars compacta and by GABA_B receptors in pars reticulata of the rat substantia nigra. Eur. J. Pharmacol., 150: 277–286.

François, C., Grabli, D., McCairn, K., Jan, C., Karachi, C., Hirsch, E.C., Feger, J. and Tremblay, L. (2004) Behavioural disorders induced by external globus pallidus dysfunction in primates II. Anatomical study. Brain, 127(Pt 9): 2055–2070.

François, C., Percheron, G., Yelnik, J. and Heyner, S. (1979) Demonstration of the existence of small local circuit neurons in the Golgi-stained primate substantia nigra. Brain Res., 172: 160–164.

François, C., Percheron, G. and Yelnik, J. (1984) Localization of nigrostriatal, nigrothalamic and nigrotectal neurons in ventricular coordinates in macaques. Neuroscience, 13: 61–76.

François, C., Tande, D., Yelnik, J. and Hirsch, E.C. (2002) Distribution and morphology of nigral axons projecting to the thalamus in primates. J. Comp. Neurol., 447: 249–260.

François, C., Yelnik, J. and Percheron, G. (1987) Golgi study of the primate substantia nigra. II Spatial organization of dendritic arborizations in relation to the cytoarchitectonic boundaries and to the striatonigral bundle. J. Comp. Neurol., 265: 473–493.

François, C., Yelnik, J., Percheron, G. and Fenelon, G. (1994) Topographic distribution of the axonal endings from the sensorimotor and associative striatum in the macaque pallidum and substantia nigra. Exp. Brain Res., 102: 305–318.

Fujimoto, K. and Kita, H. (1992) Responses of rat substantia nigra pars reticulata units to cortical stimulation. Neurosci. Lett., 142: 105–109.

Gerfen, C.R. (1985) The neostriatal mosaic. I. Compartmental organization of projections from the striatum to the substantia nigra in the rat. J. Comp. Neurol., 236: 454–476.

Gerfen, C.R. (1992a) The neostriatal mosaic: multiple levels of compartmental organization. Trends Neurosci., 15: 133–139.

Gerfen, C.R. (1992b) The neostriatal mosaic: multiple levels of compartmental organization in the basal ganglia. Annu. Rev. Neurosci., 15: 285–320.

Gerfen, C.R., Staines, W.A., Arbuthnott, G.W. and Fibiger, H.C. (1982) Crossed connections of the substantia nigra in the rat. J. Comp. Neurol., 207: 283–303.

Gerfen, C.R. and Young III, W.S. (1988) Distribution of striatonigral and striatopallidal peptidergic neurons in both patch and matrix compartments: an in situ hybridization histochemistry and fluorescent retrograde tracing study. Brain Res., 460: 161–167.

German, D.C. and Manaye, K.F. (1993) Midbrain dopaminergic neurons (nuclei A8, A9, and A10): three-dimensional reconstruction in the rat. J. Comp. Neurol., 331: 297–309.

Gonzalez-Hernandez, T. and Rodriguez, M. (2000) Compartmental organization and chemical profile of dopaminergic and GABAergic neurons in the substantia nigra of the rat. J. Comp. Neurol., 421: 107–135.

Goswell, M.J. and Sedgwick, E.M. (1971) Inhibition in the substantia nigra following stimulation of the caudate nucleus. J. Physiol., 218: 84P–85P.

Gotz, T., Kraushaar, U., Geiger, J., Lubke, J., Berger, T. and Jonas, P. (1997) Functional properties of AMPA and NMDA receptors expressed in identified types of basal ganglia neurons. J. Neurosci., 17: 204–215.

Gould, E. and Butcher, L.L. (1986) Cholinergic neurons in the rat substantia nigra. Neurosci. Lett., 63: 315–319.

Grace, A.A. and Bunney, B.S. (1985) Opposing effects of striatonigral feedback pathways on midbrain dopamine cell activity. Brain Res., 333: 271–284.

Graybiel, A.M. (1984) Neurochemically specified subsystems in the basal ganglia. Ciba. Found Symp., 107: 114–149.

Graybiel, A.M. (1986) Neuropeptides in the basal ganglia. Res. Publ. Assoc. Res. Nerv. Ment. Dis., 64: 135–161.

Graybiel, A.M. (1990) Neurotransmitters and neuromodulators in the basal ganglia. Trends Neurosci., 13: 244–254.

Graybiel, A.M. and Canales, J.J. (2001) The neurobiology of repetitive behaviors: clues to the neurobiology of Tourette syndrome. Adv. Neurol., 85: 123–131.

Graybiel, A.M. and Ragsdale Jr., C.W. (1979) Fiber connections of the basal ganglia. Prog. Brain Res., 51: 237–283.

Graybiel, A.M. and Rauch, S.L. (2000) Toward a neurobiology of obsessive-compulsive disorder. Neuron, 28: 343–347.

Groenewegen, H.J. (2003) The basal ganglia and motor control. Neural. Plast., 10: 107–120.

Groenewegen, H.J. and Berendse, H.W. (1994) Anatomical relationships between the prefrontal cortex and the basal ganglia in the rat. In: Thierry A.M., Glowinski J., Goldman-Rakic P.S. and Christen Y. (Eds.), Motor and cognitive functions of the prefrontal cortex. Springer, Berlin, pp. 51–77.

Groenewegen, H.J., Wright, C.I. and Uylings, H.B. (1997) The anatomical relationships of the prefrontal cortex with limbic structures and the basal ganglia. J. Psychopharmacol., 11: 99–106.

Grofova, I., Deniau, J.M. and Kitai, S.T. (1982) Morphology of the substantia nigra pars reticulata projection neurons intracellularly labeled with HRP. J. Comp. Neurol., 208: 352–368.

Grofova, I. and Zhou, M. (1998) Nigral innervation of cholinergic and glutamatergic cells in the rat mesopontine tegmentum: light and electron microscopic anterograde tracing and immunohistochemical studies. J. Comp. Neurol., 395: 359–379.

Gulley, R.L. and Wood, R.L. (1971) The fine structure of the neurons in the rat substantia nigra. Tissue Cell, 3: 675–690.

Gurney, K., Prescott, T.J. and Redgrave, P. (2001) A computational model of action selection in the basal ganglia. I. A new functional anatomy. Biol. Cybern., 84: 401–410.

Guyenet, P.G. and Aghajanian, G.K. (1978) Antidromic identification of dopaminergic and other output neurons of the rat substantia nigra. Brain Res., 150: 69–84.

Haber, S.N. (2003) The primate basal ganglia: parallel and integrative networks. J. Chem. Neuroanat., 26: 317–330.

Hajdu, F., Hassler, R. and Bak, I.J. (1973) Electron microscopic study of the substantia nigra and the strio-nigral projection in the rat. Z Zellforsch Mikrosk Anat., 146: 207–221.

Hammond, C., Deniau, J.M., Riszk, A. and Feger, J. (1978) Electrophysiological demonstration of an excitatory subthalamonigral pathway in the rat. Brain Res., 151: 235–244.

Hanaway, J., McConnell, J.A. and Netsky, M.G. (1970) Cytoarchitecture of the substantia nigra in the rat. Am. J. Anat., 129: 417–437.

Harting, J.K., Updyke, B.V. and Van Lieshout, D.P. (2001) The visual-oculomotor striatum of the cat: functional relationship to the superior colliculus. Exp. Brain Res., 136: 138–142.

Hattori, T. (1993) Conceptual history of the nigrostriatal dopamine system. Neurosci. Res., 16: 239–262.

Hauber, W. (1998) Involvement of basal ganglia transmitter systems in movement initiation. Prog. Neurobiol., 56: 507–540.

Hedberg, T.G., Veliskova, J., Sperber, E.F., Nunes, M.L. and Moshe, S.L. (2003) Age-related differences in NMDA/metabotropic glutamate receptor binding in rat substantia nigra. Int. J. Dev. Neurosci., 21: 95–103.

Herkenham, M. (1979) The afferent and efferent connections of the ventromedial thalamic nucleus in the rat. J. Comp. Neurol., 183: 487–517.

Hikosaka, O., Sakamoto, M. and Miyashita, N. (1993) Effects of caudate nucleus stimulation on substantia nigra cell activity in monkey. Exp. Brain Res., 95: 457–472.

Hikosaka, O., Takikawa, Y. and Kawagoe, R. (2000) Role of the basal ganglia in the control of purposive saccadic eye movements. Physiol. Rev., 80: 953–978.

Hokfelt, T., Reid, M., Herrera-Marschitz, M., Ungerstedt, U., Terenius, L., Hakanson, R., Feng, D.M. and Folkers, K. (1991) Tachykinins and related peptides in the substantia nigra and neostriatum. Ann. N. Y. Acad. Sci., 632: 192–197.

Hontanilla, B., de las Heras, S. and Gimenez-Amaya, J.M. (1994) Organization of the striatal projections from the rostral caudate nucleus to the globus pallidus, the entopeduncular nucleus, and the pars reticulata of the substantia nigra in the cat. Anat. Rec., 238: 114–124.

Hoover, J.E. and Strick, P.L. (1999) The organization of cerebellar and basal ganglia outputs to primary motor cortex as revealed by retrograde transneuronal transport of herpes simplex virus type 1. J. Neurosci., 19: 1446–1463.

Huang, K.X., Bergstrom, D.A., Ruskin, D.N. and Walters, J.R. (1998) N-methyl-D-aspartate receptor blockade attenuates D1 dopamine receptor modulation of neuronal activity in the rat substantia nigra. Synapse, 30: 18–29.

Hubert, G.W., Paquet, M. and Smith, . (2001) Differential subcellular localization of mGluR1a and mGluR5 in the rat and monkey Substantia nigra. J. Neurosci., 21: 1838–1847.

Hubert, G.W. and Smith, Y. (2004) Age-related changes in the expression of axonal and glial group I metabotropic glutamate receptor in the rat substantia nigra pars reticulata. J. Comp. Neurol., 475: 95–106.

Huerta, M.F., Van Lieshout, D.P. and Harting, J.K. (1991) Nigrotectal projections in the primate Galago crassicaudatus. Exp. Brain Res., 87: 389–401.

Hutchison, W.D., Dostrovsky, J.O., Walters, J.R., Courtemanche, R., Boraud, T., Goldberg, J. and Brown, P. (2004) Neuronal oscillations in the basal ganglia and movement disorders: evidence from whole animal and human recordings. J. Neurosci., 24: 9240–9243.

Joel, D. and Weiner, I. (1994) The organization of the basal ganglia-thalamocortical circuits: open interconnected rather than closed segregated. Neuroscience, 63: 363–379.

Joel, D. and Weiner, I. (1997) The connections of the primate subthalamic nucleus: indirect pathways and the open-interconnected scheme of basal ganglia-thalamocortical circuitry. Brain Res. Rev., 23: 62–78.

Joseph, J.P. and Boussaoud, D. (1985) Role of the cat substantia nigra pars reticulata in eye and head movements. I. Neural activity. Exp. Brain Res., 57: 286–296.

Juraska, J.M., Wilson, C.J. and Groves, P.M. (1977) The substantia nigra of the rat: a Golgi study. J. Comp. Neurol., 172: 585–600.

Kanazawa, I. and Yoshida, M. (1980) Electrophysiological evidence for the existence of excitatory fibres in the caudatonigral pathway in the cat. Neurosci. Lett., 20: 301–306.

Karabelas, A.B. and Purpura, D.P. (1980) Evidence for autapses in the substantia nigra. Brain Res., 200: 467–473.

Kawaguchi, Y., Wilson, C.J. and Emson, P.C. (1990) Projection subtypes of rat neostriatal matrix cells revealed by intracellular injection of biocytin. J. Neurosci., 10: 3421–3438.

Kemel, M.L., Desban, M., Gauchy, C., Glowinski, J. and Besson, M.J. (1988) Topographical organization of efferent projections from the cat substantia nigra pars reticulata. Brain Res., 455: 307–323.

Kita, H. (2001) Neostriatal and globus pallidus stimulation induced inhibitory postsynaptic potential in entopeduncular neurons in rat brain slice preparation. Neuroscience, 105: 871–879.

Kita, H. and Kitai, S.T. (1987) Efferent projections of the subthalamic nucleus in the rat: light and electron microscopic analysis with the PHA-L method. J. Comp. Neurol., 260: 435–452.

Kita, H. and Kitai, S.T. (1991) Intracellular study of rat globus pallidus neurons: membrane properties and responses to neostriatal, subthalamic and nigral stimulation. Brain Res., 564: 296–305.

Kita, H. and Kitai, S.T. (1994) The morphology of globus pallidus projection neurons in the rat: an intracellular staining study. Brain Res., 636: 308–319.

Kitai, S.T., Wagner, A., Precht, W. and Ono, T. (1975) Nigro-caudate and caudato-nigral relationship: an electrophysiological study. Brain Res., 85: 44–48.

Kitano, H., Tanibucchi, I. and Jinnai, K. (1998) The distribution of neurons in the susbtantia nigra pars reticulate with input from the motor, premotor and prefrontal areas of the cerebral cortex in the monkey. Brain Res., 784: 228–238.

Kolomiets, B.P., Deniau, J.M., Glowinski, J. and Thierry, A.M. (2003) Basal ganglia and processing of cortical information: functional interactions between trans-striatal and trans-subthalamic circuits in the substantia nigra pars reticulata. Neuroscience, 117: 931–938.

Kuroda, M. and Price, J.L. (1991) Ultrastructure and synaptic organization of axon terminals from brainstem structures to the mediodorsal thalamic nucleus of the rat. J. Comp. Neurol., 313: 539–552.

Korotkova, T.M., Eriksson, K.S., Haas, H.L. and Brown, R.E. (2002) Selective excitation of GABAergic neurons in the substantia nigra of the rat by orexin/hypocretin in vitro. Regul. Pept., 104: 83–89.

Langer, L.F., Jimenez-Castellanos, J. and Graybiel, A.M. (1991) The substantia nigra and its relations with the striatum in the monkey. Prog. Brain Res., 87: 81–99.

Lavin, A. and Garcia-Munoz, M. (1986) Electrophysiological changes in substantia nigra after dynorphin administration. Brain Res., 369: 298–302.

Lei, W., Jiao, Y., Del Mar, N. and Reiner, A. (2004) Evidence for differential cortical input to direct pathway versus indirect pathway striatal projection neurons in rats. J. Neurosci., 24: 8289–8299.

Levesque, M., Bedard, A., Cossette, M. and Parent, A. (2003) Novel aspects of the chemical anatomy of the striatum and its efferent projections. J. Chem. Neuroanat., 26: 271–281.

Levy, R., Hazrati, L.N., Herrero, M.T., Vila, M., Hassani, O.K., Mouroux, M., Ruberg, M., Asensi, H., Agid, Y., Feger, J., Obeso, J.A., Parent, A. and Hirsch, E.C. (1997) Re-evaluation of the functional anatomy of the basal ganglia in normal and Parkinsonian states. Neuroscience, 76: 335–343.

Lynd-Balta, E. and Haber, S.N. (1994) The organization of midbrain projections to the striatum in the primate: sensorimotor-related striatum versus ventral striatum. Neuroscience, 59: 625–640.

Mahon, S., Deniau, J.M. and Charpier, S. (2001) Relationship between EEG potentials and intracellular activity of striatal and cortico-striatal neurons: an in vivo study under different anesthetics. Cereb. Cortex, 11: 360–373.

Mahon, S., Deniau, J.M. and Charpier, S. (2003) Various synaptic activities and firing patterns in cortico-striatal and striatal neurons in vivo. J. Physiol. Paris, 97: 557–566.

Mailly, P., Charpier, S., Mahon, S., Menetrey, A., Thierry, A.M., Glowinski, J. and Deniau, J.M. (2001) Dendritic arborizations of the rat substantia nigra pars reticulata neurons: spatial organization and relation to the lamellar compartmentation of striato-nigral projections. J. Neurosci., 21: 6874–6888.

Mailly, P., Charpier, S., Menetrey, A. and Deniau, J.M. (2003) Three-dimensional organization of the recurrent axon collateral network of the substantia nigra pars reticulata neurons in the rat. J. Neurosci., 23: 5247–5257.

Mana, S. and Chevalier, G. (2001) The fine organization of nigro-collicular channels with additional observations of their relationships with acetylcholinesterase in the rat. Neuroscience, 106: 357–374.

Marino, M.J., Awad, H., Poisik, O., Wittmann, M. and Conn, P.J. (2002) Localization and physiological roles of metabotropic glutamate receptors in the direct and indirect pathways of the basal ganglia. Amino Acids, 23: 185–191.

Marti, M., Guerrini, R., Beani, L., Bianchi, C. and Morari, M. (2002) Nociceptin/orphanin FQ receptors modulate glutamate extracellular levels in the substantia nigra pars reticulate. A microdialysis study in the awake freely moving rat. Neuroscience, 112: 153–160.

Martin, L.J., Blackstone, C.D., Levey, A.I., Huganir, R.L. and Price, D.L. (1993) AMPA glutamate receptor subunits are differentially distributed in rat brain. Neuroscience, 53: 327–358.

Martin, J.L., Chesselet, M.F., Raynor, K., Gonzales, C. and Reisine, T. (1991) Differential distribution of somatostatin receptor subtypes in rat brain revealed by newly developed somatostatin analogs. Neuroscience, 41: 581–593.

Martin, L.P. and Waszczak, B.L. (1994) D1 agonist-induced excitation of substantia nigra pars reticulata neurons: mediation by D1 receptors on striatonigral terminals via a pertussis toxin-sensitive coupling pathway. J. Neurosci., 14: 4494–4506.

Martin, L.P. and Waszczak, B.L. (1996) Dopamine D2, receptor-mediated modulation of the GABAergic inhibition of substantia nigra pars reticulate neurons. Brain Res., 729: 156–169.

Martinez-Murillo, R., Villalba, R., Montero-Caballero, M.I. and Rodrigo, J. (1989) Cholinergic somata and terminals in the rat substantia nigra: an immunocytochemical study with optical and electron microscopic techniques. J. Comp. Neurol., 281: 397–415.

Maurice, N., Deniau, J.M., Glowinski, J. and Thierry, A.M. (1999) Relationships between the prefrontal cortex and the basal ganglia in the rat: physiology of the cortico-nigral circuits. J. Neurosci., 19: 4674–4681.

McBride, R.L. and Larsen, K.D. (1980) Projections of the feline globus pallidus. Brain Res., 189: 3–14.

McGeer, E.G., Staines, W.A. and McGeer, P.L. (1984) Neurotransmitters in the basal ganglia. Can. J. Neurol. Sci., 11(Suppl 1): 89–99.

McGeorge, A.J. and Faull, R.L. (1989) The organization of the projection from the cerebral cortex to the striatum in the rat. Neuroscience, 29: 503–537.

McNair, J.L., Sutin, J. and Tsubokawa, T. (1972) Suppression of cell firing in the substantia nigra by caudate nucleus stimulation. Exp. Neurol., 37: 395–411.

Middleton, F.A. and Strick, P.L. (1996) The temporal lobe is a target of output from the basal ganglia. Proc. Natl. Acad. Sci. USA, 93: 8683–8687.

Middleton, F.A. and Strick, P.L. (2002) Basal-ganglia 'projections' to the prefrontal cortex of the primate. Cereb. Cortex, 9: 926–935.

Mink, J.W. and Thach, W.T. (1993) Basal ganglia intrinsic circuits and their role in behavior. Curr. Opin. Neurobiol., 3: 950–957.

Misgeld, U. (2004) Innervation of the substantia nigra. Cell Tissue Res., 318: 107–114.

Mitchell, B.D. and Cauller, L.J. (2001) Corticocortical and thalamocortical projections to layer I of the frontal neocortex in rats. Brain Res., 921: 68–77.

Miyamoto, Y. and Jinnai, K. (1994) The inhibitory input from the substantia nigra to the mediodorsal nucleus neurons projecting to the prefrontal cortex in the cat. Brain Res., 649: 313–318.

Miyazaki, T. and Lacey, M.G. (1998) Presynaptic inhibition by dopamine of a discrete component of GABA release in rat substantia nigra pars reticulate. J. Physiol., 513: 805–817.

Montaron, M.F., Deniau, J.M., Menetrey, A., Glowinski, J. and Thierry, A.M. (1996) Prefrontal cortex inputs of the nucleus accumbens-nigro-thalamic circuit. Neuroscience, 71: 371–382.

Nakanishi, H., Kita, H. and Kitai, S.T. (1987) Intracellular study of rat substantia nigra pars reticulata neurons in an in vitro slice preparation: electrical membrane properties and response characteristics to subthalamic stimulation. Brain Res., 437: 45–55.

Nambu, A. (2004) A new dynamic model of the cortico-basal ganglia loop. Prog. Brain Res., 143: 461–466.

Nelson, E.L., Liang, C.L., Sinton, C.M. and German, D.C. (1996) Midbrain dopaminergic neurons in the mouse: computer-assisted mapping. J. Comp. Neurol., 369: 361–371.

Niijima, K. and Yoshida, M. (1982) Electrophysiological evidence for branching nigral projections to pontine reticular formation, superior colliculus and thalamus. Brain Res., 239: 279–282.

Nisenbaum, E.S. and Wilson, C.J. (1995) Potassium current responsible for inward and outward rectification in rat neostriatal spiny projection neurons. J. Neurosci., 15: 4449–4463.

Nishimura, Y., Takada, M. and Mizuno, N. (1997) Topographic distribution and collateral projections of the two major populations of nigrothalamic neurons. A retrograde labeling study in the rat. Neurosci. Res., 28: 1–9.

Oertel, W.H. and Mugnaini, E. (1984) Immunocytochemical studies of GABAergic neurons in rat basal ganglia and their relations to other neuronal systems. Neurosci. Lett., 47: 233–238.

Olazabal, U.E. and Moore, J.K. (1989) Nigrotectal projection to the inferior colliculus: horseradish peroxidase transport and tyrosine hydroxylase immunohistochemical studies in rats, cats, and bats. J. Comp. Neurol., 282: 98–118.

Paquet, M. and Smith, Y. (1996) Differential localization of AMPA glutamate receptor subunits in the two segments of the globus pallidus and the substantia nigra pars reticulata in the squirrel monkey. Eur. J. Neurosci., 8: 229–233.

Parent, A., Cote, P.Y. and Lavoie, B. (1995) Chemical anatomy of primate basal ganglia. Prog. Neurobiol., 46: 131–197.

Parent, A. and De Bellefeuille. (1983) The pallidointralaminar and pallidonigral projections in primate as studied by retrograde double-labeling method. Brain Res., 278: 11–27.

Parent, A. and Hazrati, L.N. (1994) Multiple striatal representation in primate substantia nigra. J. Comp. Neurol., 344: 305–320.

Parent, A. and Hazrati, L.N. (1995a) Functional anatomy of the basal ganglia. I. The cortico-basal ganglia-thalamo-cortical loop. Brain Res. Brain Res. Rev., 20: 91–127.

Parent, A. and Hazrati, L.N. (1995b) Functional anatomy of the basal ganglia. II. The place of subthalamic nucleus and external pallidum in basal ganglia circuitry. Brain Res. Brain Res. Rev., 20: 128–154.

Parent, A., Mackey, A., Smith, Y. and Boucher, R. (1983) The output organization of the substantia nigra in primate as revealed by a retrograde double labeling method. Brain Res. Bull., 10: 529–537.

Parent, A. and Smith, Y. (1987) Organization of efferent projections of the subthalamic nucleus in the squirrel monkey as revealed by retrograde labeling methods. Brain Res., 436: 296–310.

Paz, J.T., Deniau, J.M. and Charpier, S. (2005) Rhythmic bursting in the cortico-subthalamo-pallidal network during spontaneous genetically determined spike and wave discharges. J. Neurosci., 25: 2092–2101.

Percheron, G., Francois, C., Talbi, B., Yelnik, J. and Fenelon, G. (1996) The primate motor thalamus. Brain Res. Brain Res. Rev., 22: 93–181.

Pickel, V.M., Chan, J., Veznedaroglou, E. and Milner, T.A. (1995) Neuropeptide Y and dynorphin-immunoreactive large dense-core vesicles are strategically localized for presynaptic modulation in the hippocampal formation and substantia nigra. Synapse, 19: 160–169.

Poirier, L.J., Giguere, M. and Marchand, R. (1983) Comparative morphology of the substantia nigra and ventral tegmental area in the monkey, cat and rat. Brain Res. Bull., 11: 371–397.

Precht, W. and Yoshida, M. (1971) Blockage of caudate-evoked inhibition of neurons in the substantia nigra by picrotoxin. Brain Res., 32: 229–233.

Radnikow, G. and Misgeld, U. (1998) Dopamine D1 receptors facilitate GABA$_A$ synaptic currents in the rat substantia nigra pars reticulata. J. Neurosci., 18: 2009–2016.

Redgrave, P., Marrow, L. and Dean, P. (1992) Topographical organization of the nigrotectal projection in rat: evidence for segregated channels. Neuroscience, 50: 571–595.

Redgrave, P., Prescott, T.J. and Gurney, K. (1999). The basal ganglia: a vertebrate solution to the selection problem? Neuroscience, 89: 1009–1023.

Reiner, A., Medina, L. and Haber, S.N. (1999) The distribution of dynorphinergic terminals in striatal target regions in comparison to the distribution of substance P-containing and enkephalinergic terminals in monkeys and humans. Neuroscience, 88: 775–793.

Rhoades, R.W., Kuo, D.C., Polcer, J.D., Fish, S.E. and Voneida, T.J. (1982) Indirect visual cortical input to the deep layers of the hamster's superior colliculus via the basal ganglia. J. Comp. Neurol., 208: 239–254.

Richards, C.D., Shiroyama, T. and Kitai, S.T. (1997) Electrophysiological and immunocytochemical characterization of GABA and dopamine neurons in the substantia nigra of the rat. Neuroscience, 80: 545–557.

Rick, C.E. and Lacey, M.G. (1994) Rat substantia nigra pars reticulata neurones are tonically inhibited via GABA$_A$, but not GABA$_B$, receptors in vitro. Brain Res., 659: 133–137.

Rinvik, E. and Grofova, I. (1970) Observations on the fine structure of the substantia nigra in the cat. Exp. Brain Res., 11: 229–248.

Rinvik, E., Grofova, I. and Ottersen, O.P. (1976) Demonstration of nigrotectal and nigroreticular projections in the cat by axonal transport of proteins. Brain Res., 112: 388–394.

Rinvik, E. and Ottersen, O.P. (1993) Terminals of subthalamonigral fibres are enriched with glutamate-like immunoreactivity: an electron microscopic, immunogold analysis in the cat. J. Chem. Neuroanat., 6: 19–30.

Robledo, P. and Feger, J. (1990) Excitatory influence of rat subthalamic nucleus to substantia nigra pars reticulata and the pallidal complex: electrophysiological data. Brain Res., 518: 47–54.

Robertson, B.C., Hommer, D.W. and Skirboll, L.R. (1987) Electrophysiological evidence for a non-opioid interaction between dynorphin and GABA in the substantia nigra of the rat. Neuroscience, 23: 483–490.

Ruffieux, A. and Schultz, W. (1980) Dopaminergic activation of reticulate neurons in the substantia nigra. Nature, 285: 240–241.

Ruskin, D.N., Bergstrom, D.A., Tierney, P.L. and Walters, J.R. (2003) Correlated multisecond oscillations in firing rate in the basal ganglia: modulation by dopamine and the subthalamic nucleus. Neuroscience, 117: 427–438.

Ryan, L.J. and Sanders, D.J. (1994) Subthalamic nucleus and globus pallidus lesions alter activity in nigrothalamic neurons in rats. Brain Res. Bull., 34: 19–26.

Rye, D.B., Saper, C.B., Lee, H.J. and Wainer, B.H. (1987) Pedunculopontine tegmental nucleus of the rat: cytoarchitecture, cytochemistry, and some extrapyramidal connections of the mesopontine tegmentum. J. Comp. Neurol., 259: 483–528.

Sakai, S.T., Grofova, I. and Bruce, K. (1998) Nigrothalamic projections and nigrothalamocortical pathway to the medial agranular cortex in the rat: single- and double-labeling light and electron microscopic studies. J. Comp. Neurol., 391: 506–525.

Sakai, S.T. and Patton, K. (1993) Distribution of cerebellothalamic and nigrothalamic projections in the dog: a double anterograde tracing study. J. Comp. Neurol., 330: 183–194.

Schmitt, P., Souliere, F., Dugast, C. and Chouvet, G. (1999) Regulation of substantia nigra pars reticulata neuronal activity by excitatory amino acids. Naunyn Schmiedebergs Arch. Pharmacol., 360: 402–412.

Schwyn, R.C. and Fox, C.A. (1974) The primate substantia nigra: a Golgi and electron microscopic study. J. Hirnforsch., 15: 95–126.

Shink, E., Bevan, M.D., Bolam, J.P. and Smith, Y. (1996) The subthalamic nucleus and the external pallidum: two tightly interconnected structures that control the output of the basal ganglia in the monkey. Neuroscience, 73: 335–357.

Slaght, S.J., Paz, T., Mahon, S., Maurice, N., Charpier, S. and Deniau, J.M. (2002) Functional organization of the circuits connecting the cerebral cortex and the basal ganglia: implications for the role of the basal ganglia in epilepsy. Epileptic Disord., 4(Suppl 3): S9–S22.

Smith, Y., Bevan, M.D., Shink, E. and Bolam, J.P. (1998) Microcircuitry of the direct and indirect pathways of the basal ganglia. Neuroscience, 86: 353–387.

Smith, A.D. and Bolam, J.P. (1989) Neurons of the substantia nigra reticulata receive a dense GABA-containing input from the globus pallidus in the rat. Brain Res., 493: 160–167.

Smith, A.D. and Bolam, J.P. (1990a) The output neurons and the dopaminergic neurons of the susbtantia nigra receive a GABA-containing input from the globus pallidus in the rat. J. Comp. Neurol., 296: 47–64.

Smith, A.D. and Bolam, J.P. (1990b) The neural network of the basal ganglia as revealed by the study of synaptic connections of identified neurones. Trends Neurosci., 13: 259–265.

Smith, Y., Charara, A., Paquet, M., Kieval, J.Z., Pare, J.F., Hanson, J.E., Hubert, G.W., Kuwajima, M. and Levey, A.I. (2001) Ionotropic and metabotropic GABA and glutamate receptors in primate basal ganglia. J. Chem. Neuroanat., 22: 13–42.

Smith, Y., Hazrati, L.N. and Parent, A. (1990) Efferent projections of the subthalamic nucleus in the squirrel monkey as studied by the PHA-L anterograde tracing method. J. Comp. Neurol., 294: 306–323.

Smith, Y., Parent, A., Seguela, P. and Descarries, L. (1987) Distribution of GABA-immunoreactive neurons in the basal ganglia of the squirrel monkey (Saimiri sciureus). J. Comp. Neurol., 259: 50–64.

Standaert, D.G., Testa, C.M., Young, A.B. and Penney Jr., J.B. (1994) Organization of N-methyl-D-aspartate glutamate receptor gene expression in the basal ganglia of the rat. J. Comp. Neurol., 343: 1–16.

Steindler, D.A. and Deniau, J.M. (1980) Anatomical evidence for collateral branching of substantia nigra neurons: a combined horseradish peroxidase and [^{3}H]wheat germ agglutinin axonal transport study in the rat. Brain Res., 196: 228–236.

Steiner, H. and Gerfen, C.R. (1998) Role of dynorphin and enkephalin in the regulation of striatal output pathways and behavior. Exp. Brain Res., 123: 60–76.

Takakusaki, K., Habaguchi, T., Ohtinata-Sugimoto, J., Saitoh, K. and Sakamoto, T. (2003) Basal ganglia efferents to the brainstem centers controlling postural muscle tone and locomotion: a new concept for understanding motor disorders in basal ganglia dysfunction. Neuroscience, 119: 293–230.

Takakusaki, K., Saitoh, K., Harada, H. and Kashiwayanagi, M. (2004) Role of basal ganglia-brainstem pathways in the control of motor behaviors. Neurosci. Res., 50: 137–151.

Tepper, J.M., Martin, L.P. and Anderson, D.R. (1995) GABA$_A$ receptor-mediated inhibition of rat substantia nigra dopaminergic neurons by pars reticulata projection neurons. J. Neurosci., 15: 3092–3103.

Threlfell, S., Cragg, S.J., Kallo, I., Turi, G.F., Coen, C.W. and Greenfield, S.A. (2004) Histamine H3 receptors inhibit serotonin release in substantia nigra pars reticulata. J. Neurosci., 24: 8704–8710.

Timmerman, W. and Abercrombie, E.D. (1996) Amphetamine-induced release of dendritic dopamine in substantia nigra pars reticulate: D1-mediated behavioral and electrophysiological effects. Synapse, 23: 280–291.

Tokuno, H., Nakamura, Y., Kudo, M. and Kitao, Y. (1990) Laminar organization of the substantia nigra pars reticulata in the cat. Neuroscience, 38: 255–270.

Tokuno, H., Chiken, S., Kametani, K. and Moriizumi, T. (2002) Efferent projections from the striatal patch compartment: anterograde degeneration after selective ablation of neurons expressing mu-opioid receptor in rats. Neurosci. Lett., 332: 5–8.

Tokuno, H., Takada, M., Kondo, Y. and Mizuno, N. (1993) Laminar organization of the substantia nigra pars reticulata

in the macaque monkey, with special reference to the caudato-nigro-tectal link. Exp. Brain Res., 92: 545–548.

Totterdell, S., Bolam, J.P. and Smith, A.D. (1984) Characterization of pallidonigral neurons in the rat by a combination of Golgi impregnation and retrograde transport of horseradish peroxidase: their monosynaptic input from the neostriatum. J. Neurocytol., 13: 593–616.

Trevitt, T., Carlson, B., Correa, M., Keene, A., Morales, M. and Salamone, J.D. (2002) Interactions between dopamine D1 receptors and gamma-aminobutyric acid mechanisms in substantia nigra pars reticulata of the rat: neurochemical and behavioral studies. Psychopharmacology (Berl.), 159: 229–237.

Tse, Y.C. and Yung, K.K. (2000) Cellular expression of ionotropic glutamate receptor subunits in subpopulations of neurons in the rat substantia nigra pars reticulata. Brain Res., 854: 57–69.

Tsumori, T., Yokota, S., Ono, K. and Yasui, Y. (2003) Nigrothalamostriatal and nigrothalamocortical pathways via the ventrolateral parafascicular nucleus. Neuroreport, 14: 81–86.

Valentino, K.L., Tatemoto, K., Hunter, J. and Barchas, J.D. (1986) Distribution of neuropeptide K-immunoreactivity in the rat central nervous system. Peptides, 7: 1043–1059.

Van Der Kooy, D. and Hattori, T. (1980) Single subthalamic nucleus neurons project to both the globus pallidus and substantia nigra in rat. J. Comp. Neurol., 192: 751–768.

Vercueil, L. and Hirsch, E. (2002) Seizures and the basal ganglia: a review of the clinical data. Epileptic Disord., 4(Suppl 3): S47–S54.

Von Krosigk, M., Smith, Y., Bolam, J.P. and Smith, A.D. (1992) Synaptic organization of GABAergic inputs from the striatum and the globus pallidus onto neurons in the substantia nigra and retrorubral field which project to the medullary reticular formation. Neuroscience, 50: 531–549.

Wallmichrath, I. and Szabo, B. (2002) Cannabinoids inhibit striatonigral GABAergic neurotransmission in the mouse. Neuroscience, 113: 671–682.

Waszczak, B.L. and Walters, J.R. (1984) A physiological role for dopamine as modulator of GABA effects in substantia nigra: supersensitivity in 6-hydroxydopamine-lesioned rats. Eur. J. Pharmacol., 105: 369–373.

Waszczak, B.L. and Walters, J.R. (1986) Endogenous dopamine can modulate inhibition of substantia nigra pars reticulata neurons elicited by GABA iontophoresis or striatal stimulation. J. Neurosci., 6: 120–126.

Wichmann, T. and DeLong, M.R. (1998) Models of basal ganglia function and pathophysiology of movement disorders. Neurosurg. Clin. N. Am., 9: 223–236.

Wichmann, T. and DeLong, M.R. (2003) Pathophysiology of Parkinson's disease: the MPTP primate model of the human disorder. Ann. N. Y. Acad. Sci., 991: 199–213.

Williams, M.N. and Faull, R.L. (1988) The nigrotectal projection and tectospinal neurons in the rat A light and electron microscopic study demonstrating a monosynaptic nigral input to identified tectospinal neurons. Neuroscience, 25: 533–562.

172

Williams, D., Kuhn, A., Kupsch, A., Tijssen, M., van Bruggen, G., Speelman, H., Hotton, G., Loukas, C. and Brown, P. (2005) The relationship between oscillatory activity and motor reaction time in the parkinsonian subthalamic nucleus. Eur. J. Neurosci., 21: 249–258.

Wilson, C.J. and Kawaguchi, Y. (1996) The origins of two-state spontaneous membrane potential fluctuations of neostriatal spiny neurons. J. Neurosci., 16: 2397–2410.

Wilson, C.J., Young, S.J. and Groves, P.M. (1977) Statistical properties of neuronal spike trains in the substantia nigra: cell types and their interactions. Brain Res., 136: 243–260.

Wittmann, M., Marino, M.J., Bradley, S.R. and Conn, P.J. (2001) Activation of group III mGluRs inhibits GABAergic and glutamatergic transmission in the substantia nigra pars reticulata. J. Neurophysiol., 85: 1960–1968.

Wittmann, M., Marino, M.J. and Conn, P.J. (2002) Dopamine modulates the function of group II and group III metabotropic glutamate receptors in the substantia nigra pars reticulate. J. Pharmacol. Exp. Ther., 302: 433–441.

Wu, Y., Richard, S. and Parent, A. (2000) The organization of the striatal output system: a single-cell juxtacellular labeling study in the rat. Neurosci. Res., 38: 49–62.

Wurtz, R.H. and Hikosaka, O. (1986) Role of the basal ganglia in the initiation of saccadic eye movements. Prog. Brain Res., 64: 175–190.

Yasui, Y., Tsumori, T., Ando, A. and Domoto, T. (1995) Demonstration of axon collateral projections from the substantia nigra pars reticulata to the superior colliculus and the parvicellular reticular formation in the rat. Brain Res., 674: 122–126.

Yelnik, J., Francois, C., Percheron, G. and Heyner, S. (1987) Golgi study of the primate substantia nigra. I. Quantitative morphology and typology of nigral neurons. J. Comp. Neurol., 265: 455–472.

Yoshida, M., Nakajima, N. and Niijima, K. (1981) Effect of stimulation of the putamen on the substantia nigra in the cat. Brain Res., 217: 169–174.

You, Z.B., Saria, A., Fischer-Colbrie, R., Terenius, L., Goiny, M. and Herrera-Marschitz, M. (1996) Effects of secretogranin II-derived peptides on the release of neurotransmitters monitored in the basal ganglia of the rat with in vivo microdialysis. Naunyn Schmiedebergs Arch. Pharmacol., 354(6): 717–724.

Yung, K.K. (1998) Localization of ionotropic and metabotropic glutamate receptors in distinct neuronal elements of the rat substantia nigra. Neurochem. Int., 33: 313–326.

Yung, W.H., Hausser, M.A. and Jack, J.J. (1991) Electrophysiology of dopaminergic and non-dopaminergic neurones of the guinea-pig substantia nigra pars compacta in vitro. J. Physiol., 36: 643–667.

Zhang, J., Chiodo, L.A. and Freeman, A.S. (1994) Influence of excitatory amino acid receptor subtypes on the electrophysiological activity of dopaminergic and nondopaminergic neurons in rat substantia nigra. J. Pharmacol. Exp. Ther., 269: 313–321.

Zhang, J. and Freeman, A.S. (1994) Electrophysiological effects of cholecystokinin on neurons in rat substantia nigra pars reticulata. Brain Res., 652: 154–156.

Tepper, Abercrombie & Bolam (Eds.)
Progress in Brain Research, Vol. 160
ISSN 0079-6123

CHAPTER 10

GABAergic control of the subthalamic nucleus

Mark D. Bevan[1,*], Nicholas E. Hallworth[1] and Jérôme Baufreton[2]

[1]*Northwestern University, Department of Physiology, Feinberg School of Medicine, 303 E. Chicago Avenue, Chicago, IL 60611, USA*
[2]*Laboratoire de physiologie et physiopathologie de la signalisation cellulaire, UMR CNRS 5543, Université de Bordeaux 2, 146 Rue Léo Saignat, 33076 Bordeaux Cedex, France*

Abstract: The glutamatergic subthalamic nucleus (STN) is a key component of the basal ganglia, a group of subcortical brain nuclei important for voluntary movement and the site of dysfunction in Parkinson's disease. The rate and pattern of STN activity is precisely regulated by the reciprocally connected GABA-ergic external globus pallidus (GP_e) and glutamatergic afferents from the cortex. Subthalamic neurons possess intrinsic membrane properties that underlie the autonomous generation of action potentials and complex forms of synaptic integration. Thus, GABA acting at $GABA_A$ and/or $GABA_B$ receptors can inhibit/reset autonomous activity by deactivating postsynaptic voltage-dependent Na^+ (Na_v) channels and generate sufficient hyperpolarization for rebound burst firing, through the de-inactivation of postsynaptic voltage-dependent Ca^{2+} (Ca_v) and Na_v channels. Feedback inhibition from the GP_e can therefore paradoxically and transiently increase the efficacy of subsequent excitatory synaptic inputs, and thus enhance the response of the STN to rhythmic input from the cortex. Evidence is also provided that dopamine acting at post- and presynaptic receptors in the STN may, through actions on the integrative properties of STN neurons and activity-dependent synaptic plasticity, be critical for the patterning of STN neuronal activity in vivo. Taken together, these discoveries may be relevant for the emergence of correlated, rhythmic, burst firing in the dopamine-depleted STN of patients with PD.

Keywords: cortex; dopamine; excitation; inhibition; oscillation; Parkinson's disease; synaptic integration

Introduction

The subthalamic nucleus (STN) is a lens-shaped structure that rests on the internal capsule. In terms of its dimensions and total number of neurons, the STN is a relatively minor component of the basal ganglia (Oorschot, 1996). However, in functional terms, the STN is considerably more significant because abnormal activity of the STN is associated with profound disorders of

movement and direct manipulation of the STN can ameliorate the symptoms of such disorders (Bergman et al., 1994; Wichmann et al., 1994). Twenty years ago Kita and Kitai termed the STN the 'driving force' of the basal ganglia because it is predominantly composed of glutamatergic projection neurons that innervate the GABAergic external segment of the globus pallidus (GP_e in primates, GP in subprimates) and the basal ganglia output nuclei (the internal segment of the globus pallidus (GP_i) and the substantia nigra (SN) (Nakanishi et al., 1987a; Smith and Parent, 1988; Bevan et al., 1994). In accordance with their

*Corresponding author. Tel: +312 503 4828;
Fax: 312 503 5101; E-mail: m-bevan@northwestern.edu

DOI: 10.1016/S0079-6123(06)60010-1
173

174

proposition (which at the time was controversial), the STN has since been shown to drive, in part, the resting tonic activity of the basal ganglia and in addition, relay phasic cortical and thalamic excitation to its target nuclei during movement (Fujimoto and Kita, 1993; Maurice et al., 1998). The output of the STN is however precisely regulated by powerful feedback inhibition from the reciprocally connected GABAergic GP$_e$/GP (Shink et al., 1996). Although key to understanding both the normal and pathological operation of the basal ganglia, the interaction between the STN and GP$_e$ has been difficult to incorporate into models of basal ganglia function and dysfunction, in part, because the interaction between the STN and GP$_e$/GP neurons is complicated by their intrinsic membrane properties, which underlie unusual forms of synaptic integration (Bevan et al., 2002; Hanson et al., 2004; Baufreton et al., 2005a; Hallworth and Bevan, 2005).

Recent studies have demonstrated strong correlations between the patterns of STN activity and the symptoms of Parkinson's disease. Together with the discovery of direct dopaminergic inputs to the STN that degenerate in Parkinson's disease (PD) (Lavoie et al., 1989; Hassani et al., 1997; Francois et al., 2000; Cragg et al., 2004), there is speculation that abnormal activity in PD may be due, partly, to the loss of dopamine in the STN and abnormal patterning of the STN by GABAergic inputs from the GP$_e$ (Hallworth and Bevan, 2005; Baufreton et al., 2005a).

The primary objective of this chapter is to review the principles underlying the patterning of STN activity by GABAergic inhibition arising from the GP$_e$/GP under normal conditions and in the absence of dopamine. The following subjects will be reviewed: the major inputs to STN neurons; the intrinsic membrane properties of STN neurons; the principles underlying inhibitory synaptic integration; the influence of GABAergic inhibition on the integration of glutamatergic synaptic inputs; the regulation by dopamine of GABAergic inhibition in the STN; and finally the patterning of STN activity by GABAergic inhibition in vivo.

Afferents of the STN

GABAergic afferents of the STN

The major GABAergic input to the STN is derived from the GP$_e$/GP. The projection is ipsilateral and topographic in nature (Smith et al., 1990; Shink et al., 1996; Bevan et al., 1997) and largely reciprocated by an excitatory projection from the STN (Shink et al., 1996). Although similar functional (motor, associative and limbic) zones are reciprocally connected, a certain degree of divergence/overlap exists between functionally divergent neurons (Bevan et al., 1997; Joel and Weiner, 1997).

In rats, all GP neurons that project within the basal ganglia innervate the STN. Individual GP neurons also project to other basal ganglia via axon collaterals (Fig. 1A; Bevan et al., 1998). A similar pattern of innervation has also been described in nonhuman primates (Sato et al., 2000).

The GP projection is directed predominantly to the somata (31%), and proximal dendrites (39%) of STN neurons (Smith et al., 1990). Although only 30% of inputs are directed to distal dendrites, these inputs could be of key functional significance because they are co-aligned with cortical and thalamic inputs (Fig. 1B; Smith et al., 1990; Bevan et al., 1995) and carry functionally diverse inputs to the dendrites of STN neurons whose somata are located in distant and different functional zones of the nucleus (Bevan et al., 1997).

A minor GABAergic projection to the STN arises from the mesopontine tegmentum. Unusually, the GABAergic terminals that mediate this connection possess asymmetric synapses (Bevan and Bolam, 1995), in contrast to those arising from the GP$_e$/GP, which in common with the majority of GABAergic synapses in the nervous system are symmetric in nature (Smith et al., 1990).

Glutamatergic afferents of the STN

The major glutamatergic projections to the STN arise from motor cortical areas (primary, pre- and supplementary) (Afsharpour, 1985; Nambu et al., 1996) and the parafasicular thalamic nucleus

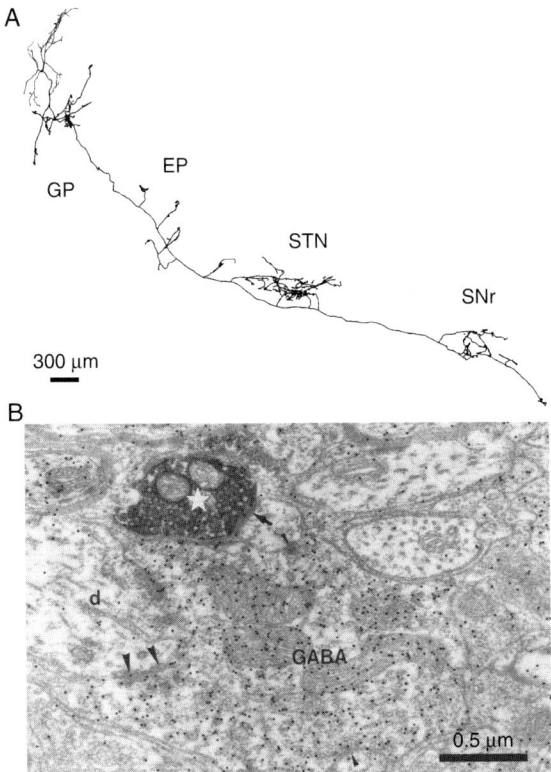

A

EP

GP

STN

SNr

300 μm

B

d

GABA

0.5 μm

Fig. 1. Microcircuitry of the pallidosubthalamic projection. (A) Individual pallidal (GP) neurons (soma and dendrites labeled in gray; axon and axon terminals labeled black) that project to the STN possess local axon collaterals and innervate other basal ganglia nuclei including the basal ganglia output nuclei, the entopeduncular nucleus (EP) and SNr. (B) A GABA-immunoreactive terminal labeled with the post-embedding immunogold technique forms symmetrical synapses (arrowheads) with the dendrites (d) and spines of subthalamic neurons. The terminal possesses the neurochemical and morphological properties of terminals derived from the pallidum. A terminal anterogradely labeled from the cortex (white star) forms an asymmetric synapse (arrow) with a spine that also receives input from the GABA-immunoreactive pallidal terminal. [Part (A) is reproduced from Bevan et al. (1998), with permission. Copyright 1998 by the *Society for Neuroscience*. Part (B) is reproduced from Bevan et al. (1995), with permission.]

(Bevan et al., 1995; Deschenes et al., 1996). The projections are ipsilateral, topographically organized, immunoreactive for glutamate and directed to the dendrites and dendritic spines of STN neurons (Fig. 1B; Bevan et al., 1995). A smaller glutamate-immunoreactive projection to the STN also arises

from the mesopontine tegmentum (Bevan and Bolam, 1995).

Dopaminergic innervation of the STN

Histochemical, immunocytochemical, tracing, pharmacological, voltammetric and electrophysiological studies have together provided definitive evidence for a direct dopaminergic pathway from the substantia nigra to the STN that degenerates in PD (Fig. 2; Brown et al., 1979; Lavoie et al., 1989; Hassani et al., 1997; Francois et al., 2000; Cragg et al., 2004). Dopaminergic terminals form conventional symmetrical synaptic contacts with all parts of STN neurons (Cragg et al., 2004). The spatial extent and timecourse of dopamine in the extracellular compartment are tightly regulated by the dopamine transporter (Cragg et al., 2004).

Electrophysiological properties of STN neurons

Autonomous pacemaker activity

In vitro and in the absence of ionotropic receptor mediated GABAergic and glutamatergic synaptic inputs, STN neurons rhythmically discharge action potentials (APs) at ~ 5–15 Hz (Fig. 3A; Bevan and Wilson, 1999; Beurrier et al., 2000; Do and Bean, 2003). This activity is driven by voltage-dependent Na^+ (Na_v) channels, which underlie subthreshold persistent and resurgent currents. The period and precision of autonomous activity is determined by small conductance Ca^{2+}-dependent K^+ (SK) channels, which underlie the so-called 'medium' duration component of the AP after-hyperpolarization (Bevan and Wilson, 1999; Hallworth et al., 2003). SK channels are functionally coupled to Ca^{2+} flowing through ω-conotoxin GVIA-sensitive $Ca_v2.2$ channels, which activate at suprathreshold voltages (Song et al., 2000; Hallworth et al., 2003). The capability of STN neurons to fire independently of synaptic input may, in part, underlie the tonic activity of STN neurons in vivo (Bergman et al., 1994; Wichmann et al., 1994). This intrinsic property is critical for the manner in which synaptic inputs are integrated (see below).

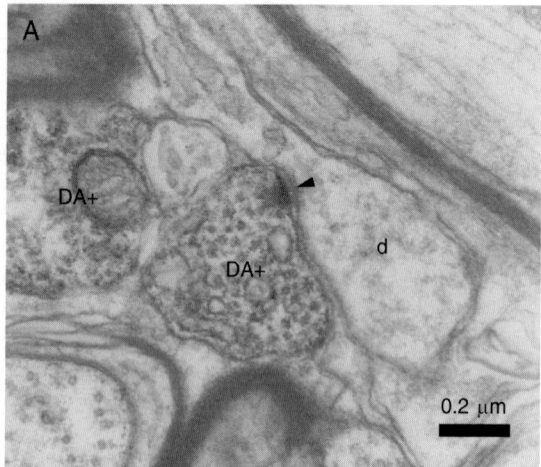

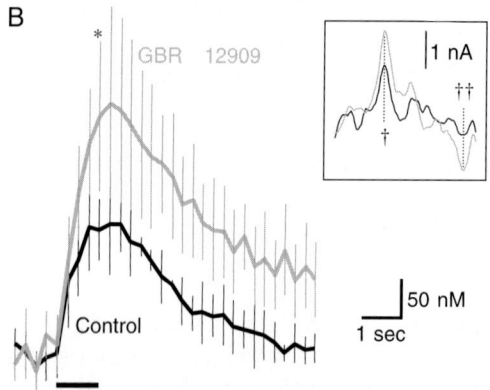

Fig. 2. Synaptic release of dopamine in the STN. (A) A dopamine-immunoreactive (DA+) axon forms a symmetrical synaptic contact (arrow) with a STN neuron dendrite (d). (B) Main panel, concentration of extracellular dopamine (black trace) in the STN following local electrical stimulation (black bar, 1 s at 50 Hz). In the presence of the dopamine transporter antagonist GBR 12909, the peak magnitude and lifetime of extracellular dopamine was enhanced (gray trace). Inset, The concentration of extra-cellular dopamine was detected using fast scan cyclic voltammetry. Examples of voltammograms under control conditions and in the presence of GBR 12909 are illustrated in the inset. [Reproduced from Cragg et al. (2004), with permission.]

Response to hyperpolarization

In vitro STN neurons respond to the termination of hyperpolarization to ~-80 mV with a rebound burst of APs (Beurrier et al., 1999; Bevan et al., 2000, 2002; Hallworth et al., 2003). STN neurons can be roughly divided into two classes with respect to their rebound activity. The majority of neurons ($\sim75\%$) exhibits rebound bursts, which last <100 ms and are driven by Ni^{2+}-sensitive Ca_v3 channels, which underlie a low-threshold Ca^{2+} spike upon which a burst of Na^+ APs rides (Fig. 3B; Beurrier et al., 1999; Bevan et al., 2000, 2002; Hallworth et al., 2003). A minority of STN neurons ($\sim25\%$) exhibit rebound bursts that last over several hundred ms. A late component of the long duration burst is driven by dihydropyridine-sensitive $Ca_v1.2$–1.3 channels, which generate plateau potentials in STN neurons (Fig. 3C; Beurrier et al., 1999; Bevan et al., 2000, 2002; Hallworth et al., 2003). Neither Ca_v3 nor $Ca_v1.2$–1.3 channels contribute greatly to autonomous activity, apparently because they are inactivated at the voltage range associated with pacemaking. However, hyperpolarization for >100 ms below ~-80 mV is sufficient for the de-inactivation of these Ca_v channels and their subsequent activation/recruitment upon repolarization (Beurrier et al., 1999; Bevan et al., 2000, 2002; Otsuka et al., 2001; Hallworth et al., 2003).

In response to hyperpolarizing current, STN neurons also exhibit a depolarizing sag in membrane potential, which is mediated by hyperpolarization-activated cyclic nucleotide gated cation channels (Fig. 3B, C). Although the role of these channels in rebound activity has not been studied, they appear to be important for restoring autonomous activity in the presence of a tonic hyperpolarizing influence (Beurrier et al., 2000).

Principles of inhibitory synaptic integration in the STN

Types of GABAR in the STN

STN neurons express the most common combination of $GABA_AR$ subunits in the brain i.e., $\alpha1$, $\beta2/3$, $\gamma2$ (Wisden et al., 1992). In accordance with this molecular composition, barbiturates and benzodiazepines enhance both the magnitude and duration of spontaneous IPSCs in STN neurons (Baufreton et al., 2001). Because benzodiazepines can enhance the amplitude of spontaneous IPSCs,

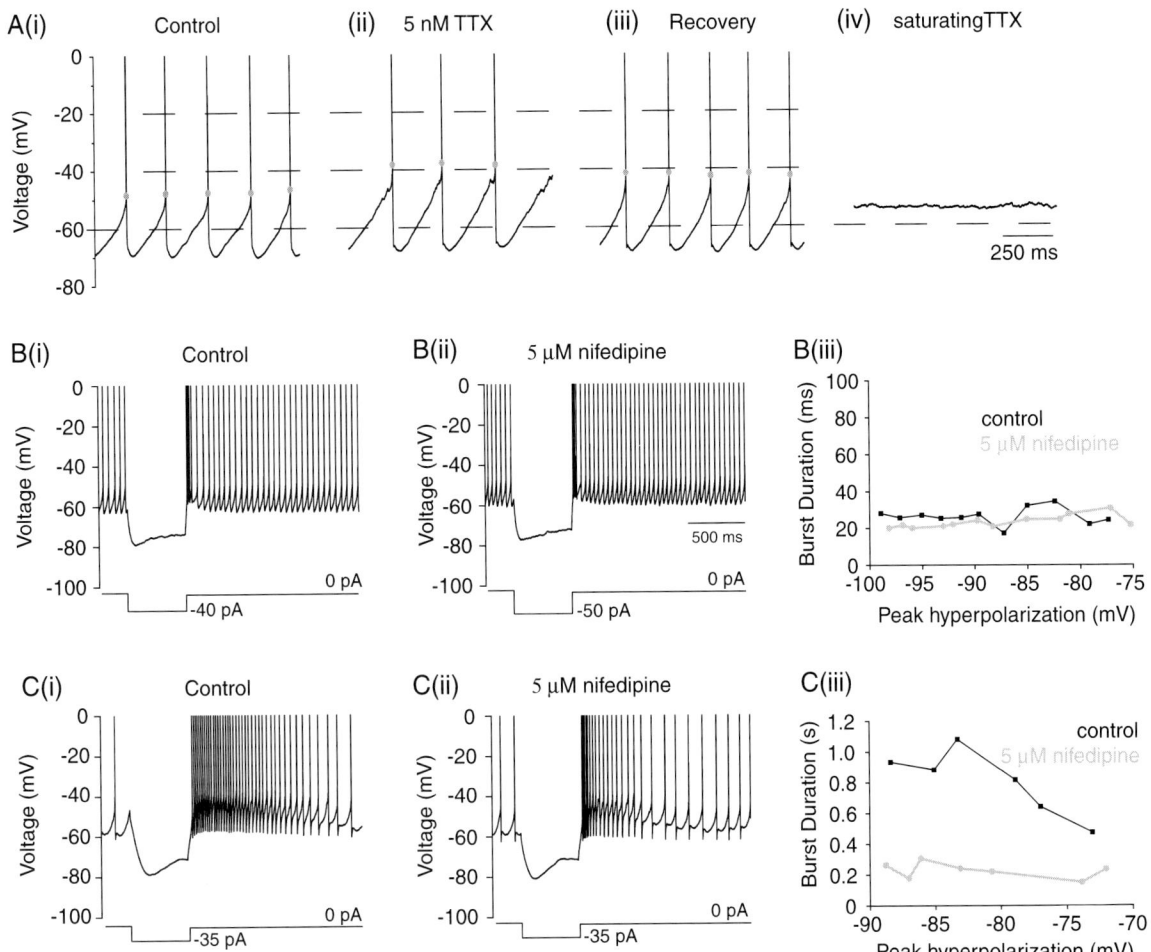

Fig. 3. Intrinsic voltage-dependent membrane properties of STN neurons. A (i), STN neurons discharge in a rhythmic fashion in the absence of ionotropic receptor mediated inputs (A (i), blocked by the bath application of APV, DNQX and GABAzine). A (ii), blockade of ~50% of Na_v channels with 5 nM TTX reduced the frequency of autonomous activity and elevated AP threshold (gray dots). A (iii), this effect was partially reversed following the removal of TTX. A (iv), Application of 1 μM TTX abolished APs, subthreshold oscillations in voltage and hyperpolarized the membrane potential below AP threshold. These data demonstrate that the autonomous activity of STN neurons is dependent on Na_v channels. B (i), the majority of STN neurons (~76%) fire short duration rebound bursts of APs following the termination of hyperpolarizing current injection. B (ii), B (iii), Short duration rebound burst firing was insensitive to the application of the dihydropyridine nifedipine. C (i), the minority of STN neurons (~25%) fire long duration rebound bursts. C (i), C (ii) The duration of these rebound bursts was reduced by nifedipine. Both types of neuron express a depolarizing sag in membrane potential in response to hyperpolarizing current injection. [Part (A) is reproduced from Baufreton et al. (2005a), with permission. Copyright 2005 by the *Society for Neuroscience*. Part (B) is reproduced from Hallworth et al. (2003), with permission. Copyright 2003 by the *Society for Neuroscience*.]

synaptic $GABA_A$Rs do not appear to be saturated during low-frequency synaptic transmission (Perrais and Ropert, 1999).

$GABA_A$R α1 subunits are concentrated at inhibitory GABAergic synapses in STN neurons. Indeed, brief electrical stimulation (1 pulse, 0.1 ms duration) of the GP or the internal capsule generates a fast IPSP/IPSC (Fig. 4A; IPSC τ_{decay}~10 ms), which is eliminated by the application of selective $GABA_A$R antagonists (Nakanishi et al., 1987b; Bevan et al., 2002; Hallworth and Bevan, 2005).

178

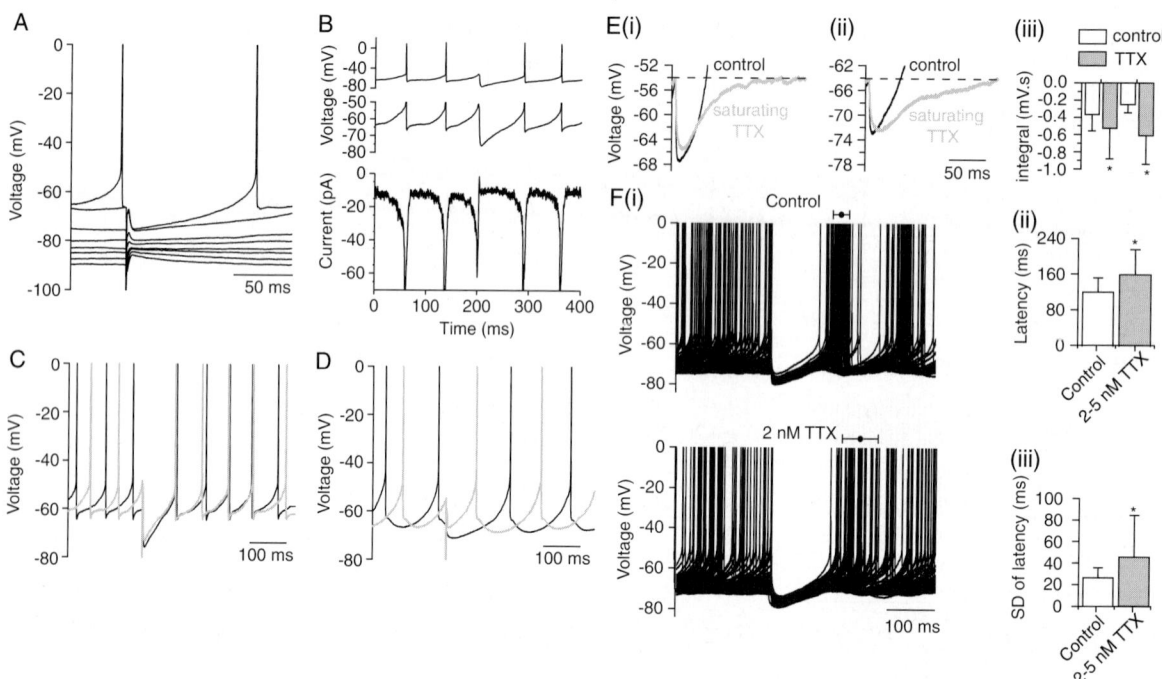

Fig. 4. Integration of individual GABA$_A$R-mediated synaptic potentials. (A) A GABA$_A$ IPSP evoked at various degrees of polarization in a gramicidin recorded STN neuron reversed at ~−80 mV. (B) In an acutely isolated STN neuron application of a GABA$_A$ IPSP waveform (upper two panels) rapidly reduced subthreshold TTX-sensitive Na$^+$ current (lower panel). (C) Large GABA$_A$ IPSPs consistently reset the phase of autonomous activity. (D) Smaller GABA$_A$ IPSPs produced more variable shifts in the phase of autonomous firing. (E) TTX-sensitive subthreshold Na$^+$ current truncated synthetic GABA$_A$ IPSPs. (F) A 20% reduction in Na$_v$ channel availability by application of 2 nM TTX disrupted the capability of a synthetic GABA$_A$ IPSP to reset autonomous firing, which led to an increase in the mean latency and variability of the AP following the dIPSP. * denotes $p < 0.05$. [Parts (A), (C), (D) are used with permission from Bevan et al. (2002). Parts (E), (F) are reproduced from Baufreton et al. (2005a), with permission. Copyright 2005 by the *Society for Neuroscience*.]

As GABA$_A$R subunits are also expressed, albeit at lower density, at extrasynaptic sites there is the potential for these receptors to be activated by GABA that 'spills over' from GABAergic synapses (Galvan et al., 2004). However, application of GABA$_A$R antagonists blocks spontaneous IPSCs but has no effect on baseline holding current, suggesting that phasic transmission at synaptic GABA$_A$Rs rather than tonic transmission at extrasynaptic receptors is the dominant mode of transmission in these neurons (Hallworth and Bevan, 2005).

Using gramicidin-based perforated patch-clamp recording, which maintains the natural transmembrane gradient of anions that permeate the GABA$_A$R, the equilibrium potential for GABA$_A$R–mediated IPSPs/IPSCs in STN was

found to be ~−83 mV (Fig. 4A; Bevan et al., 2000, 2002; Hallworth and Bevan, 2005). This voltage is 10–20 mV more hyperpolarized than the voltage traversed during autonomous activity and is also sufficiently hyperpolarized for the deinactivation of Ca$_v$ channels that mediate rebound burst activity in STN neurons, as described above (Bevan et al., 2000, 2002; Hallworth and Bevan, 2005).

STN neurons also express GABA$_B$R1 and R2 subunits but, in common with the distribution of other G protein coupled receptors (GPCRs) in the brain, these subunits are predominantly expressed at extra- and peri-synaptic sites (Charara et al., 2000; Galvan et al., 2004). GABA$_B$R1 and R2 subunits are also expressed presynaptically at excitatory and inhibitory axon terminals.

Brief high-frequency electrical stimulation (20 pulses, 0.1 ms duration, 10 ms interpulse interval) of the GP_e or the internal capsule evokes a compound IPSP with a rapid onset and slow decay in STN neurons (Hallworth and Bevan, 2005). After blockade of $GABA_A$Rs, a slow IPSP remains with a latency onset of ~135 ms and a peak hyperpolarization at ~310 ms. The slow IPSP reverses at ~−93 mV, close to the estimated reversal potential of K^+ of −97 mV. The slow IPSP is also blocked by application of the selective $GABA_B$R antagonist CGP 55845. Taken together, these observations suggest that synchronous high-frequency activation of GP_e fibers is necessary for the activation of peri- and extra-synaptic $GABA_B$Rs in STN neurons (Hallworth and Bevan, 2005).

Individual IPSPs

Individual electrically stimulated $GABA_A$R-mediated IPSPs interact with the autonomous oscillatory activity of STN neurons in a manner that is related to the amplitude of the IPSP (Bevan et al., 2002). Large IPSPs reset the phase of oscillatory activity by consistently driving the membrane potential to a level associated with the complete deactivation of the Na_v channels that underlie autonomous activity (Fig. 4B, C: Bevan et al., 2002; Baufreton et al., 2005a). Smaller IPSPs tend to reset autonomous activity in a more variable and partial fashion by less consistently and less completely deactivating pacemaker Na_v channels (Fig. 4D; Bevan et al., 2002; Baufreton et al., 2005a). Synchronous activity of presynaptic GP_e/GP neurons that presumably would underlie large IPSPs therefore has the potential to synchronize the activity of postsynaptic STN neurons. In accordance with the classical inhibitory action of GABA, the duration of the interspike interval containing the IPSP is always greater than the duration of the interspike interval associated with autonomous activity.

The kinetics of individual IPSPs is greatly influenced by Na_v channels. Synthetic $GABA_A$R-mediated IPSPs, generated through the dynamic clamp technique, are actively truncated by Na_v but not HCN or Ca_v3 channels (Fig. 4E, F). Na_v channels are rapidly and persistently activated during the decay phase of the IPSP (Fig. 4B). In contrast, the duration and degree of hyperpolarization produced by individual IPSPs appears not to be sufficient to strongly activate HCN channels and/or de-inactivate Ca_v3 channels (Baufreton et al., 2005a).

Multiple IPSPs

Tonic spontaneous activity of GP neurons, in slices that retain their axonal connections to the STN, reduces the frequency and regularity of autonomous STN activity, predominantly through the phasic activation of synaptic $GABA_A$Rs (Fig. 5; Hallworth and Bevan, 2005). Multiple $GABA_A$R-mediated and/or $GABA_B$R-mediated IPSPs evoked at various frequencies and over different periods also reduce and can completely prevent autonomous activity during the period of stimulation (Fig. 6; Bevan et al., 2002; Hallworth and Bevan, 2005). On some occasions, the degree of hyperpolarization produced by inhibition is sufficient to generate a rebound depolarization and a burst of action potentials (Bevan et al., 2002; Hallworth and Bevan, 2005). The duration of rebound bursts varies between 20 ms and several seconds depending on the intrinsic rebound properties of the postsynaptic STN neuron, as described above. Interestingly, the maximum instantaneous frequency associated with rebound burst firing relates to the frequency and number of IPSPs in STN neurons with rebound bursts < and > 100 ms, whereas only the duration of rebound activity relates to the frequency and number of IPSPs in neurons with rebound bursts in excess of 100 ms (Bevan et al., 2002). Thus rebound activity of STN neurons in response to similar patterns of inhibition is heterogeneous. Although rebound bursts may be generated in STN neurons following the synaptic activation of $GABA_A$Rs or $GABA_B$Rs, coincident activation of $GABA_A$Rs and $GABA_B$Rs produces the longest duration and/or highest frequency of rebound activity (Fig. 6; Bevan et al., 2002; Hallworth and Bevan, 2005).

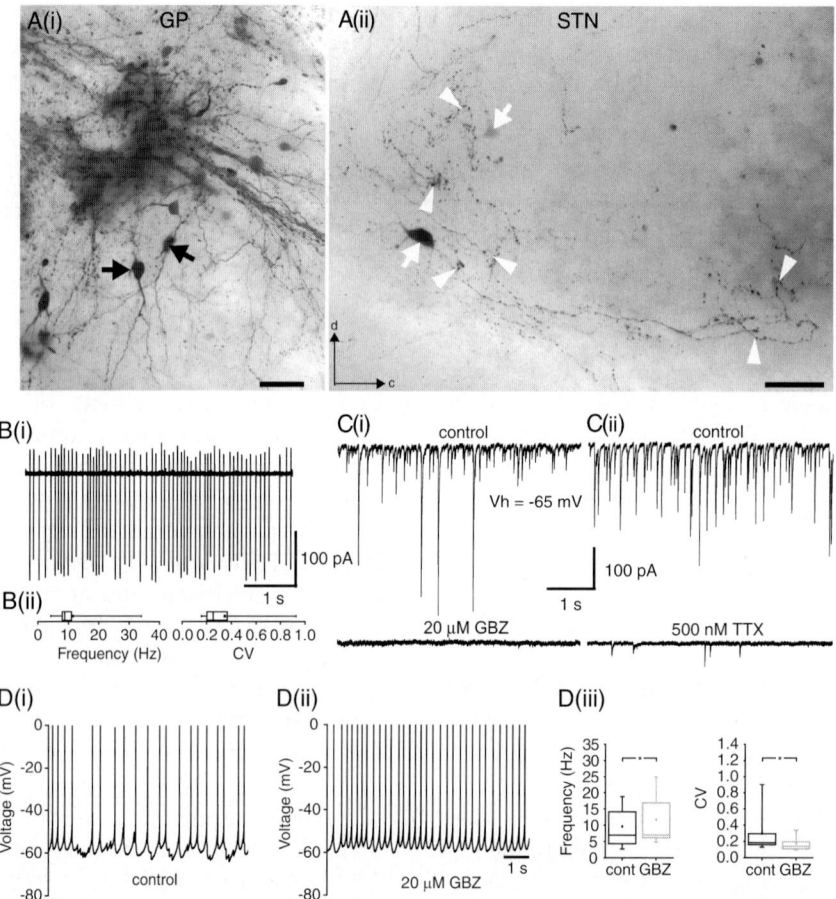

Fig. 5. Spontaneously active GP neurons pattern STN activity though GABA$_A$R-mediated IPSPs in the mouse parasagittal GP-STN slice. (A) An injection of biocytin in the GP of a parasagittal slice labeled GP neurons (A (i), arrows), their axon terminals in the STN (A (ii), arrowheads) and STN neurons that projected to the site of injection (A (ii), arrows). (B) GP neurons recorded in the cell-attached configuration were spontaneously active (B (i): example; B (ii): population data). (C) Spontaneous GABA$_A$ IPSCs recorded in STN neurons were abolished by GABAzine (GBZ; C (i)) and reduced in frequency by TTX (C (ii)). (D) Spontaneous GABA$_A$ IPSPs reduced the frequency and regularity of STN activity (D (i): example; D (ii): abolition of spontaneous IPSPs by GABAzine; D (iii): Population data, * denotes $p < 0.05$). [Parts (A)–(D) are reproduced from Hallworth and Bevan (2005), with permission. Copyright 2005 by the *Society for Neuroscience*.]

Enhancement of excitatory synaptic integration by GABAergic inhibition in the STN

Deinactivation of Na$_v$ channels by IPSPs

Following single or multiple low-frequency GABA$_A$R-mediated IPSPs, the autonomous activity of STN neurons may resume in the absence of rebound burst firing (Bevan et al., 2002; Hallworth and Bevan, 2005; Baufreton et al., 2005a). However, close inspection of APs reveals that they are briefly modified after inhibition: threshold is lowered and the maximal rate of the rise of voltage is increased (Fig. 7A; Baufreton et al., 2005a). These observations suggest that Na$_v$ channel availability is enhanced following inhibition. In order to address this possibility, current clamp recordings of autonomous activity before, during and after inhibition were utilized as voltage clamp waveforms to study the effects of autonomous activity and GABA$_A$R-mediated inhibition on Na$_v$ channel-mediated, TTX-sensitive current

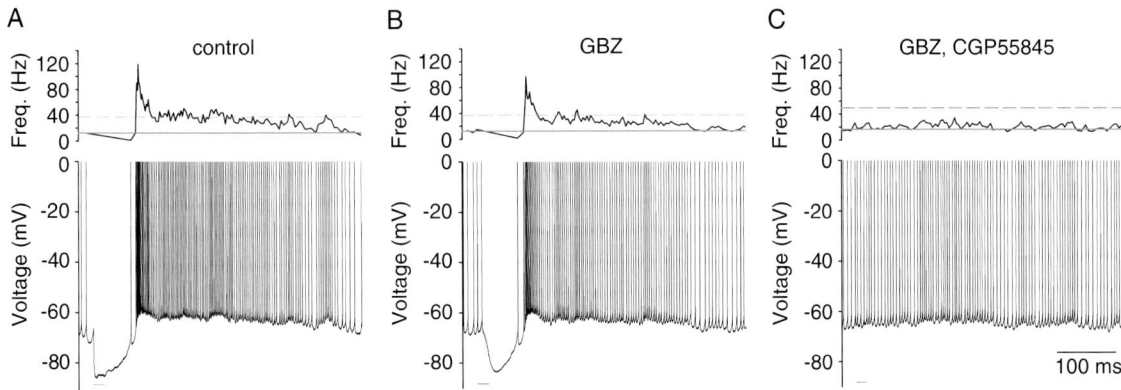

Fig. 6. GABA$_A$R and GABA$_B$R IPSPs generate rebound burst firing in STN neurons. **(A)** Electrical stimulation (10 stimuli at 100 Hz: line) inhibited activity by generating a long-lasting IPSP, which was followed by high-frequency rebound burst firing (gray line: spontaneous firing rate; dashed line 3X spontaneous firing rate). **(B)** Application of GABAzine (GBZ) blocked the early (GABA$_A$) component of the IPSP, which reduced the frequency and duration of rebound burst firing. **(C)** The additional application of the selective GABA$_B$ receptor antagonist CGP55845 blocked the late component of the IPSP and abolished rebound activity. Scale bar in C applies to all traces. [Parts (A)–(C) are reproduced from Hallworth and Bevan (2005), with permission. Copyright 2005 by the *Society for Neuroscience*.]

in acutely isolated STN neurons (Fig. 7B; Baufreton et al., 2005a). Na$_v$ channel-mediated current declined during autonomous activity to ~60% of its maximum due to an accumulation of inactivation. Subsequent IPSP waveforms led to a further but more rapid reduction in Na$_v$ channel-mediated currents due to deactivation. After IPSPs both subthreshold and suprathreshold Na$_v$ channel-mediated currents were boosted by up to 50% in the subsequent oscillatory cycle. Taken together, these data demonstrate that autonomous activity leads to the inactivation of a substantial proportion of Na$_v$ channels, which is transiently relieved by GABA$_A$R-mediated inhibition (Fig. 7B; Baufreton et al., 2005a).

Interaction of EPSPs and IPSPs

The effect of GABA$_A$R-mediated IPSPs on the integration of subsequent excitatory synaptic inputs was explored in STN neurons (Baufreton et al., 2005a). When IPSPs precede EPSPs by 20–50 ms, EPSPs generate APs with reduced latency and increased precision compared to EPSPs in isolation. When IPSPs and EPSPs are generated simultaneously, APs are generated at longer

latencies and with reduced precision compared to EPSPs alone, an action that is consistent with classical shunting inhibition. Appropriately timed, IPSPs therefore enhance the integration of subsequent excitatory synaptic inputs in STN neurons. This effect appears to be due to an enhancement of Na$_v$ channel availability following an IPSP, as described above, because blockade of ~20% of Na$_v$ channels with 2 nM TTX increases the latency and reduces the precision of APs generated after an IPSP-EPSP sequence. Furthermore, the threshold of APs generated after EPSPs alone are more depolarized than after an IPSP-EPSP sequence.

Response to rhythmic excitation in the presence and absence of feedback inhibition

In vivo, the STN can exhibit rhythmic activity, which may be coherent and driven by cortical input (Brown, 2003). In order to determine how feedback inhibition impacts the sensitivity of STN neurons to rhythmic cortical input, EPSPs were generated rhythmically at 14–18 Hz in STN neurons either through electrical stimulation or synthetic synaptic conductance injection in the absence and presence

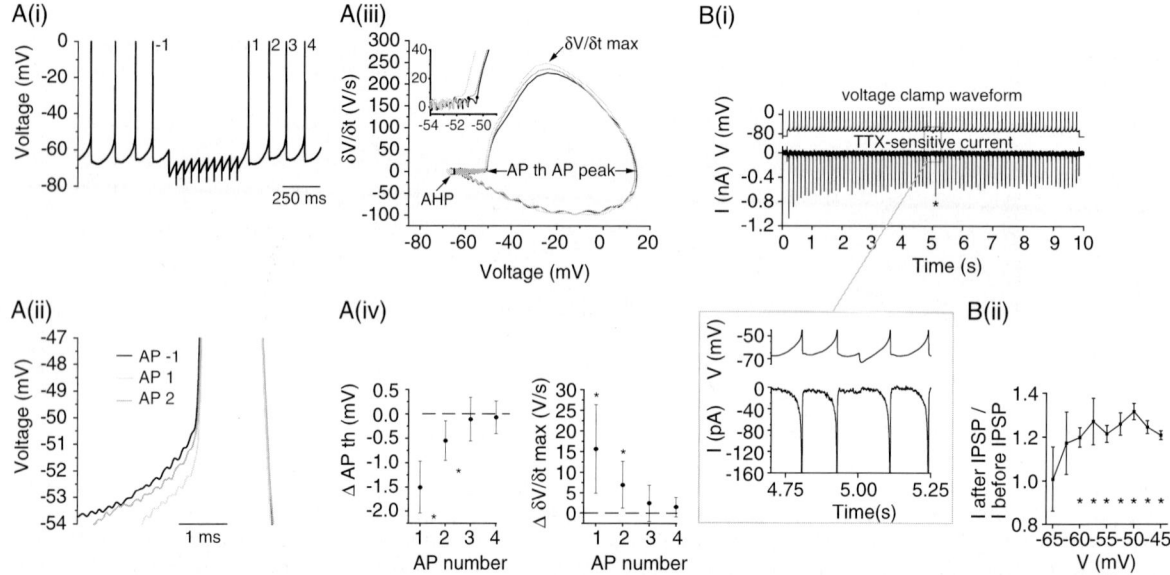

Fig. 7. GABA$_A$ IPSPs de-inactivate Na$_v$ channels in STN neurons. (A) Multiple electrically stimulated IPSPs (A (i)) lowered the threshold APs (A (ii–iv)) and increased the maximal rate of rise of APs (A (ii–iv)) generated after IPSPs. A (ii), overlaid APs generated before (AP-1) and after multiple IPSPs (AP 1–4). A (iii), Plots of $(\delta V/\delta t)/V$ illustrate the method used to measure AP threshold (AP th; defined as the first point of sustained positive 'acceleration' of voltage $((\delta V/\delta t)/\delta t)$ that was greater than 2 X S.D. of membrane noise prior to threshold) and the maximal speed $((\delta V/\delta t)$ max) of APs. Other characteristic features of APs on the phase plot, such as the spike afterhyperpolarization (AHP) and the peak of the AP (AP peak) are also noted. Inset, zoom of AP ths (dots). A (iv), Population sample of IPSP induced changes in AP th and maximal speeds of APs in 9 neurons. B (i) Autonomous spiking and inhibitory activity recorded using the perforated patch clamp technique was replayed to acutely isolated neurons as voltage clamp waveforms. This approach revealed that mean TTX-sensitive subthreshold and peak spike-associated currents declined during tonic spiking activity but were transiently increased after an IPSP in representative neuron (B(i)) and in the sample population of 13 neurons (B(ii)) compared to the mean currents flowing in the previous 5 oscillatory cycles. Expanded views (gray rectangle) of waveform and inter-spike current flowing before, during and after an IPSP demonstrate that during an IPSP Na$_v$ channel currents are reduced. After an IPSP subthreshold current was significantly increased throughout the range of voltages traversed during the interspike interval. * denotes $P < 0.05$. Abbreviations: I, current; V, voltage. [Parts (A) and (B) are reproduced from Baufreton et al. (2005a), with permission. Copyright 2005 by the *Society for Neuroscience*.]

of synthetic feedback inhibition from the GP$_e$ (Fig. 8; Baufreton et al., 2005a). In the presence of feedback inhibition, rhythmic cortical inputs drive rhythmic phase-locked firing in STN neurons. Furthermore, feedback inhibition enables intervening cortical inputs to generate APs with reduced threshold, latency and variability. Indeed when comparing the effectiveness of excitation at identical membrane potentials, EPSPs generate APs more rapidly and more reliably when they are intervened by GABA$_A$R-mediated IPSPs (Fig. 8). These data have implications for the emergence of coherent oscillatory activity in the cortex and basal ganglia during voluntary movement and in disease states (see below).

Regulation of GABAergic inhibition in the STN by dopamine

Activity-dependent synaptic plasticity/cellular excitability

In accordance with its actions in other brain regions, dopamine predominantly influences synaptic transmission in the STN through the activation of presynaptic D2-like receptors, which lower the initial probability of neurotransmitter release at GABAergic and glutamatergic synapses (Shen and Johnson, 2000). The action on GABAergic transmission is more potent than the effect on glutamatergic transmission. By reducing the initial

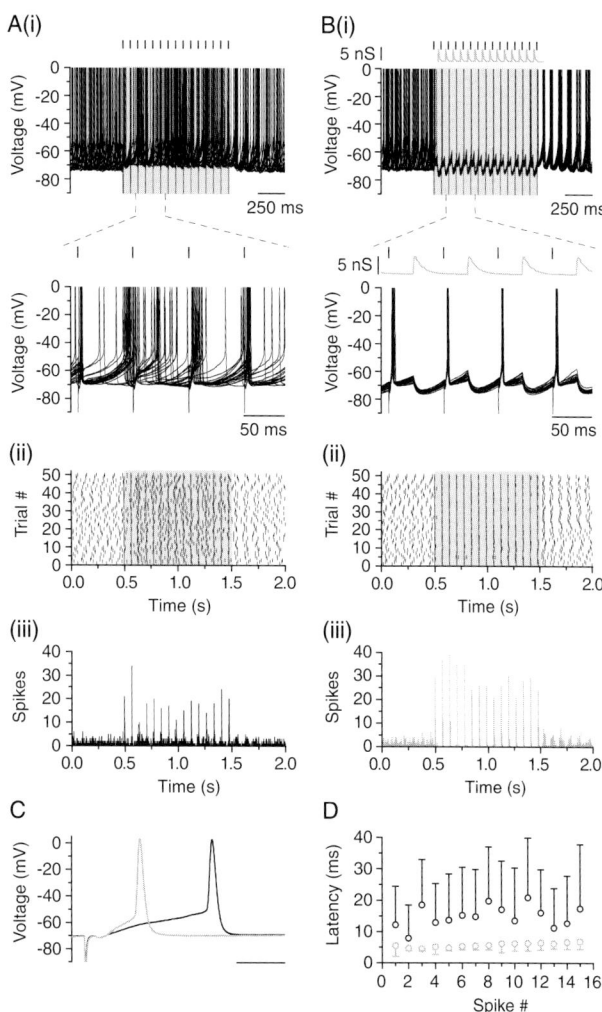

Fig. 8. Feedback inhibition enhances the coupling of action potentials to EPSPs. (A) trains of electrically stimulated EPSPs did not drive synchronized spiking in STN neurons. A (i), top panel, 20 superimposed trials during which a train of EPSPs was evoked at a frequency of 14 Hz (gray rectangle) for a period of 1 s. EPSPs were stimulated at the times marked by the vertical bars above the graph. A (i), bottom panel, enlargement of EPSP-driven action potentials. Relatively weak phase-locked activity was observed in the peristimulus raster plot (A (ii), highlighted gray box), and the peristimulus time histogram (A (iii)) during the period of evoked EPSPs. (B) Synthetic feedback inhibition enhanced EPSP-driven APs and promoted the emergence of phase-locked neuronal activity. B (i), top panel, overlay of 20 trials where EPSPs and feedback synthetic IPSPs were generated at a frequency of 14 Hz. Each synthetic IPSP ($g = 5$ nS) was injected 20 ms after each evoked EPSP for 1 s (protocol illustrated above the graph). B (i), bottom panel, enlargement of excitation-inhibition sequences revealing the precise phase-locking of EPSP-driven action potentials when feedback inhibition was incorporated. Aligned action potentials were more apparent on the peristimulus raster plots (B(ii), gray box) and the peristimulus time histogram (B(iii)). (C) Comparison of EPSPs and subsequent action potentials evoked at identical voltages in the absence (black) and presence (gray) of feedback inhibition. (D) The latency of action potential generation was reduced and its precision (S.D. of latency) improved when synthetic IPSPs intervened evoked EPSPs (gray) compared to the stimulation of EPSPs in isolation (black). [Parts (A)–(D) are reproduced from Baufreton et al. (2005a), with permission. Copyright 2005 by the *Society for Neuroscience*.]

release probability, depression due to vesicle depletion and/or receptor desensitization may be reduced and therefore the frequency at which GABAergic and glutamatergic transmission in the STN can be sustained may be elevated by dopamine (Abbott and Regehr, 2004). This observation may have significance for the emergence of rhythmic low-frequency activity in PD.

Although postsynaptic actions of dopamine on GABA and glutamate receptor-mediated transmission have not been observed thus far, the actions of dopamine on the voltage-dependent properties of postsynaptic STN neurons could be highly significant for the manner in which synaptic inputs are integrated (Baufreton et al., 2005b). D5 receptor activation can increase the duration of $Ca_v1.2$–1.3 channel-mediated component of rebound bursts through the adenylate cyclase-cAMP-protein kinase A signaling pathway (Baufreton et al., 2003), thus accentuating the heterogeneity of rebound burst firing in the STN. D2-like and D1-like receptor activation can also depolarize STN neurons and increase their rate of autonomous activity (Zhu et al., 2002a; Baufreton et al., 2005b). The D2-like receptor effect is due to the G-protein mediated closure of a voltage-independent K^+ channel (Zhu et al., 2002b), whereas the site(s) of action of D1 receptor activation is currently undefined. Given the recruitment of different sets of ion channels at hyperpolarized and depolarized potentials, the level of polarization of STN neurons appears to be critical for the manner in which GABAergic (Bevan et al., 2002) and glutamatergic synaptic inputs (Otsuka et al., 2001) are integrated.

Effects of chronic dopamine depletion

Chronic depletion of dopamine through 6-hydroxydopamine lesions of midbrain dopamine neurons increases the magnitude of $GABA_AR$-, $GABA_BR$- and NMDAR-mediated whole-cell currents in STN neurons in response to exogenous agonists (Shen and Johnson, 2005). At present it is not known whether the augmentation in currents is due to an increase in the number and/or alteration in the subunit composition of synaptic

and/or extra-synaptic receptors. It is also not known whether activity-dependent plasticity of GABAergic and glutamatergic transmission is altered by chronic dopamine depletion, although the prediction would be that synaptic depression is enhanced in the absence of dopamine.

Chronic depletion of dopamine also appears to reduce the frequency of spontaneous activity in STN neurons (Zhu et al., 2002a). However it is not known whether this is due to an alteration in the intrinsic firing properties of STN neurons and/or alterations in synaptic transmission, as described above. Based on the action of exogenously applied dopamine on the intrinsic firing properties of STN neurons, dopamine depletion would be predicted to hyperpolarize STN neurons and reduce their frequency of autonomous activity (see above).

Patterning of STN activity in vivo: proven and potential roles for GABAergic inhibition

Multiple studies provide evidence for a pivotal role of the GP_e/GP in patterning the activity of the STN in vivo. Brief motor cortical stimulation in vivo (that is meant to approximate a descending motor command) generates a stereotyped triphasic sequence of activity in STN neurons (Fujimoto and Kita, 1993; Maurice et al., 1998; Magill et al., 2004). The first response is excitatory and is mediated by the direct cortico-subthalamic pathway. In rodents it occurs with a latency of ~4–7 ms and lasts for ~5 ms. Initial excitation is followed by a brief period of reduced activity, which is mediated by feedback inhibition from the GP_e and lasts ~5 ms. The final component of the response is a second phase of excitation, which lasts for ~15 ms and is due to striatal-mediated inhibition of the GP_e, which results in brief disinhibition of the STN.

During slow-wave sleep/anesthesia and wakefulness/cortical desynchronization, GP neurons regulate the activity of STN neurons predominantly through the activation of $GABA_ARs$. Cortical slow-wave activity drives low-frequency rhythmic burst firing in STN neurons (Magill et al., 2000; Urbain et al., 2002), which is greatly increased in intensity when $GABA_ARs$ are blocked by iontophoresis of antagonists in vivo

(Urbain et al., 2002). During spontaneous wakefulness/cortical desynchronization blockade of GABA$_A$Rs in the STN also increases the frequency of STN activity (Urbain et al., 2002). Interestingly, cortical activation in intact animals increases the mean activity of both STN and GP neurons (Magill et al., 2000; Urbain et al., 2000, 2002) demonstrating that increased activity of the STN is balanced by increased activity in the reciprocally connected GP. Despite anatomical and physiological evidence for postsynaptic GABA$_B$Rs in the STN (see above), the local iontophoresis/application of GABA$_B$R antagonists has no impact on the firing of STN neurons during slow-wave sleep and wakefulness/cortical desynchronization (Urbain et al., 2002). In vitro, postsynaptic GABA$_B$Rs in the STN are only activated under conditions of enhanced GP$_e$/GP-STN transmission (Hallworth and Bevan, 2005).

Experiments in genetic absence epilepsy (GAERS) rats, provide further evidence for an intimate relationship between STN and GP activity (Paz et al., 2005). In these rats there are spontaneous cortical paroxysms in which cortico-subthalamic pyramidal neurons discharge synchronous doublets/bursts of APs every 100–200 ms. During epileptiform cortical activity, STN neurons are driven directly by cortical input. The first cortical EPSP is followed by a feedback IPSP from the GP of sufficient magnitude and duration to generate a rebound burst of action potentials. Interestingly, cortico-subthalamic neurons fire in two phases during the epileptic seizure, which are associated with the early and late phases of excitation in STN neurons. The rebound burst that follows feedback inhibition is therefore supplemented by a second coincident cortical EPSP. Interestingly, the striatum plays no apparent role in the generation of rhythmic activity in the STN-GP network in GAERS rats. Although the membrane potential of medium spiny striatal projection neurons is sculpted by the cortical rhythm, AP generation is completely suppressed by feed-forward inhibition arising from striatal GABAergic interneurons (Slaght et al., 2004). These data provide further support for the important role of the GP$_e$/GP in the expression of cortical rhythms in STN neurons (see above).

Dopamine is critical for the modulation of activity in the STN-GP$_e$/GP network. In PD patients, coherent frequency-specific activity in the cortical electroencephalogram and local field potential in the pallidum is dependent on the level of dopamine medication (Brown, 2003). In nonmedicated patients, there is coherence in the tremor frequency (4–10 Hz) and β frequency (13–30 Hz) bands. Systemic administration of a dopamine receptor agonist shifts coherent activity into the γ frequency band. Similar observations have been made in rodents that are awake. Dopamine depletion increases coherent activity between the cortex and basal ganglia in the β frequency band (Sharott et al., 2005). The close relationship between low-frequency correlated rhythmic activity in the basal ganglia of PD patients and the expression of symptoms has led to intensive research into the mechanisms underlying such activity. Because pathological correlated low-frequency rhythmic activity is most robustly expressed in the GP$_e$/GP and STN, these nuclei are thought to be the primary mediators of aberrant rhythmic activity.

Organotypic co-cultures of the STN-GP network, grown in the absence of substantia nigra dopamine neurons, exhibit spontaneous low-frequency (1–2 Hz) rhythmic activity (Plenz and Kitai, 1999). The mechanisms underlying the activity appear to be similar to those underlying spindle oscillations in the sensory thalamus during sleep (McCormick and Bal, 1997). Burst activity of GABAergic (GP) neurons generates sufficient hyperpolarization for rebound burst firing in STN neurons, which in turn drives burst firing in reciprocally connected GP neurons and the perpetuation of the low-frequency network rhythm. However studies in brain slices, which retain intact connections between the GP and the STN have failed to report spontaneous low-frequency rhythmic activity (Hallworth and Bevan, 2005; Loucif et al., 2005). Whether this failure is due to insufficient retention of the network and/or the fact that the slices were derived from intact mice and/or more normal connectivity (compared to organotypic co-cultures) remains to be determined.

In contrast, experiments in anesthetized rodents in vivo favor the cortex as a primary influence on the pattern of activity in the STN-GP_e/GP network. Under deep anesthesia the STN-GP network is patterned more powerfully by cortical slow-wave activity in dopamine-depleted animals (Magill et al., 2001). Furthermore, low-frequency rhythmic activity in the STN-GP network in dopamine-depleted animals is largely abolished by cortical ablation (Magill et al., 2001).

Several studies have provided evidence that cortical excitation of striatal-GP/GP_e neurons is enhanced by dopamine-receptor antagonists and/or dopamine depletion which in turn have important consequences for the regulation of STN activity by the GP_e/GP (Magill et al., 2001; Tseng et al., 2001; Degos et al., 2005). Systemic application of dopamine receptor antagonists enhances the second phase of excitation in STN neurons in response to brief cortical stimulation (an effect presumably due to an enhancement of the indirect pathway and more robust disinhibition of STN neurons) (Degos et al., 2005). Similarly, persistent cortical activation in dopamine-denervated animals generates more powerful inhibition of GP neurons and disinhibition of STN neurons (Magill et al., 2001). A recent modeling study has also suggested that increased activity of the striatum-GP/GP_e pathway in the dopamine-depleted/PD brain may promote the tendency of the STN-GP_e./GP network to support thalamic-like oscillatory activity by reducing lateral inhibition in the GP_e/GP (Terman et al., 2002).

The chronic loss of dopamine in the STN in PD may also promote the tendency for rhythm generation both within the STN-GP_e/GP network and in response to cortical input through actions both on activity-dependent plasticity at GP_e/GP-STN and cortical-STN synapses and the cellular excitability of STN neurons. By enhancing the release probability of synapses in the STN and hyperpolarizing STN neurons, dopamine-depletion may greatly enhance the tendency for low-frequency rhythmic burst firing in the STN-GP_e/GP network. Whether cortical and/or the STN-GP_e neurons are the primary generators of pathological rhythms in Parkinson's disease, GABAergic transmission in the STN is likely to be critical for their manifestation.

Acknowledgements

Our work is supported by NIH-NINDS Grants NS041280 (MDB) and NS040705 (MDB), the National Parkinson Foundation (NEH) and L' Association France Parkinson (JB).

References

Abbott, L.F. and Regehr, W.G. (2004) Synaptic computation. Nature, 431: 796–803.

Afsharpour, S. (1985) Topographical projections of the cerebral cortex to the subthalamic nucleus. J. Comp. Neurol., 236: 14–28.

Baufreton, J., Atherton, J.A. and Bevan, M.D. (2005a) Enhancement of excitatory synaptic integration by GABAergic inhibition in the subthalamic nucleus. J. Neurosci., 25: 8505–8517.

Baufreton, J., Garret, M., Dovero, S., Dufy, B., Bioulac, B. and Taupignon, A. (2001) Activation of GABA(A) receptors in subthalamic neurons in vitro: properties of native receptors and inhibition mechanisms. J. Neurophysiol., 86: 75–85.

Baufreton, J., Garret, M., Rivera, A., de la Calle, A., Gonon, F., Dufy, B., Bioulac, B. and Taupignon, A. (2003) D5 (not D1) dopamine receptors potentiate burst-firing in neurons of the subthalamic nucleus by modulating an L-type calcium conductance. J. Neurosci., 23: 816–825.

Baufreton, J., Zhu, Z.T., Garret, M., Bioulac, B., Johnson, S.W. and Taupignon, A.I. (2005b) Dopamine receptors set the pattern of activity generated in subthalamic neurons. FASEB J., 19: 1771–1777.

Bergman, H., Wichmann, T., Karmon, B. and DeLong, M.R. (1994) The primate subthalamic nucleus. II. Neuronal activity in the MPTP model of Parkinsonism. J. Neurophysiol., 72: 507–520.

Beurrier, C., Bioulac, B. and Hammond, C. (2000) Slowly inactivating sodium current (I(NaP)) underlies single-spike activity in rat subthalamic neurons. J. Neurophysiol., 83: 1951–1957.

Beurrier, C., Congar, P., Bioulac, B. and Hammond, C. (1999) Subthalamic nucleus neurons switch from single-spike activity to burst-firing mode. J. Neurosci., 19: 599–609.

Bevan, M.D. and Bolam, J.P. (1995) Cholinergic, GABAergic, and glutamate-enriched inputs from the mesopontine tegmentum to the subthalamic nucleus in the rat. J. Neurosci., 15: 7105–7120.

Bevan, M.D., Bolam, J.P. and Crossman, A.R. (1994) Convergent synaptic input from the neostriatum and the subthalamus onto identified nigrothalamic neurons in the rat. Eur. J. Neurosci., 6: 320–334.

Bevan, M.D., Booth, P.A., Eaton, S.A. and Bolam, J.P. (1998) Selective innervation of neostriatal interneurons by a subclass of neuron in the globus pallidus of the rat. J. Neurosci., 18: 9438–9452.

Bevan, M.D., Clark, N.P. and Bolam, J.P. (1997) Synaptic integration of functionally diverse pallidal information in the entopeduncular nucleus and subthalamic nucleus in the rat. J. Neurosci., 17: 308–324.

Bevan, M.D., Francis, C.M. and Bolam, J.P. (1995) The glutamate-enriched cortical and thalamic input to neurons in the subthalamic nucleus of the rat: convergence with GABA-positive terminals. J. Comp. Neurol., 361: 491–511.

Bevan, M.D., Magill, P.J., Hallworth, N.E., Bolam, J.P. and Wilson, C.J. (2002) Regulation of the timing and pattern of action potential generation in rat subthalamic neurons in vitro by GABA-A IPSPs. J. Neurophysiol., 87: 1348–1362.

Bevan, M.D. and Wilson, C.J. (1999) Mechanisms underlying spontaneous oscillation and rhythmic firing in rat subthalamic neurons. J. Neurosci., 19: 7617–7628.

Bevan, M.D., Wilson, C.J., Bolam, J.P. and Magill, P.J. (2000) Equilibrium potential of GABA(A) current and implications for rebound burst firing in rat subthalamic neurons in vitro. J. Neurophysiol., 83: 3169–3172.

Brown, L.L., Markman, M.H., Wolfson, L.I., Dvorkin, B., Warner, C. and Katzman, R. (1979) A direct role of dopamine in the rat subthalamic nucleus and an adjacent intrapeduncular area. Science, 206: 1416–1418.

Brown, P. (2003) Oscillatory nature of human basal ganglia activity: relationship to the pathophysiology of Parkinson's disease. Mov. Disord., 18: 357–363.

Charara, A., Heilman, T.C., Levey, A.I. and Smith, Y. (2000) Pre- and postsynaptic localization of GABA(B) receptors in the basal ganglia in monkeys. Neuroscience, 95: 127–140.

Cragg, S., Baufreton, J., Xue, Y., Bolam, P. and Bevan, M.D. (2004) Synaptic release of dopamine in the subthalamic nucleus. Eur. J. Neurosci., 20: 1788–1802.

Degos, B., Deniau, J.M., Thierry, A.M., Glowinski, J., Pezard, L. and Maurice, N. (2005) Neuroleptic-induced catalepsy: electrophysiological mechanisms of functional recovery induced by high-frequency stimulation of the subthalamic nucleus. J. Neurosci., 25: 7687–7696.

Deschenes, M., Bourassa, J., Doan, V.D. and Parent, A. (1996) A single-cell study of the axonal projections arising from the posterior intralaminar thalamic nuclei in the rat. Eur. J. Neurosci., 8: 329–343.

Do, M.T. and Bean, B.P. (2003) Subthreshold sodium currents and pacemaking of subthalamic neurons: modulation by slow inactivation. Neuron, 39: 109–120.

Francois, C., Savy, C., Jan, C., Tande, D., Hirsch, E.C. and Yelnik, J. (2000) Dopaminergic innervation of the subthalamic nucleus in the normal state, in MPTP-treated monkeys, and in Parkinson's disease patients. J. Comp. Neurol., 425: 121–129.

Fujimoto, K. and Kita, H. (1993) Response characteristics of subthalamic neurons to the stimulation of the sensorimotor cortex in the rat. Brain Res., 609: 185–192.

Galvan, A., Charara, A., Pare, J.F., Levey, A.I. and Smith, Y. (2004) Differential subcellular and subsynaptic distribution of GABA(A) and GABA(B) receptors in the monkey subthalamic nucleus. Neuroscience, 127: 709–721.

Hallworth, N.E. and Bevan, M.D. (2005) Globus pallidus neurons dynamically regulate the activity pattern of subthalamic nucleus neurons through the frequency-dependent activation of postsynaptic GABA$_A$ and GABA$_B$ receptors. J. Neurosci., 25: 6304–6315.

Hallworth, N.E., Wilson, C.J. and Bevan, M.D. (2003) Apamin-sensitive small conductance calcium-activated potassium channels, through their selective coupling to voltage-gated calcium channels, are critical determinants of the precision, pace, and pattern of action potential generation in rat subthalamic nucleus neurons in vitro. J. Neurosci., 23: 7525–7542.

Hanson, J.E., Smith, Y. and Jaeger, D. (2004) Sodium channels and dendritic spike initiation at excitatory synapses in globus pallidus neurons. J. Neurosci., 24: 329–340.

Hassani, O.K., Francois, C., Yelnik, J. and Feger, J. (1997) Evidence for a dopaminergic innervation of the subthalamic nucleus in the rat. Brain Res., 749: 88–94.

Joel, D. and Weiner, I. (1997) The connections of the primate subthalamic nucleus: indirect pathways and the open-interconnected scheme of basal ganglia-thalamocortical circuitry. Brain Res. Brain Res. Rev., 23: 62–78.

Lavoie, B., Smith, Y. and Parent, A. (1989) Dopaminergic innervation of the basal ganglia in the squirrel monkey as revealed by tyrosine hydroxylase immunohistochemistry. J. Comp. Neurol., 289: 36–52.

Loucif, K.C., Wilson, C.L., Baig, R., Lacey, M.G. and Stanford, I.M. (2005) Functional interconnectivity between the globus pallidus and the subthalamic nucleus in the mouse brain slice. J. Physiol., 567: 977–987.

Magill, P.J., Bolam, J.P. and Bevan, M.D. (2000) Relationship of activity in the subthalamic nucleus-globus pallidus network to cortical electroencephalogram. J. Neurosci., 20: 820–833.

Magill, P.J., Bolam, J.P. and Bevan, M.D. (2001) Dopamine regulates the impact of the cerebral cortex on the subthalamic nucleus-globus pallidus network. Neuroscience, 106: 313–330.

Magill, P.J., Sharott, A., Bevan, M.D., Brown, P. and Bolam, J.P. (2004) Synchronous unit activity and local field potentials evoked in the subthalamic nucleus by cortical stimulation. J. Neurophysiol., 92: 700–714.

Maurice, N., Deniau, J.M., Glowinski, J. and Thierry, A.M. (1998) Relationships between the prefrontal cortex and the basal ganglia in the rat: physiology of the corticosubthalamic circuits. J. Neurosci., 18: 9539–9546.

McCormick, D.A. and Bal, T. (1997) Sleep and arousal: thalamocortical mechanisms. Annu. Rev. Neurosci., 20: 185–215.

Nakanishi, H., Kita, H. and Kitai, S.T. (1987a) Intracellular study of rat substantia nigra *pars reticulata* neurons in an in vitro slice preparation: electrical membrane properties and response characteristics to subthalamic stimulation. Brain Res., 437: 45–55.

Nakanishi, H., Kita, H. and Kitai, S.T. (1987b) Electrical membrane properties of rat subthalamic neurons in an in vitro slice preparation. Brain Res., 437: 35–44.

Nambu, A., Takada, M., Inase, M. and Tokuno, H. (1996) Dual somatotopical representations in the primate subthalamic nucleus: evidence for ordered but reversed body-map transformations from the primary motor cortex and the supplementary motor area. J. Neurosci., 16: 2671–2683.

Oorschot, D.E. (1996) Total number of neurons in the neostriatal, pallidal, subthalamic, and substantia nigral nuclei of the rat basal ganglia: a stereological study using the cavalieri and optical dissector methods. J. Comp. Neurol., 366: 580–599.

Otsuka, T., Murakami, F. and Song, W.J. (2001) Excitatory postsynaptic potentials trigger a plateau potential in rat subthalamic neurons at hyperpolarized states. J. Neurophysiol., 86: 1816–1825.

Paz, J.T., Deniau, J.M. and Charpier, S. (2005) Rhythmic bursting in the cortico-subthalamo-pallidal network during spontaneous genetically determined spike and wave discharges. J. Neurosci., 25: 2092–2101.

Perrais, D. and Ropert, N. (1999) Effect of zolpidem on miniature IPSCs and occupancy of postsynaptic GABA$_A$ receptors in central synapses. J. Neurosci., 19: 578–588.

Plenz, D. and Kitai, S.T. (1999) A basal ganglia pacemaker formed by the subthalamic nucleus and external globus pallidus. Nature, 400: 677–682.

Sato, F., Lavallee, P., Levesque, M. and Parent, A. (2000) A single-axon tracing study of neurons of the external segment of the globus pallidus in primate. J. Comp. Neurol., 417: 17–31.

Sharott, A., Magill, P.J., Harnack, D., Kupsch, A., Meissner, W. and Brown, P. (2005) Dopamine depletion increases the power and coherence of beta-oscillations in the cerebral cortex and subthalamic nucleus of the awake rat. Eur. J. Neurosci., 21: 1413–1422.

Shen, K.Z. and Johnson, S.W. (2000) Presynaptic dopamine D2 and muscarine M3 receptors inhibit excitatory and inhibitory transmission to rat subthalamic neurones in vitro. J. Physiol., 525: 331–341.

Shen, K.Z. and Johnson, S.W. (2005) Dopamine depletion alters responses to glutamate and GABA in the rat subthalamic nucleus. Neuroreport, 16: 171–174.

Shink, E., Bevan, M.D., Bolam, J.P. and Smith, Y. (1996) The subthalamic nucleus and the external pallidum: two tightly interconnected structures that control the output of the basal ganglia in the monkey. Neuroscience, 73: 335–357.

Slaght, S.J., Paz, T., Chavez, M., Deniau, J.M., Mahon, S. and Charpier, S. (2004) On the activity of the corticostriatal networks during spike-and-wave discharges in a genetic model of absence epilepsy. J. Neurosci., 24: 6816–6825.

Smith, Y., Bolam, J.P. and Von Krosigk, M. (1990) Topographical and synaptic organization of the GABA-containing pallidosubthalamic projection in the rat. Eur. J. Neurosci., 2: 500–511.

Smith, Y. and Parent, A. (1988) Neurons of the subthalamic nucleus in primates display glutamate but not GABA immunoreactivity. Brain Res., 453: 353–356.

Song, W.J., Baba, Y., Otsuka, T. and Murakami, F. (2000) Characterization of Ca(2+) channels in rat subthalamic nucleus neurons. J. Neurophysiol., 84: 2630–2637.

Terman, D., Rubin, J.E., Yew, A.C. and Wilson, C.J. (2002) Activity patterns in a model for the subthalamopallidal network of the basal ganglia. J. Neurosci., 22: 2963–2976.

Tseng, K.Y., Kasanetz, F., Kargieman, L., Riquelme, L.A. and Murer, M.G. (2001) Cortical slow oscillatory activity is reflected in the membrane potential and spike trains of striatal neurons in rats with chronic nigrostriatal lesions. J. Neurosci., 21: 6430–6439.

Urbain, N., Gervasoni, D., Souliere, F., Lobo, L., Rentero, N., Windels, F., Astier, B., Savasta, M., Fort, P., Renaud, B., Luppi, P.H. and Chouvet, G. (2000) Unrelated course of subthalamic nucleus and globus pallidus neuronal activities across vigilance states in the rat. Eur. J. Neurosci., 12: 3361–3374.

Urbain, N., Rentero, N., Gervasoni, D., Renaud, B. and Chouvet, G. (2002) The switch of subthalamic neurons from an irregular to a bursting pattern does not solely depend on their GABAergic inputs in the anesthetic-free rat. J. Neurosci., 22: 8665–8675.

Wichmann, T., Bergman, H. and DeLong, M.R. (1994) The primate subthalamic nucleus. I. Functional properties in intact animals. J. Neurophysiol., 72: 494–506.

Wisden, W., Laurie, D.J., Monyer, H. and Seeburg, P.H. (1992) Distribution of 13 GABAA receptor subunit mRNAs in the brain. I. Telencephalon, diencephalon, mesencephalon. J. Neurosci., 12: 1040–1062.

Zhu, Z., Bartol, M., Shen, K. and Johnson, S.W. (2002a) Excitatory effects of dopamine on subthalamic nucleus neurons: in vitro study of rats pretreated with 6-hydroxydopamine and levodopa. Brain Res., 945: 31–40.

Zhu, Z.T., Shen, K.Z. and Johnson, S.W. (2002b) Pharmacological identification of inward current evoked by dopamine in rat subthalamic neurons in vitro. Neuropharmacology, 42: 772–781.

Tepper, Abercrombie & Bolam (Eds.)
Progress in Brain Research, Vol. 160
ISSN 0079-6123

CHAPTER 11

GABAergic control of substantia nigra dopaminergic neurons

James M. Tepper* and Christian R. Lee

Center for Molecular and Behavioral Neuroscience, Rutgers University, 197 University Avenue, Newark, NJ 07102, USA

Abstract: At least 70% of the afferents to substantia nigra dopaminergic neurons are GABAergic. The vast majority of these arise from the neostriatum, the external globus pallidus and the substantia nigra pars reticulata. Nigral dopaminergic neurons express both $GABA_A$ and $GABA_B$ receptors, and are inhibited by local application of $GABA_A$ or $GABA_B$ agonists in vivo and in vitro. However, in vivo, synaptic responses elicited by stimulation of neostriatal or pallidal afferents, or antidromic activation of nigral pars reticulata GABAergic projection neurons are mediated predominantly or exclusively by $GABA_A$ receptors. The clearest and most consistent role for the nigral $GABA_B$ receptor in vivo is as an inhibitory autoreceptor that presynaptically modulates $GABA_A$ synaptic responses that originate from all three principal GABAergic inputs. The firing pattern of dopaminergic neurons is also effectively modulated by GABAergic inputs in vivo. Local blockade of nigral $GABA_A$ receptors causes dopaminergic neurons to shift to a burst firing pattern regardless of the original firing pattern. This is accompanied by a modest increase in spontaneous firing rate. The GABAergic inputs from the axon collaterals of the pars reticulata projection neurons seem to be a particularly important source of a $GABA_A$ tone to the dopaminergic neurons, inhibition of which leads to burst firing. The globus pallidus exerts powerful control over the pars reticulata input, and through the latter, disynaptically over the dopaminergic neurons. Inhibition of pallidal output leads to a slight decrease in firing of the dopaminergic neurons due to disinhibition of the pars reticulata neurons whereas increased firing of pallidal neurons leads to burst firing in dopaminergic neurons that is associated with a modest increase in spontaneous firing rate and a significant increase in extracellular levels of dopamine in the neostriatum. The pallidal disynaptic disinhibitory control of the dopaminergic neurons dominates the monosynaptic inhibitory influence because of a differential sensitivity to GABA of the two nigral neuron types. Nigral GABAergic neurons are more sensitive to $GABA_A$-mediated inhibition than dopaminergic neurons, in part due to a more hyperpolarized $GABA_A$ reversal potential. The more depolarized $GABA_A$ reversal potential in the dopaminergic neurons is due to the absence of KCC2, the chloride transporter responsible for setting up a hyperpolarizing Cl^- gradient in most mature CNS neurons. The data reviewed in this chapter have made it increasingly clear that in addition to the effects that nigral GABAergic output neurons have on their target nuclei outside of the basal ganglia, local interactions between GABAergic projection neurons and dopaminergic neurons are crucially important to the functioning of the nigral dopaminergic neurons.

Keywords: IPSP; disinhibition; pars reticulata; pars compacta; burst firing; reversal potential; pallidonigral

*Corresponding author. Tel.: +973-353-1080, Ext. 3151;
Fax: +973-3531588; E-mail: tepper@axon.rutgers.edu

DOI: 10.1016/S0079-6123(06)60011-3

189

Introduction

Identification of the afferents to nigral dopaminergic neurons and study of their physiological attributes is greatly complicated by the anatomical organization of the nucleus, in particular by the morphology of the dopaminergic neurons (for recent review, see Misgeld, 2004). The cell bodies of most nigrostriatal dopaminergic neurons are situated in the pars compacta. The pars compacta is a relatively thin, disk-shaped nucleus of densely packed cells dorsal and superior to the larger, more extensive, GABAergic neuron-containing pars reticulata, thus providing a flattened sheet that covers the pars reticulata for most of its dorsolateral and anterioposteror extent (Hanaway et al., 1970). Although the majority of the dopaminergic somata in substantia nigra are in the pars compacta, there exist scattered groups of dopaminergic neurons within pars reticulata. However, the morphology and physiology of these neurons appear identical to those of the pars compacta dopaminergic neurons (Richards et al., 1997).

The dendritic organization of dopaminergic neurons contributes to the anatomical complexity of the substantia nigra. Dopaminergic somata are medium-sized, and their dendrites aspiny. Several thick but rapidly tapering dendrites emanate from dopaminergic somata and extend into the neuropil of the pars compacta (Juraska et al., 1977; Tepper et al., 1987, 1994). All dopaminergic neurons also send one or occasionally two dendrites ventrally, perpendicular to the surface of the pars compacta, deep into pars reticulata. These dendrites, up to a millimeter in length, are often the largest emitted by the neuron and it is not unusual for them to traverse the entire extent of the pars reticulata and terminate in the crus cerebri (Tepper et al., 1987). Although the long, distal regions of these dendrites receive relatively few afferents, synapses abound on the more proximal dendritic segments that are closely intermingled with the somata and dendrites of the GABAergic pars reticulata neurons (Grofova et al., 1986). Thus there is no clear anatomical distinction between terminal zones of afferents to the dopaminergic neurons and those to the GABAergic neurons. This makes standard retrograde tracing techniques of only limited value when trying to determine if certain afferents innervate dopaminergic pars compacta and/or GABAergic pars reticulata neurons.

The electrophysiological properties of dopaminergic neurons have been studied in detail, both in vivo and in vitro (for recent review, see Diana and Tepper, 2002). Almost all nigrostriatal dopaminergic neurons fire spontaneously in vivo (Tepper et al., 1984; Dai and Tepper, 1998; but see Chiodo, 1988; Floresco et al., 2003) at relatively slow rates averaging between 4 and 5 spikes/s (Bunney et al., 1973; Deniau et al., 1978; Guyenet and Aghajanian, 1978; Bunney, 1979; Grace and Bunney, 1983). The spontaneous activity exists along a continuum of firing patterns that is only loosely related to the mean rate (Wilson et al., 1977; Freeman et al., 1985; Hyland et al., 2002). In urethane-anesthetized animals the most common pattern of activity (~55% of neurons) is a random mode in which the interspike intervals are described by a Poisson-like process. The next most common pattern is a regular, pacemaker-like activity (~30%) and the least common (15%) is a slow bursting pattern (Tepper et al., 1995; Paladini and Tepper, 1999). Bursts in dopaminergic neurons in anesthetized animals are most frequently comprised of 2–8 spikes with increasing interspike intervals ranging from about 40 to well over 100 ms (Bunney et al., 1973; Grace and Bunney, 1984). The spontaneous activity is very similar in unanesthetized freely moving rats (Freeman et al., 1985; Freeman and Bunney, 1987) and the same three distinct firing patterns are evident (Hyland et al., 2002). The bursting can be sparse, with only a few two or three spike bursts occurring over a several minute period, or it can be rhythmic, lasting for several seconds or minutes. The burst firing pattern is believed to be of particular significance to the reward and/or salience signaling functions of the dopamine system (for review see Schultz, 2006).

The production of the different firing patterns, especially burst firing, in dopaminergic neurons is currently the subject of considerable study, and there is likely to be more than a single mechanism responsible (e.g., Zhang et al., 1994; Overton and Clark, 1997; Kitai et al., 1999; Waroux et al., 2005;

Ji and Shepard, 2006). As the different patterns are essentially absent in vitro (Kita et al., 1986; Grace and Onn, 1989; but see Mereu et al., 1997) afferent input is considered to be a crucial modulator of firing pattern. Considerable evidence implicates an important role for glutamatergic input and especially NMDA receptor stimulation in the burst firing pattern (Johnson et al., 1992; Overton and Clark, 1992, 1997; Chergui et al., 1993; Christoffersen and Meltzer, 1995). These conclusions are supported by several recent computational modeling studies of burst firing in dopaminergic neurons (e.g., Canavier, 1999; Amini et al., 1999; Wilson and Callaway, 2000; Medvedev et al., 2003; Komendantov et al., 2004; Kuznetsov et al., 2006). Although the ability of NMDA receptor stimulation to evoke burst firing is well established, the endogenous trigger for "spontaneous" burst firing in vivo has not been demonstrated and in vivo, the firing pattern of nigrostriatal neurons is potently modulated by blockade of GABA$_A$ receptors (Tepper et al., 1995; Paladini et al., 1999a).

GABAergic afferents to nigral dopaminergic neurons

The vast majority of inputs to pars compacta dopaminergic neurons, somewhere over 70%, are GABAergic (Bolam and Smith, 1990). These arise principally from within the basal ganglia itself, with the densest projections emanating from the neostriatum (Grofova and Rinvik, 1970; Somogyi et al., 1981; Bolam and Smith, 1990), the globus pallidus (external segment) (Grofova, 1975; Smith and Bolam, 1990) and the GABAergic neurons of the substantia nigra pars reticulata (Grace and Bunney, 1979, 1985; Nitsch and Riesenberg, 1988; Hajos and Greenfield, 1993, 1994; Tepper et al., 1995; Mailly et al., 2003).

The striatonigral pathway (direct pathway) comprises about 50% of the spiny cell efferents with the remainder (indirect pathway) projecting to the globus pallidus. Striatonigral neurons colocalize substance P and dynorphin in addition to GABA (Gerfen and Wilson, 1996). Both pars reticulata GABAergic projection neurons as well as pars compacta dopaminergic neurons receive

innervation from striatonigral neurons. The two cell populations are unlikely to receive the same set of afferent information from the striatum, however, as the major input to the GABAergic neurons comes from the striatal matrix compartment whereas that to the dopaminergic neurons, at least to their somata, arises from the patch compartment (Gerfen, 1985; Gerfen et al., 1987).

The striatonigral projection is relatively slowly conducting. It is by far the slowest of all the long-projecting GABAergic neurons in the basal ganglia (Kita and Kitai, 1991; Celada et al., 1999). The average antidromic conduction latency of striatonigral neurons in the anterior-central region of the striatum is about 10 ms corresponding to a conduction velocity around 1.4 m/s (Ryan et al., 1986). The membrane potential of the striatonigral neurons oscillates between a hyperpolarized down state during which the neurons never fire and a cortically driven, depolarized up state when firing becomes possible (Wilson, 1993). Overall, the mean firing rate is very low, in the range of 1 Hz or less (Wilson, 1993), suggesting that the effect of the striatum on the neurons of the substantia nigra is phasic and occurs only during up states.

Like the striatonigral afferents, the pallidonigral projection innervates both dopaminergic and non-dopaminergic nigral neurons (Bolam and Smith, 1990; Smith and Bolam, 1990). In contrast to the striatal input, however, the pallidonigral afferents are rapidly conducting with antidromic conduction latencies from substantia nigra of around 1 ms corresponding to a conduction velocity around 4 m/s (Kita and Kitai, 1991; Celada et al., 1999). The cells that give rise to the pallidonigral projection are among those neurons with the highest spontaneous firing rate of neurons in the basal ganglia, around 50 Hz in urethane-anesthetized rats (Celada et al., 1999).

Inputs to the dopaminergic neurons from pars reticulata GABAergic neurons arise from the local collaterals of the GABAergic output neurons (Tepper et al., 1995). These neurons are spontaneously active with a mean firing rate around 30 Hz in anesthetized rats (Celada et al., 1999) and have axonal conduction velocities similar to pallidal neurons, around 3–4 m/s (Deniau et al., 1978; Guyenet and Aghajanian, 1978).

192

At one time it was believed that the non-dopaminergic nigral neurons that were the source of the GABAergic input to the dopaminergic neurons were true local circuit neurons, leading them to be explicitly referred to as "interneurons" in the literature (e.g., Grace and Bunney, 1979, 1985; Mereu and Gessa, 1985; Araneda and Bustos, 1989; Yung et al., 1991; Johnson and North, 1992; Zhang et al., 1993; Bontempi and Sharp, 1997). However, the neuroanatomical and electrophysiological properties reported for the putative pars reticulata GABAergic interneurons are not very different from those of antidromically identified nigrothalamic and nigrotectal projection neurons (Matsuda et al., 1987; Yung et al., 1991; Lee and Tepper, 2007 but see also Grace and Bunney, 1979; Grace et al., 1980) that have been shown to send axon collaterals to the pars compacta which synapse onto dopaminergic neurons (Deniau et al., 1982; Grofova et al., 1982; Hajos and Greenfield, 1993; Tepper et al., 1995, 2002; Mailly et al., 2003). The best evidence for the existence of a nigral interneuron comes from a small population of pars compacta GABAergic neurons mapped with c-fos that are not retrogradely labeled from neostriatum (Hebb and Robertson, 2000). Virtually nothing is known about the afferent or efferent connections of these neurons however and their identity as interneurons remains to be conclusively determined. Like the dopaminergic neurons of the pars compacta, nigral GABAergic neurons also receive inhibitory input from the axon collaterals of GABAergic projection neurons (Deniau et al., 1982). At present, the balance of the evidence suggests that the bulk of the projection from the pars reticulata to the dopaminergic neurons arises from the axon collaterals of pars reticulata projection neurons, as opposed to locally projecting interneurons.

Each of the basal ganglia afferents to the substantia nigra innervates both dopaminergic and GABAergic nigral neurons with the majority of the afferents forming Gray's Type II symmetric synapses, mostly onto the dendrites of the GABAergic neurons (Rinvik and Grofova, 1970; Hattori et al., 1975; Somogyi et al., 1981; Smith and Bolam, 1990). The boutons originating from the globus pallidus and substantia nigra pars reticulata are larger than striatonigral boutons (von Krosigk et al., 1992; Tepper et al., 2002), contact proximal dendrites and somata more frequently than do the striatonigral afferents, and may innervate the dopaminergic neurons preferentially compared to the striatal inputs (Hattori et al., 1975; Smith and Bolam, 1990).

Dopaminergic substantia nigra neurons also receive GABAergic input from a number of sources outside the basal ganglia. These afferents are generally less well studied than the intrinsic basal ganglia connections. One such input originates from the superior colliculus (Comoli et al., 2003). Axons from the superior colliculus make symmetric and asymmetric synapses with both dopaminergic and GABAergic neurons in the substantia nigra (J. Boyes and J.P. Bolam, personal communication) and when stimulated, produce both inhibitory and excitatory effects in both nigral neuron types (Coizet et al., 2003; Comoli et al., 2003). A GABAergic afferent to nigral dopaminergic neurons arises from the lateral habenula, stimulation of which leads to inhibition (Bunney and Aghajanian, 1976; Christoph et al., 1986; Gao et al., 1996). Dopaminergic neurons are also inhibited in response to peripheral nociceptive stimulation (Tsai et al., 1980; Ungless et al., 2004). An additional input arises from the central nucleus of the amygdala that may preferentially innervate the pars compacta (Bunney and Aghajanian, 1976; Wallace et al., 1989, 1992; Gonzales and Chesselet, 1990; Vankova et al., 1992). The neurotransmitter used in the amygdalonigral projection is unknown but it is likely to be GABA.

Dopaminergic neurons express both $GABA_A$ and $GABA_B$ receptors somatodendritically (Bowery et al., 1987; Nicholson et al., 1992). Exogenous application of GABA, or selective $GABA_A$ or $GABA_B$ agonists, produces hyperpolarizing IPSPs in dopaminergic neurons in vitro. This is accompanied by a slowing or complete inhibition of spontaneous activity and a marked reduction in burst firing in vivo (Engberg et al., 1993). The $GABA_A$ inhibition is caused by an increase in conductance to chloride that leads to a hyperpolarization (Kaila, 1994; Gulacsi et al., 2003) whereas the $GABA_B$ inhibition is due to an increase in conductance to potassium (Lacey et al., 1988).

However, the response to endogenously released GABA evoked by afferent stimulation of striatal, pallidal, or reticulata GABAergic afferents in vivo is often complex, and depends on the type and/or intensity of stimulation.

GABAergic synaptic responses in dopaminergic neurons

In vivo recordings

In in vivo intracellular recordings, striatal stimulation produced short latency monosynaptic, chloride-mediated hyperpolarizing IPSPs in identified dopaminergic neurons (Grace and Bunney, 1985), suggesting strongly that these were $GABA_A$-mediated IPSPs even though no pharmacology was performed. Similar, but significantly larger and longer-lasting IPSPs were elicited in pars reticulata GABAergic neurons. Interestingly, the late phase of the IPSP in the GABAergic neurons was associated with a depolarization in the dopaminergic neurons (Grace and Bunney, 1985).

Extracellular in vivo recordings revealed that electrical stimulation of striatum, globus pallidus and/or pars reticulata projection neurons produced inhibition of pars compacta dopaminergic neurons that was completely blocked by local application of bicuculline or picrotoxin thus demonstrating mediation predominantly or exclusively by $GABA_A$ receptors (Paladini et al., 1999a) (Fig. 1). Local application of the highly specific and potent $GABA_B$ antagonists, saclofen or CGP 55845A, failed to block the inhibition evoked from any of these sites (Tepper et al., 1995; Paladini et al., 1999a) and in about 50% of the cases, slightly augmented the inhibition and decreased spontaneous burst firing (Paladini and Tepper, 1999; Paladini et al., 1999a). In a few cases, local application of $GABA_B$ antagonists revealed a previously unseen short latency inhibition that could subsequently be abolished by bicuculline or picrotoxin (Paladini et al., 1999a). An example of this is shown in Fig. 2. This unmasking effect is likely due to blockade of presynaptic $GABA_B$ receptors present on GABAergic afferents to substantia nigra (Giralt et al., 1990; Hausser and Yung,

1994; Shen and Johnson, 1997; Paladini et al., 1999a; Boyes and Bolam, 2003, see Misgeld, this volume), a property shared by the striatal, pallidal, and pars reticulata afferents (Paladini et al., 1999a but see Cameron and Williams, 1993). Thus, in vivo, most or all of the postsynaptic effects of striatal, pallidal, and pars reticulata GABAergic inputs in vivo appear to be mediated by $GABA_A$ receptors, with the $GABA_B$ receptors acting predominantly as presynaptic inhibitory autoreceptors.

In addition to blocking the GABAergic inhibition evoked from striatum, globus pallidus or pars reticulata, local application of $GABA_A$ antagonists exert potent effects on the spontaneous activity of dopaminergic neurons. Local application of bicuculline methiodide increases the spontaneous firing rate of dopaminergic neurons and shifts them to a robust burst pattern of firing (Tepper et al., 1995) (Fig. 3). Picrotoxin has a very similar effect on firing pattern with only a modest effect on firing rate (Paladini and Tepper, 1999) (Fig. 4), and with both drugs the effects on firing pattern were found to be independent of baseline firing rate or changes in firing rate suggesting that the mechanisms controlling the firing rate and those controlling the firing pattern are at least partially independent. It must be noted that bicuculline methiodide (and other quarternary salts, but not picrotoxin or bicuculline free base) are potent antagonists of the calcium-activated potassium channels (Johnson and Seutin, 1997; Seutin and Johnson, 1999) that underlie the long lasting spike after-hyperpolarization in dopaminergic and other neurons. Blocking this conductance also leads to increases in burst firing (Waroux et al., 2005; Ji and Shepard, 2006). Thus, the picrotoxin results are crucial for demonstrating the burst promoting effects of $GABA_A$ blockade on dopaminergic neuron firing pattern (Fig. 4).

Consistent with its lack of effects on synaptically evoked inhibition, local blockade of $GABA_B$ receptors on nigral dopaminergic neurons by application of the selective $GABA_B$ receptor antagonist, CGP 55845A or Z-hydroxysacloten, did not lead to burst firing. In fact, in about 50% of the cases, there was a modest but significant shift toward lower firing rates and more regular, less bursty firing patterns (Tepper et al., 1995; Paladini and

194

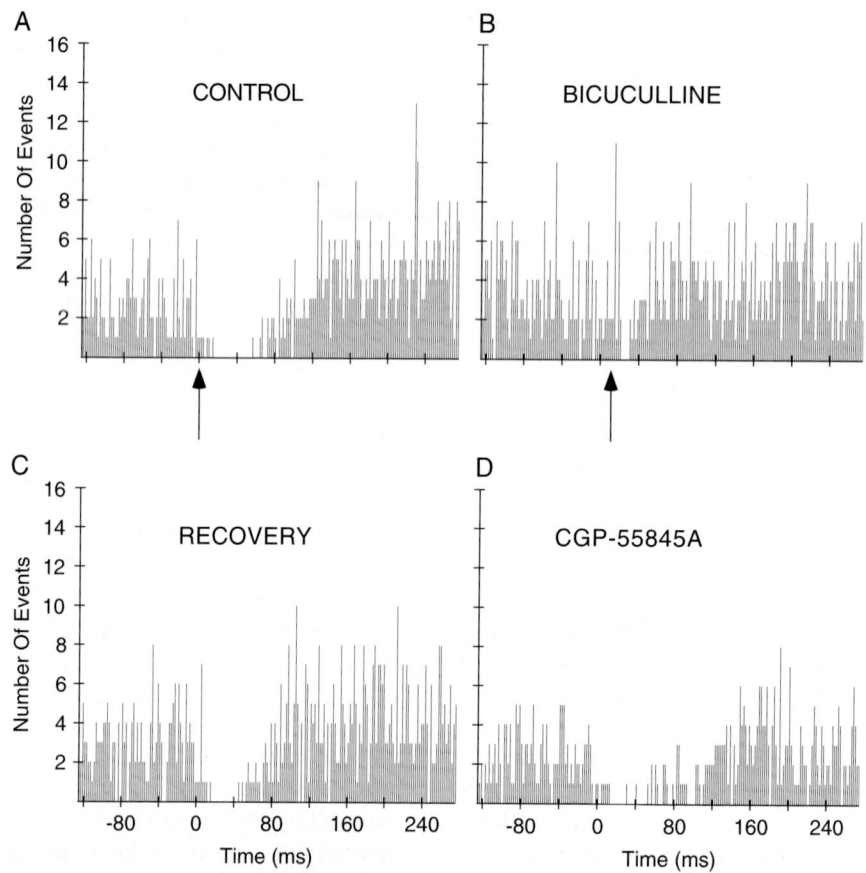

Fig. 1. Effects of globus pallidus stimulation (1.0 mA, 0.67 Hz) on a representative substantia nigra dopaminergic neuron. (A) Pallidal stimulation reliably inhibits the neuron (suppression to 10% of prestimulus firing rate, 59 ms duration). The inhibition is completely (B) and reversibly (C) blocked by bicuculline administration. (D) Application of the GABA_B antagonist, CGP-55845A, not only fails to attenuate the evokd inhibition (suppression to 18% of prestimulus firing rate) but instead *increases* the duration (82 ms duration) of inhibition relative to recovery (13% inhibition, 46 ms duration in (C) and control (A). Each PSTH consists of 200 trials. Bin width = 2 ms. [*Source*: Reprinted from Paladini (1999a) with permission from Elsevier.]

Tepper, 1999) (Fig. 3). As was the case with the effects of GABA_B antagonists on electrically evoked inhibition (Paladini et al., 1999a), these effects were interpreted to suggest that the primary locus of action of the GABA_B antagonist was pre-synaptic, leading to increased GABA release and increased postsynaptic GABA_A receptor stimulation (Paladini and Tepper, 1999).

However, there remains some controversy in the literature. While all studies agree that application of baclofen or other selective GABA_B agonists inhibit the spontaneous activity of dopaminergic neurons, early studies showed that administration of the GABA_B antagonist, CGP 35348, blocked

this effect but did not otherwise affect spontaneous activity, suggesting the absence of a GABA_B tone on dopaminergic neurons in vivo (Engberg et al., 1993). However, more recent results (Erhardt et al., 1999, 2002) indicate that administration of the GABA_B antagonist, SCH50911, or higher doses of CGP35348 did in fact lead to increases in firing rate. The reasons for these discrepancies remain unclear although it is noted that different anesthetics were used in the two series of studies (urethane in the former and chloral hydrate in the latter). It is conceivable that the different recording conditions led to differences in spontaneous GABA release such that presynaptic effects of

195

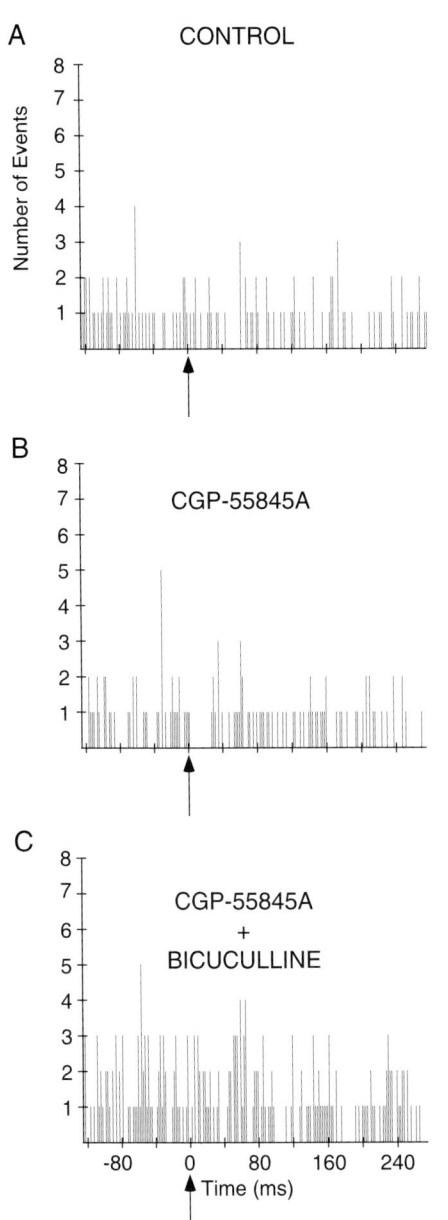

A   CONTROL

B   CGP-55845A

C   CGP-55845A
     +
     BICUCULLINE

Time (ms)

Fig. 2. Unmasking of GABA$_A$-mediated inhibition by CGP-55845A in an antidromically identified nigrostriatal dopaminergic neuron. (A) Stimulation of thalamus (1.0 mA, 0.67 Hz) has no detectable effect on the neuron. (B) Application of CGP-55845A reveals an inhibition (suppression to 0% of control for 24 ms). (C) Application of bicuculline together with CGP-55845A abolishes the unmasked inhibition. PSTH consists of 100 trials. Bin width = 2 ms. [*Source*: Reprinted from Paladini et al. (1999a) with permission from Elsevier.]

GABA$_B$ antagonists were accentuated in the former studies while postsynaptic effects were amplified in the latter. Regardless of the role of GABA$_B$ receptors, it is clear that interrupting GABA$_A$ inhibition of nigrostriatal neurons leads to phasic burst firing.

In vitro recordings

In contrast to the in vivo results, local stimulation of substantia nigra in vitro leads to both short latency and long latency (onset about 35 ms, time to peak around 150 ms; Hausser and Yung, 1994) IPSPs. The short latency IPSPs are blocked by bicuculline and picrotoxin whereas the long latency IPSPs are blocked by the selective GABA$_B$ receptor antagonists, CGP 35348 or 2-hydroxysaclofen, thereby identifying them as GABA$_A$ and GABA$_B$ IPSPs, respectively (Saitoh et al., 2004). Although GABA$_B$ IPSPs can sometimes be elicited by single pulse stimuli, they are usually not seen in response to minimal stimuli that evoke GABA$_A$ IPSPs (Hausser and Yung, 1994) and are most often reported to follow trains of stimuli at higher intensities (Johnson and North, 1992; Cameron and Williams, 1993; Hajos and Greenfield, 1993, 1994; Saitoh et al., 2004). Although spontaneous GABAergic IPSPs are frequently reported in dopaminergic neurons in vitro (e.g., Johnson and North, 1992; Hausser and Yung, 1994), these appear to be strictly GABA$_A$-mediated.

It was originally suggested that striatal-evoked GABAergic inhibition of dopaminergic neurons but not that arising from globus pallidus or substantia nigra pars reticulata, was mediated by GABA$_B$ receptors (Cameron and Williams, 1993). The evidence was that stimulation of D1 dopamine receptors in vitro selectively augmented GABA$_B$ IPSCs, and since of the striatal, pallidal, and nigral GABAergic neurons, only striatonigral neurons are known to express D1 receptors, striatonigral IPSP/Cs must be mediated predominantly or exclusively by GABA$_B$ receptors. However, more recent studies examining the effects of D1 antagonists on miniature IPSCs in the presence of TTX have revealed robust D1 modulation of GABA$_A$ IPSCs in both dopaminergic and GABAergic

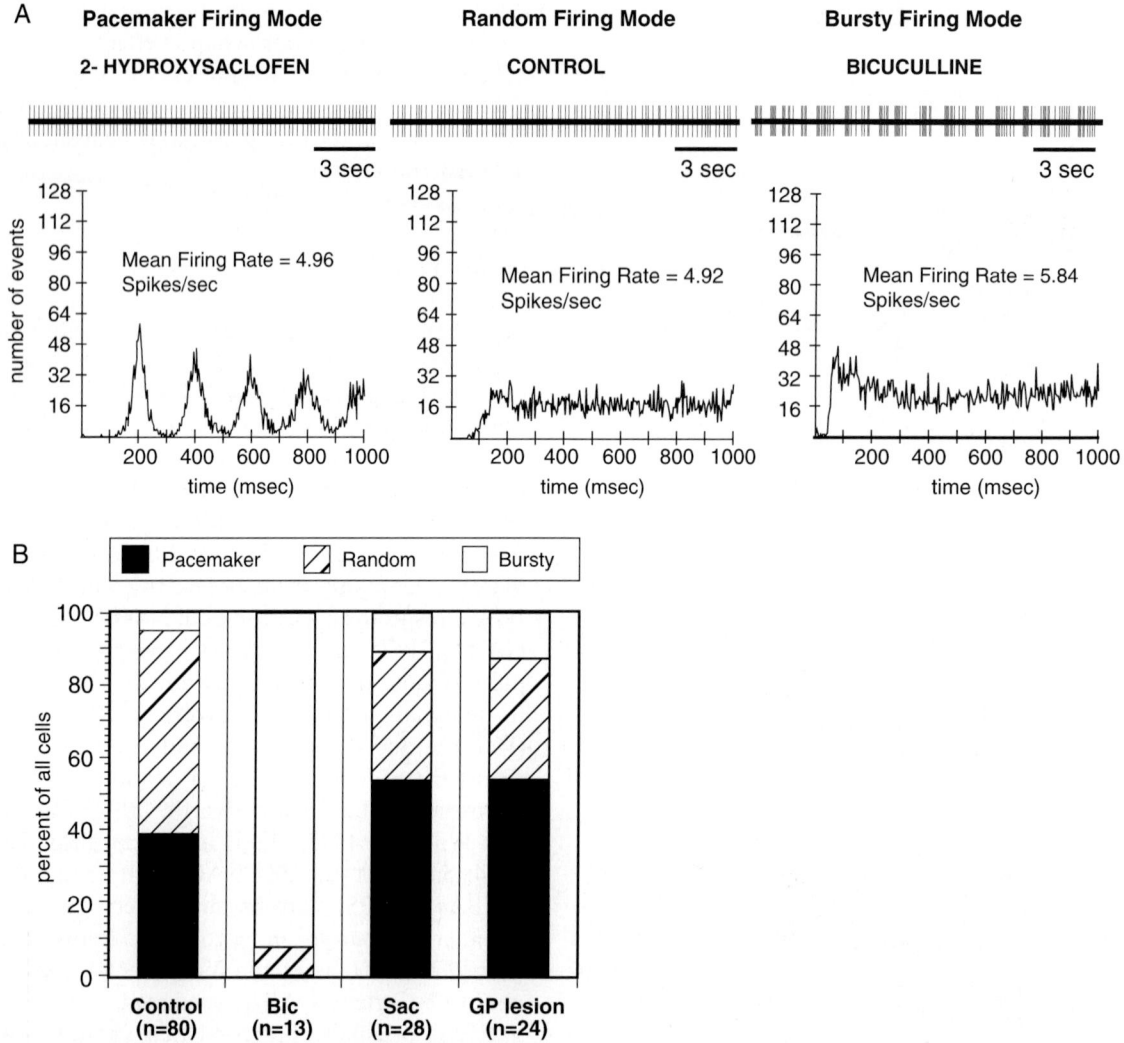

Fig. 3. Effects of bicuculline and 2-hydroxysaclofen on firing patterns of dopaminergic neurons. (A) Autocorrelograms of representative neurons exhibiting the three firing modes of dopaminergic neurons in vivo. Above each autocorrelogram is the first few seconds of the spike train used to create the autocorrelogram. The pacemaker firing neuron was obtained with a 2-hydroxysaclofen-containing micropipette, the random firing neuron with a saline micropipette, and the bursty firing neuron with a bicuculline-containing micropipette. Bin width = 3 ms. (B) Summary graph of the distribution of firing patterns recorded with control (1 M NaCl), Bic (1 M NaCl + 20 mM bicuculline methiodide) or Sac (1 mM NaCl + 20 mM 2-hydroxysaclofen) micropipettes, or with control micropipettes following kainic acid lesion of the ipsilateral globus pallidus in urethane-anesthetized rats. [*Source*: Redrawn from Tepper et al. (1995). Copyright 1995 by the *Society for Neuroscience*.]

neurons (Yanovsky et al., 2003). This suggests that the previous findings may have been due to network effects that obscured the $GABA_A$ modulation and the current view is that striatonigral and striatopallidal inhibition are mediated by both $GABA_A$ and $GABA_B$ receptors (Misgeld, 2004),

consistent with in vivo results showing blockade of striatal and pallidal evoked inhibition by bicuculline or picrotoxin (Paladini et al., 1999a).

The higher thresholds for $GABA_B$ IPSPs and the lack of spontaneous $GABA_B$ IPSPs may be related. The $GABA_B$ receptors are located

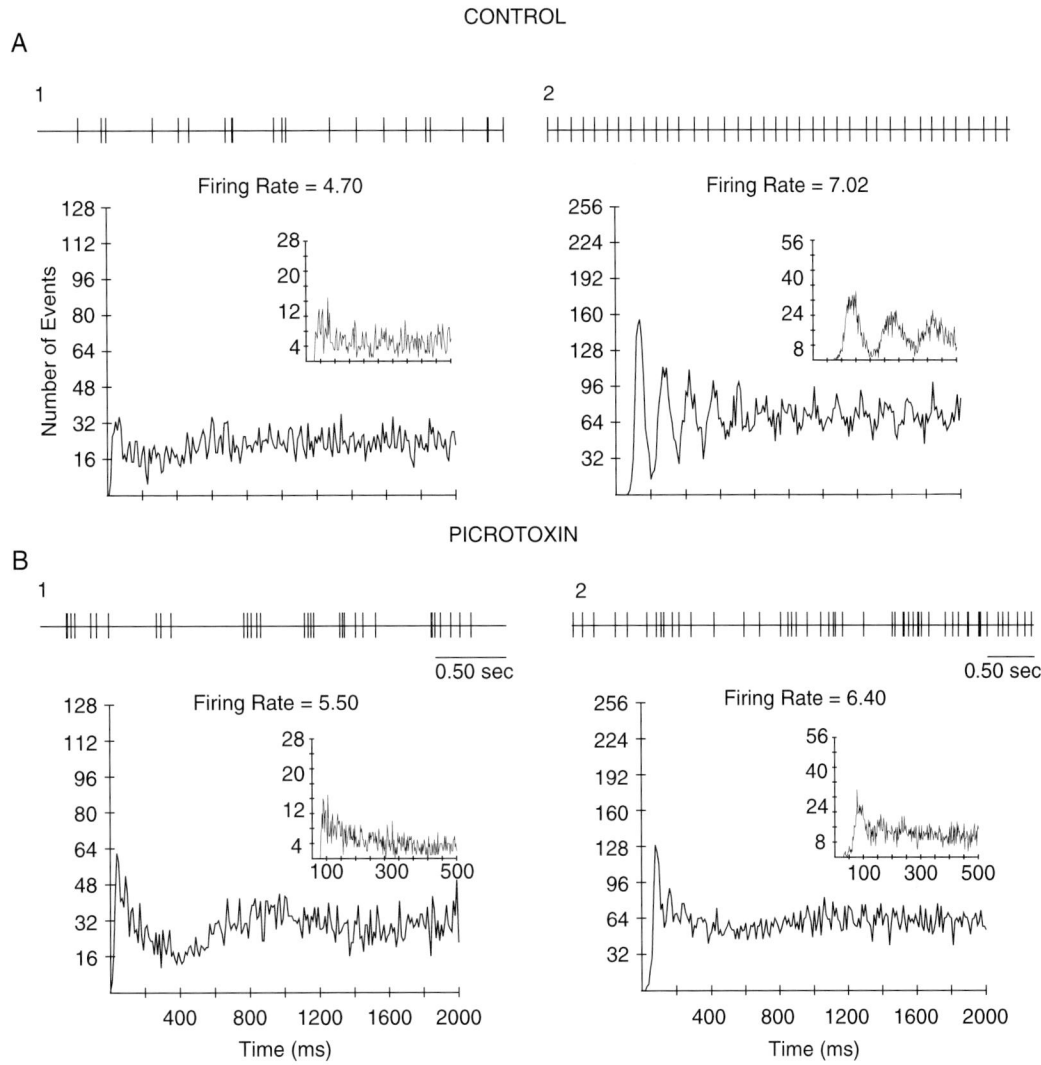

Fig. 4. Autocorrelograms showing the effects of local application of the chloride channel blocker, picrotoxin, on the firing pattern of two dopaminergic neurons. (Insets) Initial portions of the same autocorrelation at higher temporal resolution. (A) Sample spike trains and autocorrelograms of two cells before drug administration, one showing a random firing pattern (A1) and the other a pacemaker pattern (A2). (B) After application of picrotoxin, both neurons show a marked switch to a bursty firing pattern indicated by the clustering of spikes in the spike train and an initial peak with a decay to a steady-state level in the autocorrelogram. Note that the change in firing pattern is not dependent on an increase in firing rate (A2 to B2). (1,000 spikes; Bin width = 10 ms.) [*Source*: Adapted with permission from Paladini and Tepper (1999).]

extra-synaptically, at some distance from the active zone (Boyes and Bolam, 2003). Thus, relatively strong and/or prolonged presynaptic stimuli may be required to allow GABA to escape from the synaptic cleft and activate these receptors (e.g., Scanziani, 2000) compared to GABA$_A$ receptors which are believed to be located subsynaptically, in closer proximity to synaptically released GABA (Richards et al., 1987; Nicholson et al., 1992). The necessary stimulus for activation of postsynaptic GABA$_B$ receptors in the hippocampus is rhythmic, synchronous discharge in presynaptic inhibitory neurons or pharmacological block of GABA uptake (Scanziani, 2000) where GABA$_B$ receptors have been shown to have a similar extrasynaptic distribution (Lopez-Bendito et al., 2004).

As noted earlier, postsynaptic $GABA_B$ receptor-mediated events are encountered in nigral dopaminergic neurons relatively frequently in vitro, but not in vivo under most conditions. This may be due to more efficacious electrical stimulation, a reduction in GABA uptake, and/or a decrease in presynaptic GABAergic inhibition in vitro. Train stimulation of the GABAergic afferents to nigral dopaminergic neurons has thus far been ineffective at eliciting postsynaptic $GABA_B$ receptor-mediated responses in vivo (Paladini et al., 1999a) so the substrates necessary for activating postsynaptic $GABA_B$ receptors in vivo remain to be determined.

In addition, dopamine itself has been reported to have a selective presynaptic inhibitory effect on $GABA_B$ IPSPs in dopaminergic neurons (Federici et al., 2002) that could also contribute to the relative inefficiency of $GABA_B$ transmission in pars compacta dopaminergic neurons (Saitoh et al., 2004).

GABA can also affect the action of other neurotransmitters in substantia nigra, for example, glutamate. Although apparently different from spontaneous burst firing in vivo in certain important aspects, a type of burst firing can be elicited in vitro by the application of NMDA or NMDA plus apamin (Johnson et al., 1992). Burst firing induced in this way has been used to explore the effects of $GABA_A$ receptor stimulation and blockade on firing pattern in dopaminergic neurons (Paladini et al., 1999b). Under these conditions NMDA or NMDA plus apamin induced rhythmic bursting which was abolished by bath application of the $GABA_A$ agonist, isoguvacine. This was not simply due to a $GABA_A$-induced hyperpolarization, however, as hyperpolarizing NMDA-treated neurons with current injection did not eliminate the

rhythmic membrane oscillations even when hyperpolarized sufficiently to block spiking. Rather, it appears that the $GABA_A$ receptor-mediated decrease in input resistance was the critical factor for the elimination of the oscillation (Paladini et al., 1999b), consistent with predictions made from compartmental models (Canavier, 1999; Amini et al., 1999; Medvedev et al., 2003; Komendantov et al., 2004; Kuznetsov et al., 2006).

GABAergic inputs from the pars reticulata

The inputs from the pars reticulata seem to have a prepotent effect on modulating the firing pattern of dopaminergic neurons compared to those from the striatum or the globus pallidus. This conclusion is based on several independent lines of evidence. In early studies of the effects of pars reticulata neurons on dopaminergic neuron activity, it was noted that lesion of the GP produced a modest decrease in the firing rate and burst firing of dopaminergic neurons (Tepper et al., 1995), which is the same effect seen if GABAergic tone (mediated by $GABA_A$ receptors) is increased (Paladini and Tepper, 1999; Paladini et al., 1999b).

Although electrical stimulation of globus pallidus produces short latency monosynaptic IPSPs and inhibition in nigrostriatal neurons (Tepper et al., 1986; Paladini et al., 1999a), chemical stimulation of globus pallidus with bicuculline produces a paradoxical modest increase in firing rate of nigrostriatal neurons along with a significant increase in burst firing (Celada et al., 1999; Lee et al., 2004) while inhibition of globus pallidus with muscimol infusion produces modest decreases in both firing rate and burst firing (Celada et al., 1999), as shown in Fig. 5. When the same

Fig. 5. (A) Experimental design showing simultaneous dopamine microdialysis in striatum (STR), drug infusion into globus pallidus (GP) and extracellular recording in substantia nigra pars compacta (SNc). (B1-3) Representative autocorrelograms from a single dopaminergic neuron under control conditions (1), after pallidal microinjection of muscimol (2), and after pallidal microinjection of bicuculline (3). Infusion of bicuculline into the GP shifted the distribution of dopaminergic neuron firing patterns from one consisting mostly of the random firing pattern to one where the bursty firing pattern was most prominent. (C, D) The shift toward the bursty firing pattern was accompanied by a significant increase in the CV. (E) Ratemeter of a representative dopaminergic neuron following infusion of muscimol and bicuculline into the ipsilateral globus pallidus. (F) Striatal dopamine levels after injection of bicuculline into the GP. The sample collection time was 5 min beginning 1 min after the start of the bicuculline infusion. The first three samples were obtained under baseline conditions during the 15 min immediately preceding a 1 min infusion of bicuculline (BIC) into the GP as indicated by the arrow. Striatal dopamine levels increased significantly following disinhibition of the GP in the two 5 min samples immediately following drug infusion and delay.

A

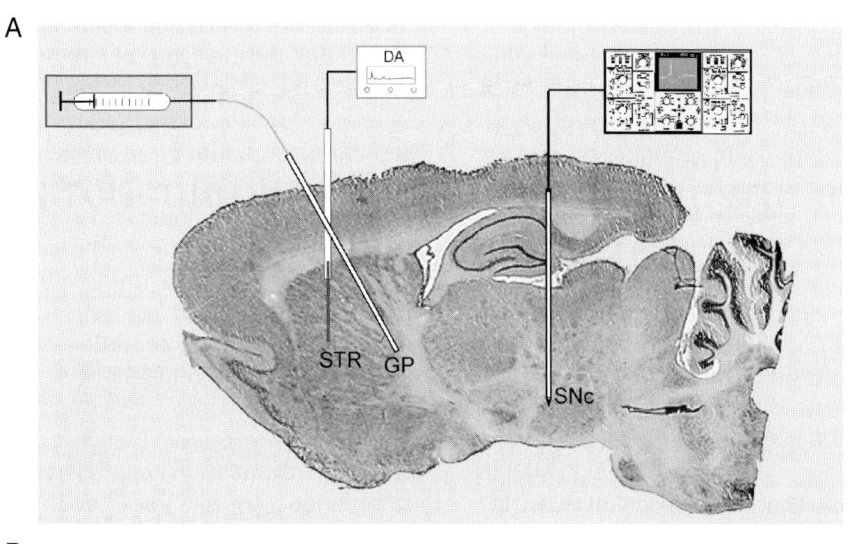

B

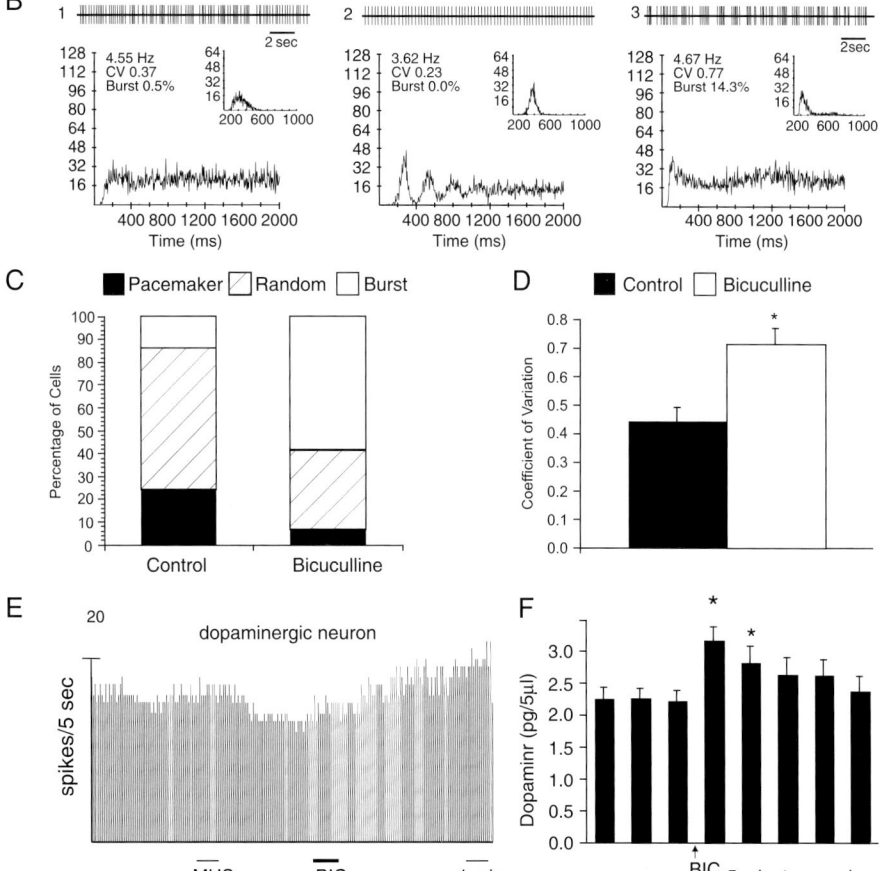

C ■ Pacemaker ▨ Random □ Burst

D ■ Control □ Bicuculline

E dopaminergic neuron

F

manipulations were made while recording from pars reticulata projection neurons, bicuculline infusions into globus pallidus produced a dramatic inhibition of the reticulata neurons, sometimes completely suppressing spontaneous activity. Pallidal muscimol infusion on the other hand led to a near doubling of the spontaneous firing rate (Celada et al., 1999). When these experiments were replicated with a microdialysis probe in striatum to measure firing pattern and dopamine overflow before and after modulation of pallidal firing rate, bicuculline infusion again produced a robust switch to burst firing accompanied by a small (11.5%) increase in firing rate but a large (45.9%) increase in dopamine overflow (Lee et al., 2004) as illustrated in Fig. 5. These results demonstrate that the pallidonigral effects on dopaminergic neurons are mediated principally through disinhibition, via a preferential inhibition of the tonically active pars reticulata projection neurons. This is consistent with neuroanatomical findings showing a particularly dense projection from the globus pallidus to the pars reticulata GABAergic neurons (Smith and Bolam, 1989).

But, how could the effects of electrical stimulation of globus pallidus be manifest as monosynaptic inhibition while the effects of chemical manipulation are seen as disinhibition? We hypothesized that the opposite effects of electrical and chemical stimulation could be attributed to differences in the sensitivity of nigrostriatal dopaminergic neurons and GABAergic pars reticulata neurons to GABA. Several studies imply that GABAergic pars reticulata neurons are more sensitive to exogenously applied GABA and/or $GABA_A$ agonists (e.g., Grace and Bunney, 1979; Waszczak et al., 1980, 1981) or to synaptically released GABA than are the dopaminergic neurons (Grace and Bunney, 1985; Celada et al., 1999). This effect is so robust that the systemic administration of the $GABA_A$ agonist, muscimol leads to an increase in the spontaneous firing rate of dopaminergic neurons (Walters and Lakoski, 1978; Grace and Bunney, 1979; Grace et al., 1980) and elicits an increase in striatal dopamine levels (Martin and Haubrich, 1978; Santiago and Westerink, 1992), despite its expected inhibitory effects on dopaminergic neuron firing. The differences observed between low-intensity electrical stimulation and chemical stimulation of GABAergic afferents as opposed to when these afferents are stimulated electrically using more standard stimulation parameters likely comes about because GABAergic afferents produce a large *synchronous* release of GABA following strong electrical stimulation that is sufficient to inhibit both the more sensitive GABAergic neurons as well as the less sensitive dopaminergic neurons. Under these conditions the monosynaptic input to the dopaminergic neurons predominates and inhibition of the nigrostriatal neurons results along with decreased burst firing. However, with chemical stimulation or low-intensity electrical stimulation (e.g., Grace and Bunney, 1985), the resulting activation and GABA release is asynchronous and preferentially inhibits the more sensitive GABAergic neurons with only minimal direct effects on the dopaminergic neurons. This leads to a disinhibition of the dopaminergic neurons as the tonically active GABAergic neurons are silenced, resulting in an increase in burst firing and a consequent increase in striatal dopamine levels despite only a minimal increase in firing rate (Lee et al., 2004).

Since low-intensity stimulation of the striatal or pallidal afferents to the substantia nigra causes disinhibition of nigral dopaminergic neurons, the GABAergic afferents that most consistently act to inhibit dopaminergic neurons in vivo are those originating from the axon collaterals of nigral GABAergic projection neurons. Under many or most in vivo conditions, the indirect action of GABA on nigral dopaminergic neurons via disinhibition predominates thus making disinhibition a fundamental process underlying signal transmission in the substantia nigra. This process also likely underlies the actions of drugs of abuse that act as $GABA_A$ agonists such as ethanol (Mereu et al., 1984; Mereu and Gessa, 1985) and benzodiazepines (Ross et al., 1982; O'Brien and White, 1987). Although all of these manipulations likely add some direct inhibition of nigral dopaminergic neurons, they remove a far greater amount by inhibiting the tonically active GABAergic neurons of the pars reticulata.

Differences in GABA$_A$ responses between dopaminergic and GABAergic neurons

There are several possible explanations for the apparently greater sensitivity to GABA of the GABAergic pars reticulata neurons compared to nigrostriatal dopaminergic neurons. Since most of the GABAergic effects in vivo are mediated by GABA$_A$ receptors, we compared GABA$_A$ functioning in the two populations of neurons. As a ligand-gated chloride channel, GABA$_A$ receptor function is dependent on cellular chloride regulatory mechanisms (Kaila, 1994). In most mature CNS neurons, a neuron-specific K^+Cl^- cotransporter (KCC2) is responsible for extruding intracellular chloride, thereby generating an inwardly directed chloride gradient which makes increases in chloride conductance hyperpolarizing (Payne et al., 1996, 2003; DeFazio et al., 2000). Light and electron microscopic immunocytochemical labeling revealed the expression of KCC2 by GABAergic pars reticulata neurons (Gulacsi et al., 2003) (Fig. 6). However, nigral dopaminergic neurons do not express KCC2 but instead express the voltage-sensitive chloride channel (ClC-2) that is absent in pars reticulata GABAergic neurons (Fig. 6). This neuron type-specific expression of Cl^- regulatory mechanisms is reflected in differences in the reversal potential for GABA$_A$ IPSPs. GABA$_A$ IPSPs evoked by local stimulation and measured with perforated patch recordings in vitro exhibit a reversal potential that is significantly more hyperpolarized than those in the dopaminergic neurons, thus accounting in part for the increased sensitivity to GABA of the nigral GABAergic neurons compared to dopaminergic neurons (Gulacsi et al., 2003) (Fig. 7).

However, even in the absence of KCC2, the GABA$_A$ reversal potential is hyperpolarizing in nigral dopaminergic neurons, indicating the presence of an active chloride extrusion mechanism. Although ClC-2 can help clear chloride in the presence of increased intracellular chloride levels, it is passive and voltage dependent and cannot create the driving force needed to account for the hyperpolarizing chloride gradient in dopaminergic neurons. The most likely candidate mechanism that can generate the appropriate driving force is

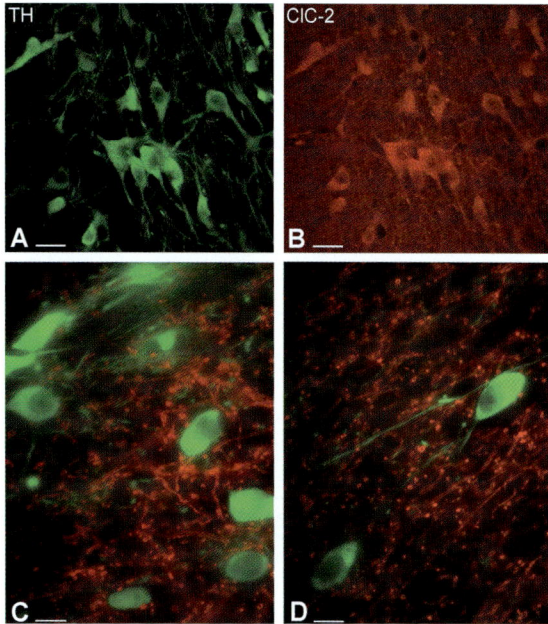

Fig. 6. Dopaminergic neurons express ClC-2 but not KCC2. (A) Green immunofluorescence labels dopaminergic neurons in the substantia nigra pars compacta neurons. (B) Same field under a different filters show red immunofluoresence for ClC-2. (C) and (D) Double immunofluoresence for KCC2 (red) and TH (green) shows that TH positive dopaminergic neurons in the pars compacta (C) and pars reticulata (D) do not express KCC2 whereas the dendrites of nigral GABAergic neurons in both regions express KCC2 but not ClC2 or TH. Scale bars 25 μm. [*Source*: Modified from Gulacsi et al. (2003). Copyright 2003, the *Society for Neuroscience*.]

the sodium-dependent anion exchanger that exchanges Cl^- for bicarbonate (NDAE; Payne et al., 2003; see Farrant and Kaila, this volume). When the IPSP reversal potential measurements were repeated in bicarbonate-free conditions that block the exchanger, the GABA$_A$ reversal potential in dopaminergic neurons was depolarized to around the resting membrane potential (-49 mV) whereas there was no significant change in the reversal potential in the GABAergic neurons (around -71 mV; Gulacsi et al., 2003) (Fig. 7). Thus, the driving force for a hyperpolarizing GABA$_A$ IPSP in dopaminergic neurons is maintained by the NDAE. However, the NDAE is much less efficient at clearing intracellular chloride than KCC2 (Kaila, 1994; Rivera et al., 1999; Payne et al., 2003), consequently nigral dopaminergic neurons

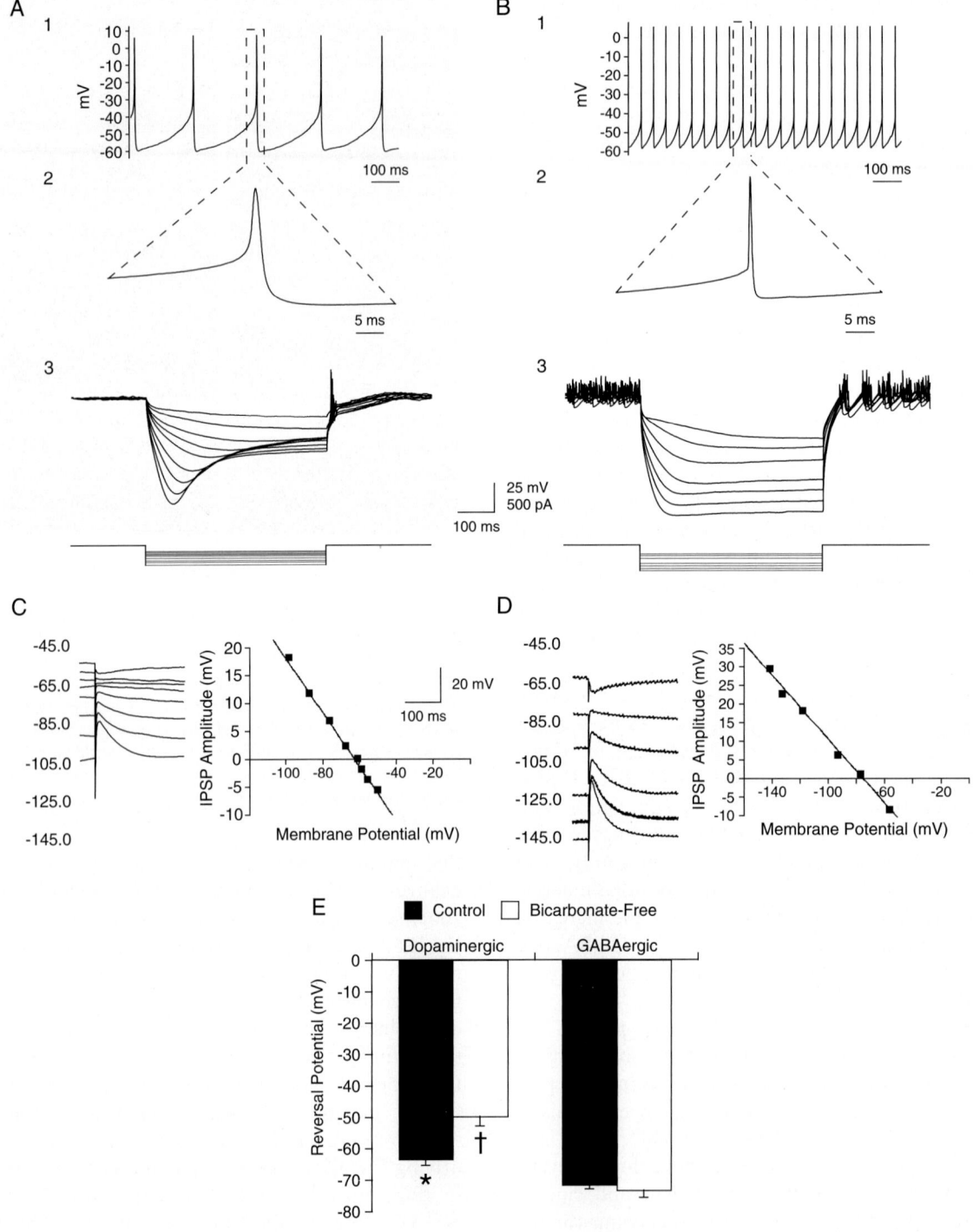

exhibit a significantly less hyperpolarized GABA$_A$ IPSPs than nigral GABAergic neurons. This leads to greater sensitivity to GABA$_A$ receptor stimulation in the pars reticulata GABAergic neurons compared to the dopaminergic neurons.

Another factor that could account for the greater efficacy of GABAergic inhibition in GABAergic reticulata neurons is the subunit composition of their GABA$_A$ receptors. Both in situ hybridization and immunocytochemical studies have identified the existence of a number of GABA$_A$ receptor subunits in the substantia nigra (Nicholson et al., 1992, 1996; Wisden et al., 1992; Fritschy and Mohler, 1995; Pirker et al., 2000; Schwarzer et al., 2001). There are some differences in the expression pattern of GABA$_A$ receptor subunits between the neurons of the substantia nigra pars reticulata and those of the substantia nigra pars compacta. A common thread which emerges from different studies using different techniques is that neurons of the substantia nigra pars reticulata express GABA$_A$ receptors with the $\alpha 1$ and $\beta 2$ receptor subunits while neurons of the substantia nigra pars compacta express receptors with the $\alpha 3$ receptor subunit (Nicholson et al., 1992, 1996; Wisden et al., 1992; Fritschy and Mohler, 1995; Guyon et al., 1999; Pirker et al., 2000; Rodriguez-Pallares et al., 2000, 2001; Schwarzer et al., 2001; Okada et al., 2004; see Goetz et al., this volume). The functional significance of this segregation is unknown but these subunit differences could contribute to the differences in sensitivity to inhibition by GABA exhibited by GABAergic and dopaminergic neurons in the substantia nigra. There are also differences in the overall density of GABA$_A$ receptors with GABA/benzodiazepine immunoreactivity being much lower in the pars compacta than in the pars reticulata, consistent with a relatively reduced GABA sensitivity of the dopaminergic neurons (Nicholson et al., 1992).

Conclusions

The advances made in the past decade have shed light on several key elements underlying the afferent control of nigrostriatal dopaminergic neurons. Chief among them are GABAergic afferents from other parts of the basal ganglia that make up more than 70% of the inputs to nigral dopaminergic neurons.

Though it has been appreciated for some time that disinhibition is the major mechanism for signaling among GABAergic neurons in the basal ganglia (e.g., Chevalier et al., 1985; Deniau and Chevalier, 1985; Chevalier and Deniau, 1990), it is only more recently that this mode of control has been shown to apply to the nigrostriatal neurons as well. A crucial first step in understanding the neuronal interactions that control the activity of nigral dopaminergic neurons was the discovery that nigral GABAergic projection neurons strongly inhibit the dopaminergic neurons through their axon collaterals. It is now clear that disinhibition of the globus pallidus can lead to inhibition of nigral GABAergic projection neurons and a subsequent disinhibition of nigral dopaminergic neurons that is manifest primarily as an increase in burst firing. The burst firing results in a significant increase in striatal dopamine levels over those resulting from normal spontaneous firing in the pacemaker and random modes.

The seemingly contradictory excitation of dopaminergic neurons in vivo by GABA or systemic administration of GABA agonists can now be understood as a consequence of disinhibition

---

Fig. 7. The GABA$_A$ IPSP reversal potential is more hyperpolarized in GABAergic than in dopaminergic nigral neurons. (A1, B1) Perforated patch recordings of electrophysiologically identified substantia nigra pars compacta dopaminergic (A1) and pars reticulata GABAergic neuron (B1). (A2, B2) Higher sweep speed reveals the considerably longer duration action potential and large and longer duration spike after-hyperpolarization in the dopaminergic neuron (A2) compared to the GABAergic neuron (B2). (A3, B3) The dopaminergic neuron (A3) exhibits a marked sag in response to hyperpolarizing current pulses due to a slowly activating I_h and inward rectification whereas the GABAergic neuron displays little or no sag and a linear I–V relation. (C, D) Representative GABA$_A$-mediated IPSPs evoked locally in normal Ringer's solution in the presence of CGP 35348, CNQX and APV at different membrane potentials (left panels) reveal a more hyperpolarized reversal potential in the GABAergic neuron than in the dopaminergic neuron (right panels). (E) Summary bar graphs of effects of cell type and inhibition of NDAE on GABA$_A$ IPSPs in nigral neurons. * = significantly different from GABAergic neuron, $p < 0.01$.[†]

resulting from a more potent GABAergic inhibition of the pars reticulata GABAergic neurons than the dopaminergic neurons due, in part, to differences in the $GABA_A$ IPSP reversal potential. This difference is also ultimately responsible, at least in part, for a number of seemingly paradoxical excitations of nigrostriatal dopaminergic neurons by drugs acting at other receptors including the NMDA and mu opioid receptors (Hommer and Pert, 1983; Zhang et al., 1992) as well as the effects of ethanol and benzodiazepines discussed above (Ross et al., 1982; Mereu et al., 1984; Mereu and Gessa, 1985; O'Brien and White, 1987).

Understanding the activity and interactions of neurons in the substantia nigra is a necessary prerequisite to understanding the functioning of the basal ganglia in both normal and pathological behaviors. It has become increasingly clear that in addition to the effects that nigral GABAergic projection neurons have on their target nuclei outside of the basal ganglia, local interactions between GABAergic projection neurons and dopaminergic neurons are crucially important to the functioning of the substantia nigra and of the basal ganglia as a whole.

Acknowledgments

We thank Fulva Shah for expert technical assistance. Supported by a grant from the National Institute for Neurological Diseases and Stroke (NS-34865) to J.M.T.

References

Amini, B., Clark, J.W. and Canavier, C.C. (1999) Calcium dynamics underlying pacemaker-like and burst firing oscillations in midbrain dopaminergic neurons: a computational study. J. Neurophysiol., 82: 2249–2261.

Araneda, R. and Bustos, G. (1989) Modulation of dendritic release of dopamine by N-Methyl-D-aspartate receptors in rat substantia nigra. J. Neurochem., 52: 962–970.

Bolam, J.P. and Smith, Y. (1990) The GABA and substance P input to dopaminergic neurones in the substantia nigra of the rat. Brain Res., 529: 57–78.

Bontempi, B. and Sharp, F.R. (1997) Systemic morphine-induced Fos protein in the rat striatum and nucleus accumbens is regulated by μ opioid receptors in the substantia nigra and ventral tegmental area. J. Neurosci., 17: 8596–8612.

Bowery, N.G., Hudson, A.L. and Price, G.W. (1987) $GABA_A$ and $GABA_B$ receptor site distribution in the rat central nervous system. Neuroscience, 20: 365–383.

Boyes, J. and Bolam, J.P. (2003) The subcellular localization of $GABA_B$ receptor subunits in the rat substantia nigra. Eur. J. Neurosci., 18: 3279–3293.

Bunney, B.S. (1979) The electrophysiological pharmacology of midbrain dopaminergic systems. In: Horn A.S., Korf J. and Westerink B.H.C. (Eds.), The Neurobiology of Dopamine. Academic Press, New York, pp. 417–452.

Bunney, B.S. and Aghajanian, G.K. (1976) The precise localization of nigral afferents in the rat as determined by a retrograde tracing technique. Brain Res., 117: 423–435.

Bunney, B.S., Walters, J.R., Roth, R.H. and Aghajanian, G.K. (1973) Dopaminergic neurons: effect of antipsychotic drugs and amphetamines on single cell activity. J. Pharmacol. Exp. Ther., 185: 560–571.

Cameron, D.L. and Williams, J.T. (1993) Dopamine D1 receptors facilitate transmitter release. Nature, 366: 344–347.

Canavier, C.C. (1999) Sodium dynamics underlying burst firing and putative mechanisms for the regulation of the firing pattern in midbrain dopamine neurons: a computational approach. J. Comput. Neurosci., 6: 49–69.

Celada, P., Paladini, C.A. and Tepper, J.M. (1999) GABAergic control of rat substantia nigra dopaminergic neurons: role of globus pallidus and substantia nigra pars reticulata. Neuroscience, 89: 813–825.

Chergui, K., Charlety, P.J., Akaoka, H., Saunier, C.F., Brunet, J.-L., Buda, M., Svensson, T.H. and Chouvet, G. (1993) Tonic activation of NMDA receptors causes spontaneous burst discharge of rat midbrain dopamine neurons in vivo. Eur. J. Neurosci., 5: 137–144.

Chevalier, G. and Deniau, J.M. (1990) Disinhibition as a basic process in the expression of striatal functions. Trends Neurosci., 13: 277–280.

Chevalier, G., Vacher, S., Deniau, J.M. and Desban, M. (1985) Disinhibition as a basic process in the expression of striatal functions. I. The striato-nigral Influence on tecto-spinal/ tecto-diencephalic neurons. Brain Res., 334: 215–226.

Chiodo, L.A. (1988) Dopamine-containing neurons in the mammalian central nervous system: electrophysiology and pharmacology. Neurosci. Biobehav. Rev., 12: 49–91.

Christoffersen, C.L. and Meltzer, L.T. (1995) Evidence for N-methyl-D-aspartate and AMPA subtypes of the glutamate receptor on substantia nigra dopamine neurons: Possible preferential role for N-methyl-D-aspartate receptors. Neuroscience, 67: 373–381.

Coizet, V., Comoli, E., Westby, G.W. and Redgrave, P. (2003) Phasic activation of substantia nigra and the ventral tegmental area by chemical stimulation of the superior colliculus: an electrophysiological investigation in the rat. Eur. J. Neurosci., 17: 28–40.

Comoli, E., Coizet, V., Boyes, J., Bolam, J.P., Canteras, N.S., Quirk, R.H., Overton, P.G. and Redgrave, P. (2003) A direct projection from superior colliculus to substantia nigra for detecting salient visual events. Nat. Neurosci., 6: 974–980.

Christoph, G.R., Leonzio, R.J. and Wilcox, K.S. (1986) Stimulation of the lateral habenula inhibits dopamine-containing neurons in the substantia nigra and ventral tegmental area of the rat. J. Neurosci., 6: 613–619.

Dai, M. and Tepper, J.M. (1998) Do silent dopaminergic neurons exist in rat substantia nigra in vivo? Neuroscience, 85(4): 1089–1099.

DeFazio, R.A., Keros, S., Quick, M.W. and Hablitz, J.J. (2000) Potassium-coupled chloride cotransport controls intracellular chloride in rat neocortical pyramidal neurons. J. Neurosci., 20: 8069–8076.

Deniau, J.M. and Chevalier, G. (1985) Disinhibition as a basic process in the expression of striatal functions. II. The striatonigral influence on thalamocortical cells of the ventromedial thalamic nucleus. Brain Res., 334: 227–233.

Deniau, J.M., Hammond, C., Riszk, A. and Feger, J. (1978) Electrophysiological properties of identified output neurons of the rat substantia nigra (pars compacta and pars reticulata): evidences for the existence of branched neurons. Exp. Brain Res., 32: 409–422.

Deniau, J.M., Kitai, S.T., Donoghue, J.P. and Grofova, I. (1982) Neuronal Interactions in the substantia nigra pars reticulata through axon collaterals of the projection neurons. Exp. Brain Res., 47: 105–113.

Diana, M. and Tepper, J.M. (2002) Electrophysiological pharmacology of mesencephalic dopaminergic neurons. In: DiChiara G. (Ed.), Handbook of Experimental Pharmacology, Vol. 54/II. Springer, Berlin, pp. 1–61.

Engberg, G., Kling-Petersen, T. and Nissbrandt, H. (1993) GABAB-receptor activation alters the firing pattern of dopamine neurons in the rat substantia nigra. Synapse, 15: 229–238.

Erhardt, S., Mathe, J.M., Chergui, K., Engberg, G. and Svensson, T.H. (2002) GABA$_B$ receptor-mediated modulation of the firing pattern of ventral tegmental area dopamine neurons in vivo. Naunyn Schmiedebergs Arch. Pharmacol., 365: 173–180.

Erhardt, S., Nissbrandt, H. and Engberg, G. (1999) Activation of nigral dopamine neurons by the selective GABA(B)-receptor antagonist SCH 50911. J. Neural. Transm., 106: 383–394.

Federici, M., Natoli, S., Bernardi, G. and Mercuri, N.B. (2002) Dopamine selectively reduces GABA$_B$ transmission onto dopaminergic neurones by an unconventional presynaptic action. J. Physiol., 540: 119–128.

Floresco, S.B., West, A.R., Ash, B., Moore, H. and Grace, A.A. (2003) Afferent modulation of dopamine neuron firing differentially regulates tonic and phasic dopamine transmission. Nat. Neurosci., 6: 968–973.

Freeman, A.S. and Bunney, B.S. (1987) Activity of A9 and A10 dopaminergic neurons in unrestrained rats: further characterization and effects of apomorphine and cholecystokinin. Brain Res., 405: 46–55.

Freeman, A.S., Meltzer, L.T. and Bunney, B.S. (1985) Firing properties of substantia nigra dopaminergic neurons in freely moving rats. Life Sci., 36: 1983–1994.

Fritschy, J.M. and Mohler, H. (1995) GABA$_A$-receptor heterogeneity in the adult rat brain: differential regional and cellular distribution of seven major subunits. J. Comp. Neurol., 359: 154–194.

Gao, D.M., Hoffman, D. and Benabid, A.L. (1996) Simultaneous recording of spontaneous activities and nociceptive responses from neurons in the pars compacta of substantia nigra and in the lateral habenula. Eur. J. Neurosci., 8: 1474–1478.

Gerfen, C.R. (1985) The neostriatal mosaic. I. Compartmental organization of projections from the striatum to the substantia nigra in the rat. J. Comp. Neurol., 236: 454–476.

Gerfen, C.R., Herkenham, M. and Thibault, J. (1987) The neostriatal mosaic: II. Patch- and matrix-directed mesostriatal dopaminergic and non-dopaminergic systems. J. Neurosci., 7: 3915–3934.

Gerfen, C.R. and Wilson, C.J. (1996) The basal ganglia. In: Swanson L.W., Bjorklund A. and Hokfelt T. (Eds.), Handbook of Chemical Neuroanatomy (12th Edition). Elsevier Science Pub, Amsterdam, pp. 371–468.

Giralt, M.T., Bonanno, G. and Raiteri, M. (1990) GABA terminal autoreceptors in the pars compacta and in the pars reticulata of the rat substantia nigra are GABA$_B$. Eur. J. Pharmacol., 175: 137–144.

Gonzales, C. and Chesselet, M.F. (1990) Amygdalonigral pathway: an anterograde study in the rat with phaseolus vulgaris leucoagglutinin (PHA-L). J. Comp. Neurol., 297: 182–200.

Grace, A.A. and Bunney, B.S. (1979) Paradoxical GABA excitation of nigral dopaminergic cells: Indirect mediation through reticulata inhibitory neurons. Eur. J. Pharmacol., 59: 211–218.

Grace, A.A. and Bunney, B.S. (1983) Intracellular and extracellular electrophysiology of nigral dopaminergic neurons-1. Identification and characterization. Neuroscience, 2: 301–315.

Grace, A.A. and Bunney, B.S. (1984) The control of firing pattern in the nigral dopamine neurons: Burst firing. J. Neurosci., 4: 2877–2890.

Grace, A.A. and Bunney, B.S. (1985) Opposing effects of striatonigral feedback pathways on midbrain dopamine cell activity. Brain Res., 333: 271–284.

Grace, A.A., Hommer, D.W. and Bunney, B.S. (1980) Peripheral and striatal influences on nigral dopamine cells: Mediation by reticulata neurons. Brain Res. Bull, 5: 105–109.

Grace, A.A. and Onn, S.-P. (1989) Morphology and electrophysiological properties of immunocytochemically identified rat dopamine neurons recorded in vitro. J. Neurosci., 9: 3463–3481.

Grofova, I. (1975) The identification of striatal and pallidal neurons projecting to substantia nigra. An experimental study by means of retrograde axonal transport of horseradish peroxidase. Brain Res., 91: 286–291.

Grofova, I. and Rinvik, E. (1970) An experimental electron microscopic study on the striatonigral projection in the cat. Exp. Brain Res., 11: 249–262.

Grofova, I., Deniau, J.M. and Kitai, S.T. (1982) Morphology of the substantia nigra pars reticulata projection neurons intracellularly labeled with HRP. J. Comp. Neurol., 208: 352–368.

Grofova, I., Kita, H. and Kitai, S.T. (1986) Patterns of synaptic contacts upon intracellularly labeled nigrostriatal neurons in the rat. Soc. Neurosci. Abstr., 12: 1542.

Gulacsi, A., Lee, C.R., Sik, A., Viitanen, T., Kaila, K., Tepper, J.M. and Freund, T.F. (2003) Cell type-specific differences in chloride-regulatory mechanisms and GABA$_A$ receptor-mediated inhibition in rat substantia nigra. J. Neurosci., 23: 8237–8246.

Guyenet, P.G. and Aghajanian, G.K. (1978) Antidromic identification of dopaminergic and other output neurons of the rat substantia nigra. Brain Res., 150: 69–84.

Guyon, A., Laurent, S., Paupardin-Tritsch, D., Rossier, J. and Eugene, D. (1999) Incremental conductance levels of GABA$_A$ receptors in dopaminergic neurones of the rat substantia nigra pars compacta. J. Physiol., 516: 719–737.

Hajos, M. and Greenfield, S.A. (1993) Topographic heterogeneity of substantia nigra neurons: Diversity in intrinsic membrane properties and synaptic inputs. Neuroscience, 55: 919–934.

Hajos, M. and Greenfield, S.A. (1994) Synaptic connections between pars compacta and pars reticulata neurones: Electrophysiological evidence for functional modules within the substantia nigra. Brain Res., 660: 216–224.

Hanaway, J., McConnell, J.A. and Netsky, M.G. (1970) Cytoarchitecture of the substantia nigra in the rat. Am. J. Anat., 129: 417–438.

Hattori, T., Fibiger, H.C. and McGeer, P.L. (1975) Demonstration of a pallido-nigral projection innervating dopaminergic neurons. J. Comp. Neurol., 162: 487–504.

Hausser, M.A. and Yung, W.H. (1994) Inhibitory synaptic potentials in guinea-pig substantia nigra dopamine neurones in vitro. J. Physiol., 479(Pt 3): 401–422.

Hebb, M.O. and Robertson, H.A. (2000) Identification of a subpopulation of substantia nigra pars compacta gamma-aminobutyric acid neurons that is regulated by basal ganglia activity. J. Comp. Neurol., 416: 30–44.

Hommer, D.W. and Pert, A. (1983) The actions of opiates in the rat substantia nigra: an electrophysiological analysis. Peptides, 4: 603–608.

Hyland, B.I., Reynolds, J.N.J., Hay, J., Perk, C.G. and Miller, R. (2002) Firing modes of midbrain dopamine cells in the freely moving rat. Neuroscience, 114: 475–492.

Ji, H. and Shepard, P.D. (2006) SK Ca^{2+}-activated K$^+$ channel ligands alter the firing pattern of dopamine-containing neurons in vivo. Neuroscience.

Johnson, S.W. and North, R.A. (1992) Two types of neurone in the rat ventral tegmental area and their synaptic inputs. J. Physiol. (Lond.), 450: 455–468.

Johnson, S.W. and Seutin, V. (1997) Bicuculline methiodide potentiates NMDA-dependent burst firing in rat dopamine neurons by blocking apamin-sensitive Ca^{2+}-activated K$^+$ currents. Neurosci. Lett., 231: 13–16.

Johnson, S.W., Seutin, V. and North, R.A. (1992) Burst firing in dopamine neurons induced by N-methyl-D-aspartate: role of electrogenic sodium pump. Science, 258: 665–667.

Juraska, J.M., Wilson, C.J. and Groves, P.M. (1977) The substantia nigra of the rat: A golgi study. J. Comp. Neurol., 172: 585–599.

Kaila, K. (1994) Ionic basis of GABA$_A$ receptor channel function in the nervous system. Prog. Neurobiol., 42: 489–537.

Kita, T., Kita, H. and Kitai, S.T. (1986) Electrical membrane properties of rat substantia nigra compacta neurons in an in vitro slice preparation. Brain Res., 372: 21–30.

Kita, H. and Kitai, S.T. (1991) Intracellulur study of rat globus pallidus neurons: membrane properties and responses to neostriatal, subthalamic and nigral stimulation. Brain Res., 569: 296–305.

Kitai, S.T., Shepard, P.D., Callaway, J.C. and Scroggs, R. (1999) Afferent modulation of dopamine neuron firing patterns. Curr. Opin. Neurobiol., 9: 690–697.

Komendantov, A.O., Komendantova, O.G., Johnson, S.W. and Canavier, C.C. (2004) A modeling study suggests complementary roles for GABA$_A$ and NMDA receptors and the SK channel in regulating the firing pattern in midbrain dopamine neurons. J. Neurophysiol., 91: 346–357.

Kuznetsov, A.S., Kopell, N.J. and Wilson, C.J. (2006) Transient high-frequency firing in a coupled-oscillator model of the mesencephalic dopaminergic neuron. J. Neurophysiol., 95: 932–947.

Lacey, M.G., Mercuri, N.B. and North, R.A. (1988) On the potassium conductance increase activated by GABA$_B$ and dopamine D$_2$ receptors in rat substantia nigra neurones. J. Physiol., 401: 437–453.

Lee, C.R., Abercrombie, E.D. and Tepper, J.M. (2004) Pallidal control of substantia nigra dopaminergic neuron firing pattern and its relation to extracellular neostriatal dopamine levels. Neuroscience, 129: 481–489.

Lee, C.R. and Tepper, J.M. (2007) Morphological and physiological properties of parvalbumin and calretinin containing GABAergic neurons in the substantia nigra. J. Comp. Neurol., 500: 958–972.

Lopez-Bendito, G., Shigemoto, R., Kulik, A., Vida, I., Fairen, A. and Lujan, R. (2004) Distribution of metabotropic GABA receptor subunits GABA$_{B1a/b}$ and GABA$_{B2}$ in the rat hippocampus during prenatal and postnatal development. Hippocampus, 14: 836–848.

Matsuda, Y., Fujimura, K. and Yoshida, S. (1987) Two types of neurons in the substantia nigra pars compacta studied in a slice preparation. Neurosci. Res., 5: 172–179.

Mailly, P., Charpier, S., Menetrey, A. and Deniau, J.M. (2003) Three-dimensional organization of the recurrent axon collateral network of the substantia nigra pars reticulata neurons in the rat. J. Neurosci., 23: 5247–5257.

Martin, G.E. and Haubrich, D.R. (1978) Striatal dopamine release and contraversive rotation elicited by intranigrally applied muscimol. Nature, 275: 230–231.

Medvedev, G.S., Wilson, C.J., Callaway, J.C. and Kopell, N. (2003) Dendritic synchrony and transient dynamics in a coupled oscillator model of the dopaminergic neuron. J. Comput. Neurosci., 15: 53–69.

Mereu, G., Fadda, F. and Gessa, G.L. (1984) Ethanol stimulates the firing rate of nigral dopaminergic neurons in unanesthetized rats. Brain Res., 292: 63–69.

Mereu, G. and Gessa, G.L. (1985) Low doses of ethanol inhibit the firing of neurons in the substantia nigra, pars reticulata: a GABAergic effect? Brain Res., 360: 325–330.

Mereu, G., Lilliu, V., Casula, A., Vargiu, P.F., Diana, M., Musa, A. and Gessa, G.L. (1997) Spontaneous bursting activity of dopaminergic neurons in midbrain slices from immature rats: Role of N-methyl-D-aspartate receptors. Neuroscience, 77: 1029–1036.

Misgeld, U. (2004) Innervation of the substantia nigra. Cell Tissue Res., 318: 107–114.

Nicholson, L.F., Faull, R.L., Waldvogel, H.J. and Dragunow, M. (1992) The regional, cellular and subcellular localization of GABA$_A$/benzodiazepine receptors in the substantia nigra of the rat. Neuroscience, 50: 355–370.

Nicholson, L.F., Waldvogel, H.J. and Faull, R.L. (1996) GABA$_A$ receptor subtype changes in the substantia nigra of the rat following quinolinate lesions in the striatum: a correlative in situ hybridization and immunohistochemical study. Neuroscience, 74: 89–98.

Nitsch, C. and Riesenberg, R. (1988) Immunocytochemical demonstration of GABAergic synaptic connections in rat substantia nigra after different lesions of the striatonigral projection. Brain Res., 461: 127–142.

O'Brien, D.P. and White, F.J. (1987) Inhibition of non-dopamine cells in the ventral tegmental area by benzodiazepines: relationship to A10 dopamine cell activity. Eur. J. Pharmacol., 142: 343–354.

Okada, H., Matsushita, N. and Kobayashi, K. (2004) Identification of GABA$_A$ receptor subunit variants in midbrain dopaminergic neurons. J. Neurochem., 89: 7–14.

Overton, P. and Clark, D. (1992) Iontophoretically administered drugs acting at the N-methyl-D-aspartate receptor modulate burst firing in A9 dopamine neurons in the rat. Synapse, 10: 131–140.

Overton, P.G. and Clark, D. (1997) Burst firing in midbrain dopaminergic neurons. Brain Res. Rev., 25: 312–334.

Paladini, C.A., Celada, P. and Tepper, J.M. (1999a) Striatal, pallidal, and pars reticulata evoked inhibition of nigrostriatal dopaminergic neurons is mediated by GABA$_A$ receptors in vivo. Neuroscience, 89: 799–812.

Paladini, C.A., Iribe, Y. and Tepper, J.M. (1999b) GABA$_A$ receptor stimulation blocks NMDA-induced bursting of dopaminergic neurons in vitro by decreasing input resistance. Brain Res., 832: 145–151.

Paladini, C.A. and Tepper, J.M. (1999) GABA$_A$ and GABA$_B$ antagonists differentially affect the firing pattern of substantia nigra dopaminergic neurons in vivo. Synapse, 32: 165–176.

Payne, J.A., Rivera, C., Voipio, J. and Kaila, K. (2003) Cation-chloride co-transporters in neuronal communication, development and trauma. Trends Neurosci., 26: 199–206.

Payne, J.A., Stevenson, T.J. and Donaldson, L.F. (1996) Molecular characterization of a putative K-Cl cotransporter in rat brain. A neuronal-specific isoform. J. Biol. Chem., 271: 16245–16252.

Pirker, S., Schwarzer, C., Wieselthaler, A., Sieghart, W. and Sperk, G. (2000) GABA$_A$ receptors: immunocytochemical distribution of 13 subunits in the adult rat brain. Neuroscience, 101: 815–850.

Richards, J.G., Schoch, P., Haring, P., Takacs, B. and Mohler, H. (1987) Resolving GABAA/benzodiazepine receptors: cellular and subcellular localization in the CNS with monoclonal antibodies. J. Neurosci., 7: 1866–1886.

Richards, C.D., Shiroyama, T. and Kitai, S.T. (1997) Electrophysiological and immunocytochemical characterization of GABA and dopamine neurons in the substantia nigra of the rat. Neuroscience, 80: 545–557.

Rinvik, E. and Grofova, I. (1970) Observations on the fine structure of the substantia nigra in the cat. Exp. Brain Res., 11: 229–248.

Rivera, C., Voipio, J., Payne, J.A., Ruusuvuori, E., Lahtinen, H., Lamsa, K., Pirvola, U., Saarma, M. and Kaila, K. (1999) The K$^+$/Cl$^-$ co-transporter KCC2 renders GABA hyperpolarizing during neuronal maturation. Nature, 397: 251–255.

Rodriguez-Pallares, J., Caruncho, H.J., Lopez-Real, A., Wojcik, S., Guerra, M.J. and Labandeira-Garcia, J.L. (2001) Rat brain cholinergic, dopaminergic, noradrenergic and serotonergic neurons express GABA$_A$ receptors derived from the alpha3 subunit. Recept. Chan., 7: 471–478.

Rodriguez-Pallares, J., Caruncho, H.J., Munoz, A., Guerra, M.J. and Labandeira-Garcia, J.L. (2000) GABA$_A$ receptor subunit expression in intrastriatal ventral mesencephalic transplants. Exp. Brain Res., 135: 331–340.

Ross, R.J., Waszczak, B.L., Lee, E.K. and Walters, J.R. (1982) Effects of benzodiazepines on single unit activity in the substantia nigra pars reticulata. Life Sci., 31: 1025–1035.

Ryan, L.J., Young, S.J. and Groves, P.M. (1986) Substantia nigra stimulation evolved antidromic responses in rat neostriatum. Exp. Brain Res., 63: 449–460.

Saitoh, K., Isa, T. and Takakusaki, K. (2004) Nigral GABAergic inhibition upon mesencephalic dopaminergic cell groups in rats. Eur. J. Neurosci., 19: 2399–2409.

Santiago, M. and Westerink, B.H. (1992) The role of GABA receptors in the control of nigrostriatal dopaminergic neurons: dual-probe microdialysis study in awake rats. Eur. J. Pharmacol., 219: 175–181.

Scanziani, M. (2000) GABA spillover activates postsynaptic GABA$_B$ receptors to control rhythmic hippocampal activity. Neuron, 25: 673–681.

Schwarzer, C., Berresheim, U., Pirker, S., Wieselthaler, A., Fuchs, K., Sieghart, W. and Sperk, G. (2001) Distribution of the major gamma-aminobutyric acid(A) receptor subunits in the basal ganglia and associated limbic brain areas of the adult rat. J. Comp. Neurol., 433: 526–549.

Schultz, W. (2006) Behavioral theories and the neurophysiology of reward. Annu. Rev. Psychol., 57: 87–115.

Seutin, V. and Johnson, S.W. (1999) Recent advances in the pharmacology of quaternary salts of bicuculline. Trends Pharmacol. Sci., 20: 268–270.

Shen, K.Z. and Johnson, S.W. (1997) Presynaptic GABA$_B$ and adenosine A$_1$ receptors regulate synaptic transmission to rat substantia nigra reticulata neurones. J. Physiol., 505: 153–163.

Smith, Y. and Bolam, J.P. (1989) Neurons of the substantia nigra reticulata receive a dense GABA-containing input from the globus pallidus in the rat. Brain Res., 493: 160–167.

Smith, Y. and Bolam, J.P. (1990) The output neurones and the dopaminergic neurones of the substantia nigra receive a GABA-containing input from the globus pallidus in the rat. J. Comp. Neurol., 296: 47–64.

Somogyi, P., Bolam, J.P., Totterdell, S. and Smith, A.D. (1981) Monosynaptic input from the nucleus accumbens–ventral striatum region to retrogradely labelled nigrostriatal neurones. Brain Res., 217: 245–263.

208

Tepper, J.M., Celada, P., Iribe, Y. and Paladini, C. (2002) Afferent control of nigral dopaminergic neurons: the role of GABAergic inputs. In: Graybiel A., et al. (Eds.), The Basal Ganglia VI. Kluwer Academic/Plenum Publishers, New York, pp. 641–651.

Tepper, J.M., Damlama, M. and Trent, F. (1994) Postnatal changes in the distribution and morphology of rat substantia nigra dopaminergic neurons. Neuroscience, 60(2): 469–477.

Tepper, J.M., Martin, L.P. and Anderson, D.R. (1995) GABA$_A$ receptor-mediated inhibition of rat substantia nigra dopaminergic neurons by pars reticulata projection neurons. J. Neurosci., 15: 3092–3103.

Tepper, J.M., Nakamura, S., Young, S.J. and Groves, P.M. (1984) Autoreceptor-mediated changes in dopaminergic terminal excitability: effects of striatal drug infusions. Brain Res., 309: 317–333.

Tepper, J.M., Sawyer, S.F. and Groves, P.M. (1987) Electrophysiologically identified nigral dopaminergic neurons intracellularly labeled with HRP: light-microscopic analysis. J. Neurosci., 7(9): 2794–2806.

Tepper, J.M., Sawyer, S.F., Young, S.J. and Groves, P.M. (1986) Intracellular recording and HRP staining of rat nigral neurons. Soc Neurosci. Abstr., 1542.

Tsai, C., Nakamura, S. and Iwama, K. (1980) Inhibition of neuronal activity of the substantia nigra by noxious stimuli and its modification by the caudate nucleus. Brain Res., 195: 299–311.

Ungless, M.A., Magill, P.J. and Bolam, J.P. (2004) Uniform inhibition of dopamine neurons in the ventral tegmental area by aversive stimuli. Science, 303: 2040–2042.

Vankova, M., Arluison, M., Leviel, V. and Tramu, G. (1992) Afferent connections of the rat substantia nigra pars lateralis with special reference to peptide-containing neurons of the amygdalo-nigral pathway. J. Chem. Neuroanat., 5: 39–50.

von Krosigk, M., Smith, Y., Bolam, J.P. and Smith, A.D. (1992) Synaptic organization of GABAergic inputs from the striatum and the globus pallidus onto neurons in the substantia nigra and retrorubral field which project to the medullary reticular formation. Neuroscience, 50: 531–549.

Wallace, D.M., Magnuson, D.J. and Gray, T.S. (1989) The amygdalo-brainstem pathway: selective innervation of dopaminergic, noradrenergic and adrenergic cells in the rat. Neurosci. Lett., 97: 252–258.

Wallace, D.M., Magnuson, D.J. and Gray, T.S. (1992) Organization of amygdaloid projections to brainstem dopaminergic, noradrenergic, and adrenergic cell groups in the rat. Brain Res. Bull., 28: 447–454.

Walters, J.R. and Lakoski, J.M. (1978) Effect of muscimol on single unit activity of substantia nigra dopamine neurons. Eur. J. Pharmacol., 47: 469–471.

Waroux, O., Massotte, L., Alleva, L., Graulich, A., Thomas, E., Liegeois, J.F., Scuvee-Moreau, J. and Seutin, V. (2005) SK channels control the firing pattern of midbrain dopaminergic neurons in vivo. Eur. J. Neurosci., 22: 3111–3121.

Waszczak, B.L., Bergstrom, D.A. and Walters, J.R. (1981) Single unit responses of substantia nigra and globus pallidus neurons to GABA agonist and antagonist drugs. Adv. Biochem. Psychopharmacol., 30: 79–94.

Waszczak, B.L., Eng, N. and Walters, J.R. (1980) Effects of muscimol and picrotoxin on single unit activity of substantia nigra neurons. Brain Res., 188: 185–197.

Wilson, C.J. (1993) The generation of natural firing patterns in neostriatal neurons. In: Arbuthnott G.W. and Emson P.C. (Eds.), Progress in Brain Research. Elsevier Science Publishers B.V., Amsterdam, pp. 277–297.

Wilson, C.J. and Callaway, J.C. (2000) Coupled oscillator model of the dopaminergic neuron of the substantia nigra. J. Neurosci., 83: 3084–3100.

Wilson, C.J., Young, S.J. and Groves, P.M. (1977) Statistical properties of neuronal spike trains in the substantia nigra: Cell types and their interactions. Brain Res., 136: 243–260.

Wisden, W., Laurie, D.J., Monyer, H. and Seeburg, P.H. (1992) The distribution of 13 GABA$_A$ receptor subunit mRNAs in the rat brain. I. Telencephalon, diencephalon, mesencephalon. J. Neurosci., 12: 1040–1062.

Yanovsky, Y., Mades, S. and Misgeld, U. (2003) Retrograde signaling changes the venue of postsynaptic inhibition in rat substantia nigra. Neuroscience, 122: 317–328.

Yung, W.H., Hausser, M.A. and Jack, J.J.B. (1991) Electrophysiology of dopaminergic and non-dopaminergic neurones of the guinea-pig substantia nigra pars compacta in vitro. J. Physiol. (Lond.), 436: 643–667.

Zhang, J., Chiodo, L.A. and Freeman, A.S. (1992) Electrophysiological effects of MK-801 on rat nigrostriatal and mesoaccumbal dopaminergic neurons. Brain Res., 590: 153–163.

Zhang, J., Chiodo, L.A. and Freeman, A.S. (1993) Effects of phencyclidine, MK-801 and 1,3-di(2-tolyl)guanidine on non-dopaminergic midbrain neurons. Eur. J. Pharmacol., 230: 371–374.

Zhang, J., Chiodo, L.A. and Freeman, A.S. (1994) Influence of excitatory amino acid receptor subtypes on the electrophysiological activity of dopaminergic and nondopaminergic neurons in rat substantia nigra. J. Pharmacol. Exp. Ther., 269: 313–321.

Tepper, Abercrombie & Bolam (Eds.)
Progress in Brain Research, Vol. 160
ISSN 0079-6123
2007 Elsevier B.V.

CHAPTER 12

GABAergic output of the basal ganglia

O. Hikosaka*

Laboratory of Sensorimotor Research, National Eye Institute, National Institute of Health, 49 Convent Drive, Bldg. 49, Rm. 2A50, Bethesda, MD 20892-4435, USA

Abstract: Using GABAergic outputs from the SNr or GP_i, the basal ganglia exert inhibitory control over several motor areas in the brainstem which in turn control the central pattern generators for the basic motor repertoire including eye–head orientation, locomotion, mouth movements, and vocalization. These movements are by default kept suppressed by tonic rapid firing of SNr/GP_i neurons, but can be released by a selective removal of the tonic inhibition. Derangement of the SNr/GP_i outputs leads to either an inability to initiate movements (akinesia) or an inability to suppress movements (involuntary movements). Although the spatio-temporal patterns of individual movements are largely innate and fixed, it is essential for survival to select appropriate movements and arrange them in an appropriate order depending on the context, and this is what the basal ganglia presumably do. To achieve such a goal, however, the basal ganglia need to be trained to optimize their outputs with the aid of cortical inputs carrying sensorimotor and cognitive information and dopaminergic inputs carrying reward-related information. The basal ganglia output to the thalamus, which is particularly developed in primates, provides the basal ganglia with an advanced ability to organize behavior by including the motor skill mechanisms in which new movement patterns can be created by practice. To summarize, an essential function of the basal ganglia is to select, sort, and integrate innate movements and learned movements, together with cognitive and emotional mental operations, to achieve purposeful behaviors. Intricate hand–finger movements do not occur in isolation; they are always associated with appropriate motor sets, such as eye–head orientation and posture.

Keywords: substantia nigra pars reticulata; internal segment of globus pallidus; caudate nucleus; putamen; selection of behavior; saccadic eye movement; superior colliculus; memory-guided behavior; sequential procedure; reward; involuntary movement

Outputs of the basal ganglia

The output of the basal ganglia is issued largely (but not exclusively) from the substantia nigra pars reticulata (SNr) and the internal segment of the globus pallidus (GP_i) (Carpenter, 1981). The output neurons in the SNr and GP_i share two important features: (1) they are GABAergic and inhibitory

(Uno and Yoshida, 1975; Di Chiara et al., 1979; Yoshida and Omata, 1979), and (2) they fire tonically and rapidly (DeLong and Georgopoulos, 1981). The tonic firing is present even in vitro slice preparations (Nakanishi et al., 1987). This means that the brain areas that receive inputs from the basal ganglia are, by default, under a strong tonic inhibition. This is very important when we consider the functions and dysfunctions of the basal ganglia.

However, a mere tonic inhibition would not be useful for controlling behavior; some mechanisms

*Corresponding author. Tel.: +301-402-7959;
Fax: +301-402-0511; E-mail: oh@lsr.nei.nih.gov

DOI: 10.1016/S0079-6123(06)60012-5

that modulate the level of the tonic inhibition would be necessary. In fact, the activity of SNr and GP_i neurons does change (decrease or increase) when the animal is alert and behaving (Hikosaka and Wurtz, 1983a; Anderson and Horak, 1985). Anatomical and electrophysiological studies have suggested that these changes in neuronal activity are caused by inputs from other basal ganglia nuclei. A decrease in SNr or GP_i neuronal activity may be caused by a direct input from the striatum (caudate nucleus and putamen) which is, again, GABAergic and inhibitory (Yoshida and Precht, 1971; Hikosaka et al., 1993a). An increase in SNr or GP_i neuronal activity may be caused by excitatory inputs from the subthalamic nucleus (STN) (Nakanishi et al., 1987; Robledo and Féger, 1990) or a decrease in inhibitory inputs from the external segment of the globus pallidus (GP_e) (Smith and Bolam, 1989).

These mechanisms give the basal ganglia a potential for controlling other brain areas by decreasing or increasing the inhibitions. An attractive hypothesis is that a major function of the basal ganglia *is the selection of appropriate behavior* (Hikosaka et al., 1993b; Mink, 1996; Nambu et al., 2002). Unwanted behaviors would be suppressed by maintaining or increasing the SNr/GP_i-induced inhibition while desired behaviors would be released by decreasing or removing the SNr/GP_i-induced inhibition. This hypothesis is consistent with movement disorders observed in patients with basal ganglia dysfunctions. A symptom commonly observed in basal ganglia patients is *involuntary movement*, which appears in various forms, such as tremor, dyskinesia, dystonia, chorea, athetosis, and ballism (Denny-Brown, 1968). Such involuntary movements could be caused by a reduction or interruption of the SNr/GP_i-induced inhibition. This is supported by single unit recording studies in such patients (Vitek et al., 1999; Starr et al., 2005) or experimental animal models (Wichmann et al., 1999; Raz et al., 2000). On the other hand, patients with basal ganglia dysfunction commonly have difficulty in initiating purposeful movements (akinesia) and, if initiated, the movements tend to be slow and small (hypokinesia). This type of movement disorder could be caused by an insufficient removal (or disinhibition) of the SNr/GP_i-induced inhibition (Burbaud et al., 1998).

The evidence based on clinical observations (as described above) for *the selection hypothesis* indicates a potential function of the basal ganglia, but not *the* function of the basal ganglia. In other words, it is still unclear whether the basal ganglia actually use the inhibition/disinhibition mechanisms to select purposeful behaviors (but see Aron and Poldrack, 2006). A real test of the selection hypothesis requires the recording of neuronal activity in the basal ganglia while the animal is performing purposeful behaviors.

In this chapter I will discuss the mechanisms and functions of the GABAergic output of the basal ganglia in relation to motor control. I will first discuss a series of studies using saccadic eye movement as a behavioral measure and then discuss other kinds of body movements. My discussion will be largely focused on the output of the SNr.

An example of basal ganglia-controlled functions — Saccadic eye movement

A big advantage of using saccadic eye movement as a behavioral measure is that its neuronal mechanisms have been studied extensively so that the basal ganglia mechanisms can be studied more or less independently from motor execution mechanisms. Let me first describe what aspects of saccadic eye movement are controlled by motor execution mechanisms in the brainstem, which is situated downstream to the basal ganglia.

Saccadic eye movement (or saccade) is an extremely fast and simultaneous rotation of the both eyes (typically $400°/s$) (Becker, 1989). It is present in most vertebrates but is highly developed in primates (Easter, 1975). Its function is *orienting*, that is, to change the line of sight from one position to another (Grantyn et al., 2004). It occurs fairly frequently in primates, typically 2–3 times per second. Between saccades are the periods of visual fixation during which visual information on an object of interest is acquired through the fovea (the most sensitive part of the retina). Saccades must be fast because no or little visual information can be acquired while the eyes are rotating and consequently visual images are sweeping over the retina (Judge et al., 1980). Saccadic eye movements are often accompanied by saccadic head movements (Guitton and Volle, 1987), which is more common among non-primate mammals (Fuller, 1985) and nonmammalian vertebrates (Du Lac and Knudsen, 1990).

The physical parameters of saccadic eye movement are controlled by the neuronal networks in the brainstem (Fuchs et al., 1985) and the cerebellum (Keller, 1989). For example, the fast rotation of the eyes is enabled by a high-frequency burst of spikes in *burst neurons* in the reticular formation, and this burst is enabled by the removal of tonic inhibition originating from *omnipause neurons* in the pontine raphe nucleus (Evinger et al., 1982). The burst and omnipause neurons are controlled by the superior colliculus (SC) (Raybourn and Keller, 1977), which receives direct inputs from the retina (Sparks, 1986). The SC (which may be called the optic tectum) plays a pivotal role in orienting behavior, which is crucial for survival (Goodale and Murison, 1975). Orienting of the eye-head-body is prerequisite for visually capturing an object (through the fovea) and/or physically capturing the object (using the mouth, hand, or other body parts). In the SC are spatial maps for visual, auditory, and somotosensory information, and these sensory maps are in register with a motor map for orienting (Meredith et al., 1992). This organization serves a basic mechanism of behavioral orienting such that an object of interest, detected as one (or more than one) of the sensory information, is mapped onto a unique location in the SC, which then is converted to a vector (direction and magnitude) of eye and/or head movement (Sparks, 1986).

The involvement of the basal ganglia in the behavioral orienting was first suggested by the discovery of the efferent connection of the SNr to the SC (Rinvik et al., 1976; Jayaraman et al., 1977; Faull and Mehler, 1978; Graybiel, 1978; Vincent et al., 1978) (Fig. 1(B)). The discovery provided a good opportunity to study the function of the basal ganglia because the neuronal mechanisms downstream to the SC had already been understood to a great degree, as described above. Other

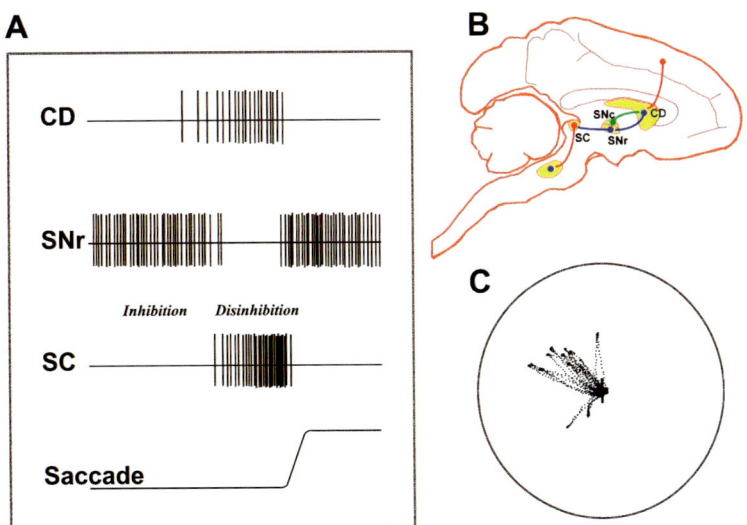

Fig. 1. Basal ganglia mechanism for control of saccadic eye movement. (A) A cardinal saccade mechanism in the basal ganglia. The SC is normally inhibited by rapid firing of GABAergic neurons in the SNr. The tonic inhibition can be interrupted by GABAergic inputs from the caudate nucleus. This disinhibition, together with excitatory cortical inputs, allows SC neurons to fire in burst, which leads to a saccade to a contralateral location. (B) Simplified neural circuits in the basal ganglia for control of saccadic eye movement in a parasagittal view of the macaque brain. Red, blue, green lines indicate excitatory, inhibitory, and modulatory connections. (C) Involuntary eye movement of a monkey after muscimol injection into right SNr, shown as trajectories of saccades during 2 s fixation periods. The monkey was unable to keep fixating at the central spot of light. Adapted with permission from (Hikosaka and Wurtz, 1985b).

efferent connections of the basal ganglia are either far more complex (as seen in the connections to the thalamus) or much less studied (as seen in the connections to brainstem areas other than the SC). Reinforcing the importance of the SNr–SC connection is the hypothesis stating that this connection is phylogenetically the best preserved one among the basal ganglia efferents (Marín et al., 1998; Reiner et al., 1998). This suggests that the essential or primary functions of the basal ganglia can be discerned by studying the SNr–SC connection in relation to saccadic eye movement.

Single unit recordings from the SNr in monkeys (Hikosaka and Wurtz, 1983a, b, c, d) and cats (Joseph and Boussaoud, 1985) performing saccade tasks provided firm evidence for the oculomotor role of the basal ganglia. In the following section I will summarize the results obtained in a series of studies on trained monkeys. Neurons related to saccadic eye movement were found mainly in the dorsolateral part of the SNr (Hikosaka and Wurtz, 1983a). Their typical response is a pause in firing in relation to events related to the preparation of saccades (Fig. 1(A)). The pause may occur as a visual response to a spot of light to which the monkey makes a saccade, as a pre-saccadic response, or as a sustained response during a delay period before the saccade (Hikosaka and Wurtz, 1983b, c). Most SNr neurons have response fields, usually centered in the contralateral hemifield, such that saccades into the response field are associated with stronger responses. Some SNr neurons increased firing in relation to saccades (Handel and Glimcher, 1999; Sato and Hikosaka, 2002).

Many of the saccade-related SNr neurons were activated antidromically by stimulating the intermediate layer of the SC where saccadic burst neurons are concentrated (Hikosaka and Wurtz, 1983d). This result suggested that the saccade-related SNr neurons have synaptic connections with saccade-related neurons in the SC, which was confirmed by anatomical studies (May and Hall, 1984; Karabelas and Moschovakis, 1985; Williams and Faull, 1988). There was a mirror-image relationship in activity between the antidromically activated SNr neurons and SC neurons recorded at the stimulation sites: the SNr neurons decrease firing while the SC neurons increase firing at similar time periods during the preparation of saccades with similar directions and amplitudes (Fig. 1(A)). This result is consistent with the fact that the SNr–SC connection is inhibitory (Vincent et al., 1978; Chevalier et al., 1981).

These results support the *selection* function of the basal ganglia. Saccade-related SNr neurons are highly active, exerting a strong inhibition on saccade-related SC neurons, unless the saccade to a particular location is planned or executed. It is as if a gate were closed for the saccade. In fact, saccade burst neurons in the SC are inactive, rarely emitting spikes unless a saccade is about to be executed. Such quiescence may be maintained by the SNr–SC tonic inhibition. On the other hand, once SNr neurons stop firing, the strong inhibition would be removed at once, giving a strong drive for SC neurons to fire. It is as if the gate were opened.

A more straightforward demonstration of the SNr–SC gating function was obtained by the experimental blockade of the SNr–SC inhibition. When a small amount of $GABA_A$ antagonist, bicuculline, was injected in the saccade-related region of the SC, the monkey became unable to keep fixating on a visual stimulus and made saccades incessantly to the side contralateral to the injection site (Hikosaka and Wurtz, 1985a). A likely explanation of such involuntary saccades is that the GABAergic SNr–SC inhibition was blocked by the GABA antagonist. Alternatively, the involuntary saccade might have been induced by the blockade of the effects of GABAergic interneurons within the SC. A support for the first hypothesis was obtained by an injection of $GABA_A$ agonist, muscimol, in the SNr, which induced similar involuntary saccades to the contralateral side (Hikosaka and Wurtz, 1985b) (Fig. 1(C)). Since SNr neurons express an extremely high concentration of $GABA_A$ receptors (Fahn and Cote, 1968; Okada et al., 1971) reflecting abundant GABAergic inputs from the striatum (caudate nucleus and putamen) and the external segment of the GP_e (Smith and Bolam, 1991), the injected muscimol would bind to the $GABA_A$ receptors and suppress the rapid and tonic firing. This chemical manipulation would effectively remove the SNr–SC inhibition. The fact that the monkey developed involuntary saccades indicates that the SC is under the influence of excitatory inputs which are capable of triggering spike activity in SC saccadic neurons, which however can be prevented by the SNr–SC inhibition.

These experimental results provide an insight into the motor control mechanisms. The excitatory inputs to the SC, which are normally gated by the SNr–SC inhibition, arise from many brain areas including the cerebral cortex and the cerebellum (Hikosaka et al., 2000). Among the cerebral cortical areas are the frontal eye field (FEF) (Bruce and Goldberg, 1985), supplementary eye field (SEF) (Schlag and Schlag-Rey, 1987), and the lateral intraparietal area (LIP) (Barash et al., 1991). All of these cortical areas contain many neurons that fire before saccades, although there seem some differences in their relationship to behavioral context (Coe et al., 2002). These results suggest that there is not a single area in the brain that dictates the initiation of saccadic eye movement. Individual cerebral cortical areas (and cerebellar areas as well) can contribute to saccade initiation, but only partially, by sending excitatory signals to the SC.

Such parallel excitatory inputs to the SC alone would not be suitable for the selection of behavior. Once some of these excitatory inputs are turned on simultaneously, their effects would be added up and reach the threshold for firing in SC neurons. The outcome would be uncontrollable emission of saccadic motor outputs, as observed when the SNr–SC inhibition was blocked. The SNr–SC tonic inhibition is capable of preventing the excitatory inputs from triggering off spike activity in SC neurons, and this is a necessary condition for selection. Selection would then be done by a pause in activity of SNr neurons, which is caused by GABAergic inhibitory inputs from the caudate nucleus (Hikosaka et al., 1993a) (Fig. 1(A)).

Functions of basal ganglia GABAergic outputs

So far I have been describing a neural mechanism of behavioral selection. However, for the mechanism to be functional, it must be operated appropriately. One way to answer this question is to examine the behavioral context in which the mechanism is deployed. An experimental objective

would then be to examine whether the key neurons constituting the mechanism (in this case SNr neurons) change their activity selectively in that context. This is a difficult task because the experimenter has to create the particular context by training the animal using a particular task and then test the key neurons. Experiments using this approach have revealed at least three kinds of behavioral context in which the basal ganglia appear to deploy its selection mechanism: (1) memory-guided behavior, (2) sequential procedure, and (3) reward-oriented behavior.

The selection of memory-guided behavior by the basal ganglia was first suggested by the discovery of neurons in the SNr that pause (Hikosaka and Wurtz, 1983c) and neurons in the caudate nucleus that burst (Hikosaka et al., 1989a) selectively or preferentially with memory-guided saccades, as opposed to visually guided saccades. These results suggest that the SNr–SC inhibition is removed more readily when saccades are planned to an invisible, but remembered, stimulus. The lack of this function should lead to a preferential impairment of memory-guided saccades. Indeed, patients of Parkinson's disease and Huntington's disease may have difficulty in making memory-guided saccades compared with visually guided saccades (Crawford et al., 1989; Nakamura et al., 1994).

The deficits in memory-guided saccades may be relevant to a phenomenon often observed in patients with basal ganglia dysfunctions: parkinsonian patients are able to move in response to sensory stimulation, but otherwise are unable to initiate a movement (Cooke et al., 1978; Glickstein and Stein, 1991; Azulay et al., 2002). For example, parkinsonian patients, who have great difficulties in walking on a flat plane floor, could initiate and keep walking if there are visible steps on the floor. Such a peculiar phenomenon is usually described as a selective impairment in internally guided behavior. However, the concept of internally guided behavior is rather obscure. Its nature may, at least partly, be memory-guided behavior.

The second context in which the basal ganglia may play an important role is sequential procedure. A goal-directed learned behavior is often composed of a fixed sequence of elementary movements. To drink a glass of juice, for example, you

214

first go to the refrigerator, open its door, find a bottle of orange juice, retrieve it, open its cap, and pour the content into a glass. The whole sequence is organized smoothly, each movement predicting a next movement. Many neurons are found in the basal ganglia, especially the caudate and putamen, which become active after one event until a next predicted event (Hikosaka et al., 1989b; Kermadi and Joseph, 1995; Miyachi et al., 1997; Meyer-Luehmann et al., 2002; Fujii and Graybiel, 2005), as if they linked one elementary behavior to another. Some studies have shown that patients with basal ganglia dysfunctions may have difficulties in executing such sequential and learned movements smoothly (Stern et al., 1983). Human functional imaging studies have revealed activation in the basal ganglia when the subject performs such sequential movement (Lehericy et al., 2005) as well as nonmovement tasks (Tinaz et al., 2006).

One important aspect of sequential procedure is that it is acquired through repeated practice, and once acquired, it becomes automatic or habitual (Hikosaka et al., 1995).Several lines of evidence suggest that the basal ganglia are necessary for acquisition and/or maintenance of such habitual and sequential behavior (Barnes et al., 2005; Yin and Knowlton, 2006). For example, a series of studies in which monkeys were trained to press buttons in a fixed order have shown that the anterior part of the striatum (caudate and putamen) is necessary for learning of new sequences whereas the middle part of the putamen is necessary for the execution of learned sequences (Miyachi et al., 1997; Miyachi et al., 2002). In this task, learning occurs in both eye movements and hand movements (Miyashita et al., 1996).

The third context, reward-oriented behavior, has been emphasized recently. A goal-directed sequential behavior is typically culminated with the attainment of reward. Many striatal (caudate and putamen) neurons exhibit tonic activity before the reward delivery (Hikosaka et al., 1989b; Schultz et al., 1992; Hollerman et al., 1998). Similar reward-related activation has been found in the human basal ganglia (Delgado et al., 2000; Knutson et al., 2000; O'Doherty et al., 2002). Visual or saccadic activity of neurons in the caudate (Kawagoe et al., 1998; Watanabe et al., 2003b) and the SNr

(Sato and Hikosaka, 2000) is strongly enhanced or depressed depending on whether the current trial will end with a big reward or a small reward. This was demonstrated using a saccade task in which the amount of reward delivered after a correct saccade is big for one direction, but small for the other directions; the task is called one-direction-reward task (1DR). The reward-dependent modulation of caudate neuronal activity is so strong that the neuron's response field can be changed completely depending on the rewarded location (Kawagoe et al., 1998). A group of caudate neurons exhibit sustained activity before a saccade depending on the rewarded location (Lauwereyns et al., 2002; Takikawa et al., 2002a). The reward-modulated activity of caudate neurons is transmitted to the SC (Ikeda and Hikosaka, 2003) through the SNr, leading to a biased tendency of saccades, namely, facilitation of saccades to the big-reward position (i.e., shorter latencies) and inhibition of saccades to the small-reward position (i.e., longer latencies) (Takikawa et al., 2002b; Watanabe et al., 2003a).

Further experiments suggest that the reward-dependent modulation of caudate neuronal activity is, at least partly, caused by inputs from dopamine neurons. Schultz and colleagues have shown that dopamine neurons, which are located in the substantia nigra pars compacta (SNc) and the ventral tegmental area (VTA), respond to unexpected reward or a sensory event that indicate the upcoming delivery of reward (Schultz, 1998, 2002). In the 1DR saccade task, dopamine neurons exhibit a short burst of spikes in response to the big-reward-indicating target and a decrease in spike activity in response to the small-reward-indicating targets, regardless of the direction of the big reward (Kawagoe et al., 2004). In other words, the visual-saccadic activity of most caudate neurons is enhanced when it is associated with a burst of dopamine release in the caudate, but is depressed when it is associated with a reduction of dopamine release (Hikosaka et al., 2006). The causal relationship of dopamine release in the caudate, activity of caudate projection neurons, and saccadic eye movement is suggested by an experiment in which dopamine D1 or D2 antagonist was injected in the caudate (Nakamura and Hikosaka, 2006). The reward-dependent modulation of saccade latencies was attenuated by the injection of D1 antagonist, but enhanced by the D2 antagonist. This suggests that the reward-modulation of saccadic eye movement is caused by the changes in neuronal activity in the caudate which depends on the reward-dependent changes in dopaminergic inputs to caudate projection neurons (Hikosaka et al., 2006).

Note that such reward-dependent modulation is not unique to the basal ganglia. Neuronal activity in many areas in the cerebral cortex is modulated by expected reward in monkeys (Watanabe, 1996; Platt and Glimcher, 1999; Tremblay and Schultz, 2000; Kobayashi et al., 2002; Roesch and Olson, 2003; Sugrue et al., 2004; Amiez et al., 2006) and in humans (O'Doherty et al., 2001; Ramnani and Miall, 2003; McClure et al., 2004). For example, activity of neurons in the frontal eye field can be modulated strongly by expected reward or expected reward position, although the modulation is not so drastic as in the caudate (Ding and Hikosaka, 2006). Nonetheless, the evidence described above certainly indicates that the source of reward-oriented behavior is at least partly in the basal ganglia.

Basal ganglia GABAergic control of basic movements

So far I have been discussing the function of the basal ganglia in relation to the control of saccadic eye movement, which is mediated by the SNr–SC connection. It is then tempting to hypothesize that other kinds of body movements or behaviors may also be controlled directly by the basal ganglia. Noteworthy in this respect is the fact that the basal ganglia send their outputs to many areas in the midbrain and the brainstem, which contain neural mechanisms of specific motor behaviors. Although this idea may be found in a long-standing concept of the extrapyramidal motor system (Jung and Hassler, 1960), recent studies using more controlled paradigms have provided strong support for this hypothesis.

A powerful technique to study this issue is reversible blockade of neuronal activity, typically by injecting muscimol (GABA$_A$ agonist) into localized areas within the SNr or GP$_i$ (Hikosaka and Wurtz,

1983e). Since a great majority of basal ganglia output neurons in the SNr and GP$_i$ are GABAergic and exhibit tonic rapid firing, the basal ganglia control of motor behaviors is likely to use the same mechanisms as used for the control of saccadic eye movement, namely, tonic inhibition and disinhibition. Injected muscimol would inhibit surrounding neurons in the SNr or GP$_i$, remove the tonic inhibition on the target structures, and release body movements or other behaviors that are controlled by the neurons in the target structures. For example, muscimol injections in the rat SNr induced a wide variety of body movements, including movements of the head, mouth, and upper and lower limbs in addition to saccadic eye movements (Sakamoto and Hikosaka, 1989). Subsequent studies have demonstrated that different subregions in the SNr or GP$_i$ have control over different kinds of motor behaviors, as shown below (Fig. 2).

Locomotion and posture

It is known that the oscillatory pattern of locomotion is generated by intricately connected

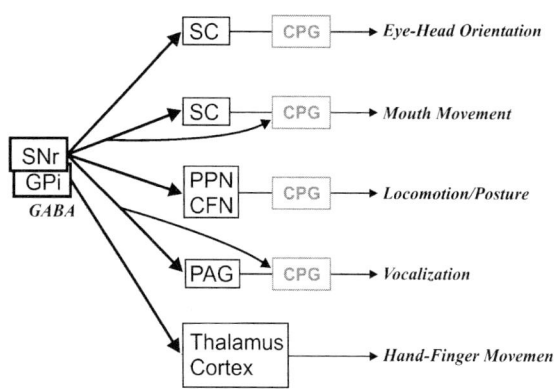

Fig. 2. Body movements controlled by the basal ganglia. With their projections to the brainstem motor areas and thalamus, GABAergic neurons in the SNr and GP$_i$ keep suppressing various body movements but remove the suppression selectively to fulfill biological and motivational needs. The output of each brainstem motor area is mediated by a central pattern generator (CPG) located in the reticular formation or the spinal cord where a fixed spatiotemporal pattern of specific movement is generated. SC: superior colliculus; PPN: pedunculopontine nucleus; CFN: cuneiform nucleus; PAG: periaqueductal gray.

neurons in the spinal cord (Grillner, 1985). Such locomotor movements, together with an increase in muscle tone, are initiated by electrical or chemical stimulation of an area at the junction of the midbrain and the pons, which approximately corresponds to the cuneiform nucleus and is often called *midbrain locomotor region* or MLR (Grillner and Shik, 1973; Mori, 1987). In contrast, stimulation of the pedunculopontine nucleus (PPN), which is adjacent to the MLR, suppresses postural muscle tone (Takakusaki et al., 2003). The MLR- and PPN-mediated effects are mediated by their efferent pathways through the pontine and medullary reticular formation, and the effects remain even after being disconnected from the forebrain (i.e., decerebrate preparation) (Takakusaki et al., 2004a). Taken together, MLR and PPN play pivotal roles in locomotion and posture.

The role of the GABAergic outputs of the basal ganglia in the control of locomotion and posture was first suggested by Garcia-Rill (1986). It has been known that the mesopontine tegmentum, including the MLR and the PPN, receives a substantial amount of inputs from the basal ganglia, especially the SNr (Beckstead et al., 1979; Garcia-Rill et al., 1983; Spann and Grofova, 1991). Electrical stimulation of the SNr induces monosynaptic IPSPs in PPN neurons (Noda and Oka, 1986; Kang and Kitai, 1990). In a series of experiments, Takakusaki and colleagues demonstrated that the basal ganglia control locomotion and posture using the SNr–GABAergic output (Takakusaki et al., 2004a) (Fig. 2). They found that electrical stimulation at the lateral part of the SNr blocked the PPN-induced muscle tone suppression, whereas stimulation at the medial part of the SNr suppressed the MLR-induced locomotion (Takakusaki et al., 2003). The heterogeneous effects of SNr-stimulation suggest that the SNr is capable of controlling locomotion and posture using different channels depending on a given context.

A cardinal symptom of Parkinson's disease is gait disturbance in which the patients have difficulties in initiating or terminating walking (Azulay et al., 2002). It has been shown that the level of the GABAergic outputs of the basal ganglia is abnormally increased in Parkinson's patients (Miller and DeLong, 1987; DeLong, 1990; Filion and Tremblay, 1991). The studies by Takakusaki and colleagues suggest that the gait disturbance of Parkinson's patients is caused by an abnormal increase in the SNr-induced inhibition of the MLR. On the other hand, muscle rigidity, which is characterized by excessive muscle tone, may be due to an abnormally increased inhibition of the PPN which otherwise would contribute to muscle relaxation. This pathological mechanism may also account for changes in muscle tone observed in other basal ganglia disorders. Dystonia, which is characterized by focal and involuntary changes in muscle tone, posture, or movements, may be caused by a reduction or derangement of the GABAergic outputs of the basal ganglia to the PPN (Starr et al., 2005). Hypotonia, which is associated with Huntington's chorea, may be caused by a decreased SNr–PPN inhibition. To summarize, these studies on the MLR and the PPN have provided a clear and straightforward view on the locomotor function of the basal ganglia.

Important issues remain to be solved, however. First, the experiments on the MLR and the PPN have been done using quadrupedal animals (cats and rats), and it is unknown whether the same mechanisms exist in bipedal humans. Second, it is unclear what drives or dictates the SNr-induced control of locomotion and posture. One possibility is that the nucleus accumbens (or the ventral striatum), which projects to the SNr (Lynd-Balta and Haber, 1994), controls locomotion (Brudzynski and Mogenson, 1985). On the other hand, the ventral striatum is now known to carry reward-related motivational signals in rodents (Nicola et al., 2004), monkeys (Schultz et al., 1992), and humans (Knutson et al., 2001; Pagnoni et al., 2002). These results together suggest that the ventral striatum, by sending signals to the SNr, controls reward-oriented locomotion. This scheme (ventral striiatum-SNr-MLR/PPN) appears to parallel the caudate-SNr-SC mechanism, which controls reward-oriented saccadic eye movement. In fact, an essential part of reward-seeking behavior is to approach the place where reward may be available. The basal ganglia may control the two elements of reward-seeking behavior, orienting and locomotion, using separate mechanisms (Goodale and Murison, 1975).

A sustained decrease (or loss) of muscle tone occurs during rapid eye movement (REM) sleep. Several lines of evidence suggest that the PPN and surrounding regions are responsible for the REM-associated muscle atonia (Rye, 1997; Datta and Siwek, 2002). Using decerebrate cats Takakusaki and colleagues created the behavioral state that mimics REM (with atonia) by stimulating the PPN, and showed that additional stimulation at the lateral part of the SNr abolished the PPN-induced REM (Takakusaki et al., 2004b). This study suggested that the SNr is capable of controlling REM with its GABAergic connection to the PPN. This mechanism may be involved in sleep disorders that are often observed in parkinsonian patients (Larsen and Tandberg, 2001).

Mouth movements

The central pattern generator for rhythmical mouth movements is located in the medial part of the medullary reticular formation (Chandler and Tal, 1986; Nozaki et al., 1986). This area contains premotor interneurons that are involved in various types of mouth movements (Nozaki et al., 1993), and is directly controlled by the mouth area of the motor cortex (Nozaki et al., 1986). That the basal ganglia also control mouth movements has been suggested by electrical and chemical stimulation studies (Nakamura et al., 1990). Using chronically decorticated rats (Inchul et al., 2005) found that electrical stimulation at the ventral part of the SNr induced rhythmic activity in orofacial muscles and that an injection of GABA in the same region induced sustained elevation of muscle activity. These effects could be mediated by the SNr–GABAergic connection to the medullary reticular formation (Von Krosigk and Smith, 1991; Yasui et al., 1992). Another line of evidence suggests that mouth movements are mediated by the SNr–SC connection (Fig. 2): Apomorphine-induced mouth movements are suppressed by ablation of the SC (Chandler and Goldberg, 1984), mouth movements can be evoked by $GABA_A$ receptor blockade in the lateral deep layer of the SC (Adachi et al., 2003), and the lateral SC is thought to be critical for mouth movements (biting and licking)

directed at noxious stimuli applied to the limb (Wang and Redgrave, 1997). The mouth movement region in the SC may extend to the mesencephalic reticular formation (Hashimoto et al., 1989), which may also receive inputs from the SNr (Beckstead, 1983; Yasui et al., 1994). The basal ganglia control of mouth movements is further supported by neuronal activity accompanied by natural mouth movements found in the SNr (Mora et al., 1977; Joseph et al., 1985; Nishino et al., 1985) and other basal ganglia nuclei (Mittler et al., 1994; Masuda et al., 2001).

Mouth movements include biting, licking, chewing, and swallowing, which constitute a basic means of food and liquid intake. Following orienting movements (including saccadic eye movement) and locomotion (discussed above), mouth movements would complete reward-seeking behavior. These considerations suggest that an essential function of the basal ganglia is to control reward-seeking behavior by organizing and sequencing multiple body movements.

Vocalization

Vocalization, which involves mouth movements, may also be controlled by the basal ganglia. A large body of evidence comes from clinical observations of basal ganglia patients. Vocalization is weak in Parkinson's patients (Goberman and Elmer, 2005). Dysarthria is common among patients with corticobasal degeneration (Ozsancak et al., 2000). Patients with Tourette syndrome may burst out with grunts or obscene words, in addition to abrupt and ritualistic movements of various body parts. Imaging and postmortem studies have shown that the size of the basal ganglia nuclei is abnormally small in Tourette patients (Leckman et al., 1991; Peterson et al., 1993; Singer et al., 1993). Dopamine antagonists or agonists may be used for treatment, suggesting an involvement of the basal ganglia. A recent postmortem study indicated that the number of parvalbumin-positive GABAergic neurons is abnormal in the GP_i and the caudate (Kalanithi et al., 2005).

Vocalization involves coordinated activity in respiratory, laryngeal, and orofacial muscle groups.

Similarly to orienting, locomotion, and mouth movements, vocalization is likely to be controlled by a central pattern generator which is located in the nucleus retroambiguus (Holstege, 1989; Zhang et al., 1995) or in the parvocellular reticular formation around the nucleus ambiguous (Kirzinger and Jürgens, 1991; Hage and Jürgens, 2006), or both. This vocal pattern generator is controlled by the periaqueductal gray (PAG) (Zhang et al., 1995), which is known to control vocalization (Jürgens and Pratt, 1979; Larson and Kistler, 1986; Zhang et al., 1994). The PAG is known to be innervated by the SNr (Hopkins and Niessen, 1976; Cebrian et al., 2005; Castellan-Baldan et al., 2006), and this might be responsible for the basal ganglia control of vocalization (Von Krosigk and Smith, 1991) (Fig. 2). However, there has been no study, to my knowledge, that tested this hypothesis.

The role of the basal ganglia in vocalization has been studied in relation to bird song. In the birdbrain is an area called X, which is equivalent to the striatum and the GP_i combined in mammals (Farries and Perkel, 2002). Area X contains a large number of GABAergic neurons, similarly to the SNr/GP_i and the striatum, and is essential for the bird to learn to sing species-specific songs. As in the mammalian striatum, inputs from the hyperstriatum (corresponding to the mammalian cerebral cortex) to neurons in area X are modulated presynaptically by dopamine (Ding et al., 2003) and undergo synaptic plasticity (LTP) in the presence of dopamine (Ding and Perkel, 2004). These studies on the role of the basal ganglia in bird song should motivate more research in the mammalian brain.

Similarly to orienting, locomotion, and mouth movements, vocalization is a basic motor behavior whose patterns are generated by distinct neural circuits in the brainstem. On the other hand, vocalization is used for expression of emotions and social communications (Gil-da-Costa et al., 2004; Poremba et al., 2004), which require complicated context-dependent selections. The basal ganglia mechanisms utilizing their GABAergic outputs to the brainstem vocalization centers may play important roles in such selections.

Hand-finger movements

A majority of basal ganglia outputs is directed to several nuclei in the thalamus including nucleus ventralis lateralis (VL), nucleus ventralis anterior (VA), nucleus centrum medianum (CM), nucleus parafascicularis (Pf), and nucleus medialis dorsalis (MD) (Graybiel and Ragsdale, 1979; Carpenter, 1981; Parent, 1990). These thalamic nuclei are mutually connected with different areas in the cerebral cortex (Colwell, 1975) and in addition project to the striatum (caudate and putamen) (Beckstead, 1984; Sadikot et al., 1992; Mcfarland and Haber, 2001). In both cases the information sent out of the basal ganglia may be returned to the basal ganglia forming loop circuits: cortex-striatum-SNr/GP_i-thalamus-cortex and striatum-SNr/GP_i-thalamus-striatum. A dominant hypothesis is that the loop circuits are parceled into several divisions, which correspond to different functions such as sensorimotor, oculomotor, cognitive, and limbic functions (Alexander et al., 1986). While the presence of such loop circuits is likely, it is unclear whether the different loop circuits are independent. An alternative hypothesis is that signals can be relayed from one functional loop to another like a spiral (Haber et al., 2000). The latter type of information processing may be advantageous in integrating different kinds of information (e.g., cognitive and emotional), which is required for purposeful sbehaviors. The general scheme of the basal ganglia outputs to the thalamus described above suggest that the basal ganglia are capable of contributing to almost all kinds of functions.

A prominent motor function that is likely to be controlled by the basal ganglia output to the thalamus, not their output to the brainstem structures, is motor skill (Fig. 2). Motor skills are likely to be controlled by the mechanisms located in the motor cortices, cerebellum, basal ganglia, or these areas combined (Hikosaka et al., 1999; Hikosaka et al., 2002), not by the central pattern generators located in the brainstem or spinal cord. A distinct feature of motor skills is that they are acquired through repeated practice, and many studies suggest that the basal ganglia are involved in the learning process and possibly in the storage of the procedural skill memory as well (Miyachi et al., 1997; Hikosaka et al., 1999; Miyachi et al., 2002; Lehericy et al., 2005). These results suggest that, among the different lines of basal ganglia outputs, the output to the thalamus is likely to be involved in motor skills. With its output to the thalamus the GP$_i$ is deeply involved in motor skills, whereas the SNr is preferentially involved in cognitive or emotive functions (Sidibe et al., 2002). An outstanding example of motor skills is skilled hand–finger movement, which is particularly developed in primates (Wiesendanger, 1999). Another motor skill largely unique to humans is speech, which seems to require the intact basal ganglia (Damasio, 1983; Ullman, 2001).

Pathological mechanisms of GABAergic outputs of the basal ganglia

I suggested previously that the basal ganglia GABAergic output acts as a gate for motor signals such that there should be no motor output as long as the gate is closed. For this gating function to work properly, the level of the GABAergic output must, by default, be maintained at a steady level. A subtle change in this mechanism may cause serious problems in motor behavior. An example is found in Parkinson's disease. It has repeatedly been shown that the average level of the basal ganglia GABAergic output is higher in Parkinson's patients compared with control subjects. The enhanced GABAergic output would excessively suppress the target areas including the SC, MLR, PPN, thalamo-cortical circuits, and possibly mouth movement and vocalization centers, leading to akinesia or hypokinesia. On the other hand, dyskinesias induced by l–DOPA (dopamine precursor) or apomorphine (dopamine agonist) are associated with a reduction of the basal ganglia GABAergic outputs (Nevet et al., 2004). The depressed GABAergic output would be insufficient to suppress the motor centers mentioned above, which leads to excessive body movements, dyskinesia.

Further experiments support these conclusions. If the GABAergic output is reduced by injecting muscimol (GABA$_A$ agonist) in the SNr of MPTP-induced parkinsonian monkeys, their body movements are improved (Wichmann et al., 2001). In normal monkeys injections of muscimol (GABA agonist) in the GP$_i$, which should reduce the GABA output, induces choreiform movements, while injections of bicuculline (GABA$_A$ antagonist) in the GP$_i$, which should block incoming GABA-induced inhibitions, induced hypokinesia (Burbaud et al., 1998). Bergmann and colleagues have also shown that the GABAergic output is not stable but fluctuates in an oscillatory manner in MPTP-induced parkinsonian monkeys, but not in control monkeys (Bar-Gad et al., 2004). This should lead to oscillatory firing of motor neurons in the above areas, which may induce tremor or other involuntary movements.

Some mechanisms have been suggested for the deranged GABAergic output of the basal ganglia. GABAergic neurons in the SNr/GP$_i$ are highly sensitive to GABAergic inputs that arise from the GP$_e$ or caudate/putamen. It has been shown that the level of GABAergic outputs of the GP$_e$ is lower in Parkinson's patients, which might be a cause of an increase in GABAergic outputs of the SNr/GP$_i$. However, this would not be an ultimate explanation because the primary cause of Parkinson's disease is a loss of dopamine neurons in the SNc. Some studies have shown that dopamine released or delivered in the SNr can cause short-term and long-term changes in the activity of GABAergic output neurons (Waszczak and Walters, 1986; Miyazaki and Lacey, 1998; Trevitt et al., 2001) which may be paralleled by dyskinesia (Vila et al., 1996; Katz et al., 2005).

References

Adachi, K., Hasegawa, M., Ikeda, H., Sato, M., Koshikawa, N. and Cools, A.R. (2003) The superior colliculus contains a discrete region involved in the control of jaw movements: role of GABAA receptors. Eur. J. Pharmacol., 464: 147–154.

Alexander, G.E., DeLong, M.R. and Strick, P.L. (1986) Parallel organization of functionally segregated circuits linking basal ganglia and cortex. Annu. Rev. Neurosci., 9: 357–381.

Amiez, C., Joseph, J.P. and Procyk, E. (2006) Reward encoding in the monkey anterior cingulate cortex. Cereb. Cortex, 16: 1040–1055.

220

Anderson, M.E. and Horak, F.B. (1985) Influence of the globus pallidus on arm movements in monkeys. III. Timing of movement-related information. J. Neurophysiol., 54: 433–448.

Aron, A.R. and Poldrack, R.A. (2006) Cortical and subcortical contributions to Stop signal response inhibition: role of the subthalamic nucleus. J. Neurosci., 26: 2424–2433.

Azulay, J.P., Mesure, S., Amblard, B. and Pouget, J. (2002) Increased visual dependence in Parkinson's disease. Percept. Mot. Skills, 95: 1106–1114.

Barash, S., Bracewell, R.M., Fogassi, L., Gnadt, J.W. and Andersen, R.A. (1991) Saccade-related activity in the lateral intraparietal area. I. Temporal properties; comparison with area 7a. J. Neurophysiol., 66: 1095–1108.

Bar-Gad, I., Elias, S., Vaadia, E. and Bergman, H. (2004) Complex locking rather than complete cessation of neuronal activity in the globus pallidus of a 1-methyl-4-phenyl-1,2,3, 6-tetrahydropyridine-treated primate in response to pallidal microstimulation. J. Neurosci., 24: 9410–9419.

Barnes, T.D., Kubota, Y., Hu, D., Jin, D.Z. and Graybiel, A.M. (2005) Activity of striatal neurons reflects dynamic encoding and recoding of procedural memories. Nature, 437: 1158–1161.

Becker, W. (1989) Metrics. In: Wurtz R.H. and Goldberg M.E. (Eds.), The Neurobiology of Saccadic Eye Movements. Elsevier, Amsterdam, pp. 13–67.

Beckstead, R.M. (1983) Long collateral branches of substantia nigra pars reticulata axons to thalamus, superior colliculus and reticular formation in monkey and cat. Multiple retrograde neuronal labeling with fluorescent dyes. Neurosci., 10: 767–779.

Beckstead, R.M. (1984) The thalamostriatal projection in the cat. J. Comp. Neurol., 223: 313–346.

Beckstead, R.M., Domesick, V.B. and Nauta, W.J.H. (1979) Efferent connections of the substantia nigra and ventral tegmental area in the rat. Brain Res., 175: 191–217.

Bruce, C.J. and Goldberg, M.E. (1985) Primate frontal eye fields. I. Single neurons discharging before saccades. J. Neurophysiol., 53: 603–635.

Brudzynski, S.M. and Mogenson, G.J. (1985) Association of the mesencephalic locomotor region with locomotor activity induced by injections of amphetamine into the nucleus accumbens. Brain Res, 334: 77–84.

Burbaud, P., Bonnet, B., Guehl, D., Lagueny, A. and Bioulac, B. (1998) Movement disorders induced by gamma-aminobutyric agonist and antagonist injections into the internal globus pallidus and substantia nigra pars reticulata of the monkey. Brain Res, 780: 102–107.

Carpenter, M.B. (1981) Anatomy of the corpus striatum and brain stem integrating systems. In: Brooks V.B. (Ed.), The Nervous System. American Physiological Society, Bethesda, MD, pp. 947–995.

Castellan-Baldan, L., da Costa Kawasaki, M., Ribeiro, S.J., Calvo, F., Correa, V.M. and Coimbra, N.C. (2006) Topographic and functional neuroanatomical study of GABAergic disinhibitory striatum-nigral inputs and inhibitory nigrocollicular pathways: neural hodology recruiting the substantia nigra, pars reticulata, for the modulation of the neural activity in the inferior colliculus involved with panic-like emotions. J. Chem. Neuroanat., 32: 1–27.

Cebrian, C., Parent, A. and Prensa, L. (2005) Patterns of axonal branching of neurons of the substantia nigra pars reticulata and pars lateralis in the rat. J. Comp. Neurol., 492: 349–369.

Chandler, S.H. and Goldberg, L.J. (1984) Differentiation of the neural pathways mediating cortically induced and dopaminergic activation of the central pattern generator (CPG) for rhythmical jaw movements in the anesthetized guinea pig. Brain Res, 323: 297–301.

Chandler, S.H. and Tal, M. (1986) The effects of brain stem transections on the neuronal networks responsible for rhythmical jaw muscle activity in the guinea pig. J. Neurosci., 6: 1831–1842.

Chevalier, G., Thierry, A.M., Shibazaki, T. and Féger, J. (1981) Evidence for a GABAergic inhibitory nigrotectal pathway in the rat. Neurosci. Lett., 21: 67–70.

Coe, B., Tomihara, K., Matsuzawa, M. and Hikosaka, O. (2002) Visual and anticipatory bias in three cortical eye fields of the monkey during an adaptive decision-making task. J. Neurosci., 22: 5081–5090.

Colwell, S.A. (1975) Thalamocortical-corticothalamic reciprocity: a combined anterograde-retrograde tracer technique. Brain Res, 92: 443–449.

Cooke, J.D., Brown, J.D. and Brooks, V.B. (1978) Increased dependence on visual information for movement control in patients with Parkinson's disease. Can. J. Neurol. Sci., 5: 413–415.

Crawford, T.J., Henderson, L. and Kennard, C. (1989) Abnormalities of nonvisually-guided eye movements in Parkinson's disease. Brain, 112: 1573–1586.

Damasio, A.R. (1983) Language and the basal ganglia. Trends Neurosci, 6: 442–444.

Datta, S. and Siwek, D.F. (2002) Single cell activity patterns of pedunculopontine tegmentum neurons across the sleep-wake cycle in the freely moving rats. J. Neurosci. Res., 70: 611–621.

Delgado, M.R., Nystrom, L.E., Fissell, C., Noll, D.C. and Fiez, J.A. (2000) Tracking the hemodynamic responses to reward and punishment in the striatum. J. Neurophysiol., 84: 3072–3077.

DeLong, M.R. (1990) Primate models of movement disorders of basal ganglia origin. Trends Neurosci, 13: 281–285.

DeLong, M.R. and Georgopoulos, A.P. (1981) Motor functions of the basal ganglia. In: Brooks V.B. (Ed.), The Nervous System. American Physiological Society, Bethesda, MD, pp. 1017–1061.

Denny-Brown, D. (1968). Clinical symptomatology of diseases of the basal ganglia. In: Vinken, P.J. and Bruyn, G.W. (eds.), Diseases of the Basal Ganglia. Amsterdam, North Holland, pp. 133–172.

Di Chiara, G., Porceddu, M.L., Morelli, M., Mulas, M.L. and Gessa, G.L. (1979) Evidence for a GABAergic projection from the substantia nigra to the ventromedial thalamus and to the superior colliculus of the rat. Brain Res, 176: 273–284.

Ding, L. and Hikosaka, O. (2006) Comparison of reward modulation in the frontal eye field and caudate of the macaque. J. Neurosci., 26: 6695–6703.

Ding, L. and Perkel, D.J. (2004) Long-term potentiation in an avian basal ganglia nucleus essential for vocal learning. J. Neurosci., 24: 488–494.

Ding, L., Perkel, D.J. and Farries, M.A. (2003) Presynaptic depression of glutamatergic synaptic transmission by D1-like dopamine receptor activation in the avian basal ganglia. J. Neurosci., 23: 6086–6095.

Du Lac, S. and Knudsen, E.I. (1990) Neural maps of head movement vector and speed in the optic tectum of the barn owl. J. Neurophysiol., 63: 131–146.

Easter, S.S. (1975) The time course of saccadic eye movements in goldfish. Vision Res, 15: 405–409.

Evinger, C., Kaneko, C.R.S. and Fuchs, A.F. (1982) Activity of omnipause neurons in alert cats during saccadic eye movements and visual stimuli. J. Neurophysiol., 47: 827–844.

Fahn, S. and Cote, L. (1968) Regional distribution of g-aminobutyric acid (GABA) in brain of the rhesus monkey. J. Neurochem., 15: 209–213.

Farries, M.A. and Perkel, D.J. (2002) A telencephalic nucleus essential for song learning contains neurons with physiological characteristics of both striatum and globus pallidus. J. Neurosci., 22: 3776–3787.

Faull, R.L.M. and Mehler, W.R. (1978) The cells of origin of nigrotectal, nigrothalamic and nigrostriatal projections in the rat. Neuroscience, 3: 989–1002.

Filion, M. and Tremblay, L. (1991) Abnormal spontaneous activity of globus pallidus neurons in monkeys with MPTP-induced parkinsonism. Brain Res, 547: 142–151.

Fuchs, A.F., Kaneko, C.R.S. and Scudder, C.A. (1985) Brainstem control of saccadic eye movements. Annu. Rev. Neurosci., 8: 307–337.

Fujii, N. and Graybiel, A.M. (2005) Time-varying covariance of neural activities recorded in striatum and frontal cortex as monkeys perform sequential-saccade tasks. Proc. Natl. Acad. Sci. USA, 102: 9032–9037.

Fuller, J.H. (1985) Eye and head movements in the pigmented rat. Vision Res, 25: 1121–1128.

Garcia-Rill, E. (1986) The basal ganglia and the locomotor regions. Brain Res, 396: 47–63.

Garcia-Rill, E., Skinner, R.D., Jackson, M.B. and Smith, M.M. (1983) Connections of the mesencephalic locomotor region (MLR) I. Substantia nigra afferents. Brain Res. Bull., 10: 57–62.

Gil-da-Costa, R., Braun, A., Lopes, M., Hauser, M.D., Carson, R.E., Herscovitch, P. and Martin, A. (2004) Toward an evolutionary perspective on conceptual representation: species-specific calls activate visual and affective processing systems in the macaque. Proc. Natl. Acad. Sci. USA, 101: 17516–17521.

Glickstein, M. and Stein, J. (1991) Paradoxical movement in Parkinson's disease. Trends Neurosci, 14: 480–482.

Goberman, A.M. and Elmer, L.W. (2005) Acoustic analysis of clear versus conversational speech in individuals with Parkinson disease. J. Commun. Disord., 38: 215–230.

Goodale, M.A. and Murison, R.C. (1975) The effects of lesions of the superior colliculus on locomotor orientation and the orienting reflex in the rat. Brain Res, 88: 243–261.

Grantyn, A., Moschovakis, A.K. and Kitama, T. (2004) Control of orienting movements: role of multiple tectal projections to the lower brainstem. Prog. Brain Res., 143: 423–438.

Graybiel, A.M. (1978) Organization of the nigrotectal connection: an experimental tracer study in the cat. Brain Res, 143: 339–348.

Graybiel, A.M. and Ragsdale, C.W. (1979) Fiber connections of the basal ganglia. In: Cuenod M., Kreutzberg G.W. and Bloom F.E. (Eds.), Development of Chemical Specificity of Neurons. Elsevier, Amsterdam, pp. 239–283.

Grillner, S. (1985) Neurobiological bases of rhythmic motor acts in vertebrates. Science, 228: 143–149.

Grillner, S. and Shik, M.L. (1973) On the descending control of the lumbosacral spinal cord from the 'mesencephalic locomotor region'. Acta Physiol. Scand., 87: 320–333.

Guitton, D. and Volle, M. (1987) Gaze control in humans: eye-head coordination during orienting movements to targets within and beyond the oculomotor range. J. Neurophysiol., 58: 427–459.

Haber, S.N., Fudge, J.L. and McFarland, N.R. (2000) Striatonigrostriatal pathways in primates form an ascending spiral from the shell to the dorsolateral striatum. J. Neurosci., 20: 2369–2382.

Hage, S.R. and Jürgens, U. (2006) On the role of the pontine brainstem in vocal pattern generation: a telemetric single-unit recording study in the squirrel monkey. J. Neurosci., 26: 7105–7115.

Handel, A. and Glimcher, P.W. (1999) Quantitative analysis of substantia nigra pars reticulata activity during a visually guided saccade task. J. Neurophysiol., 82: 3458–3475.

Hashimoto, N., Katayama, T., Ishiwata, Y. and Nakamura, Y. (1989) Induction of rhythmic jaw movements by stimulation of the mesencephalic reticular formation in the guinea pig. J. Neurosci., 9: 2887–2901.

Hikosaka, O., Matsumura, M., Kojima, J. and Gardiner, T.W. (1993b) Role of basal ganglia in initiation and suppression of saccadic eye movements. In: Mano N., Hamada I. and DeLong M.R. (Eds.), Role of the Cerebellum and Basal Ganglia in Voluntary Movement. Elsevier, Amsterdam, pp. 213–219.

Hikosaka, O., Nakamura, K. and Nakahara, H. (2006) Basal Ganglia orient eyes to reward. J. Neurophysiol., 95: 567–584.

Hikosaka, O., Nakahara, H., Rand, M.K., Sakai, K., Lu, X., Nakamura, K., Miyachi, S. and Doya, K. (1999) Parallel neural networks for learning sequential procedures. Trends Neurosci, 22: 464–471.

Hikosaka, O., Nakamura, K., Sakai, K. and Nakahara, H. (2002) Central mechanisms of motor skill learning. Curr. Opin. Neurobiol., 12: 217–222.

Hikosaka, O., Rand, M.K., Miyachi, S. and Miyashita, K. (1995) Learning of sequential movements in the monkey: process of learning and retention of memory. J. Neurophysiol., 74: 1652–1661.

Hikosaka, O., Sakamoto, M. and Miyashita, N. (1993a) Effects of caudate nucleus stimulation on substantia nigra cell activity in monkey. Exp. Brain Res., 95: 457–472.

Hikosaka, O., Sakamoto, M. and Usui, S. (1989a) Functional properties of monkey caudate neurons. I. Activities related to saccadic eye movements. J. Neurophysiol., 61: 780–798.

Hikosaka, O., Sakamoto, M. and Usui, S. (1989b) Functional properties of monkey caudate neurons. III. Activities related to expectation of target and reward. J. Neurophysiol., 61: 814–832.

Hikosaka, O., Takikawa, Y. and Kawagoe, R. (2000) Role of the basal ganglia in the control of purposive saccadic eye movements. Physiol. Rev., 80: 953–978.

Hikosaka, O. and Wurtz, R.H. (1983a) Visual and oculomotor functions of monkey substantia nigra pars reticulata. I. Relation of visual and auditory responses to saccades. J. Neurophysiol., 49: 1230–1253.

Hikosaka, O. and Wurtz, R.H. (1983b) Visual and oculomotor functions of monkey substantia nigra pars reticulata. II. Visual responses related to fixation of gaze. J. Neurophysiol., 49: 1254–1267.

Hikosaka, O. and Wurtz, R.H. (1983c) Visual and oculomotor functions of monkey substantia nigra pars reticulata. III. Memory-contingent visual and saccade responses. J. Neurophysiol., 49: 1268–1284.

Hikosaka, O. and Wurtz, R.H. (1983d) Visual and oculomotor functions of monkey substantia nigra pars reticulata. IV. Relation of substantia nigra to superior colliculus. J. Neurophysiol., 49: 1285–1301.

Hikosaka, O. and Wurtz, R.H. (1983e) Effects on eye movements of a GABA agonist and antagonist injected into monkey superior colliculus. Brain Res, 272: 368–372.

Hikosaka, O. and Wurtz, R.H. (1985a) Modification of saccadic eye movements by GABA-related substances. I. Effect of muscimol and bicuculline in the monkey superior colliculus. J. Neurophysiol., 53: 266–291.

Hikosaka, O. and Wurtz, R.H. (1985b) Modification of saccadic eye movements by GABA-related substances. II. Effects of muscimol in monkey substantia nigra pars reticulata. J. Neurophysiol., 53: 292–308.

Hollerman, J.R., Tremblay, L. and Schultz, W. (1998) Influence of reward expectation on behavior-related neuronal activity in primate striatum. J. Neurophysiol., 80: 947–963.

Holstege, G. (1989) Anatomical study of the final common pathway for vocalization in the cat. J. Comp. Neurol., 284: 242–252.

Hopkins, D.A. and Niessen, L.W. (1976) Substantia nigra projections to the reticular formation, superior colliculus and central gray in the rat, cat and monkey. Neurosci. Lett., 2: 253–259.

Ikeda, T. and Hikosaka, O. (2003) Reward-dependent gain and bias of visual responses in primate superior colliculus. Neuron, 39: 693–700.

Inchul, P., Amano, N., Satoda, T., Murata, T., Kawagishi, S., Yoshino, K. and Tanaka, K. (2005) Control of orofacio-lingual movements by the substantia nigra pars reticulata: high-frequency electrical microstimulation and GABA microinjection findings in rats. Neuroscience, 134: 677–689.

Jayaraman, A., Batton, R.R.I. and Carpenter, M.B. (1977) Nigrotectal projections in the monkey: an autoradiographic study. Brain Res, 135: 147–F152.

Joseph, J.P. and Boussaoud, D. (1985) Role of the cat substantia nigra pars reticulata in eye and head movements. I. Neural activity. Exp. Brain Res., 57: 286–296.

Joseph, J.P., Boussaoud, D. and Biguer, B. (1985) Activity of neurons in the cat substantia nigra pars reticulata during drinking. Exp. Brain Res., 60: 375–379.

Judge, S.J., Wurtz, R.H. and Richmond, B.J. (1980) Vision during saccadic eye movements. I. Visual interactions in striate cortex. J. Neurophysiol., 43: 1133–1155.

Jung, R. and Hassler, R. (1960) The extrapyramidal motor system. In: Field J., Magoun H.W. and Hall V.E. (Eds.), Neurophysiology. American Phsiological Society, Washington, DC, pp. 863–927.

Jürgens, U. and Pratt, R. (1979) Role of the periaqueductal grey in vocal expression of emotion. Brain Res, 167: 367–378.

Kalanithi, P.S., Zheng, W., Kataoka, Y., Difiglia, M., Grantz, H., Saper, C.B., Schwartz, M.L., Leckman, J.F. and Vaccarino, F.M. (2005) Altered parvalbumin-positive neuron distribution in basal ganglia of individuals with Tourette syndrome. Proc. Natl. Acad. Sci. USA, 102: 13307–13312.

Kang, Y. and Kitai, S.T. (1990) Electrophysiological properties of pedunculopontine neurons and their postsynaptic responses following stimulation of substantia nigra reticulata. Brain Res, 535: 79–95.

Karabelas, A.B. and Moschovakis, A.K. (1985) Nigral inhibitory termination on efferent neurons of the superior colliculus: an intracellular horseradish peroxidase study in the cat. J. Comp. Neurol., 239: 309–329.

Katz, J., Nielsen, K.M. and Soghomonian, J.J. (2005) Comparative effects of acute or chronic administration of levodopa to 6-hydroxydopamine-lesioned rats on the expression of glutamic acid decarboxylase in the neostriatum and GABA$_A$ receptors subunits in the substantia nigra, pars reticulata. Neuroscience, 132: 833–842.

Kawagoe, R., Takikawa, Y. and Hikosaka, O. (1998) Expectation of reward modulates cognitive signals in the basal ganglia. Nat. Neurosci., 1: 411–416.

Kawagoe, R., Takikawa, Y. and Hikosaka, O. (2004) Reward-predicting activity of dopamine and caudate neurons — a possible mechanism of motivational control of saccadic eye movement. J. Neurophysiol., 91: 1013–1024.

Keller, E.L. (1989) The cerebellum. In: Wurtz R.H. and Goldberg M.E. (Eds.), The Neurobiology of Saccadic Eye Movements. Elsevier, Amsterdam, pp. 391–411.

Kermadi, I. and Joseph, J.P. (1995) Activity in the caudate nucleus of monkey during spatial sequencing. J. Neurophysiol., 74: 911–933.

Kirzinger, A. and Jürgens, U. (1991) Vocalization-correlated single-unit activity in the brain stem of the squirrel monkey. Exp. Brain Res., 84: 545–560.

Knutson, B., Adams, C.M., Fong, G.W. and Hommer, D. (2001) Anticipation of increasing monetary reward selectively recruits nucleus accumbens. J. Neurosci., 21: RC159: 1–5.

Knutson, B., Westdorp, A., Kaiser, E. and Hommer, D. (2000) FMRI visualization of brain activity during a monetary incentive delay task. Neuroimage, 12: 20–27.

Kobayashi, S., Lauwereyns, J., Koizumi, M., Sakagami, M. and Hikosaka, O. (2002) Influence of reward expectation on visuospatial processing in macaque lateral prefrontal cortex. J. Neurophysiol., 87: 1488–1498.

Larsen, J.P. and Tandberg, E. (2001) Sleep disorders in patients with Parkinson's disease: epidemiology and management. CNS Drugs, 15: 267–275.

Larson, C.R. and Kistler, M.K. (1986) The relationship of periaqueductal gray neurons to vocalization and laryngeal EMG in the behaving monkey. Exp. Brain Res., 63: 596–606.

Lauwereyns, J., Watanabe, K., Coe, B. and Hikosaka, O. (2002) A neural correlate of response bias in monkey caudate nucleus. Nature, 418: 413–417.

Leckman, J.F., Knorr, A.M., Rasmusson, A.M. and Cohen, D.J. (1991) Basal ganglia research and Tourette's syndrome. Trends Neurosci, 14: 94.

Lehericy, S., Benali, H., Van de Moortele, P.F., Pelegrini-Issac, M., Waechter, T., Ugurbil, K. and Doyon, J. (2005) Distinct basal ganglia territories are engaged in early and advanced motor sequence learning. Proc. Natl. Acad. Sci. USA, 102: 12566–12571.

Lynd-Balta, E. and Haber, S.N. (1994) Primate striatonigral projections: a comparison of the sensorimotor-related striatum and the ventral striatum. J. Comp. Neurol., 345: 562–578.

Marín, O., Smeets, W.J.A.J. and González, A. (1998) Evolution of the basal ganglia in tetrapods: a new perspective based on recent studies in amphibians. Trends Neurosci, 21: 487–494.

Masuda, Y., Kato, T., Hidaka, O., Matsuo, R., Inoue, T., Iwata, K. and Morimoto, T. (2001) Neuronal activity in the putamen and the globus pallidus of rabbit during mastication. Neurosci. Res., 39: 11–19.

May, P.J. and Hall, W.C. (1984) Relationships between the nigrotectal pathway and the cells of origin of the predorsal bundle. J. Comp. Neurol., 226: 357–376.

McClure, S.M., York, M.K. and Montague, P.R. (2004) The neural substrates of reward processing in humans: the modern role of FMRI. Neuroscientist, 10: 260–268.

McFarland, N.R. and Haber, S.N. (2001) Organization of thalamostriatal terminals from the ventral motor nuclei in the macaque. J. Comp. Neurol., 429: 321–336.

Meredith, M.A., Wallace, M.T. and Stein, B.E. (1992) Visual, auditory and somatosensory convergence in output neurons of the cat superior colliculus: multisensory properties of the tecto-reticulo-spinal projection. Exp. Brain Res., 88: 181–186.

Meyer-Luehmann, M., Thompson, J.F., Berridge, K.C. and Aldridge, J.W. (2002) Substantia nigra pars reticulata neurons code initiation of a serial pattern: implications for natural action sequences and sequential disorders. Eur. J. Neurosci., 16: 1599–1608.

Miller, W.C. and DeLong, M.R. (1987) Altered tonic activity of neurons in the globus pallidus and subthalamic nucleus in the primate MPTP model of parkinsonism. In: Carpenter M.B.

and Jayaraman A. (Eds.), The Basal Ganglia II. Plenum, New York, pp. 415–427.

Mink, J.W. (1996) The basal ganglia: focused selection and inhibition of competing motor programs. Prog. Neurobiol., 50: 381–425.

Mittler, T., Cho, J., Peoples, L.L. and West, M.O. (1994) Representation of the body in the lateral striatum of the freely moving rat: single neurons related to licking. Exp. Brain Res., 98: 163–167.

Miyachi, S., Hikosaka, O. and Lu, X. (2002) Differential activation of monkey striatal neurons in the early and late stages of procedural learning. Exp. Brain Res., 146: 122–126.

Miyachi, S., Hikosaka, O., Miyashita, K., Karádi, Z. and Rand, M.K. (1997) Differential roles of monkey striatum in learning of sequential hand movement. Exp. Brain Res., 115: 1–5.

Miyashita, K., Rand, M.K., Miyachi, S. and Hikosaka, O. (1996) Anticipatory saccades in sequential procedural learning in monkeys. J. Neurophysiol., 76: 1361–1366.

Miyazaki, T. and Lacey, M.G. (1998) Presynaptic inhibition by dopamine of a discrete component of GABA release in rat substantia nigra pars reticulata. J. Physiol., 513(Pt 3): 805–817.

Mora, F., Mogenson, G.J. and Rolls, E.T. (1977) Activity of neurons in the region of the substantia nigra during feeding in the monkey. Brain Res, 133: 267–276.

Mori, S. (1987) Integration of posture and locomotion in acute decerebrate cats and in awake, freely moving cats. Prog. Neurobiol. (Oxf.), 28: 161–195.

Nakamura, K. and Hikosaka, O. (2006) Role of dopamine in the primate caudate nucleus in reward modulation of saccades. J. Neurosci., 26: 5360–5369.

Nakamura, S., Muramatsu, S. and Yoshida, M. (1990) Role of the basal ganglia in manifestation of rhythmical jaw movement in rats. Brain Res, 535: 335–338.

Nakamura, T., Bronstein, A.M., Lueck, C., Marsden, C.D. and Rudge, P. (1994) Vestibular, cervical and visual remembered saccades in Parkinson's disease. Brain, 117: 1423–1432.

Nakanishi, H., Kita, H. and Kitai, S.T. (1987) Intracellular study of rat substantia nigra pars reticulata neurons in an in vitro slice preparation: electrical membrane properties and response characteristics to subthalamic stimulation. Brain Res, 437: 45–55.

Nambu, A., Tokuno, H. and Takada, M. (2002) Functional significance of the cortico-subthalamo-pallidal 'hyperdirect' pathway. Neurosci. Res., 43: 111–117.

Nevet, A., Morris, G., Saban, G., Fainstein, N. and Bergman, H. (2004) Discharge rate of substantia nigra pars reticulata neurons is reduced in non-parkinsonian monkeys with apomorphine-induced orofacial dyskinesia. J. Neurophysiol., 92: 1973–1981.

Nicola, S.M., Yun, I.A., Wakabayashi, K.T. and Fields, H.L. (2004) Cue-evoked firing of nucleus accumbens neurons encodes motivational significance during a discriminative stimulus task. J. Neurophysiol., 91: 1840–1865.

Nishino, H., Ono, T., Fukuda, M. and Sasaki, K. (1985) Monkey substantia nigra (pars reticulata) neuron discharges during operant feeding. Brain Res, 334: 190–193.

Noda, T. and Oka, H. (1986) Distribution and morphology of tegmental neurons receiving nigral inhibitory inputs in the cat: an intracellular HRP study. J. Comp. Neurol., 244: 254–266.

Nozaki, S., Iriki, A. and Nakamura, Y. (1986) Localization of central rhythm generator involved in cortically induced rhythmical masticatory jaw-opening movement in the guinea pig. J. Neurophysiol., 55: 806–825.

Nozaki, S., Iriki, A. and Nakamura, Y. (1993) Trigeminal premotor neurons in the bulbar parvocellular reticular formation participating in induction of rhythmical activity of trigeminal motoneurons by repetitive stimulation of the cerebral cortex in the guinea pig. J. Neurophysiol., 69: 595–608.

O'Doherty, J., Kringelbach, M.L., Rolls, E.T., Hornak, J. and Andrews, C. (2001) Abstract reward and punishment representations in the human orbitofrontal cortex. Nat. Neurosci., 4: 95–102.

O'Doherty, J.P., Deichmann, R., Critchley, H.D. and Dolan, R.J. (2002) Neural responses during anticipation of a primary taste reward. Neuron, 33: 815–826.

Okada, Y., Nitsch-Hassler, C., Kim, J.S., Bak, I.J. and Hassler, R. (1971) Role of g-aminobutyric acid (GABA) in the extrapyramidal motor system. I. Regional distribution of GABA in rabbit, guinea pig and baboon CNS. Exp. Brain Res., 13: 514–518.

Ozsancak, C., Auzou, P. and Hannequin, D. (2000) Dysarthria and orofacial apraxia in corticobasal degeneration. Mov. Disord., 15: 905–910.

Pagnoni, G., Zink, C.F., Montague, P.R. and Berns, G.S. (2002) Activity in human ventral striatum locked to errors of reward prediction. Nat. Neurosci., 5: 97–98.

Parent, A. (1990) Extrinsic connections of the basal ganglia. Trends Neurosci, 13: 254–258.

Peterson, B., Riddle, M.A., Cohen, D.J., Katz, L.D., Smith, J.C., Hardin, M.T. and Leckman, J.F. (1993) Reduced basal ganglia volumes in Tourette's syndrome using three-dimensional reconstruction techniques from magnetic resonance images. Neurology, 43: 941–949.

Platt, M.L. and Glimcher, P.W. (1999) Neural correlates of decision variables in parietal cortex. Nature, 400: 233–238.

Poremba, A., Malloy, M., Saunders, R.C., Carson, R.E., Herscovitch, P. and Mishkin, M. (2004) Species-specific calls evoke asymmetric activity in the monkey's temporal poles. Nature, 427: 448–451.

Ramnani, N. and Miall, R.C. (2003) Instructed delay activity in the human prefrontal cortex is modulated by monetary reward expectation. Cereb. Cortex, 13: 318–327.

Raybourn, M.S. and Keller, E.L. (1977) Colliculoreticular organization in primate oculomotor system. J. Neurophysiol., 40: 861–878.

Raz, A., Vaadia, E. and Bergman, H. (2000) Firing patterns and correlations of spontaneous discharge of pallidal neurons in the normal and the tremulous 1-methyl-4-phenyl-1,2,3, 6-tetrahydropyridine vervet model of parkinsonism. J. Neurosci., 20: 8559–8571.

Reiner, A., Medina, L. and Veenman, C.L. (1998) Structural and functional evolution of the basal ganglia in vertebrates. Brain Res. Brain Res. Rev., 28: 235–285.

Rinvik, E., Grofova, I. and Ottersen, O.P. (1976) Demonstration of nigrotectal and nigroreticular projections in the cat by axonal transport of proteins. Brain Res, 112: 388–394.

Robledo, P. and Féger, J. (1990) Excitatory influence of rat subthalamic nucleus to substantia nigra pars reticulata and the pallidal complex: electrophysiological data. Brain Res, 518: 47–54.

Roesch, M.R. and Olson, C.R. (2003) Impact of expected reward on neuronal activity in prefrontal cortex, frontal and supplementary eye fields and premotor cortex. J. Neurophysiol., 90: 1766–1789.

Rye, D.B. (1997) Contributions of the pedunculopontine region to normal and altered REM sleep. Sleep, 20: 757–788.

Sadikot, A.F., Parent, A. and François, C. (1992) Efferent connections of the centromedian and parafascicular thalamic nuclei in the squirrel monkey: a PHA-L study of subcortical projections. J. Comp. Neurol., 315: 137–159.

Sakamoto, M. and Hikosaka, O. (1989) Eye movements induced by microinjection of GABA agonist in the rat substantia nigra pars reticulata. Neurosci. Res., 6: 216–233.

Sato, M. and Hikosaka, O. (2000) Reward-related modulation of spatial information in substantia nigra pars reticulata neurons for subsequent saccades. Soc Neurosci. Abstr., 26: 682.

Sato, M. and Hikosaka, O. (2002) Role of primate substantia nigra pars reticulata in reward-oriented saccadic eye movement. J. Neurosci., 22: 2363–2373.

Schlag, J. and Schlag-Rey, M. (1987) Evidence for a supplementary eye field. J. Neurophysiol., 57: 179–200.

Schultz, W. (1998) Predictive reward signal of dopamine neurons. J. Neurophysiol., 80: 1–27.

Schultz, W. (2002) Getting formal with dopamine and reward. Neuron, 36: 241–263.

Schultz, W., Apicella, P., Scarnati, E. and Ljungberg, T. (1992) Neuronal activity in monkey ventral striatum related to the expectation of reward. J. Neurosci., 12: 4595–4610.

Sidibe, M., Pare, J.F. and Smith, Y. (2002) Nigral and pallidal inputs to functionally segregated thalamostriatal neurons in the centromedian/parafascicular intralaminar nuclear complex in monkey. J. Comp. Neurol., 447: 286–299.

Singer, H.S., Reiss, A.L., Brown, J.E., Aylward, E.H., Shih, B., Chee, E., Harris, E.L., Reader, M.J., Chase, G.A., Bryan, R.N., et al. (1993) Volumetric MRI changes in basal ganglia of children with Tourette's syndrome. Neurology, 43: 950–956.

Smith, Y. and Bolam, J.P. (1989) Neurons of the substantia nigra reticulata receive a dense GABA-containing input from the globus pallidus in the rat. Brain Res, 493: 160–167.

Smith, Y. and Bolam, J.P. (1991) Convergence of synaptic inputs from the striatum and the globus pallidus onto identified nigrocollicular cells in the rat: a double anterograde labelling study. Neuroscience, 44: 45–73.

Spann, B.M. and Grofova, I. (1991) Nigropedunculopontine projection in the rat: an anterograde tracing study with Phaseolus vulgaris-leucoagglutinin (PHA-L). J. Comp. Neurol., 311: 375–388.

Sparks, D.L. (1986) Translation of sensory signals into commands for control of saccadic eye movements: role of primate superior colliculus. Physiol. Rev., 66: 118–171.

Starr, P.A., Rau, G.M., Davis, V., Marks Jr., W.J., Ostrem, J.L., Simmons, D., Lindsey, N. and Turner, R.S. (2005) Spontaneous pallidal neuronal activity in human dystonia: comparison with Parkinson's disease and normal macaque. J. Neurophysiol., 93: 3165–3176.

Stern, Y., Mayeux, R., Rosen, J. and Ilson, J. (1983) Perceptual motor dysfunction in Parkinson's disease: a deficit in sequential and predictive voluntary movement. J. Neurol. Neurosur. Psychiatry, 46: 145–151.

Sugrue, L.P., Corrado, G.S. and Newsome, W.T. (2004) Matching behavior and the representation of value in the parietal cortex. Science, 304: 1782–1787.

Takakusaki, K., Habaguchi, T., Ohtinata-Sugimoto, J., Saitoh, K. and Sakamoto, T. (2003) Basal ganglia efferents to the brainstem centers controlling postural muscle tone and locomotion: a new concept for understanding motor disorders in basal ganglia dysfunction. Neuroscience, 119: 293–308.

Takakusaki, K., Saitoh, K., Harada, H. and Kashiwayanagi, M. (2004a) Role of basal ganglia-brainstem pathways in the control of motor behaviors. Neurosci. Res., 50: 137–151.

Takakusaki, K., Saitoh, K., Harada, H., Okumura, T. and Sakamoto, T. (2004b) Evidence for a role of basal ganglia in the regulation of rapid eye movement sleep by electrical and chemical stimulation for the pedunculopontine tegmental nucleus and the substantia nigra pars reticulata in decerebrate cats. Neuroscience, 124: 207–220.

Takikawa, Y., Kawagoe, R. and Hikosaka, O. (2002a) Reward-dependent spatial selectivity of anticipatory activity in monkey caudate neurons. J. Neurophysiol., 87: 508–515.

Takikawa, Y., Kawagoe, R., Itoh, H., Nakahara, H. and Hikosaka, O. (2002b) Modulation of saccadic eye movements by predicted reward outcome. Exp. Brain Res., 142: 284–291.

Tinaz, S., Schendan, H.E., Schon, K. and Stern, C.E. (2006) Evidence for the importance of basal ganglia output nuclei in semantic event sequencing: an fMRI study. Brain Res, 1067: 239–249.

Tremblay, L. and Schultz, W. (2000) Reward-related neuronal activity during go-nogo task performance in primate orbitofrontal cortex. J. Neurophysiol., 83: 1864–1876.

Trevitt, J.T., Carlson, B.B., Nowend, K. and Salamone, J.D. (2001) Substantia nigra pars reticulata is a highly potent site of action for the behavioral effects of the D1 antagonist SCH 23390 in the rat. Psychopharmacology (Berl.), 156: 32–41.

Ullman, M.T. (2001) A neurocognitive perspective on language: the declarative/procedural model. Nat. Rev. Neurosci., 2: 717–726.

Uno, M. and Yoshida, M. (1975) Monosynaptic inhibition of thalamic neurons produced by stimulation of the pallidal nucleus in cats. Brain Res, 99: 377–380.

Vila, M., Herrero, M.T., Levy, R., Faucheux, B., Ruberg, M., Guillen, J., Luquin, MR., Guridi, J., Javoy-Agid, F., Agid, Y., Obeso, J.A. and Hirsch, E.C. (1996) Consequences of nigrostriatal denervation on the gamma-aminobutyric acidic neurons of substantia nigra pars reticulata and superior colliculus in parkinsonian syndromes. Neurology, 46: 802–809.

Vincent, S.R., Hattori, T. and McGeer, E.G. (1978) The nigrotectal projection: a biochemical and ultrastructural characterization. Brain Res, 151: 159–164.

Vitek, J.L., Chockkan, V., Zhang, J.Y., Kaneoke, Y., Evatt, M., DeLong, M.R., Triche, S., Mewes, K., Hashimoto, T. and Bakay, R.A. (1999) Neuronal activity in the basal ganglia in patients with generalized dystonia and hemiballismus. Ann. Neurol., 46: 22–35.

Von Krosigk, M. and Smith, A.D. (1991) Descending projections from the substantia nigra and retrorubral field to the medullary and pontomedullary reticular formation. Eur. J. Neurosci., 3: 260–273.

Wang, S. and Redgrave, P. (1997) Microinjections of muscimol into lateral superior colliculus disrupt orienting and oral movements in the formalin model of pain. Neuroscience, 81: 967–988.

Waszczak, B.L. and Walters, J.R. (1986) Endogenous dopamine can modulate inhibition of substantia nigra pars reticulata neurons elicited by GABA iontophoresis or striatal stimulation. J. Neurosci., 6: 120–126.

Watanabe, M. (1996) Reward expectancy in primate prefrontal neurons. Nature, 382: 629–632.

Watanabe, K., Lauwereyns, J. and Hikosaka, O. (2003a) Effects of motivational conflicts on visually elicited saccades in monkeys. Exp. Brain Res., 152: 361–367.

Watanabe, K., Lauwereyns, J. and Hikosaka, O. (2003b) Neural correlates of rewarded and unrewarded eye movements in the primate caudate nucleus. J. Neurosci., 23: 10052–10057.

Wichmann, T., Bergman, H., Starr, P.A., Subramanian, T., Watts, R.L. and DeLong, M.R. (1999) Comparison of MPTP-induced changes in spontaneous neuronal discharge in the internal pallidal segment and in the substantia nigra pars reticulata in primates. Exp. Brain Res., 125: 397–409.

Wichmann, T., Kliem, M.A. and DeLong, M.R. (2001) Antiparkinsonian and behavioral effects of inactivation of the substantia nigra pars reticulata in hemiparkinsonian primates. Exp. Neurol., 167: 410–424.

Wiesendanger, M. (1999) Manual dexterity and the making of tools – an introduction from an evolutionary perspective. Exp. Brain Res., 128: 1–5.

Williams, M.N. and Faull, R.L. (1988) The nigrotectal projection and tectospinal neurons in the rat. A light and electron microscopic study demonstrating a monosynaptic nigral input to identified tectospinal neurons. Neuroscience, 25: 533–562.

Yasui, Y., Nakano, K., Nakagawa, Y., Kayahara, T., Shiroyama, T. and Mizuno, N. (1992) Non-dopaminergic neurons in the substantia nigra project to the reticular formation around the trigeminal motor nucleus in the rat. Brain Res, 585: 361–366.

Yasui, Y., Tsumori, T., Ando, A., Domoto, T., Kayahara, T. and Nakano, K. (1994) Descending projections from the superior colliculus to the reticular formation around the motor trigeminal nucleus and the parvicellular reticular formation of the medulla oblongata in the rat. Brain Res, 656: 420–426.

Yin, H.H. and Knowlton, B.J. (2006) The role of the basal ganglia in habit formation. Nat. Rev. Neurosci., 7: 464–476.

Yoshida, M. and Omata, S. (1979) Blocking by picrotoxin of nigra-evoked inhibition of neurons of ventromedial nucleus of the thalamus. Experientia, 35: 794.

Yoshida, M. and Precht, W. (1971) Monosynaptic inhibition of neurons in the substantia nigra by caudate-nigral fibers. Brain Res, 32: 225–228.

Zhang, S.P., Bandler, R. and Davis, P.J. (1995) Brain stem integration of vocalization: role of the nucleus retroambigualis. J. Neurophysiol., 74: 2500–2512.

Zhang, S.P., Davis, P.J., Bandler, R. and Carrive, P. (1994) Brain stem integration of vocalization: Role of the midbrain periaqueductal gray. J. Neurophysiol., 72: 1337–1356.

GABAergic Neurotransmission in the Basal Ganglia

Tepper, Abercrombie & Bolam (Eds.)
Progress in Brain Research, Vol. 160
ISSN 0079-6123

CHAPTER 13

Localization of GABA receptors in the basal ganglia

Justin Boyes[2] and J. Paul Bolam[1,*]

[1]*MRC Anatomical Neuropharmacology Unit, Department of Pharmacology, University of Oxford, Mansfield Road, Oxford, UK*
[2]*School of Life and Health Sciences, Aston University, Aston Triangle, Birmingham, UK*

Abstract: The majority of neurons in the basal ganglia utilize GABA as their principal neurotransmitter and, as a consequence, most basal ganglia neurons receive extensive GABAergic inputs derived from multiple sources. In order to understand the diverse roles of GABA in the basal ganglia it is necessary to define the precise localization of GABA receptors in relation to known neuron subtypes and known afferents. In this chapter, we summarize data on the ultrastructural localization of ionotropic $GABA_A$ receptors and metabotropic $GABA_B$ receptors in the basal ganglia. In each of the regions of the basal ganglia that have been studied, $GABA_A$ receptor subunits are located primarily at symmetrical synapses formed by GABAergic boutons, where they display a several-hundred-fold enrichment over extrasynaptic sites. In contrast, $GABA_B$ receptors are widely distributed at synaptic and extrasynaptic sites on both presynaptic and postsynaptic membranes. Presynaptic $GABA_B$ receptors are localized on striatopallidal, striatonigral and pallidonigral afferent terminals, as well as glutamatergic terminals derived from the cortex, thalamus and subthalamic nucleus. It is concluded that fast GABA transmission mediated by $GABA_A$ receptors in the basal ganglia occurs primarily at synapses whereas GABA transmission mediated by $GABA_B$ receptors is more complex, involving receptors located at presynaptic, postsynaptic and extrasynaptic sites.

Keywords: $GABA_A$ receptor; $GABA_B$ receptor; striatum; globus pallidus; substantia nigra

The inhibitory amino acid, γ-aminobutyric acid (GABA), plays a critical role in the neuronal networks of the basal ganglia. Most of the major classes of neurons in the basal ganglia utilize GABA as a neurotransmitter. Thus, neurons of the globus pallidus (GP), basal ganglia output neurons in the entopeduncular nucleus (EP) (or internal segment of the GP in primates; GP_i) and the substantia nigra pars reticulata (SNr) are GABAergic (Smith and Bolam, 1990; Smith et al., 1998; Bolam et al., 2000). Within the striatum, spiny projection neurons are GABAergic (Smith and Bolam, 1990; Smith et al., 1998) as are populations of GABAergic interneurons (Bolam et al., 1983, 1985) that express different calcium-binding proteins (Cowan et al., 1990; Kita et al., 1990; Kubota et al., 1993; Clarke and Bolam, 1997) or synthesize nitric oxide (Kubota et al., 1993). Thus, by virtue of the high degree of interconnectivity of basal ganglia structures, each region receives abundant GABAergic afferents (Smith et al., 1998) (see Chapter 1).

In view of the widespread distribution of GABA and its receptors in the basal ganglia and its critical role in basal ganglia function, it is important to characterize the localization and composition of

---

*Corresponding author. E-mail: paul.bolam@pharm.ox.ac.uk

DOI: 10.1016/S0079-6123(06)60013-7

229

GABA receptors in relation to the known synaptic circuitry of the basal ganglia. In this chapter we summarize data on the ultrastructural localization of the two major classes of GABA receptors, ionotropic GABA$_A$ receptors and metabotropic GABA$_B$ receptors, in the basal ganglia. The primary objectives are several-fold. First, to summarize data relating to GABA receptor location at the cellular level. Second, to characterize the subcellular localization of GABA receptors in different nuclei of the basal ganglia. More specifically, to determine their spatial relationship to synaptic specializations and to identify the nature of the axon terminals that are associated with GABA receptors. Finally, to summarize what is known about the co-expression of different receptor subunits at individual synapses.

Technical issues

Many approaches are available to define the localization of neurotransmitter receptors at the regional, cellular and subcellular levels, including ligand binding, in situ hybridization, various molecular techniques and immunocytochemistry. The latter is the method of choice for the precise localization of receptors. Of the different immunocytochemical techniques that are available each has its relative advantages and disadvantages.

- Fluorescence immunolabelling has a high resolution, particularly when used in combination with confocal microscopy, and has the advantage of the ease and rapidity with which the labelling can be carried out. In addition, because of the large diversity of fluorescent markers available, multiple immunolabelling can be carried out. The major disadvantage is the impermanence of the fluorescent marker and the inability to take the material directly to the electron microscopic level.
- Immunoperoxidase labelling is a highly sensitive technique that has a resolution at the light microscopic level similar to that of fluorescence labelling, but is slightly more complicated to perform. However, the reaction products are permanent and suitable for

electron microscopic analysis. Furthermore, multiple labelling can be performed using different chromogens. The major disadvantage of the peroxidase method for the localization of antigens at the electron microscopic level is that, although the resolution is high, the peroxidase reaction endproduct diffuses from the site of synthesis until it adheres to a membrane or other subcellular structure. Thus, although an antigen can be localized at the level of a subcellular structure, e.g., spine, dendrite or axon, immunoperoxidase labelling is not suitable for the localization of a receptor, or indeed any antigen, at the subcellular or subsynaptic level.

- The issue of diffusion of the reaction product is overcome by the use of particulate markers, the most commonly used being colloidal gold conjugated to the secondary antibody, i.e., immunogold labelling. The main advantages of the immunogold labelling techniques are their excellent spatial resolution and the ease of quantification and, as such, are the techniques of choice to precisely localize an antigen at the subcellular level.
- Immunogold labelling can be performed in two ways, *pre-embedding* and *post-embedding*, which refers to whether the immunolabelling is performed before or after, respectively, the tissue is embedded in a resin suitable for electron microscopy. Pre-embedding immunogold labelling is a highly sensitive technique but suffers from the fact that very small gold particles (about 1 nm in diameter) need to be used to enable the reagent to penetrate into the tissue. These small gold particles are difficult to resolve in the electron microscope and so require intensification by reaction with silver salts. The resulting silver-gold particles are clearly visible in the electron microscope but their increased size (and variability in size) results in some loss of spatial resolution (Figs. 1C–G; 2A, C–F; 4D–G). Another limitation of the pre-embedding method is the apparent restricted access of the immunogold reagents into synapses, thus false-negative labelling at synaptic sites is common (Baude et al., 1993; Bernard et al., 1997; Kulik et al., 2002).

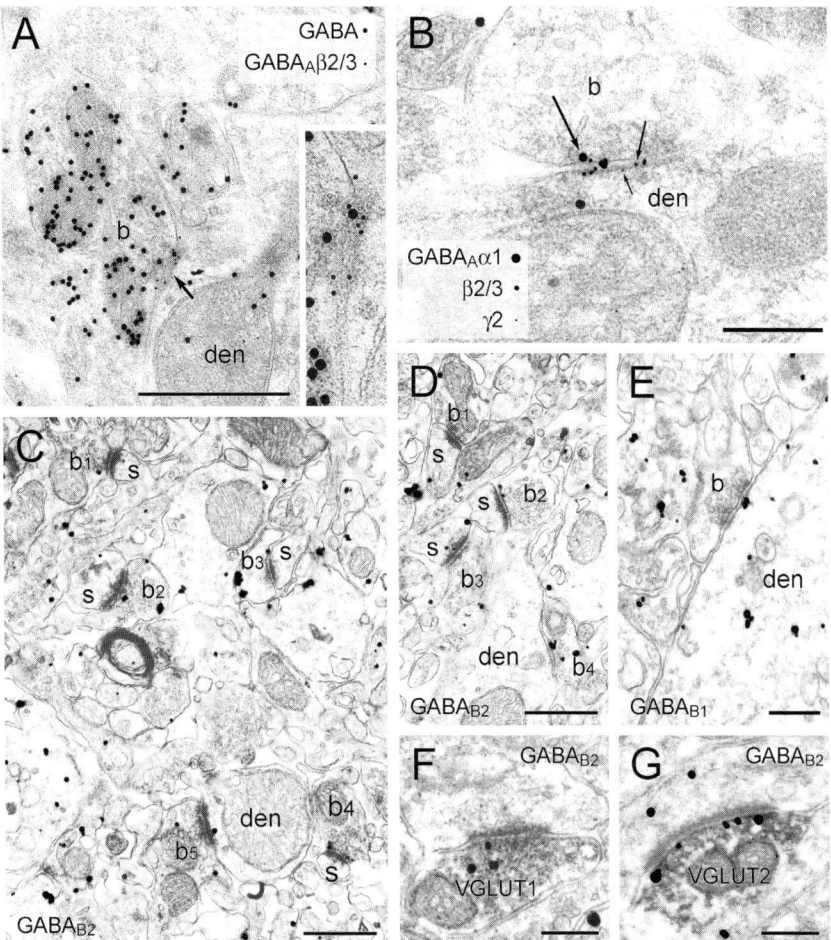

Fig. 1. Localization of GABA receptors in the striatum. (A) Post-embedding immunolabelling for GABA (20 nm gold particles) and the β2/3 subunits of the GABA$_A$ receptor (10 nm gold particles). The strongly GABA-positive bouton (*b*) forms a synapse (*arrow*) with the dendrite (*den*) of a presumed medium spiny neuron. The synapse is positive for the β2/3 subunits of the GABA$_A$ receptor as revealed by the smaller gold particles (*insert*). (B) Co-localization of GABA$_A$ receptor subunits at a synapse between a bouton (*b*) and a dendrite (*den*). The section was immunolabelled for three subunits of the GABA$_A$ receptor using different-sized gold particles. All three subunits, α1 (20 nm gold particles; *large arrow*), β2/3 (10 nm gold particles; *medium arrow*) and γ2 (5 nm gold particles; *small arrow*), are localized at the synapse. (C) Pre-embedding immunolabelling for GABA$_{B2}$ in the striatum. Boutons labelled for GABA$_{B2}$ form asymmetric synapses with spines (*s*) (*b1–b4*) and a dendrite (*den*) (*b5*). Gold particles in labelled boutons are associated with the presynaptic specialization, extrasynaptic sites on the membrane and intracellular sites. Immunolabelling is also associated with postsynaptic structures. (D) Pre- and postsynaptic immunolabelling for GABA$_{B2}$ at axospinous synapses. Gold particles are associated with the extrasynaptic membrane (*b1*), the presynaptic membrane (*b2*) or the postsynaptic spine (*b3*). Bouton, *b4*, is in symmetrical synaptic contact with a dendrite (*den*) and has GABA$_{B2}$ immunolabelling associated with the presynaptic membrane. (E) Immunolabelling for GABA$_{B1}$ on the postsynaptic membrane at a symmetrical synapse between a bouton (*b*) and a dendrite (*den*). (F) and (G) Presynaptic GABA$_{B2}$ labelling of glutamatergic boutons identified by immunoreactivity for the vesicular glutamate transporters, VGLUT1 and VGLUT2 (immunoperoxidase). The VGLUT1-positive bouton is probably derived from the cortex and the VGLUT2-positive bouton from the thalamus. In both cases, gold particles are associated with the presynaptic membrane. Scale bars: (A) 0.5 μm; (B) 0.2 μm; (C)–(G), 0.25 μm. [*Source*: Panels (A) and (B) were modified from Fujiyama et al. (2000) and (C)–(G) were modified from Lacey et al. (2005).]

Table 1. Summary of in situ hybridization studies of GABA receptors subunits in different regions of the rat basal ganglia

	GABA$_A$										GABA$_B$	
	α1	α2	α3	α4	β1	β2	β3	γ1	γ2	γ3	R1	R2
Striatum	(+)	+ + +	+	+ +	(+)	(+)	+ +	(+)	+	+	+ +	0
GP	+ + +	+ +	+	0	(+)	+ + +	+	+ +	+	0	+	+
EP	+ + +	(+)	0	0	0	+ + +	0	0	+	0	+	+
STN	+ + +	0	0	0	0	+ +	(+)	0	+ +	(+)	+ +	+ +
SNr	+ + +	(+)	+	0	0	+ +	(+)	(+)	+	+	+	+
SNc	(+)	0	+	+	(+)	(+)	+	(+)	+	(+)	+ + +	+ +

Note: 0 (not detectable) — + + + (very strong signal).
Source: Data for GABA$_A$ receptor subunits derived from Chadha et al. (2000) and Wisden et al. (1992); data for GABA$_B$ receptor subunits derived from Bischoff et al. (1999), Durkin et al. (1999), Johnston and Duty (2003), Liang et al. (2000) and Lu et al. (1999).

Post-embedding immunolabelling (i.e., that performed on electron microscope sections) overcomes the limitations of the pre-embedding method. The tissue sections, in effect, have no depth, so problems of penetration of reagents are overcome. As a result larger-sized gold particles that do not require silver intensification can be used and the superb spatial resolution is maintained. In addition, the problem of access of the reagents into synapses is obviated. Furthermore, multiple immunolabelling can be achieved using different-sized gold particles (Figs. 1A, B; 2B; 3; 4A–C; 5). A major limitation of the post-embedding technique is that antigenicity for proteins and peptides is generally not well maintained on conventionally prepared electron microscope sections (i.e., osmium-treated and epoxy resin-embedded). As a consequence, it is necessary to use freeze-substituted, low temperature-embedded, non-osmium treated tissue (Figs. 1A, B; 2B; 3; 4A–C; 5). However, not all antigens are maintained and ultrastructural preservation is not optimal.

- It should be noted that recent studies have used freeze-fracture replica immunolabelling to analyse quantitatively the distribution of membrane proteins, including receptors, with superb resolution and high sensitivity (Hagiwara et al., 2005; Kulik et al., 2006).

In reality, to define the localization of a particular receptor at regional, cellular and subcellular levels requires the application of multiple immunocytochemical techniques (e.g., see Baude et al., 1993; Bernard et al., 1997). Furthermore data derived from these multiple immunolabelling studies at the cellular level should be validated by comparison to data derived from in situ hybridization studies (see Table 1). However the data discussed in this chapter will be primarily from tissue labelled by the pre-embedding immunogold method and post-embedding immunolabelling on freeze-substituted, Lowicryl-embedded tissue.

Localization of GABA receptors in the striatum

GABA$_A$ receptors

Given the cellular heterogeneity that exists in the striatum, it is unsurprising that the expression of GABA$_A$ receptor subunits is also markedly heterogeneous (Table 2). Immunolabelling for the α1 subunit is confined to GABAergic interneurons that express parvalbumin or calretinin, whereas cholinergic interneurons express the α3 subunit (Caruncho et al., 1996; Waldvogel et al., 1997, 1998, 1999, 2004). The major cell type in the striatum, the medium spiny projection neurons, display immunoreactivity for both α2 and α3 subunits (Waldvogel et al., 1999). The distribution of the β2/3 and γ2 subunits is considerably more homogeneous, likely reflecting their association with GABA$_A$ receptors containing different α subunits.

Table 2. Summary of immunohistochemical studies of GABA receptor subunits in different regions of the rat basal ganglia

	GABA$_A$												GABA$_B$	
	α1	α2	α3	α4	α5	β1	β2	β3	γ1	γ2	γ3	δ	R1	R2
Striatum	+	+ + +	+	+ +	+	+	+	+ + +	(+)	+ +	(+)	+	+ +	+ +
GP	+ + +	+	+	(+)	(+)	0	+ + +	(+)	+ +	+ +	(+)	0	+	+
EP	+ + +	+	+	0	0	(+)	+ + +	(+)	+	+ + +	0	0	+	−
STN	+ +	+	+	0	(+)	0	+	+	0	+ +	(+)	(+)	+ +	−
SNr	+ + +	0	(+)	0	(+)	(+)	+ +	0	+	+ + +	+	0	+	+
SNc	+	(+)	+ +	(+)	+	(+)	+	0	+	+ +	+ +	+	+ + +	+ +

Note: 0 (not detectable) — + + + (very strong staining);—not reported.
Source: Data for GABA$_A$ receptor subunits derived from Fritschy and Mohler (1995), Pirker et al. (2000) and Schwarzer et al. (2001); data for GABA$_B$ receptor subunits derived from Margeta-Mitrovic et al. (1999) and Charles et al. (2001).

Data on the subcellular distribution of GABA$_A$ receptors is restricted to the α1, β2/3 and γ2 subunits. At the ultrastructural level, immunogold particles coding for these subunits are widely distributed in the striatum when analysed using the post-embedding method (Figs. 1A, B), with the most prominent labelling being at symmetrical synapses. Although it is difficult to unequivocally ascribe the labelling to the presynaptic or postsynaptic membrane in this type of preparation, the weight of evidence from in situ hybridization and pre-embedding labelling in the striatum and other regions of the basal ganglia is that GABA$_A$ receptors are associated primarily, if not exclusively, with postsynaptic structures. The majority of GABA$_A$ receptor-positive synapses in the striatum express immunolabelling for the β2/3 subunit, with much smaller proportions positive for α1 and γ2 subunits. This reflects the ubiquitous nature of the β subunits in striatal GABA$_A$ receptors, but may also relate to the quality of the primary antibodies used in the experiments. Overall, the receptor immunolabelling is closely associated with membranes and also with a variety of intracellular organelles including saccules of endoplasmic reticulum and the nuclear envelope. This intracellular labelling presumably represents receptors undergoing synthesis, trafficking to and from the plasma membrane or degradation. In β2/3-immunolabelled tissue, more than half of the gold particles associated with the membrane are found at symmetrical synapses (Figs. 1A, B). Based on assumptions about the proportion of striatal

neuronal membranes that is occupied by symmetrical synapses, these figures suggest that there is a 220- to 440-fold enrichment of GABA$_A$ receptor immunolabelling at synapses compared to extrasynaptic sites (Fujiyama et al., 2000). β2/3-immunopositive synapses are located on dendritic shafts (of both presumed spiny projection neurons and aspiny interneurons), spines and perikarya. At least two populations of terminals give rise to the β2/3 subunit-positive synapses, those that express high level of GABA (about 60%) and those that express low or undetectable levels of GABA (about 40%). The former have been proposed to be the terminals of GABA interneurons or terminals derived from the GPe and the latter have been proposed to be the local terminals of striatal spiny projection neurons (Fujiyama et al., 2000). Co-localization of α1, β2/3 and γ2 subunits is observed at individual symmetrical synapses in the striatum (Fig. 1B). Co-localization for all three subunits occurs at 12% of the synapses, co-localization of α1 and β2/3 subunits at 22% and co-localization of β2/3 and γ2 subunits at 14% (Fujiyama et al., 2000). These figures again reflect the relative expression of the different subunits in the striatum but may also be influenced by the quality of the antibodies used.

GABA$_B$ receptors

The pattern of immunoreactivity for GABA$_B$ receptor subunits has been described at the light microscopic level in rats, monkeys and humans, revealing little difference between species. Immunoreactivity for $GABA_{B1}$ and $GABA_{B2}$ is relatively homogeneous throughout the striatum, with the vast majority of striatal neurons expressing both subunits (Charara et al., 2000, 2004; Charles et al., 2001; Waldvogel et al., 2004). Double-labelling studies have demonstrated that labelling for $GABA_B$ receptor subunits is associated with all the major subpopulations of striatal neurons (Yung et al., 1999; Ng and Yung, 2001a; Waldvogel et al., 2004). Medium spiny neurons typically show the weakest labelling for $GABA_{B1}$, with the strongest labelling being found in the large cholinergic interneurons (Waldvogel et al., 2004).

The ultrastructural localization of $GABA_B$ receptor subunits in the striatum has been analysed in monkeys using the immunoperoxidase method (Charara et al., 2000, 2004) and in rats using the pre-embedding immunogold method (Lacey et al., 2005). In contrast to $GABA_A$ receptors, immunolabelling for $GABA_{B1}$ and $GABA_{B2}$ is associated with *both* presynaptic and postsynaptic structures (Figs. 1C–G). Thus, in addition to perikarya, dendrites and spines, labelling for both subunits is detected in unmyelinated axons and axon terminals forming symmetrical or asymmetrical synapses. Although quantitative analyses reveal that a larger proportion of the immunogold labelling for $GABA_{B1}$ and $GABA_{B2}$ is associated with dendrites and spines than with presynaptic structures, labelling in axons and axon terminals is more frequently located on the plasma membrane. Thus, in presynaptic structures, 30–40% of gold particles are associated with the membrane, compared to about 20% in postsynaptic structures (Lacey et al., 2005), suggesting that a larger proportion of the presynaptic labelling represents functional receptors.

Immunolabelling for $GABA_{B1}$ and $GABA_{B2}$ is particularly prominent in axon terminals forming asymmetrical synapses. In labelled terminals, immunogold particles associated with the membrane are localized at extrasynaptic sites and at the presynaptic active zone (Figs. 1C, D). These $GABA_B$ receptor-positive terminals have been identified as glutamatergic terminals derived from the cortex or the thalamus on the basis of their expression of the vesicular glutamate transporters VGLUT1 and VGLUT2, respectively (Fujiyama et al., 2004) (Figs. 1F, G). The presence of $GABA_B$ heteroreceptors on corticostriatal terminals is consistent with the inhibitory effects of $GABA_B$ receptor activation on corticostriatal glutamatergic transmission (Calabresi et al., 1990, 1991; Nisenbaum et al., 1992, 1993) and the reduced $GABA_B$ binding following lesions of the corticostriatal pathway (Moratalla and Bowery, 1991). The postsynaptic targets of the $GABA_B$-immunopositive terminals forming asymmetrical synapses are mostly dendritic spines of medium spiny neurons (Figs. 1C, D) and, occasionally, dendritic shafts, which are likely to belong to both medium spiny neurons and interneurons (Lacey et al., 2005) (Fig. 1C).

Labelling for $GABA_{B1}$ and $GABA_{B2}$ is also seen in terminals forming symmetrical synapses, primarily with dendritic shafts (Fig. 1D). As with the terminals forming asymmetrical synapses, immunogold particles are localized at extrasynaptic sites on the membrane and at the presynaptic active zone (Fig. 1D). There are several possible origins of these terminals, including the local collaterals of spiny projection neurons and GABA-ergic striatal interneurons. An interesting possibility is that these terminals originate from the axons of dopaminergic nigrostriatal neurons, which give rise to symmetrical synaptic contacts within the striatum. In this way, GABA could directly modulate the release of dopamine from nigrostriatal terminals. However, further studies are required to establish the precise nature of these terminals. Immunolabelling for $GABA_{B1}$ and $GABA_{B2}$ is also found at the postsynaptic membrane specialization at symmetrical synapses (Fig. 1C), where the receptors are presumably co-localized with $GABA_A$ receptors.

In summary, presynaptic $GABA_B$ heteroreceptors are in a position to modulate the two major excitatory inputs to striatal spiny projection neurons arising in the cortex and thalamus. In addition, presynaptic $GABA_B$ autoreceptors are present on the terminals of spiny projection neurons and/or striatal GABAergic interneurons. Furthermore, it is likely that GABA may also affect the excitability of striatal neurons through postsynaptic $GABA_B$ receptors.

Localization GABA receptors in the globus pallidus (internal and external)

GABA$_A$ receptors

A similar pattern of expression for GABA$_A$ receptor subunits is seen in the GPe and the GPi/EP at the light microscopic level. Neurons and dendritic processes in both regions label strongly for the α1, β2/3 and γ2 subunits (Fritschy and Mohler, 1995; Waldvogel et al., 1998, 1999; Schwarzer et al., 2001). In humans, these include neurons that express either or both of the calcium-binding proteins parvalbumin and calretinin (Waldvogel et al., 1999, 2004). Immunostaining for the α3 subunit has been observed in all species examined. In contrast to the striatum, however, there is little or no expression of the α2 subunit in either pallidal region (see Table 2).

Consistent with observations in other regions of the basal ganglia, immunolabelling for GABA$_A$ receptor subunits is largely localized on membranes and is concentrated at symmetrical synapses (Figs. 2A, B; 3) (Somogyi et al., 1996; Fujiyama et al., 2003; Charara et al., 2005). In the rat, the majority of terminals forming β2/3 subunit-positive synapses in the GP and EP are immunoreactive for GABA (Fig. 3) (Fujiyama et al., 2003). Many of these have morphological characteristics that are consistent with terminals derived from the striatum. Additionally, a smaller proportion has the characteristics of terminals derived from neurons of the GP (see Smith et al., 1998). Multiple labelling studies have revealed that the α1, β2/3 and γ2 subunits co-localize at symmetrical synapses in both regions (Figs. 2B and 3B) (Somogyi et al., 1996; Fujiyama et al., 2003).

GABA$_B$ receptors

Immunolabelling for GABA$_B$ receptor subunits in the GPe is similar in both rats and primates. Labelling for both GABA$_{B1}$ and GABA$_{B2}$ is associated with cell bodies and proximal dendrites of virtually all neurons, as well as the surrounding neuropil, with the intensity of cellular immunolabelling stronger for GABA$_{B1}$ (Charara et al., 2000, 2004, 2005; Chen et al., 2004).

The subcellular distributions of GABA$_B$ receptor subunits have been characterized by pre-embedding immunogold labelling in the GP in rats (Chen et al., 2004) and the GPe/GPi in monkeys (Charara et al., 2005). Immunolabelling for GABA$_{B1}$ and GABA$_{B2}$ is found in both presynaptic and postsynaptic elements throughout the GP (Figs. 2C–F). In the rat, labelling for both subunits is found in terminals forming symmetrical, putative GABAergic synapses, including striatopallidal terminals (Chen et al., 2004) (Figs. 2F, G). Gold particles are localized at extrasynaptic sites, as well as the presynaptic active zone. In addition, labelling is also associated with the presynaptic membrane (Fig. 2D) in terminals forming asymmetrical synapses, the majority of which express VGLUT2 and are thus likely to be derived from the subthalamic nucleus (STN) (Chen et al., 2004). Immunolabelling for GABA$_{B1}$ and GABA$_{B2}$ in GP dendrites is mainly at intracellular sites or at extrasynaptic sites on the plasma membrane (Chen et al., 2004; Charara et al., 2005), however, labelling for both subunits is also observed at the postsynaptic specialization at symmetrical synapses, including striatopallidal synapses, and at the edge of the postsynaptic density at glutamatergic synapses (Figs. 2D–F).

Localization GABA receptors in the substantia nigra

GABA$_A$ receptors

The pattern of expression of GABA$_A$ receptor subunits in the SNr resembles that seen in the GP, with strong expression of α1, β2/3 and γ2 subunits. In contrast to all other regions of the basal ganglia, neurons in the substantia nigra pars compacta (SNc), presumed to be dopaminergic, display low immunoreactivity for both α1 and α2 subunits, but label relatively strongly for the α3 subunit. Other GABA$_A$ subunits expressed in the SNc include α4, β2, γ2 and γ3 (Fritschy and Mohler, 1995; Nicholson et al., 1996; Pirker et al., 2000; Schwarzer et al., 2001) (see Table 2).

As is the case in the striatum and GP, immunolabelling for GABA$_A$ receptor subunits (α1, β2/3

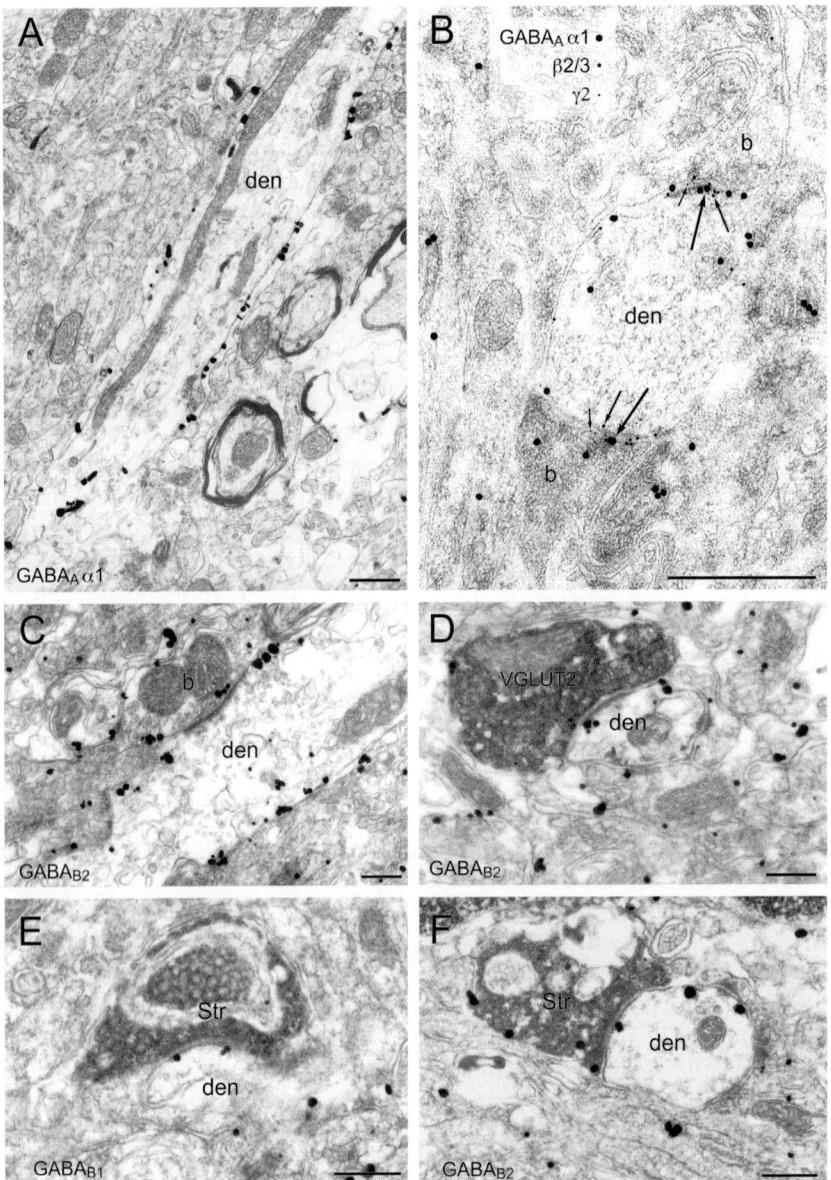

Fig. 2. Localization of GABA receptors in the globus pallidus (GP$_e$). (A) Pre-embedding immunolabelling for the α1 subunit of the GABA$_A$ receptor. Note that most of the labelling is associated with the external surface of the dendritic (*den*) membrane as the antibody was raised against an extracellular epitope of the subunit. (B) Two boutons (*b*) forming GABA$_A$ receptor-positive synapses with a dendritic shaft (*den*). Both synapses are immunopositive for the α1 (20 nm gold, *large arrows*), β2/3 (10 nm gold, *medium arrows*) and the γ2 subunits (5 nm gold, *small arrows*). (C) Pre-embedding immunogold labelling for GABA$_{B2}$. A large proportion of the labelling is associated with membranes, some is associated with the presynaptic membrane of a bouton (*b*) forming an asymmetrical synapse with a dendrite (*den*). (D) GABA$_{B2}$ immunolabelling at the edge of the postsynaptic specialization of a glutamatergic bouton (VGLUT2-positive, immunoperoxidase) forming an asymmetrical synapse. (E) and (F) Anterogradely labelled striatal boutons (*Str*) forming symmetrical synapses with dendrites (*den*). In E, the postsynaptic membrane is immunopositive for GABA$_{B1}$ and in F, both the postsynaptic membrane and the striatal bouton are labelled for GABA$_{B2}$. Scale bars: 0.25 μm. [*Source*: Panel (A), Boyes, Bolam and Stanford (*unpublished*), panel (B), Fujiyama and Bolam (*unpublished*) and panels (C)–(F) were modified from Chen et al. (2004).]

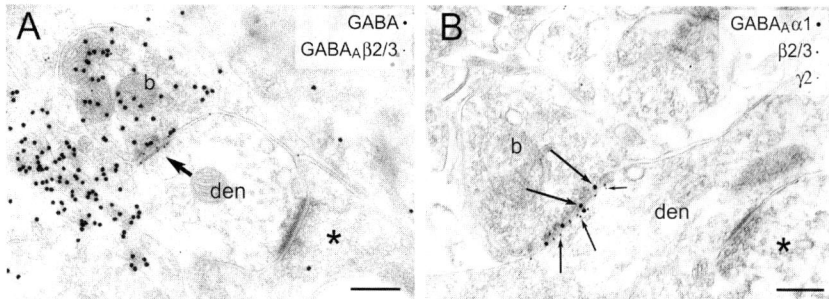

Fig. 3. Localization of GABA$_A$ receptors in the entopeduncular nucleus. (A) Post-embedding immunolabelling for GABA (20 nm gold particles) and the β2/3 subunits of the GABA$_A$ receptor (10 nm gold particles). A bouton (*b*) that is strongly immunopositive for GABA is in symmetrical synaptic contact with a dendrite (*den*). The synapse is positive for the β2/3 subunits (*arrow*). Note the second bouton (*asterisk*) that is negative for GABA and forms an asymmetrical synapse that is unlabelled. (B) GABA$_A$ receptor-positive synapse formed between a bouton (*b*) and a dendrite (*den*). This section was immunolabelled by the post-embedding method to reveal different subunits of the GABA$_A$ receptor using different-sized gold particles as markers. All three subunits, α1 (20 nm gold particles; *large arrows*), β2/3 (10 nm gold particles; *medium arrows*) and γ2 (5 nm gold particles; *small arrow*) are localized at the synapse. Note the second bouton (*asterisk*) that forms an asymmetric synapse that is unlabelled. Scale bars: 0.25 μm. Data from Fujiyama and Bolam, unpublished.

and γ2) in the substantia nigra (SN) is concentrated at symmetrical synapses (Figs. 4A–C) (Fujiyama et al., 2002) and terminals establishing GABA$_A$ subunit-positive synapses are immunoreactive for GABA (Fig. 4A). There are at least three sources of GABAergic terminals in the SN, the striatum, the GP and the local collaterals of SNr output neurons. In both the SNc and SNr, the morphological features of terminals forming receptor-positive synapses are consistent with those derived from the striatum (b2 in Fig. 4A) and the GP (b1 in Fig. 4A) (see Smith et al., 1998; Fujiyama et al., 2002). Direct evidence that β2/3 subunit-positive synapses are formed by terminals derived from the striatum comes from labelling of striatal terminals (by anterograde degeneration) combined with the receptor immunolabelling (Fig. 4C). Multiple immunolabelling has revealed that at least some of symmetrical synapses in both SNc and SNr co-express the α1, β2/3 and γ2 subunits of the GABA$_A$ receptor (Fig. 4B). Some of the terminals forming synapses positive for all three subunits have been identified as originating in the striatum by anterograde degeneration (Fujiyama et al., 2002). It is interesting to note that following quinolinic acid lesions of the striatum in the rat, which is a model of Huntington's disease, there is an up-regulation of immunolabelling for GABA$_A$ β2/3 subunits (but not glutamate receptors) at synapses in the SN (Fujiyama et al., 2002).

GABA$_B$ receptors

Immunocytochemical studies in rats and primates indicate that GABA$_{B1}$ and GABA$_{B2}$ are expressed by neurons in both the SNc and SNr, but with a marked difference between the two regions (Table 2). Thus, presumed dopaminergic neurons in the SNc are strongly immunoreactive for GABA$_{B1}$ and, to a lesser extent, GABA$_{B2}$. In contrast, cells in the SNr display only light staining for GABA$_{B1}$ and GABA$_{B2}$, with immunoreactivity mostly restricted to the neuropil (Margeta-Mitrovic et al., 1999; Charara et al., 2000, 2004; Ng and Yung, 2000, 2001b; Charles et al., 2001; Boyes and Bolam, 2003).

The subcellular distributions of GABA$_B$ receptor subunits have been described in monkeys using the immunoperoxidase method (Charara et al., 2000, 2004) and in rats using the pre-embedding immunogold method (Boyes and Bolam, 2003). The patterns of immunolabelling for GABA$_{B1}$ and GABA$_{B2}$ in the SN closely resemble those seen in the GP, with both subunits localized in presynaptic and postsynaptic elements, and most labelling at intracellular sites or associated with extrasynaptic sites on the plasma membrane. In addition, labelling for both subunits is associated with the pre- and postsynaptic membranes of symmetrical, putative GABAergic synapses, including those formed by striatonigral and pallidonigral terminals (Figs. 4F, G). Labelling is also localized on the presynaptic

238

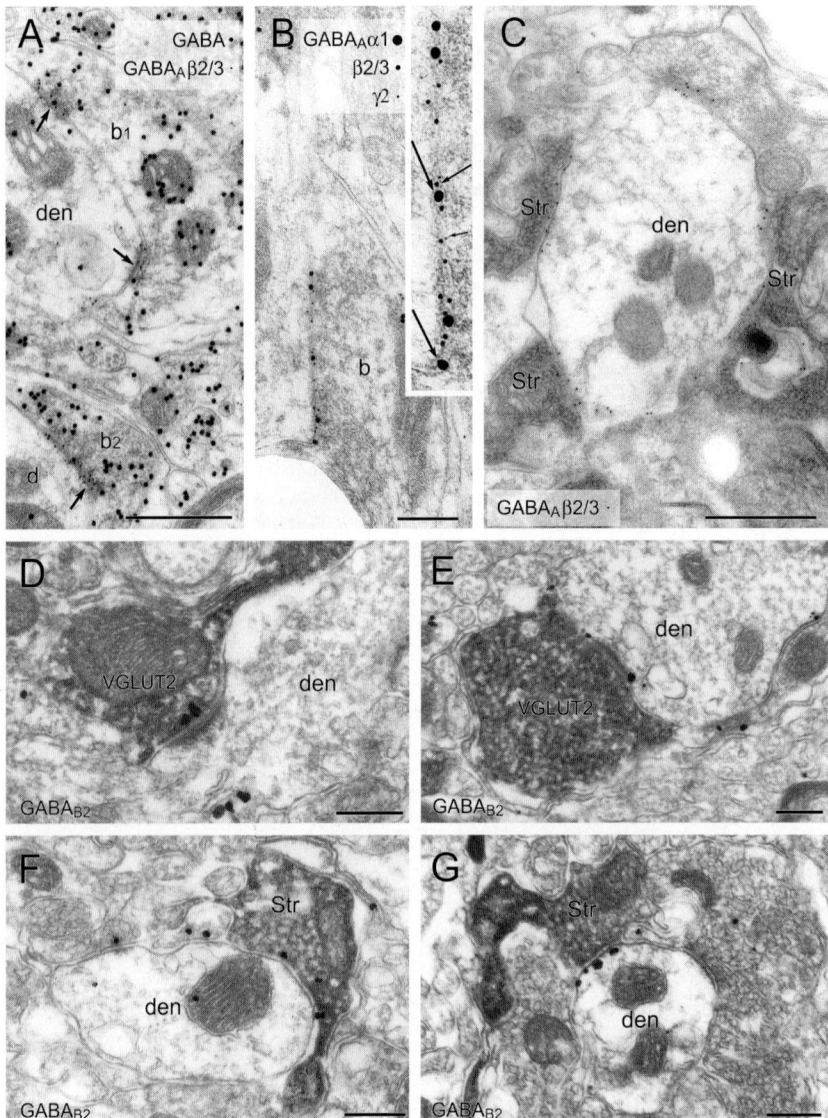

Fig. 4. Localization of GABA receptors in the substantia nigra. (A)–(C) Post-embedding immunolabelling for GABA$_A$ receptor subunits. (A) Section of the SNc immunolabelled to reveal both GABA (20 nm gold particles) and the β2/3 subunit of the GABA$_A$ receptor (10 nm gold particles). Two boutons, b1 and b2, are GABA-positive and form GABA$_A$ receptor-positive synapses (*arrows*). Bouton b1, has features in common terminals derived from the globus pallidus (GP) and b2 has features in common with terminals derived from the striatum. (B) A bouton (*b*) in the SNr forming an axodendritic synapse that is immunopositive (see inset) for the α1 (20 nm gold, *large arrow*), β2/3 (10 nm gold, *medium arrow*) and γ2 (5 nm gold, *small arrow*) subunits of the GABA$_A$ receptor. (C) Degenerating boutons derived from the striatum (*Str*) forming GABA$_A$ β2/3 subunit-positive synapses with a dendrite (*den*) in the SNr. (D) and (E) Pre-embedding GABA$_{B2}$ immunolabelling on the presynaptic membrane (D) or at the edge of the postsynaptic density (E) of glutamatergic synapses (identified by immunoperoxidase labelling for VGLUT2) in the SNc. (F) and (G) Pre-embedding GABA$_{B2}$ immunolabelling on the presynaptic (F) or postsynaptic (G) membrane of anterogradely labelled (peroxidase labelling) striatal terminals (*Str*)–in the SNr. Scale bars: (A)–(C) 0.5 μm; (D)–(G), 0.25 μm. [*Source*: Panels (A)–(C) were modified from Fujiyama et al. (2002) and (D)–(G) were modified from Boyes and Bolam (2003).]

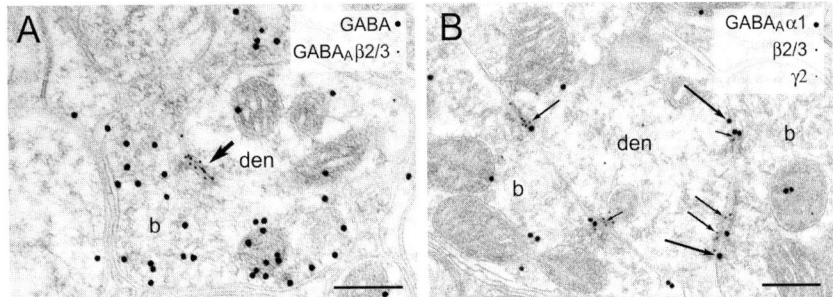

Fig. 5. Localization of GABA$_A$ receptors in the subthalamic nucleus. (A) Post-embedding immunolabelling for the GABA$_A$ β2/3 subunits (10 nm gold particles) at a synapse (*arrow*) between a GABA-positive bouton (20 nm gold particles) (*b*) and a dendrite (*den*). (B) Two boutons (*b*) forming GABA$_A$ receptor-positive synapses with a dendrite (d). The boutons have the typical appearance of those derived from the globus pallidus (see Smith et al., 1998) and form synapses that are positive for α1 (20 nm gold particles, *large arrows*), β2/3 (10 nm gold particles, *medium arrows*) and γ2 (5 nm gold particles, *small arrows*). Note that the immunolabelling is concentrated at the active zones of synapses. Note also the non-specific labelling associated with mitochondria. Scale bars: 0.25 μm. Data from Fujiyama and Bolam, *unpublished*.

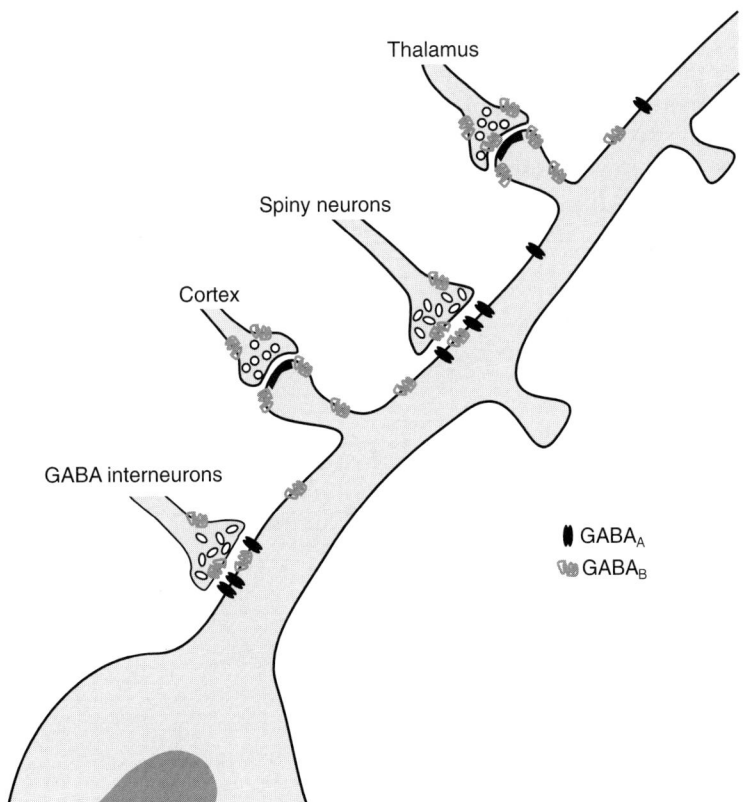

Fig. 6. Summary of the distribution of GABA$_A$ and GABA$_B$ receptors in spiny striatal neurons. GABA$_A$ receptors are primarily associated with synapses on the dendritic shafts formed by GABA-positive terminals of GABA interneurons, the local collaterals of spiny neurons and possibly, terminals from the globus pallidus. Extrasynaptic GABA$_A$ receptors are present but their density is much lower than at synapses. GABA$_B$ receptors, in contrast, are located at both presynaptic and postsynaptic sites and are associated with both presumed GABAergic terminals of GABA interneurons and local collaterals of spiny neurons and with glutamatergic terminals from the cortex and thalamus. A higher proportion of GABA$_B$ receptors are located at extrasynaptic sites.

240

membrane (Fig. 4D) and at the edge of the postsynaptic density of asymmetrical (Fig. 4E) synapses identified as glutamatergic by immunolabelling for VGLUT2. These terminals are most likely derived from the STN. Consistent with light microscopic observations, strong labelling for $GABA_{B1}$ is found in nigrostriatal, presumed dopaminergic, neurons in the SNc (Boyes and Bolam, 2003).

The widespread distribution of $GABA_B$ subunits in the SNc and SNr suggests that $GABA_B$-mediated effects in these regions are likely to be complex, involving presynaptic auto- and heteroreceptors and postsynaptic receptors on different populations of SN neurons.

Localization GABA receptors in the subthalamic nucleus

The localization of both $GABA_A$ and $GABA_B$ receptors in the STN is the least extensively studied in the basal ganglia (Charara et al., 2000; Fujiyama et al., 2003; Galvan et al., 2004). The overall findings are essentially similar to those in other regions of the basal ganglia. Thus, $GABA_A$ receptor subunits are selectively enriched at symmetrical

synapses (Fig. 5) formed by GABA-positive terminals (Fig. 5A) that are likely to be derived from the GPe (Fujiyama et al., 2003; Galvan et al., 2004). At least some of these synapses co-express the $\alpha1$, $\beta2/3$ and $\gamma2$ subunits (Fig. 5B) (Fujiyama et al., 2003). Immunolabelling for $GABA_B$ receptors is widely expressed in neuronal perikarya and dendrites in the STN and at the ultrastructural level localized at both presynaptic and postsynaptic sites. Particularly strong labelling is associated with a sub-population of terminals forming asymmetrical synapses, presumably derived from the thalamus or cortex where the labelling was detected on the presynaptic and postsynaptic membranes and at perisynaptic sites in the postsynaptic structure (Charara et al., 2000; Galvan et al., 2004).

Summary and conclusions

- In all regions of the basal ganglia, $GABA_A$ receptor subunits are localized at symmetrical synapses formed by GABA-containing terminals. $GABA_A$ receptor subunits are also expressed at extrasynaptic sites, albeit at much lower densities than at synapses (Figs. 6 and 7).

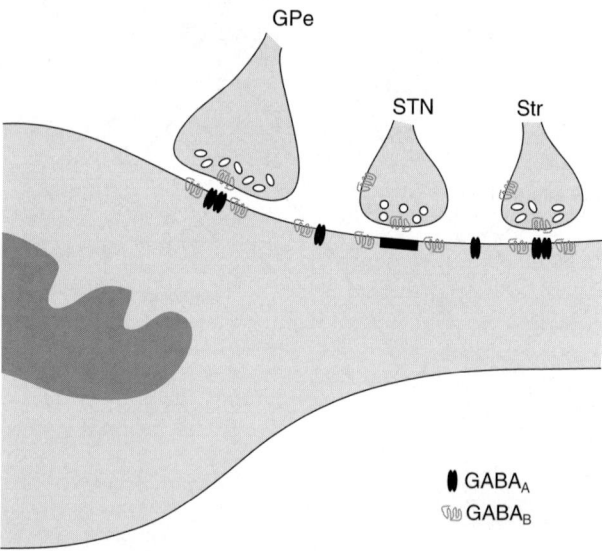

Fig. 7. Summary of the distribution of $GABA_A$ and $GABA_B$ receptors in the pallidal complex and substantia nigra. $GABA_A$ receptors are mainly located on the postsynaptic membrane at synapses formed by striatal and pallidal terminals. Extrasynaptic $GABA_A$ receptors are present but their density is much lower than at synapses. $GABA_B$ receptors, in contrast, are located at both presynaptic and postsynaptic sites and are associated with striatal, pallidal and glutamatergic terminals derived from the STN and/or the thalamus. A higher proportion of $GABA_B$ receptors are located at extrasynaptic sites.

- Synapses expressing the α1, β2/3 and γ2 subunits of the GABA$_A$ receptors are widely distributed in the networks of the basal ganglia.
- GABA$_B$ receptors are more widely distributed than GABA$_A$ receptors, including both presynaptic and postsynaptic sites. GABA$_B$ receptor subunits are mainly expressed at extrasynaptic sites, although a proportion of them are located at symmetrical synapses where they are presumably co-localized with GABA$_A$ receptors (Figs. 6 and 7).
- GABA$_B$ receptors are located presynaptically on striatopallidal, striatonigral and pallidonigral terminals, as well as on glutamatergic terminals derived from the cortex, thalamus and/or STN (Figs. 6 and 7).
- It is concluded that fast GABAergic neurotransmission within the basal ganglia is mediated by synaptic GABA$_A$ receptors, whereas GABA$_B$ receptors mediate more complex and subtle effects of GABA, acting as autoreceptors and heteroreceptors to modulate the release of neurotransmitters, and influencing the excitability of neurons through postsynaptic receptors.

Abbreviations

α1 subunit	alpha 1 subunit of the type A γ-aminobutyric acid receptor
β2/3 subunit	beta 2 and 3 subunits of the type A γ-aminobutyric acid receptor
γ2 subunit	gamma 2 subunit of the type A γ-aminobutyric acid receptor
EP	entopeduncular nucleus
GABA	γ-aminobutyric acid
GABA$_A$	type A γ-aminobutyric acid receptor
GABA$_B$	type B γ-aminobutyric acid receptor
GABA$_{B1}$	Subunit 1 of the type B γ-aminobutyric acid receptor
GABA$_{B2}$	Subunit 2 of the type B γ-aminobutyric acid receptor
GP	globus pallidus
GPi	internal segment of the globus pallidus
GPe	external segment of the globus pallidus
SN	substantia nigra
SNc	substantia nigra *pars compacta*
SNr	substantia nigra *pars reticulata*
STN	subthalamic nucleus
VGLUT1	vesicular glutamate transporter 1
VGLUT2	vesicular glutamate transporter 2

Acknowledgments

The authors' own work described in this manuscript was funded by the Medical Research Council, U.K. Many thanks to Ben Micklem for the preparation of the figures.

References

Baude, A., Nusser, Z., Roberts, J.D.B., Mulvihill, E., McIlhinney, R.A.J. and Somogyi, P. (1993) The metabotropic glutamate receptor (mGluR1α) is concentrated at perisynaptic membrane of neuronal subpopulations as detected by immunogold reaction. Neuron, 11: 771–787.

Bernard, V., Somogyi, P. and Bolam, J.P. (1997) Cellular, subcellular, and subsynaptic distribution of AMPA-type glutamate receptor subunits in the neostriatum of the rat. J. Neurosci, 17: 819–833.

Bischoff, S., Leonhard, S., Reymann, N., Schuler, V., Shigemoto, R., Kaupmann, K. and Bettler, B. (1999) Spatial distribution of GABA$_B$ R1 receptor mRNA and binding sites in the rat brain. J. Comp. Neurol., 412: 1–16.

Bolam, J.P., Booth, P.A.C., Hanley, J.J. and Bevan, M.D. (2000) Synaptic organisation of the basal ganglia. J. Anat., 196: 527–542.

Bolam, J.P., Clarke, D.J., Smith, A.D. and Somogyi, P. (1983) A type of aspiny neuron in the rat neostriatum accumulates (^{3}H) γ aminobutyric acid: combination of Golgi-staining, autoradiography and electron microscopy. J. Comp. Neurol., 213: 121–134.

Bolam, J.P., Powell, J.F., Wu, J.-Y. and Smith, A.D. (1985) Glutamate decarboxylase-immunoreactive structures in the rat neostriatum. A correlated light and electron microscope study including a combination of Golgi-impregnation with immunocytochemistry. J. Comp. Neurol., 237: 1–20.

Boyes, J. and Bolam, J.P. (2003) The subcellular localization of GABA$_B$ receptor subunits in the rat substantia nigra. Eur. J. Neurosci., 18: 3279–3293.

Calabresi, P., Mercuri, N.B., De Murtas, M. and Bernardi, G. (1990) Endogenous GABA mediates presynaptic inhibition of spontaneous and evoked excitatory synaptic potentials in the rat neostriatum. Neurosci. Lett., 118: 99–102.

242

Calabresi, P., Mercuri, N.B., De Murtas, M. and Bernardi, G. (1991) Involvement of GABA systems in feedback regulation of glutamate- and GABA-mediated synaptic potentials in rat neostriatum. J. Physiol., 440: 581–599.

Caruncho, H.J., Liste, I. and Labandiera-Garcia, J.L. (1996) GABA$_A$ receptor α1-subunit-immunopositive neurons in the rat striatum. Brain Res., 722: 185–189.

Chadha, A., Dawson, L.G., Jenner, P.G. and Duty, S. (2000) Effect of unilateral 6-hydroxydopamine lesions of the nigrostriatal pathway on GABA$_A$ receptor subunit gene expression in the rodent basal ganglia and thalamus. Neuroscience, 95: 119–126.

Charara, A., Galvan, A., Kuwajima, M., Hall, R.A. and Smith, Y. (2004) An electron microscope immunocytochemical study of GABA$_B$ R2 receptors in the monkey basal ganglia: a comparative analysis with GABA$_B$ R1 receptor distribution. J. Comp. Neurol., 476: 65–79.

Charara, A., Heilman, C., Levey, A.I. and Smith, Y. (2000) Pre- and postsynaptic localization of GABA$_B$ receptors in the basal ganglia in monkeys. Neuroscience, 95: 127–140.

Charara, A., Pare, J.F., Levey, A.I. and Smith, Y. (2005) Synaptic and extrasynaptic GABA-A and GABA-B receptors in the globus pallidus: an electron microscopic immunogold analysis in monkeys. Neuroscience, 131: 917–933.

Charles, K.J., Evans, M.L., Robbins, M.J., Calver, A.R., Leslie, R.A. and Pangalos, M.N. (2001) Comparative immunohistochemical localisation of GABA$_{B1a}$, GABA$_{B1b}$ and GABA$_{B2}$ subunits in rat brain, spinal cord and dorsal root ganglion. Neuroscience, 106: 447–467.

Chen, L., Boyes, J., Yung, W.H. and Bolam, J.P. (2004) Sub-cellular localization of GABA$_B$ receptor subunits in rat globus pallidus. J. Comp. Neurol., 474: 340–352.

Clarke, N.P. and Bolam, J.P. (1997) Colocalization of neurotransmitters in the basal ganglia of the rat. Brit. J. Pharmacol., 120: 281P.

Cowan, R.L., Wilson, C.J., Emson, P.C. and Heizmann, C.W. (1990) Parvalbumin-containing GABAergic interneurons in the rat neostriatum. J. Comp. Neurol., 302: 197–205.

Durkin, M.M., Gunwaldsen, C.A., Borowsky, B., Jones, K.A. and Branchek, T.A. (1999) An in situ hybridization study of the distribution of the GABA$_{B2}$ protein mRNA in the rat CNS. Brain Res. Mol. Brain Res., 71: 185–200.

Fritschy, J.M. and Mohler, H. (1995) GABA$_A$-receptor heterogeneity in the adult rat brain: differential regional and cellular distribution of seven major subunits. J. Comp. Neurol., 359: 154–194.

Fujiyama, F., Fritschy, J.M., Stephenson, F.A. and Bolam, J.P. (2000) Synaptic localization of GABA$_A$ receptor subunits in the striatum of the rat. J. Comp. Neurol., 416: 158–172.

Fujiyama, F., Fritschy, J.M., Stephenson, F.A. and Bolam, J.P. (2003) Synaptic localization of GABA$_A$ receptor subunits in the basal ganglia of the rat. In: Graybiel A., DeLong M. and Kitai S. (Eds.), Basal Ganglia VI. Plenum Press, New York, pp. 631–640.

Fujiyama, F., Kuramoto, E., Okamoto, K., Hioki, K., Furuta, T., Zhou, L., Nomura, S. and Kaneko, T. (2004) Presynaptic localization of an AMPA-type glutamate receptor in corticostriatal and thalamostriatal axon terminals. Eur. J. Neurosci., 20: 3322–3330.

Fujiyama, F., Stephenson, F.A. and Bolam, J.P. (2002) Synaptic localization of GABA$_A$ receptor subunits in the substantia nigra of the rat: effects of quinolinic acid lesions of the striatum. Eur. J. Neurosci., 15: 1961–1975.

Galvan, A., Charara, A., Pare, J.F., Levey, A.I. and Smith, Y. (2004) Differential subcellular and subsynaptic distribution of GABA$_A$ and GABA$_B$ receptors in the monkey subthalamic nucleus. Neuroscience, 127: 709–721.

Hagiwara, A., Fukazawa, Y., Deguchi-Tawarada, M., Ohtsuka, T. and Shigemoto, R. (2005) Differential distribution of release-related proteins in the hippocampal CA3 area as revealed by freeze-fracture replica labeling. J. Comp. Neurol., 489: 195–216.

Johnston, T. and Duty, S. (2003) Changes in GABA$_B$ receptor mRNA expression in the rodent basal ganglia and thalamus following lesion of the nigrostriatal pathway. Neuroscience, 120: 1027–1035.

Kita, H., Kosaka, T. and Heizmann, C.W. (1990) Parvalbumin-immunoreactive neurons in the rat neostriatum: a light and electron microscopic study. Brain Res., 536: 1–15.

Kubota, Y., Mikawa, S. and Kawaguchi, Y. (1993) Neostriatal GABAergic interneurones contain NOS, calretinin or parvalbumin. Neuroreport, 5: 205–208.

Kulik, A., Nakadate, K., Nyiri, G., Notomi, T., Malitschek, B., Bettler, B. and Shigemoto, R. (2002) Distinct localization of GABA$_B$ receptors relative to synaptic sites in the rat cerebellum and ventrobasal thalamus. Eur. J. Neurosci., 15: 291–307.

Kulik, A., Vida, I., Fukazawa, Y., Guetg, N., Kasugai, Y., Marker, C.L., Rigato, F., Bettler, B., Wickman, K., Frotscher, M. and Shigemoto, R. (2006) Compartment-dependent colocalization of Kir3.2-containing K$^+$ channels and GABA$_B$ receptors in hippocampal pyramidal cells. J. Neurosci., 26: 4289–4297.

Lacey, C.J., Boyes, J., Gerlach, O., Chen, L., Magill, P.J. and Bolam, J.P. (2005) GABA$_B$ receptors at glutamatergic synapses in the rat striatum. Neuroscience, 136: 1083–1095.

Liang, F., Hatanaka, Y., Saito, H., Yamamori, T. and Hashikawa, T. (2000) Differential expression of γ-aminobutyric acid type B receptor-1a and -1b mRNA variants in GABA and non-GABAergic neurons of the rat brain. J. Comp. Neurol., 416: 475–495.

Lu, X.Y., Ghasemzadeh, M.B. and Kalivas, P.W. (1999) Regional distribution and cellular localization of γ-aminobutyric acid subtype 1 receptor mRNA in the rat brain. J. Comp. Neurol., 407: 166–182.

Margeta-Mitrovic, M., Mitrovic, I., Riley, R.C., Jan, L.Y. and Basbaum, A.I. (1999) Immunohistochemical localization of GABA$_B$ receptors in the rat central nervous system. J. Comp. Neurol., 405: 299–321.

Moratalla, R. and Bowery, N.G. (1991) Chronic lesion of corticostriatal fibers reduces GABA$_B$ but not GABA$_A$ binding in rat caudate putamen: an autoradiographic study. Neurochem. Res., 16: 309–315.

Ng, T.K.Y. and Yung, K.K.L. (2000) Distinct cellular distribution of GABA$_B$ R1 and GABA$_A$ α1 receptor immunoreactivity in the rat substantia nigra. Neuroscience, 99: 65–76.

Ng, T.K.Y. and Yung, K.K.L. (2001a) Differential expression of GABA$_B$ R1 and GABA$_B$ R2 receptor immunoreactivity in neurochemically identified neurons of the rat neostriatum. J. Comp. Neurol., 433: 458–470.

Ng, T.K.Y. and Yung, K.K.L. (2001b) Subpopulations of neurons in rat substantia nigra display GABA$_B$ R2 receptor immunoreactivity. Brain Res., 920: 210–216.

Nicholson, L.F.B., Waldvogel, H.J. and Faull, R.L.M. (1996) GABA$_A$ receptor subtype changes in the substantia nigra of the rat following quinolinate lesions in the striatum: a correlative in situ hybridization and immunohistochemical study. Neuroscience, 74: 89–98.

Nisenbaum, E.S., Berger, T.W. and Grace, A.A. (1992) Presynaptic modulation by GABA$_B$ receptors of glutamatergic excitation and GABAergic inhibition of neostriatal neurons. J. Neurophysiol., 67: 477–481.

Nisenbaum, E.S., Berger, T.W. and Grace, A.A. (1993) Depression of glutamatergic and GABAergic synaptic responses in striatal spiny neurons by stimulation of presynaptic GABA$_B$ receptors. Synapse, 14: 221–242.

Pirker, S., Schwarzer, C., Wieselthaler, A., Sieghart, W. and Sperk, G. (2000) GABA$_A$ receptors: immunocytochemical distribution of 13 subunits in the adult rat brain. Neuroscience, 101: 815–850.

Schwarzer, C., Berresheim, U., Pirker, S., Wieselthaler, A., Fuchs, K., Sieghart, W. and Sperk, G. (2001) Distribution of the major γ-aminobutyric acid$_A$ receptor subunits in the basal ganglia and associated limbic brain areas of the adult rat. J. Comp. Neurol., 433: 526–549.

Smith, A.D. and Bolam, J.P. (1990) The neural network of the basal ganglia as revealed by the study of synaptic connections of identified neurones. Trends Neurosci., 13: 259–265.

Smith, Y., Bevan, M.D., Shink, E. and Bolam, J.P. (1998) Microcircuitry of the direct and indirect pathways of the basal ganglia. Neuroscience, 86: 353–387.

Somogyi, P., Fritschy, J.M., Benke, D., Roberts, J.D.B. and Sieghart, W. (1996) The γ_2 subunit of the GABA$_A$ receptor is concentrated in synaptic junctions containing the α_1 and $\beta_{2/3}$ subunits in hippocampus, cerebellum and globus pallidus. Neuropharmacology, 35: 1425–1444.

Waldvogel, H.J., Billinton, A., White, J.H., Emson, P.C. and Faull, R.L. (2004) Comparative cellular distribution of GABA$_A$ and GABA$_B$ receptors in the human basal ganglia: immunohistochemical colocalization of the α_1 subunit of the GABA$_A$ receptor, and the GABA$_B$ R1 and GABA$_B$ R2 receptor subunits. J. Comp. Neurol., 470: 339–356.

Waldvogel, H.J., Fritschy, J.M., Mohler, H. and Faull, R.L.M. (1998) GABA$_A$ receptors in the primate basal ganglia: an autoradiographic and a light and electron microscopic immunohistochemical study of the α_1 and $\beta_{2/3}$ subunits in the baboon brain. J. Comp. Neurol., 397: 297–325.

Waldvogel, H.J., Kubota, Y., Fritschy, J.M., Mohler, H. and Faull, R.L.M. (1999) Regional and cellular localisation of GABA$_A$ receptor subunits in the human basal ganglia: an autoradiographic and immunohistochemical study. J. Comp. Neurol., 415: 313–340.

Waldvogel, H.J., Kubota, Y., Trevallyan, S.C., Kawaguchi, Y., Fritschy, J.M., Mohler, H. and Faull, R.L.M. (1997) The morphological and chemical characteristics of striatal neurons immunoreactive for the α_1-subunit of the GABA$_A$ receptor in the rat. Neuroscience, 80: 775–792.

Wisden, W., Laurie, D.J., Monyer, H. and Seeburg, P.H. (1992) The distribution of 13 GABA$_A$ receptor subunit messenger mRNAs in the rat brain. 1. Telencephalon, diencephalon, mesencephalon. J. Neurosci., 12: 1040–1062.

Yung, K.K.L., Ng, T.K.Y. and Wong, C.K. (1999) Subpopulations of neurons in the rat neostriatum display GABA$_B$ R1 receptor immunoreactivity. Brain Res., 830: 345–352.

Tepper, Abercrombie & Bolam (Eds.)
Progress in Brain Research, Vol. 160
ISSN 0079-6123

CHAPTER 14

Presynaptic modulation of GABA release in the basal ganglia

Ulrich Misgeld*, Geoffrey Drew and Yevgenij Yanovsky

Institute of Physiology and Pathophysiology and Interdisciplinary Center for Neurosciences (IZN), University of Heidelberg, Im Neuenheimer Feld 326, D-69120 Heidelberg, Germany

Abstract: Presynaptic receptors provide plasticity to GABAergic synapses in the basal ganglia network, in which GABA neurons outnumber all other neurons. Presynaptic receptors, mostly of the metabotropic type, enhance or reduce the strength of synaptic inhibition and are activated by ligands being released from the GABA terminals themselves (autoreceptors) or by ligands coming from other sources (heteroreceptors), including the target neurons innervated by the GABA terminals. The latter mechanism, termed retrograde signaling, is given particular emphasis as far as it occurs in substantia nigra.

Keywords: autoreceptors; heteroreceptors; metabotropic receptors; retrograde signaling; feed-back

Introduction

Chains of GABAergic neurons connected in series constitute a characteristic principle of network organization in the basal ganglia. Versatility in the synaptic functioning of these chains is achieved, in part, through short- or long-term changes in neurotransmitter release from presynaptic terminals (i.e., presynaptic plasticity). Presynaptic receptors located at the terminals of inhibitory GABAergic neurons provide a high degree of flexibility to synaptic signaling by increasing or decreasing the strength of coupling between a GABAergic neuron and its target neuron. Furthermore, by regulating distinct intracellular mechanisms within the terminal, presynaptic receptors provide specificity to presynaptic plasticity. The presence of presynaptic receptors is well established for many GABAergic

pathways within the basal ganglia. They are found prominently at the terminals of medium spiny neurons (MSNs), the major GABAergic neuron type projecting from the neostriatum and connecting to most other nuclei of the basal ganglia (see Wilson, Chapter 6 this volume). GABA release from pallidonigral and pallidosubthalamic neurons is also modulated through presynaptic receptors. GABAergic terminals within the basal ganglia express receptor types for all neurotransmitters that are released within these nuclei, including glutamate, dopamine, acetylcholine, serotonin and GABA itself. In addition, neuromodulatory peptides and substances originating from metabolic cascades, e.g., adenosine, have been found to activate presynaptic receptors at GABAergic basal ganglia synapses (Mori et al., 1996; Fassio et al., 2000; Shindou et al., 2001).

Both ionotropic and metabotropic receptors have been implicated in presynaptic regulation of neurotransmitter release in the central nervous system. It is thought that activation of ionotropic

*Corresponding author. Tel.: +49-6221-544558;
Fax: +49-6221-546364; E-mail: ulrich.misgeld@physiologie.uni-heidelberg.de

DOI: 10.1016/S0079-6123(06)60014-9

245

receptors leads to transient changes in transmitter release, whereas metabotropic receptor activation is capable of inducing more enduring changes in transmitter release. Changes in neurotransmitter release following presynaptic G-protein-coupled receptor activation occur through two broad types of mechanisms. These are the modulation of Ca^{2+} and/or K^+ channels and actions directly on the release machinery. One of the most common forms of presynaptic plasticity is inhibition produced via the activation of such G-protein-coupled receptors. Inhibition of voltage-dependent Ca^{2+} channels located on synaptic terminals is the major reason for presynaptic inhibition. K^+ channel activation in the terminal can also contribute to presynaptic inhibition by shunting the action potential, thereby further inhibiting Ca^{2+} influx. Presynaptic inhibition may also occur at sites downstream to Ca^{2+} entry through the modulation of one or more steps in synaptic vesicle cycling. Often a transmitter can depress release both by reducing Ca^{2+} influx and by acting at a site downstream to Ca^{2+} influx (for a review see Malenka and Siegelbaum, 2001).

Presynaptic receptors almost certainly present an important target for drugs used in the treatment of motor disturbances as they occur, for example, in Parkinson's disease. It is more difficult to address the question of their physiological significance at a particular synaptic connection, however, since the circumstances in which they are engaged under physiological conditions in vivo remain poorly understood. Nevertheless, substantial information has been gained from both in vivo and in vitro studies regarding principles of presynaptic plasticity within the central nervous system. It is clear from these studies that the specific role for any given presynaptic receptor is highly dependent upon the structure, history and active state of the synapse. Presynaptic receptor-mediated regulation of neurotransmitter release can be initiated via at least three distinct routes: (1) activation of autoreceptors by synaptically released GABA or other co-released ligands, (2) activation of heteroreceptors, either by spillover of transmitter from neighboring synapses/glia or by volume transmission and (3) retrograde signaling mediated by ligands released from the postsynaptic neuron. In this chapter, we discuss these principles as they relate to GABAergic synapses within the basal ganglia, with special emphasis on retrograde signaling. For this purpose, we focus predominantly on electrophysiological studies in which presynaptic mechanisms can be directly inferred, for example, through analysis of isolated 'miniature' synaptic currents (Fig. 1A) and/or paired-pulse ratios. We have included neurochemical data obtained, for example, from studies measuring transmitter release evoked by K^+ or electrical stimulation only if they are helpful in the discussion of electrophysiological data.

Autoreceptors

A possible functional role is immediately obvious for autoreceptors that are activated by ligands released from the very same terminals on which they reside. Whereas, to date, there is little compelling evidence for a functional role of ionotropic autoreceptors on GABA terminals within the basal ganglia system, an involvement of the metabotropic $GABA_B$ autoreceptor in inhibitory synaptic function is well established. Activation of this receptor by synaptically released GABA exerts a negative feedback control by inhibiting subsequent GABA release. In addition to the well-known inhibition of Ca^{2+} channels (see Emson, Chapter 4, this volume), $GABA_B$ receptors directly inhibit the machinery for GABA release downstream from Ca^{2+} influx in substantia nigra (SN; Fig. 1; Misgeld, 2004; Giustizieri et al., 2005). Furthermore, activation of an inwardly rectifying K^+ conductance (Misgeld et al., 1995; Radnikow et al., 2001) may contribute to presynaptic inhibition by hyperpolarizing the membrane.

$GABA_B$ receptors can be found at almost all GABA terminals within the basal ganglia, including those of MSNs. The amplitude of inhibitory postsynaptic potentials (IPSPs) electrically evoked in MSNs from neostriatal slices is reduced by the selective $GABA_B$ receptor agonist baclofen in a concentration-dependent manner (0.5–100 µM). This inhibition in IPSP amplitude is associated with a relative increase in IPSP amplitude in response to the second of two closely paired stimuli (paired-pulse ratio), a phenomenon that is thought

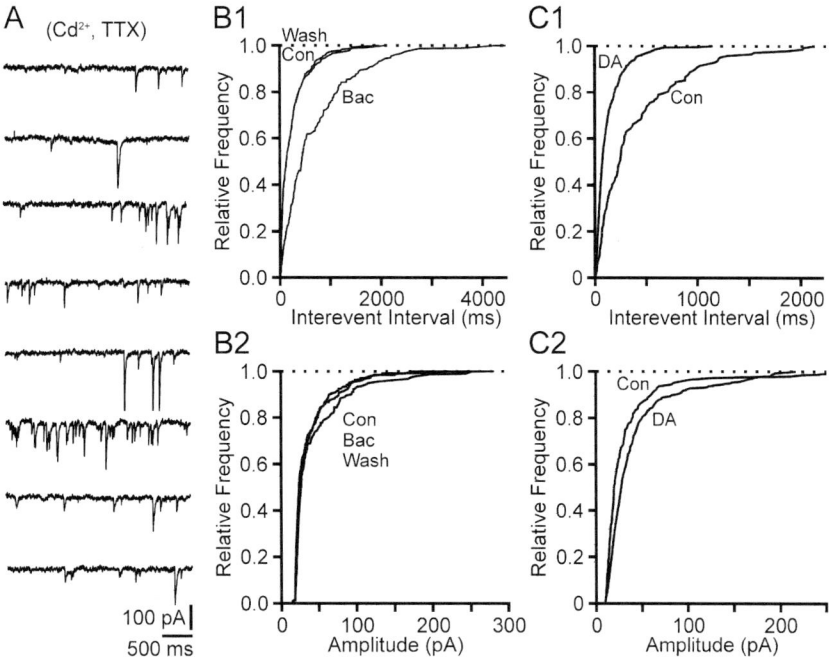

Fig. 1. GABA$_B$ receptor- and dopamine D$_1$ receptor-mediated modulation of GABA release in dopamine neurons. A: mIPSCs were recorded with high Cl$^-$ concentration (137 mM) in the pipette solution. Na$^+$ spikes were blocked by tetrodotoxin (TTX, 1 μM) and voltage-activated Ca^{2+} channels by Cd^{2+} (50 μM). Miniature IPSCs that can be blocked by GABA$_A$ receptor antagonists (not shown) reflect spontaneous vesicular GABA release. B: Plots of the cumulative distribution of the interevent intervals (B1) and of the amplitudes (B2) in control (Con), R-baclofen (Bac, 5 μM) and after washout of the drug (Wash) reveal a rightward shift in the interevent interval distribution. The plots indicate that the selective GABA$_B$ receptor agonist baclofen inhibits GABA release, because it reduces the frequency, but not the amplitude of mIPSCs. An activation of K$^+$ channels through postsynaptic GABA$_B$ receptors was prevented by the inclusion of Cs$^+$ and QX314 in the patch pipette solution (modified from Misgeld, 2004). C: Applying the same test to dopamine (DA, 3 μM) in presence of a D$_2$ receptor antagonist revealed a potentiation of GABA release by the dopamine D$_1$ receptor agonist (modified from Yanovsky et al., 2003).

to reflect changes in transmitter release probability (Thomson, 2000). In addition, baclofen does not alter the somato-dendritic membrane properties of MSNs. Therefore, the actions of baclofen on single evoked IPSPs and on the paired-pulse ratio within the neostriatum are likely to occur at a site presynaptic to MSNs, either on the terminal region of MSN axon collaterals or on interposed non-spiny GABAergic neurons (Nisenbaum et al., 1992, 1993). Similarly, in pallidal neurons, baclofen reduces the frequency, but has no effect on the amplitude, of spontaneous miniature inhibitory postsynaptic currents (mIPSCs; Kaneda and Kita, 2005), a result, which also strongly suggests a presynaptic site of action. The GABA$_B$ receptors responsible for this effect could reside on striatopallidal terminals in rat globus pallidus

(Chen et al., 2004) and/or on collaterals of pallidal neurons (Boyes and Bolam, 2003).

Comparable findings have been reported for neurons in the pars compacta of substantia nigra (SNc, mostly dopaminergic neurons) and in pars reticulata (SNr, mostly GABAergic neurons). In slices, baclofen reduces the amplitude of IPSP/Cs evoked in SNc and SNr neurons by electrical stimulation of striatonigral or pallidonigral fibers, whilst a possible involvement of postsynaptic receptors in these studies is minimized by using Cs$^+$ and QX314 in the recording pipette solution to block K$^+$ conductance in the recorded neurons (Häusser and Yung, 1994; Stanford and Lacey, 1996; Shen and Johnson, 1997). Furthermore, baclofen reduces the frequency of mIPSCs in SNc neurons without changing their amplitude

(Fig. 1B; Misgeld, 2004; Giustizieri et al., 2005). Similarly, in slices of the subthalamic nucleus, baclofen reduces the amplitude of evoked IPSCs, increases the paired-pulse ratio of evoked IPSCs and reduces mIPSC frequency but not amplitude (Shen and Johnson, 2001; Chen and Yung, 2005). Again, these results strongly suggest activation of presynaptic $GABA_B$ receptors.

Studies on GABA release induced by electrical or K^+ stimulation in slices or synaptosomal fractions (Florán et al., 1988; Giralt et al., 1990; Waldmeier et al., 1992; for a review of the older literature see Bowery, 1993) support the inhibitory role of $GABA_B$ autoreceptors throughout the basal ganglia. All GABAergic inhibitory inputs to SN, including fibers originating in the neostriatum, the pallidum and from GABAergic neurons in the pars reticulata, appear to be under the control of the negative feedback mediated by $GABA_B$ autoreceptors. These $GABA_B$ autoreceptors may limit excessive inhibition by GABAergic systems (Davies et al., 1990). Whilst the in vitro preparation may be of limited use for characterizing tonic activation of $GABA_B$ receptors by ambient GABA, there is compelling evidence for a tonic inhibition of GABA release by presynaptic $GABA_B$ autoreceptors in SN in vivo (Paladini et al., 1999).

Activation of $GABA_B$ autoreceptors is not a requirement for paired-pulse depression. Various other mechanisms regulate the strength of synaptic connections through their influence on the quantity and the reliability of transmitter release into the synaptic cleft following an incoming train of action potentials. Short-term depression, for example, can be due to reduced transmitter release resulting from changes intrinsic to the presynaptic terminal, such as inactivation of presynaptic voltage-gated Na^+ or Ca^{2+} channels, a transient depletion of docked vesicles or a transient decrease in release probability of the releasable pool of vesicles. Thus, at synapses that have a high initial release probability, synaptic depression predominates (Malenka and Siegelbaum, 2001). There is the possibility that not all GABA terminals in the basal ganglia express functional $GABA_B$ receptors. Indeed, pharmacological analysis of the reduction of intrastriatally evoked IPSPs in MSNs by $GABA_B$ receptor agonists has revealed evidence for the existence of two discrete populations of GABAergic fibers, one of which is regulated by presynaptic $GABA_B$ receptors and one of which is not (Seabrook et al., 1991). To further address this possibility, we applied 'minimal' intrastriatal stimulation to evoke IPSPs in an 'all-or-none' fashion in MSNs (Radnikow et al., 1997). This technique restricts synaptic input to only one or a few afferent fibers. These responses were compared with IPSPs evoked in the same neurons by compound stimulation, which recruits a much higher number of fibers. At an interstimulus interval of 300 ms, we observed paired-pulse depression of IPSP responses to both minimal (Fig. 2B) and compound stimulation. A selective, high affinity $GABA_B$ receptor antagonist (CGP55845A, 0.5 μM) had little effect on paired-pulse depression of responses evoked by compound stimulation, but abolished paired-pulse depression evoked by minimal stimulation (Fig. 2B, C), suggesting the existence of both $GABA_B$ receptor-dependent and -independent mechanisms for paired-pulse depression. In a separate series of experiments, we then eliminated the synaptic contribution of $GABA_B$ receptor-expressing fibers to the evoked IPSP by tonically inhibiting GABA release at these terminals with baclofen. Under these conditions, the residual IPSP displayed a strong paired-pulse depression in response to both minimal and compound stimulation. Whereas the paired-pulse depression evoked by compound stimulation was then decreased by application of the $GABA_B$ receptor antagonist, the paired-pulse depression observed with minimal stimulation remained unchanged. From these studies we were able to conclude that GABAergic fibers lacking $GABA_B$ receptors were those mediating stronger paired-pulse depression than those carrying $GABA_B$ receptors. Based on the fact that axon terminals of MSNs in SN carry functional $GABA_B$ receptors, we suggested that the intrastriatal fibers possessing $GABA_B$ receptors were the axon terminals of MSN collaterals, whereas the fibers without $GABA_B$ receptors, which exhibited strong paired-pulse depression, were axon terminals of local GABAergic neurons in the neostriatum (Fig. 2A; Radnikow et al., 1997). Similarly, in SN dopamine neurons, we observed that most IPSCs evoked by minimal stimulation were abolished by baclofen

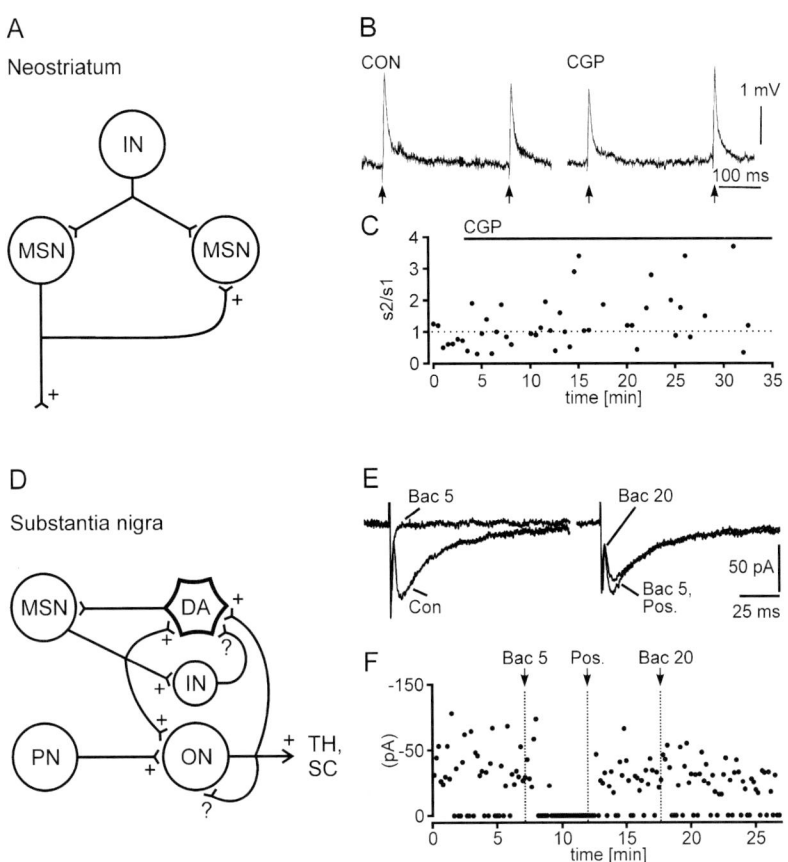

Fig. 2. Autoinhibition by presynaptic GABA$_B$ receptors in neostriatum and SN. A: The schematic drawing of the intrastriatal GABAergic network denotes the presumed location of GABA$_B$ receptors on GABAergic terminals (+) originating from MSN, but not from IN. B: Paired-pulse depression in the responses evoked by minimal intrastriatal stimulation in a MSN in the presence of glutamate antagonists at a time interval of 300 ms is converted to paired-pulse facilitation by the application of the high affinity GABA$_B$ receptor antagonist CGP55845A (CGP, 0.5 µM). The responses that were depolarizing were recorded with sharp micro-electrodes filled with 3 M KCl. They were blocked by the GABA$_A$ receptor antagonist bicuculline (30 µM). Thus, electrical stimulation activated in this example neostriatal GABAergic fibers with GABA$_B$ autoreceptors (modified from Radnikow et al., 1997). C: The scatter plot of the same responses illustrates the increase in the ratio of the response to the second stimulation over the response to the first stimulation (s2/s1). D: The schematic drawing of the GABAergic input and output to and from SN denotes the terminals that are equipped with GABA$_B$ autoreceptors (+) and those for which it is yet unclear (?). MSN = medium spiny neuron, PN = pallidal neuron, DA = dopamine neuron in pars compacta, IN = interneuron in pars compacta, ON = output neuron in pars reticulata, TH = thalamus, SC = superior colliculus. E: A GABA$_A$ receptor-mediated inward current (IPSC) was evoked by minimal stimulation in a dopamine neuron (Con) and blocked by R-baclofen (5 µM Bac, Bac 5) indicating that fibers with GABA$_B$ receptors were stimulated. After repositioning the stimulating electrode (Pos.), an IPSC was evoked in the same cell in the presence of R-baclofen. R-Baclofen, even at a concentration of 20 µM (Bac 20) had no effect indicating that fibers without GABA$_B$ receptors were stimulated. F: Scatter plot to illustrate the time course of the inhibition of the response by R-baclofen. Each dot in the plot represents the amplitude of consecutively evoked IPSCs (unpublished observation).

(5 µM). However, by making small adjustments to the location of the stimulating electrode in these experiments, we were able to identify a subpopulation of inputs through which minimal stimulation elicited IPSCs that were insensitive even to much higher concentrations of baclofen (Fig. 2E; Radnikow and Misgeld, unpublished observations). In this case, however, the possible origin of the baclofen-insensitive GABAergic terminals is not clear (Fig. 2D, F).

An intriguing, but as yet untested possibility is that there is a positive feedback regulation of GABA release in basal ganglia that is mediated by the neuropeptide neurotensin. Immunohistochemical studies in cats have revealed that most of the neurotensin-containing neurons in the neostriatum send massive projections to the globus pallidus, the pars compacta and the pars lateralis of SN, but sparse projections to the pars reticulata (Sugimoto and Mizuno, 1987). Neurotensin-related neuropeptide, LANT6, is colocalized with GABA and parvalbumin in a subpopulation of striatal interneurons, as well as in numerous pallidal and nigral reticulata neurons (Reiner and Anderson, 1993). Neurotensin increases striatal (O'Connor et al., 1992) and pallidal GABA release (Ferraro et al., 1998), and is released from terminals of striatonigral and striatopallidal neurons inside the SN (Frankel et al., 2005). Thus, it is tempting to speculate that neurotensin could enhance GABA release at very low concentrations, as might occur during conditions of low neuronal activity, whereas conditions of high neuronal activity would engage negative feedback mediated through GABA$_B$ autoreceptors.

Heteroreceptors

A considerable number of presynaptic receptors that react to ligands other than those liberated by GABA terminals (heteroreceptors) have been found to modulate GABA release within the basal ganglia system. There is no doubt that such receptors are relevant to understanding the actions of drugs used in the treatment of motor disturbances resulting from dysfunction of basal ganglia networks. To assign a functional role to presynaptic heteroreceptors, however, requires some knowledge about the origin and nature of the signals involved in their activation. Unfortunately, in this regard, many questions remain to be answered. The issue is further complicated because, in many cases, the same receptors operate both pre- and postsynaptically. Indeed, they can even have opposing effects, as in the case of receptors of the dopamine D$_1$ family, which may activate MSNs on the one hand but strengthen their inhibition

through axon collaterals of neighboring MSNs on the other. In general, it is often better known what presynaptic heteroreceptors are capable of doing than what they actually do. Finally, some apparently direct presynaptic effects may be mediated instead via postsynaptic receptors and release of a secondary transmitter(s), e.g., endocannabinoids (eCBs), which then act in a retrograde fashion to modulate GABA release at the presynaptic terminal (Centonze et al., 2004a). Rather than listing all the heteroreceptors that have been found to modulate GABA release in the basal ganglia, here we describe a few examples that illustrate possible presynaptic functions for these receptors.

Dopamine and its receptors play a very important role in basal ganglia network functions, as highlighted by the multiple motor disturbances resulting from failures in this transmitter system. Despite their unequivocal importance, the relative contributions made by presynaptic vs. postsynaptic dopamine receptors to basal ganglia functions remain largely unknown. Hence, dopamine receptors located on the terminals of GABAergic neurons are of as much potential interest as at any other location within the basal ganglia. Dopamine is released from non-synaptic release sites and hence does not act in a strict point-to-point manner, but rather by diffusion through the extracellular space ('volume transmission'). GABA release from the intrastriatal axon terminals of MSN collaterals is enhanced by activation of presynaptic receptors of the dopamine D$_1$ family and reduced by activation of presynaptic receptors of the dopamine D$_2$ family. As a consequence, these receptors have a strong impact on short-term plasticity. Thus, paired-pulse depression predominates at terminals under the influence of D$_1$ receptors, whereas paired-pulse facilitation predominates at terminals in which presynaptic D$_2$ receptors are active (Guzmán et al., 2003). The opposing actions mediated by D$_1$ and D$_2$ receptors have been substantiated by studies examining the influence of selective agonists for these receptors on background and evoked GABA release in the basal ganglia (Florán et al., 1990, 1997, 2002, 2004, 2005). Under the influence of dopamine itself, the D$_2$ receptor-mediated effect appears to prevail in the neostriatum (Shen and Johnson, 2000;

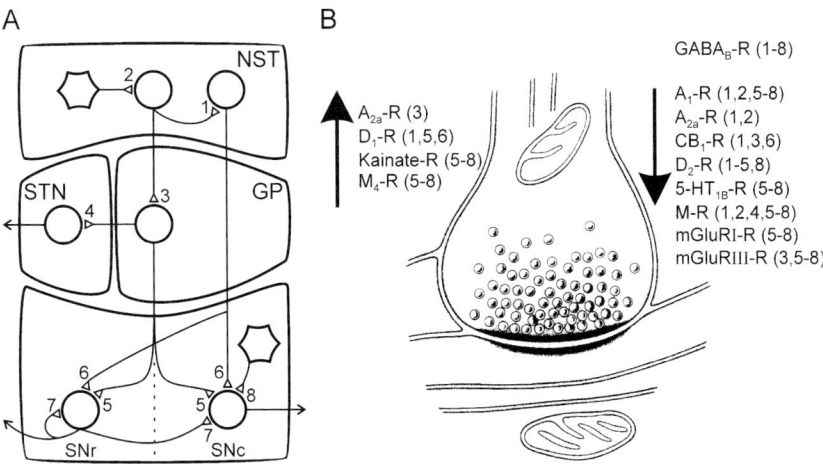

Fig. 3. A: Scheme of GABAergic connections within the basal ganglia. Numbers refer to presynaptic receptors listed under B. NST = Neostriatum, STN = Subthalamic nucleus, GP = Globus pallidus, SNr = Substantia nigra pars reticulata, SNc = Substantia nigra pars compacta. B: Scheme of a GABAergic synapse with presynaptic receptors enhancing (upward arrow) and reducing GABA release (downward arrow). Except the GABA$_B$ receptor, all others are heteroreceptors. An example for an ionotropic receptor is the kainate receptor all others are metabotropic receptors. A$_1$-R = adenosine receptor subtype (Centonze et al., 2001; Florán et al., 2002), A$_{2a}$-R = adenosine receptor subtype (enhancing: Mayfield et al., 1993; Kurokawa et al., 1994; Mayfield et al., 1996; Mori et al., 1996; Zahniser et al., 2000; Shindou et al., 2001; Florán et al., 2005; reducing: Chergui et al., 2000), CB$_1$-R = cannabinoid receptor subtype (Szabo et al., 1998; Wallmichrath and Szabo, 2002a, b; Yanovsky et al., 2003; Centonze et al., 2004a; Köfalvi et al., 2005; Caillé and Parsons, 2006; Engler et al., 2006), D$_1$-R = dopamine receptor subtype (Radnikow and Misgeld, 1998; Florán et al., 2002; Trevitt et al., 2002; Guzmán et al., 2003; Yanovsky et al., 2003), D$_2$-R = dopamine receptor subtype (Florán et al., 1997; Shen and Johnson, 2000; Cooper and Stanford, 2001; Centonze et al., 2002; Guzmán et al., 2003; Valenti et al., 2003; Centonze et al., 2004a, b; Florán et al., 2004; Florán et al., 2005; Salgado et al., 2005), GABA$_B$-R = GABA autoreceptor (Nisenbaum et al., 1992; Nisenbaum et al., 1993; Häuser and Yung, 1994; Stanford and Lacey, 1996; Radnikow et al., 1997; Chan et al., 1998; Paladini et al., 1999; Radnikow et al., 2001; Giustizieri et al., 2005; Kaneda and Kita, 2005), 5-HT$_{1B}$-R = serotonin receptor subtype (Johnson et al., 1992; Stanford and Lacey, 1996), kainate receptor = ionotropic glutamate receptor (Nakamura et al., 2003), M-R = muscarinic receptor, not specified (Marchi et al., 1990; Raiteri et al., 1990; Grillner et al., 2000; Shen and Johnson, 2000; Koós and Tepper, 2002; Perez-Rosello et al., 2005), M$_4$-R = muscarinic receptor subtype (Kayadjanian et al., 1994), mGluRI-R and mGluRIII-R = metabotropic glutamate receptor subtypes (mGluRI: Marino et al., 2001; mGluRIII: Wittmann et al., 2001; Matsui and Kita, 2003; Valenti et al., 2003; Giustizieri et al., 2005).

Momiyama and Koga, 2001). In the SN, in which activation of D$_1$ receptors elevates GABA release (Fig. 1C; Radnikow and Misgeld, 1998), endogenous dopamine exerts a tonic effect that increases the amplitude of evoked IPSPs (Yanovsky et al., 2003). Moreover, endogenous tonic activation of nigral D$_1$ receptors has been reported to influence motor behavior in rats (Trevitt et al., 2001). Another example of a presynaptic receptor that enhances GABA release in the basal ganglia is the kainate receptor, an ionotropic receptor found on synaptic terminals apposed to acutely isolated dopamine neurons (Nakamura et al., 2003).

Receptor-mediated enhancement of GABA release appears to be the exception rather than the rule, however. Most heteroreceptors with G-protein requirement reduce GABA release. In addition to the dopamine D$_2$ receptors already mentioned, adenosine receptors, muscarinic receptors, metabotropic glutamate receptors and cannabinoid CB$_1$ receptors belong to this list (Fig. 3; for references see Fig. 3 legend). Some receptors exert mixed or opposing effects depending on which subtypes are involved. Muscarinic M$_4$ receptors, for example, enhance GABA release from GABAergic terminals impinging upon dopamine neurons, whereas all other muscarinic receptor subtypes appear to reduce release (Fig. 3; Kayadjanian et al., 1994; but see Grillner et al., 2000). Alternatively, the action of a particular presynaptic heteroreceptor can vary between different synapses. Adenosine A$_{2a}$ receptors, for example, enhance GABA release at pallidal synapses but reduce it at intrastriatal synapses (Fig. 3; Kirk and Richardson, 1994; Kurokawa et al., 1994; Mori et al., 1996; Zahniser et al., 2000; Florán et al., 2005).

As far as specific GABAergic pathways are concerned, a wealth of literature describes the presynaptic modulation of transmitter release from terminals of MSNs in the neostriatum, pallidum and SN. In contrast, little if anything is known about modulation of GABA release from local neurons in these structures. One synaptic mechanism in which presynaptic heteroreceptors are known to play a significant role is retrograde signaling. In the SN, two opposing effects – a dopamine D_1 receptor-mediated enhancement of GABA release and a cannabinoid CB_1 receptor-mediated suppression of GABA release – result from postsynaptic release of dopamine and eCBs, respectively. As a synaptic mechanism, retrograde signaling offers the intriguing possibility of being able to regulate input selection in a network by establishing a switch from one input to another. Given the wealth of recent literature in this field and the focus of our own studies in SN, we will discuss retrograde activation of presynaptic receptors in the basal ganglia in greater detail below.

Retrograde signaling

Retrograde signaling through presynaptic heteroreceptors provides a mechanism by which a postsynaptic neuron can modulate its own synaptic input. As mentioned already, retrograde signaling through dopamine D_1 receptors and cannabinoid CB_1 receptors modulates $GABA_A$ receptor-mediated inhibition in SN in opposite directions. Endogenous dopamine enhances, whereas eCBs reduce $GABA_A$ receptor-mediated postsynaptic inhibition. The intranigral presynaptic targets for dopamine are MSN and pallidonigral terminals, whereas eCBs target only MSN terminals. The vast majority of these GABAergic terminals innervate the inhibitory output neurons of the SN (Paladini et al., 1999; Lévesque and Parent, 2005).

Dopaminergic neurons of the SNc release dopamine not only from their terminals in the neostriatum but also from the dendrites they extend

into the SNr (Geffen et al., 1976; Korf et al., 1976; Cheramy et al., 1981). Extrasynaptic dopamine release from the latter location has properties that are consistent with a vesicular mode of release (Jaffe et al., 1998) and may be supported by dendritic backfiring (Häusser et al., 1995). Possible intranigral targets for the released dopamine include D_2 receptors located on the dopaminergic neurons themselves, which may mediate autoinhibition via their coupling to inwardly rectifying K^+ channels (Innis and Aghajanian, 1987; for a review, see Smith and Kieval, 2000), and D_1 type receptors located on GABAergic terminals. Studies on rodent SN slices have shown that selective D_1 receptor agonists (e.g., SKF38393 or dopamine in the presence of the D_2 receptor antagonist sulpiride) enhance GABA release (Florán et al., 2002) and increase the frequency, but not the amplitude of $GABA_A$ receptor-mediated mIPSCs in both dopamine (Fig. 1; Yanovsky et al., 2003) and GABA neurons (Radnikow and Misgeld, 1998). The increase in IPSC frequency could be mimicked by forskolin, suggesting a cAMP-dependent mechanism was involved. A selective D_1 receptor antagonist, SCH23390, had no effect on mIPSCs when applied alone. Collectively, these findings indicate a presynaptic enhancement of GABA release by D_1 receptor activation, which is in good agreement with anatomical data demonstrating an exclusive location of D_1 receptors on striatonigral and pallidonigral afferents within the SN (Barone et al., 1987; Harrison et al., 1990; Mansour et al., 1991; Jaber et al., 1996).

To study the impact of this enhanced GABA release on evoked IPSCs, we applied minimal stimulation in order to limit the number of fibers activated and hence minimize conductance changes that would otherwise be imposed by the evoked IPSC and would impair voltage control over the synaptic current. In addition, this type of stimulation helps to avoid contamination of the response by the co-release of other modulators. Taking these precautions, we found that D_1 receptor activation increases the amplitude of most evoked IPSCs (Radnikow and Misgeld, 1998; but see Miyazaki and Lacey, 1998). This response to D_1 receptor activation could be demonstrated more readily in GABA neurons than in dopamine neurons (Yanovsky et al., 2003). One reason for this observation could be that a greater number of GABAergic inputs from the forebrain terminate on GABA neurons in SNr than on dopamine neurons in SNc (Tepper et al., 1995). The stimulatory effect of D_1 receptor activation is blocked by the selective D_1 receptor antagonist SCH23390 and enhanced by the noradrenaline-uptake inhibitor imipramine.

We then investigated whether dopamine release induced by depolarization of dopamine neurons would result in D_1 receptor-mediated enhancement of evoked IPSCs recorded in the same neurons (Yanovsky et al., 2003). The predominant response to depolarization of the postsynaptic neuron was, on the contrary, a depolarization-induced suppression of inhibition (DSI, see below). The DSI observed under these conditions, however, was significantly smaller than DSI recorded in the presence of a D_1 receptor antagonist, indicating an additional, opposing, contribution of D_1 receptors. Unfortunately, we were unable to demonstrate D_1 receptor-mediated enhancement of inhibition after DSI was reduced using pharmacological tools.

Application of the D_1 receptor antagonist SCH23390 (1–3 µM) alone produced a consistent reduction in the amplitude of evoked IPSCs. However, as mentioned, this effect was not accompanied by changes in the amplitude or frequency of mIPSCs. This lack of effect on mIPSCs might have been due to a reduction in vesicular dopamine release resulting from a suppression of action potentials by tetrodotoxin in the dopamine neurons (Jaffe et al., 1998). Our conclusion, with respect to the influence of dopamine on GABAergic transmission in SN, is that D_1 receptor activation enhances GABA release from axon terminals of MSNs and of pallidonigral fibers. Ambient dopamine tonically stimulates D_1 receptors on these GABA terminals, thereby enhancing GABA release. In the case of dopamine neurons, this provides a good example of retrograde signaling, whereby postsynaptic release of dopamine leads to activation of presynaptic dopamine heteroreceptors located on the terminals of GABAergic afferents innervating dopamine neurons. However, the retrograde spread of released dopamine can also reach synapses on GABAergic neurons that have

the dual role of transmitting signals out of the basal ganglia and of inhibiting dopamine neurons. Therefore, dopamine can reach synapses across short distances that do not directly inhibit dopamine neurons. This is made possible by the interposition of GABAergic neurons in the net formed by dopamine dendrites in SNr. According to experiments performed in vivo, somatodendritic dopamine release, in turn, is controlled by the GABA input itself (Cobb and Abercrombie, 2002, 2003).

A second retrograde messenger system within the basal ganglia that is gaining increasing interest is the endocannabinoid family. Most of the effects produced by eCBs in the central nervous system are mediated by the G-protein-coupled CB_1 receptor, which is highly expressed in GABAergic MSNs. CB_1 receptor agonists suppress IPSCs evoked in SN neurons, an effect that is blocked by CB_1 receptor antagonists (Wallmichrath and Szabo, 2002a, b; Yanovsky et al., 2003; Mátyás et al., 2006). That these agonists act through CB_1 receptors has been confirmed by the use of hippocampal slices from CB_1 receptor knockout mice (Hájos et al., 2000; Hájos and Freund, 2002). Whereas CB_1 receptor agonists reduce the frequency of mIPSCs in hippocampal slices (Wilson and Nicoll, 2001), they do not affect tetrodotoxin-sensitive spontaneous IPSCs or mIPSCs in SN slices (Wallmichrath and Szabo, 2002a, b). One likely explanation for this difference is that there is an abundance of GABA release sites of intrinsic origin within SN that do not carry CB_1 receptors. Nevertheless, the effect is presynaptic, since localization of CB_1 receptors within SN is restricted to presynaptic sites. Thus, mRNA for CB_1 receptors is highly expressed in the neostriatum, but essentially absent from SNr (Egertová and Elphick, 2000). Moreover, immunohistochemical techniques have shown that the CB_1 receptor protein is localized exclusively presynaptically to the membrane of terminals and preterminal axons (Mátyás et al., 2006). Interestingly, the intensity of CB_1 receptor staining was found to be far greater for preterminal axons than for terminals. The majority of CB_1 receptors were found in the axonal membranes of striatonigral fibers that originate from MSNs. Both the CB_1 receptor agonist WIN55,212-2 and the CB_1 receptor antagonist AM251 impaired action

254

potential generation in MSN axons, but this occurred via a CB$_1$-independent mechanism, since these effects were also observed in CB$_1$ knockout animals. The effect may have been due to nonselective actions of these drugs on the input resistance of MSNs, which could result from interactions with the lipid-protein interface of membrane proteins. We concluded that CB$_1$ receptors located in the axonal membrane are not involved in modulating action potential propagation but, more likely, are being transported to synaptic terminals (Mátyás et al., 2006).

Further evidence for localization of functional CB$_1$ receptors at presynaptic terminals comes from studies on DSI in SN neurons (Wallmichrath and Szabo, 2002b; Yanovsky et al., 2003). In this paradigm, eCBs are released upon postsynaptic depolarization of the recorded neuron. Both GABA and dopamine neurons are capable of releasing eCBs using this method (Yanovsky et al., 2003). The resulting transient suppression of evoked IPSCs is prevented by appropriate CB$_1$ receptor antagonists. In addition, we could strengthen DSI by coapplication of low concentrations (0.1–5 μM) of muscarinic receptor agonists. Indeed, under these conditions, even spontaneous IPSCs were transiently suppressed (Yanovsky et al., 2003).

As well as blocking agonist effects, the CB$_1$ receptor antagonist alone was capable of increasing the amplitude of evoked IPSCs, suggesting a tonic activation of CB$_1$ receptors on GABAergic terminals originating from MSNs. Interestingly, it has recently been suggested that ryanodine receptors, which regulate internal Ca^{2+} mobilization, play a key role in the depolarization-induced production of eCBs in the hippocampus (Isokawa and Alger, 2006). Since we found that both GABA- and dopamine neurons produce eCBs upon depolarization (Yanovsky et al., 2003) and that, in both types of neuron, spontaneous ryanodine receptor openings are prominent (Yanovsky et al., 2005), these results suggest that ryanodine receptors might play an important role in setting a tonic endocannabinoid tone in SN.

Interactions may exist between the dopaminergic and the endocannabinoid retrograde signaling systems at the level of the presynaptic terminal. Direct CB$_1$ receptor-mediated effects of eCBs on dopaminergic neurons are unlikely, since the dopaminergic neurons in the SNc do not express CB$_1$ receptors (Mátyás et al., 2006). Presynaptic CB$_1$ receptor activation may disinhibit GABAergic striatonigral, pallidonigral or SNr neurons. The CB$_1$ receptor-mediated tonic

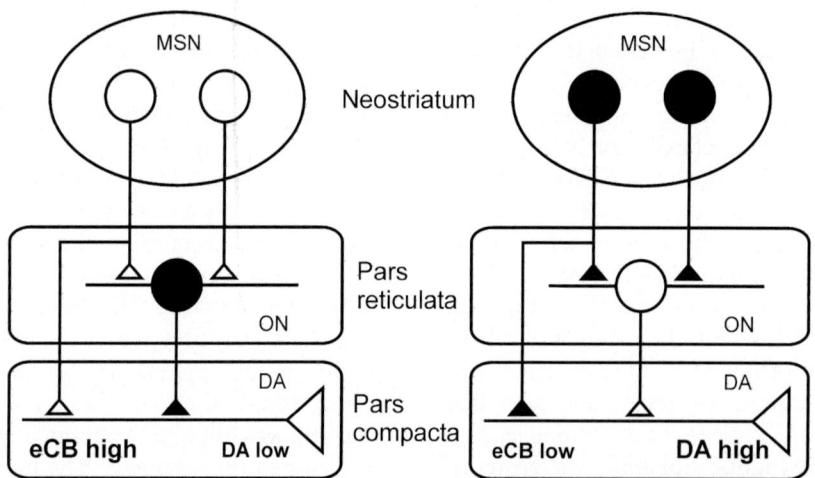

Fig. 4. Hypothesis for the cannabinoid CB$_1$ and dopamine D$_1$ receptor-mediated tone in SN. In the double inhibitory link connecting neostriatal medium spiny neurons (MSN) to dopamine neurons (DA) inhibition through GABAergic output neurons (ON) prevails if the endocannabinoid tone activating CB$_1$ receptors on MSN terminals dominates (left side). Vice versa, inhibition through MSN prevails if the dopamine tone activating D$_1$ receptors on MSN terminals dominates (right side).

disinhibition of SNr neurons has a counterpart in a tonic enhancement of inhibition of SNr neurons that is mediated by presynaptic dopamine D_1 receptors. Thus, when the CB_1 receptor-mediated tone prevails, inhibition in SN is conveyed primarily by GABA neurons in SNr and, vice versa, when the D_1 receptor-mediated tone prevails, inhibition in SN is dominated by the striatal and pallidal input (Fig. 4; Yanovsky et al., 2003). Disinhibition of SNr neurons may also occur via an indirect pathway when eCBs are released in the neostriatum. Thus, disinhibition of striatopallidal neurons will enhance inhibition of pallidonigral cells that in turn, will reduce pallidal inhibition of SN neurons. Changes in the inhibition of dopaminergic neurons will alter their dopamine release (Tepper et al., 1995; Cobb and Abercrombie, 2002, 2003) and, thereby, the dopamine receptor-mediated tone in SN and neostriatum.

Concluding remarks

Presynaptic receptors provide versatility to the basal ganglia network in which GABA neurons and $GABA_A$ receptor-mediated synaptic mechanisms play a dominant role. Negative feedback through presynaptic $GABA_B$ receptors can reduce the level of inhibition at each individual GABAergic connection, in particular under the condition of an overactive output. Presynaptic heteroreceptors allow for the fine-tuning of postsynaptic inhibition by sources not directly influenced by the basal ganglia network or the synaptic connection itself. Retrograde signaling, for which eCB- and dopamine-mediated mechanisms are prominent examples, has the capacity to regulate the input of the target neuron. Last but not least, balanced ambient levels of neurotransmitters and/or neuromodulators for which presynaptic receptors are available can accomplish the fine-tuning of $GABA_A$ receptor-mediated synaptic transmission over extended time periods.

Abbreviations

DSI   depolarization-induced suppression of inhibition
eCBs   endocannabinoids
IPSCs inhibitory postsynaptic currents
IPSPs inhibitory postsynaptic potentials
mIPSCs miniature inhibitory postsynaptic currents
MSNs   medium spiny neurons
SN   substantia nigra
SNc   substantia nigra pars compacta
SNr   substantia nigra pars reticulata

Acknowledgments

The authors are grateful to Ms Andrea Lewen for excellent editorial assistance. Data reported in this chapter were from studies funded by the German Research Society (DFG).

References

Barone, P., Tucci, I., Parashos, S.A. and Chase, T.N. (1987) D-1 dopamine receptor changes after striatal quinolinic acid lesion. Eur. J. Pharmacol., 138: 141–145.

Bowery, N.G. (1993) $GABA_B$ receptor pharmacology. Ann. Rev. Pharmacol. Toxicol., 33: 109–147.

Boyes, J. and Bolam, J.P. (2003) The subcellular localization of $GABA_B$ receptor subunits in the rat substantia nigra. Eur. J. Neurosci., 18: 3279–3293.

Caillé, S. and Parsons, L.H. (2006) Cannabinoid modulation of opiate reinforcement through the ventral striatopallidal pathway. Neuropsychopharmacology, 31: 804–813.

Centonze, D., Battista, N., Rossi, S., Mercuri, N.B., Finazzi-Agrò, A., Bernardi, G., Calabresi, P. and Maccarrone, M. (2004a) A critical interaction between dopamine D2 receptors and endocannabinoids mediates the effects of cocaine on striatal GABAergic transmission. Neuropsychopharmacology, 29: 1488–1497.

Centonze, D., Gubellini, P., Usiello, A., Rossi, S., Tscherter, A., Bracci, E., Erbs, E., Tognazzi, N., Bernardi, G., Pisani, A., Calabresi, P. and Borrelli, E. (2004b) Differential contribution of dopamine D2S and D2L receptors in the modulation of glutamate and GABA transmission in the striatum. Neuroscience, 129: 157–166.

Centonze, D., Picconi, B., Baunez, C., Borrelli, E., Pisani, A., Bernardi, G. and Calabresi, P. (2002) Cocaine and amphetamine depress striatal GABAergic synaptic transmission through D2 dopamine receptors. Neuropsychopharmacology, 26: 164–175.

Centonze, D., Saulle, E., Pisani, A., Bernardi, G. and Calabresi, P. (2001) Adenosine-mediated inhibition of striatal GABAergic synaptic transmission during in vitro ischaemia. Brain, 124: 1855–1865.

Chan, P.K.Y., Leung, C.K.S. and Yung, W. (1998) Differential expression of pre- and postsynaptic $GABA_B$ receptors in rat substantia nigra pars reticulata neurones. Eur. J. Pharmacol., 349: 187–197.

Chen, L., Boyes, J., Yung, W. and Bolam, J.P. (2004) Subcellular localization of GABA$_B$ receptor subunits in rat globus pallidus. J. Comp. Neurol., 474: 340–352.

Chen, L. and Yung, W. (2005) Tonic activation of presynaptic GABA$_B$ receptors on rat pallidosubthalamic terminals. Acta Pharmacol. Sin., 26: 10–16.

Cheramy, A., Leviel, V. and Glowinski, J. (1981) Dendritic release of dopamine in the substantia nigra. Nature, 289: 537–543.

Chergui, K., Bouron, A., Normand, E. and Mulle, C. (2000) Functional GluR6 kainate receptors in the striatum: indirect downregulation of synaptic transmission. J. Neurosci., 20: 2175–2182.

Cobb, W.S. and Abercrombie, E.D. (2002) Distinct roles for nigral GABA and glutamate receptors in the regulation of dendritic dopamine release under normal conditions and in response to systemic haloperidol. J. Neurosci., 22: 1407–1413.

Cobb, W.S. and Abercrombie, E.D. (2003) Relative involvement of globus pallidus and subthalamic nucleus in the regulation of somatodendritic dopamine release in substantia nigra is dopamine-dependent. Neuroscience, 119: 777–786.

Cooper, A.J. and Stanford, I.M. (2001) Dopamine D2 receptor mediated presynaptic inhibition of striatopallidal GABA$_A$ IPSCs in vitro. Neuropharmacology, 41: 62–71.

Davies, C.H., Davies, S.N. and Collingridge, G.L. (1990) Paired-pulse depression of monosynaptic GABA-mediated inhibitory postsynaptic responses in rat hippocampus. J. Physiol. (Lond.), 424: 513–531.

Egertová, M. and Elphick, M.R. (2000) Localisation of cannabinoid receptors in the rat brain using antibodies to the intracellular C-terminal tail of CB$_1$. J. Comp. Neurol., 422: 159–171.

Engler, B., Freiman, I., Urbanski, M. and Szabo, B. (2006) Effects of exogenous and endogenous cannabinoids on GABA-ergic neurotransmission between the caudate-putamen and the globus pallidus in the mouse. J. Pharmacol. Exp. Ther., 316: 608–617.

Fassio, A., Evans, G., Grisshammer, R., Bolam, J.P., Mimmack, M. and Emson, P.C. (2000) Distribution of the neurotensin receptor NTS1 in the rat CNS studied using an amino-terminal directed antibody. Neuropharmacology, 39: 1430–1442.

Ferraro, L., Antonelli, T., O'Connor, W.T., Fuxe, K., Soubrié, P. and Tanganelli, S. (1998) The striatal neurotensin receptor modulates striatal and pallidal glutamate and GABA release: functional evidence for a pallidal glutamate-GABA interaction via the pallidal-subthalamic nucleus loop. J. Neurosci., 18: 6977–6989.

Florán, B., Aceves, J., Sierra, A. and Martinez-Fong, D. (1990) Activation of D$_1$ dopamine receptors stimulates the release of GABA in the basal ganglia of the rat. Neurosci. Lett., 116: 136–140.

Florán, B., Barajas, C., Florán, L., Erlij, D. and Aceves, J. (2002) Adenosine A1 receptors control dopamine D1-dependent [^{3}H]GABA release in slices of substantia nigra pars

reticulata and motor behavior in the rat. Neuroscience, 115: 743–751.

Florán, B., Florán, L., Erlij, D. and Aceves, J. (2004) Dopamine D4 receptors inhibit depolarization-induced [^{3}H]GABA release in the rat subthalamic nucleus. Eur. J. Pharmacol., 498: 97–102.

Florán, B., Florán, L., Sierra, A. and Aceves, J. (1997) D2 receptor-mediated inhibition of GABA release by endogenous dopamine in the rat globus pallidus. Neurosci. Lett., 237: 1–4.

Florán, B., Gonzalez, B., Florán, L., Erlij, D. and Aceves, J. (2005) Interactions between adenosine A$_{2a}$ and dopamine D2 receptors in the control of [^{3}H]GABA release in the globus pallidus of the rat. Eur. J. Pharmacol., 520: 43–50.

Florán, B., Silva, I., Nava, C. and Aceves, J. (1988) Presynaptic modulation of the release of GABA by GABA$_A$ receptors in pars compacta and by GABA$_B$ receptors in pars reticulata of the rat substantia nigra. Eur. J. Pharmacol., 150: 277–286.

Frankel, P.S., Hoonakker, A.J., Hanson, G.R., Bush, L., Keefe, K.A. and Alburges, M.E. (2005) Differential neurotensin responses to low and high doses of methamphetamine in the terminal regions of striatal efferents. Eur. J. Pharmacol., 522: 47–54.

Geffen, L.B., Jessell, T.M., Cuello, A.C. and Iversen, L.L. (1976) Release of dopamine from dendrites in rat substantia nigra. Nature, 260: 258–260.

Giralt, M.T., Bonanno, G. and Raiteri, M. (1990) GABA terminal autoreceptors in the pars compacta and in the pars reticulata of the rat substantia nigra are GABA$_B$. Eur. J. Pharmacol., 175: 137–144.

Giustizieri, M., Bernardi, G., Mercuri, N.B. and Berretta, N. (2005) Distinct mechanisms of presynaptic inhibition at GABAergic synapses of the rat substantia nigra pars compacta. J. Neurophysiol., 94: 1992–2003.

Grillner, P., Berretta, N., Bernardi, G., Svensson, T.H. and Mercuri, N.B. (2000) Muscarinic receptors depress GABA-ergic synaptic transmission in rat midbrain dopamine neurons. Neuroscience, 96: 299–307.

Guzmán, J.N., Hernández, A., Galarraga, E., Tapia, D., Laville, A., Vergara, R., Aceves, J. and Bargas, J. (2003) Dopaminergic modulation of axon collaterals interconnecting spiny neurons of the rat striatum. J. Neurosci., 23: 8931–8940.

Hájos, N. and Freund, T.F. (2002) Pharmacological separation of cannabinoid sensitive receptors on hippocampal excitatory and inhibitory fibers. Neuropharmacology, 43: 503–510.

Hájos, N., Katona, I., Naiem, S.S., Mackie, K., Ledent, C., Mody, I. and Freund, T.F. (2000) Cannabinoids inhibit hippocampal GABAergic transmission and network oscillations. Eur. J. Neurosci., 12: 3239–3249.

Harrison, M.B., Wiley, R.G. and Wooten, G.F. (1990) Selective localization of striatal D$_1$ receptors to striatonigral neurons. Brain Res., 528: 317–322.

Häusser, M., Stuart, G., Racca, C. and Sakmann, B. (1995) Axonal initiation and active dendritic propagation of action potentials in substantia nigra neurons. Neuron, 15: 637–647.

Häusser, M.A. and Yung, W.H. (1994) Inhibitory synaptic potentials in guinea-pig substantia nigra dopamine neurones in vitro. J. Physiol. (Lond.), 479: 401–422.

Innis, R.B. and Aghajanian, G.K. (1987) Pertussis toxin blocks autoreceptor-mediated inhibition of dopaminergic neurons in rat substantia nigra. Brain Res., 411: 139–143.

Isokawa, M. and Alger, B.E. (2006) Ryanodine receptor regulates endogenous cannabinoid mobilization in the hippocampus. J. Neurophysiol., 95: 3001–3011.

Jaber, M., Robinson, S.W., Missale, C. and Caron, M.G. (1996) Dopamine receptors and brain function. Neuropharmacology, 35: 1503–1519.

Jaffe, E.H., Marty, A., Schulte, A. and Chow, R.H. (1998) Extrasynaptic vesicular transmitter release from the somata of substantia nigra neurons in rat midbrain slices. J. Neurosci., 18: 3548–3553.

Johnson, S.W., Mercuri, N.B. and North, R.A. (1992) 5-hydroxytryptamine$_{1B}$ receptors block the GABA$_B$ synaptic potential in rat dopamine neurons. J. Neurosci., 12: 2000–2006.

Kaneda, K. and Kita, H. (2005) Synaptically released GABA activates both pre- and postsynaptic GABA$_B$ receptors in the rat globus pallidus. J. Neurophysiol., 94: 1104–1114.

Kayadjanian, N., Gioanni, H., Ménetrey, A. and Besson, M.J. (1994) Muscarinic receptor stimulation increases the spontaneous [^{3}H]GABA release in the rat substantia nigra through muscarinic receptors localized on striatonigral terminals. Neuroscience, 63: 989–1002.

Kirk, I.P. and Richardson, P.J. (1994) Adenosine A$_{2a}$ receptor-mediated modulation of striatal [^{3}H]GABA and [^{3}H]acetylcholine release. J. Neurochem., 62: 960–966.

Köfalvi, A., Rodrigues, R.J., Ledent, C., Mackie, K., Vizi, E.S., Cunha, R.A. and Sperlágh, B. (2005) Involvement of cannabinoid receptors in the regulation of neurotransmitter release in the rodent striatum: a combined immunochemical and pharmacological analysis. J. Neurosci., 25: 2874–2884.

Koós, T. and Tepper, J.M. (2002) Dual cholinergic control of fast-spiking interneurons in the neostriatum. J. Neurosci., 22: 529–535.

Korf, J., Zieleman, M. and Westerink, B.H.C. (1976) Dopamine release in substantia nigra? Nature, 260: 257–258.

Kurokawa, M., Kirk, I.P., Kirkpatrick, K.A., Kase, H. and Richardson, P.J. (1994) Inhibition by KF17837 of adenosine A2A receptor-mediated modulation of striatal GABA and ACh release. Br. J. Pharmacol., 113: 43–48.

Lévesque, M. and Parent, A. (2005) The striatofugal fiber system in primates: a reevaluation of its organization based on single-axon tracing studies. Proc. Natl. Acad. Sci. USA, 102: 11888–11893.

Malenka, R.C. and Siegelbaum, S.A. (2001) Synaptic plasticity: diverse targets and mechanisms for regulating synaptic efficacy. In: Cowan W.M., Südhof T.C. and Stevens C.F. (Eds.), Synapses. The Johns Hopkins University Press, Baltimore, Maryland, pp. 393–453.

Mansour, A., Meador-Woodruff, J.H., Zhou, Q.-Y., Civelli, O., Akil, H. and Watson, S.J. (1991) A comparison of D$_1$ receptor binding and mRNA in rat brain using receptor

autoradiographic and in situ hybridization techniques. Neuroscience, 45: 359–371.

Marchi, M., Sanguineti, P. and Raiteri, M. (1990) Muscarinic receptors mediate direct inhibition of GABA release from rat striatal nerve terminals. Neurosci. Lett., 116: 347–351.

Marino, M.J., Wittmann, M., Bradley, S.R., Hubert, G.W., Smith, Y. and Conn, P.J. (2001) Activation of group I metabotropic glutamate receptors produces a direct excitation and disinhibition of GABAergic projection neurons in the substantia nigra pars reticulata. J. Neurosci., 21: 7001–7012.

Matsui, T. and Kita, H. (2003) Activation of group III metabotropic glutamate receptors presynaptically reduces both GABAergic and glutamatergic transmission in the rat globus pallidus. Neuroscience, 122: 727–737.

Mátyás, F., Yanovsky, Y., Mackie, K., Kelsch, W., Misgeld, U. and Freund, T.F. (2006) Subcellular localization of type 1 cannabinoid receptors in the rat basal ganglia. Neuroscience, 137: 337–361.

Mayfield, R.D., Larson, G., Orona, R.A. and Zahniser, N.R. (1996) Opposing actions of adenosine A$_{2a}$ and dopamine D$_2$ receptor activation on GABA release in the basal ganglia: evidence for an A$_{2a}$/D$_2$ receptor interaction in globus pallidus. Synapse, 22: 132–138.

Mayfield, R.D., Suzuki, F. and Zahniser, N.R. (1993) Adenosine A$_{2a}$ receptor modulation of electrically evoked endogenous GABA release from slices of rat globus pallidus. J. Neurochem., 60: 2334–2337.

Misgeld, U. (2004) Innervation of the substantia nigra. Cell Tissue Res., 318: 107–114.

Misgeld, U., Bijak, M. and Jarolimek, W. (1995) A physiological role for GABA$_B$ receptors and the effects of baclofen in the mammalian central nervous system. Prog. Neurobiol., 46: 423–462.

Miyazaki, T. and Lacey, M.G. (1998) Presynaptic inhibition by dopamine of a discrete component of GABA release in rat substantia nigra pars reticulata. J. Physiol. (Lond.), 513: 805–817.

Momiyama, T. and Koga, E. (2001) Dopamine D$_2$-like receptors selectively block N-type Ca^{2+} channels to reduce GABA release onto rat striatal cholinergic interneurones. J. Physiol. (Lond.), 533: 479–492.

Mori, A., Shindou, T., Ichimura, M., Nonaka, H. and Kase, H. (1996) The role of adenosine A$_{2A}$ receptors in regulating GABAergic synaptic transmission in striatal medium spiny neurons. J. Neurosci., 16: 605–611.

Nakamura, M., Jang, I.-S., Ishibashi, H., Watanabe, S. and Akaike, N. (2003) Possible roles of kainate receptors on GABAergic nerve terminals projecting to rat substantia nigra dopaminergic neurons. J. Neurophysiol., 90: 1662–1670.

Nisenbaum, E.S., Berger, T.W. and Grace, A.A. (1992) Presynaptic modulation by GABA$_B$ receptors of glutamatergic excitation and GABAergic inhibition of neostriatal neurons. J. Neurophysiol., 67: 477–481.

Nisenbaum, E.S., Berger, T.W. and Grace, A.A. (1993) Depression of glutamatergic and GABAergic synaptic responses in striatal spiny neurons by stimulation of presynaptic GABA_B receptors. Synapse, 14: 221–242.

O'Connor, W.T., Tanganelli, S., Ungerstedt, U. and Fuxe, K. (1992) The effects of neurotensin on GABA and acetylcholine release in the dorsal striatum of the rat: an in vivo microdialysis study. Brain Res., 573: 209–216.

Paladini, C.A., Celada, P. and Tepper, J.M. (1999) Striatal, pallidal, and pars reticulata evoked inhibition of nigrostriatal dopaminergic neurons is mediated by GABA_A receptors in vivo. Neuroscience, 89: 799–812.

Perez-Rosello, T., Figueroa, A., Salgado, H., Vilchis, C., Tecuapetla, F., Guzman, J.N., Galarraga, E. and Bargas, J. (2005) Cholinergic control of firing pattern and neurotransmission in rat neostriatal projection neurons: role of Ca_V2.1 and Ca_V2.2 Ca^{2+} channels. J. Neurophysiol., 93: 2507–2519.

Radnikow, G. and Misgeld, U. (1998) Dopamine D_1 receptors facilitate GABA_A synaptic currents in the rat substantia nigra pars reticulata. J. Neurosci., 18: 2009–2016.

Radnikow, G., Rohrbacher, J. and Misgeld, U. (1997) Heterogeneity in use-dependent depression of inhibitory postsynaptic potentials in the rat neostriatum in vitro. J. Neurophysiol., 77: 427–434.

Radnikow, G., Titz, S., Mades, S., Bäurle, J. and Misgeld, U. (2001) γ-Aminobutyric acid_B autoreceptors in substantia nigra and neostriatum of the weaver mutant mouse. Neurosci. Lett., 299: 81–84.

Raiteri, M., Marchi, M., Paudice, P. and Pittaluga, A. (1990) Muscarinic receptors mediating inhibition of γ-aminobutyric acid release in rat corpus striatum and their pharmacological characterization. J. Pharmacol. Exp. Ther., 254: 496–501.

Reiner, A. and Anderson, K.D. (1993) Co-occurrence of γ-aminobutyric acid, parvalbumin and the neurotensin-related neuropeptide LANT6 in pallidal, nigral and striatal neurons in pigeons and monkeys. Brain Res., 624: 317–325.

Salgado, H., Tecuapetla, F., Perez-Rosello, T., Perez-Burgos, A., Perez-Garci, E., Galarraga, E. and Bargas, J. (2005) A reconfiguration of Ca_V2 Ca^{2+} channel current and its dopaminergic D_2 modulation in developing neostriatal neurons. J. Neurophysiol., 94: 3771–3787.

Seabrook, G.R., Howson, W. and Lacey, M.G. (1991) Subpopulations of GABA-mediated synaptic potentials in slices of rat dorsal striatum are differentially modulated by presynaptic GABA_B receptors. Brain Res., 562: 332–334.

Shen, K.-Z. and Johnson, S.W. (1997) Presynaptic GABA_B and adenosine A_1 receptors regulate synaptic transmission to rat substantia nigra reticulata neurones. J. Physiol. (Lond.), 505: 153–163.

Shen, K.-Z. and Johnson, S.W. (2000) Presynaptic dopamine D_2 and muscarine M_3 receptors inhibit excitatory and inhibitory transmission to rat subthalamic neurones in vitro. J. Physiol. (Lond.), 525: 331–341.

Shen, K.-Z. and Johnson, S.W. (2001) Presynaptic GABA_B receptors inhibit synaptic inputs to rat subthalamic neurons. Neuroscience, 108: 431–436.

Shindou, T., Mori, A., Kase, H. and Ichimura, M. (2001) Adenosine A_{2A} receptor enhances GABA_A-mediated IPSCs in the rat globus pallidus. J. Physiol. (Lond.), 532: 423–434.

Smith, Y. and Kieval, J.Z. (2000) Anatomy of the dopamine system in the basal ganglia. Trends Neurosci., 23: S28–S33.

Stanford, I.M. and Lacey, M.G. (1996) Differential actions of serotonin, mediated by 5-HT_{1B} and 5-HT_{2C} receptors, on GABA-mediated synaptic input to rat substantia nigra pars reticulata neurons in vitro. J. Neurosci., 16: 7566–7573.

Sugimoto, T. and Mizuno, N. (1987) Neurotensin in projection neurons of the striatum and nucleus accumbens, with reference to coexistence with enkephalin and GABA: an immunohistochemical study in the cat. J. Comp. Neurol., 257: 383–395.

Szabo, B., Dörner, L., Pfreundtner, C., Nörenberg, W. and Starke, K. (1998) Inhibition of GABAergic inhibitory postsynaptic currents by cannabinoids in rat corpus striatum. Neuroscience, 85: 395–403.

Tepper, J.M., Martin, L.P. and Anderson, D.R. (1995) GABA_A receptor-mediated inhibition of rat substantia nigra dopaminergic neurons by pars reticulata projection neurons. J. Neurosci., 15: 3092–3103.

Thomson, A.M. (2000) Facilitation, augmentation and potentiation at central synapses. Trends Neurosci., 23: 305–312.

Trevitt, T., Carlson, B., Correa, M., Keene, A., Morales, M. and Salamone, J.D. (2002) Interactions between dopamine D1 receptors and γ-aminobutyric acid mechanisms in substantia nigra pars reticulata of the rat: neurochemical and behavioral studies. Psychopharmacology (Berl.), 159: 229–237.

Trevitt, J.T., Carlson, B.B., Nowend, K. and Salamone, J.D. (2001) Substantia nigra pars reticulata is a highly potent site of action for the behavioral effects of the D1 antagonist SCH 23390 in the rat. Psychopharmacology (Berl.), 156: 32–41.

Valenti, O., Marino, M.J., Wittmann, M., Lis, E., DiLella, A.G., Kinney, G.G. and Conn, P.J. (2003) Group III metabotropic glutamate receptor-mediated modulation of the striatopallidal synapse. J. Neurosci., 23: 7218–7226.

Waldmeier, P.C., Stöcklin, K. and Feldtrauer, J.-J. (1992) Systemic administration of baclofen and the GABA_B antagonist, CGP 35348, does not affect GABA, glutamate or aspartate in microdialysates of the striatum of conscious rats. Naunyn-Schmiedeberg's Arch. Pharmacol., 345: 548–552.

Wallmichrath, I. and Szabo, B. (2002a) Analysis of the effect of cannabinoids on GABAergic neurotransmission in the substantia nigra pars reticulata. Naunyn-Schmiedeberg's Arch. Pharmacol., 365: 326–334.

Wallmichrath, I. and Szabo, B. (2002b) Cannabinoids inhibit striatonigral GABAergic neurotransmission in the mouse. Neuroscience, 113: 671–682.

Wilson, R.I. and Nicoll, R.A. (2001) Endogenous cannabinoids mediate retrograde signalling at hippocampal synapses. Nature, 410: 588–592.

Wittmann, M., Marino, M.J., Bradley, S.R. and Conn, P.J. (2001) Activation of group III mGluRs inhibits GABAergic and glutamatergic transmission in the substantia nigra pars reticulata. J. Neurophysiol., 85: 1960–1968.

Yanovsky, Y., Mades, S. and Misgeld, U. (2003) Retrograde signaling changes the venue of postsynaptic inhibition in rat substantia nigra. Neuroscience, 122: 317–328.

Yanovsky, Y., Zhang, W. and Misgeld, U. (2005) Two pathways for the activation of small-conductance potassium channels in neurons of substantia nigra pars reticulata. Neuroscience, 136: 1027–1036.

Zahniser, N.R., Simosky, J.K., Mayfield, R.D., Negri, C.A., Hanania, T., Larson, G.A., Kelly, M.A., Grandy, D.K., Rubinstein, M., Low, M.J. and Fredholm, B.B. (2000) Functional uncoupling of adenosine A_{2A} receptors and reduced response to caffeine in mice lacking dopamine D_2 receptors. J. Neurosci., 20: 5949–5957.

Tepper, Abercrombie & Bolam (Eds.)
Progress in Brain Research, Vol. 160
ISSN 0079-6123
2007 Elsevier B.V.

CHAPTER 15

Development of striatal fast-spiking GABAergic interneurons

Marie-Françoise Chesselet[1,*], Joshua L. Plotkin[1], Nanping Wu[2] and Michael S. Levine[2]

[1]*Departments of Neurology and Neurobiology, The David Geffen School of Medicine at UCLA, Los Angeles, CA 90095-1769, USA*
[2]*Mental Retardation Research Center, The David Geffen School of Medicine at UCLA, Los Angeles, CA 90095-1769, USA*

Abstract: Fast-spiking GABAergic interneurons represent a very small portion of striatal neurons, yet they play a critical role in modulating cortical input and mediating inhibition of striatal medium-sized spiny projection neurons. Considering their pivotal role in the adult striatum, it is of importance to determine when during development these neurons acquire their characteristic properties and function. In this review we describe recent work from our laboratories indicating that fast-spiking GABAergic interneurons are under stronger cortical control than efferent neurons at postnatal day 12 but mature considerably between postnatal days 12–19 in the rat striatum. During this time period, their molecular development is under the control of GABAergic and cholinergic mechanisms. Thus, fast-spiking interneurons are poised to influence striatal function and perhaps development during the postnatal period in rats, and their properties could be influenced by commonly used pharmacological agents during a protracted developmental window. These findings point to the need for future research to better understand the functional maturation of this critical population of striatal GABAergic neurons, and the consequences of abnormal maturation of these cells.

Keywords: basal ganglia; acetylcholine; rat; corticostriatal; organotypic cultures

Introduction

After it was realized that contrary to previous beliefs, medium-sized spiny efferent neurons comprise more than 90% of striatal neurons, less attention was focused on the sparse interneurons present in the striatum. In the last few years, however, it has been realized that these interneurons play important roles in striatal circuitry, and these roles differ for each class of interneuron. Similar to the striatal efferent neurons, three of the four known classes of

interneurons (excluding cholinergic interneurons), parvalbumin-positive, nitric oxide synthase (NOS)/nicotinamide adenine dinucleotide phosphate-diaphorase (NADPH-diaphorase)-positive and calretinin-positive interneurons contain some level of γ-aminobutyric acid (GABA) (Cowan et al., 1990; Kawaguchi, 1993; Kawaguchi et al., 1995). Parvalbumin interneurons, however, occupy a special position in that they contain very high levels of GABA, GABA synthesizing enzymes, especially glutamic acid decarboxylase (Mr 67,000: GAD67) and the GABA transporter (Chesselet and Robbins, 1989; Augood et al., 1995). These molecular characteristics suggest that they may

*Corresponding author. Tel.: 310-267-1781; Fax: 310-267-1786;
E-mail: Mchesselet@mednet.ucla.edu

DOI: 10.1016/S0079-6123(06)60015-0

synthesize and release high levels of GABA, and indeed, it has been shown that they belong to the category of fast-spiking GABAergic interneurons also found in other brain regions (Kawaguchi et al., 1987; Kawaguchi, 1993; Kawaguchi and Kubota, 1993). Similar to these cells, in addition to high levels of GAD67 and parvalbumin, they express the Shaw-like potassium channel Kv3.1 and are linked by GAP junctions containing connexin 36 (Chesselet and Robbins, 1989; Kawaguchi, 1993; Lenz et al., 1994; Belluardo et al., 2000).

In the striatum, fast-spiking GABAergic interneurons receive direct inputs from the cerebral cortex and synapse on striatal efferent neurons (Bennett and Bolam, 1994). Although they represent less than 1% of the total number of neurons in the dorsal striatum (Luk and Sadikot, 2001), they play a critical role in striatal information processing because they mediate cortical feed-forward inhibition of medium-sized spiny neurons, the main efferent cells of the striatum (Kita et al., 1990; Bennett and Bolam, 1994; Plenz and Aertsen, 1996; Plenz and Kitai, 1998; Koos and Tepper, 1999; Tepper et al., 2004).

Electrophysiological properties of striatal fast-spiking GABAergic interneurons during postnatal development

Embryonic birth-dating studies have revealed that striatal fast-spiking GABAergic interneurons are born slightly earlier than most medium-sized spiny efferent neurons (Bayer, 1984; Sadikot and Sasseville, 1997). Earlier studies of their postnatal maturation relied on immunohistochemical detection of parvalbumin, a characteristic molecular marker of these neurons in the striatum (Cowan et al., 1990; Lenz et al., 1994). Based on this index, fast-spiking GABAergic interneurons mature during the second and third postnatal weeks in rat pups (Schlosser et al., 1999). This corresponds to the maturation period of morphological and physiological properties of medium-sized spiny efferent neurons (Tepper and Trent, 1993; Colwell et al., 1998; Tepper et al., 1998; Uryu et al., 1999; Hurst et al., 2001). Although evidence indicated that striatal fast-spiking GABAergic interneurons morphologically

mature earlier than efferent neurons (Tepper et al., 1998; Schlosser et al., 1999), it was unknown whether their functional maturation parallels or precedes that of medium-sized spiny neurons.

Electrophysiological studies performed at postnatal day 16 in rat pups show that striatal fast-spiking GABAergic interneurons already have the ability to fire at extremely high frequencies with little adaptation, narrow action potentials (AP) and rapid large amplitude afterhyperpolarizations (AHP) at that early age (Kawaguchi, 1993; Koos and Tepper, 1999). In view of their critical contribution to striatal physiology, it is of interest to determine how these properties mature during this critical postnatal phase of striatal development (Aaron and Chesselet, 1989; Szele et al., 1994; Butler et al., 1998), which, in the rat, also includes the major period of corticostriatal synaptogenesis (Hattori and McGeer, 1973; Sharpe and Tepper, 1998; Uryu et al., 1999).

The electrophysiological properties of fast-spiking GABAergic interneurons just described are believed to critically depend on their molecular make-up, in particular the expression of the Kv3.1 potassium channel (Yokoyama et al., 1989; Kanemasa et al., 1995; Perney and Kaczmarek, 1997; Wang et al., 1998). Similarly, their ability to handle high intracellular calcium resulting from their fast-spiking, and to release correspondingly high amounts of GABA, is thought to depend on their high levels of expression of parvalbumin and GAD67, respectively. Therefore, in a first set of studies, we examined the level of expression of mRNA encoding these proteins in rat striatum at various times during postnatal development, with single cell in situ hybridization histochemistry. In the adult striatum, expression of high levels of these mRNAs is specific for fast-firing GABAergic interneurons and can be used to identify them. For example, levels of expression of GAD67 mRNA are approximately tenfold higher in fast-spiking interneurons than in medium-sized spiny efferent neurons (Soghomonian et al., 1992). Parvalbumin and Kv3.1 mRNA are not routinely detected by in situ hybridization histochemistry in other striatal cells (Lenz et al., 1994).

Sections of dorsal striatum from rat pups killed at postnatal day 14 (P14), P18 and P22 were processed for in situ hybridization histochemistry with cRNA probes labeled with ^{35}S-UTP and subsequently processed for emulsion autoradiography to permit single cell resolution and quantification (Chesselet et al., 1987; Chesselet, 1996). We observed a 68% increase in the relative expression of parvalbumin mRNA per cell from P14 to P22 with an intermediate level at P18 (Plotkin et al., 2005). Similar to findings with parvalbumin immunohistochemistry (Schlosser et al., 1999), the number of parvalbumin mRNA-positive striatal cells gradually increased from P14 to P22. Kv3.1 mRNA expression also increased 40% during the same time period (Plotkin et al., 2005). In contrast, GAD67 mRNA levels did not change between P14 and P22 (Fig. 1). These molecular data suggest a differential regulation of mRNA encoding the principal enzyme of GABA synthesis in these neurons, compared to mRNA encoding proteins that may be more critical for their physiological properties. The progressive increase of the latter suggests that the functional properties of these proteins mature during the third postnatal week, and prompted us to examine the functional development of fast-spiking GABAergic interneurons during the same period of postnatal development, in comparison to medium-sized spiny neurons.

The changes in intrinsic electrophysiological properties and responses to cortical afferents were determined in striatal slices from rat pups aged P12–14 and P19–23 (Plotkin et al., 2005). Interneurons were initially identified under infrared differential interference contrast (IR-DIC) optics by their size, shape and morphology, and confirmed to be interneurons by their electrophysiological characteristics, which were already quite distinct even at the younger age. Indeed, we found with patch clamp recordings that at both P12–14 and P19–23 fast-spiking interneurons fired at high frequencies with little adaptation and had narrower APs with larger AHPs than medium-sized spiny neurons (Fig. 2), and generally similar to those previously reported in P16–20 Wistar rat pups (Kawaguchi, 1993). At P19–23, fast-spiking interneurons had input resistance and qualitative properties comparable to those observed by others (Bracci et al., 2002; Koos and Tepper, 1999) in P25–40 Wistar rats and P17–32 Sprague-Dawley

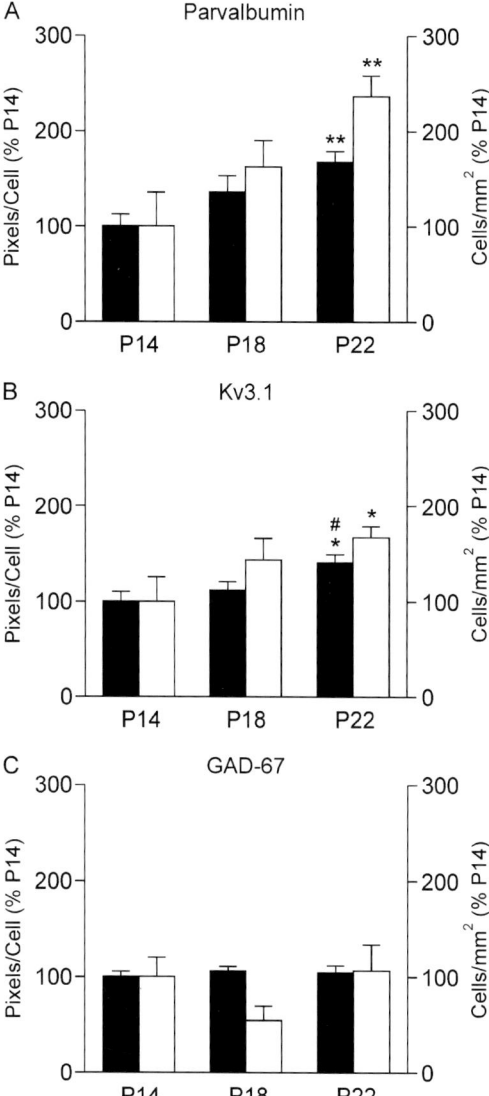

Fig. 1. Relative mRNA levels in FS at P14, P18 and P22. Quantitative in situ hybridization results showing relative mRNA levels (measured as silver grain area per cell; black bars) and the number of cells expressing mRNA per mm^2 (white bars) for (A) parvalbumin, (B) Kv3.1 and (C) GAD-67 in the dorsolateral striatum. There is a significant increase in both the mRNA expression per cell and the number of mRNA-expressing cells for parvalbumin and Kv3.1 between P14–P22. There is no significant increase in either mRNA levels or cell numbers for GAD-67. All statistics were performed on absolute numbers. ** indicates a significant difference ($p<0.01$) from the P14 value; * indicates a significant difference ($p<0.05$) from the P14 value; # indicates a significant difference ($p<0.05$) from the P18 value using Fisher's LSD post hoc tests. FS = fast-spiking interneurons. Figure taken from Plotkin et al. (2005).

264

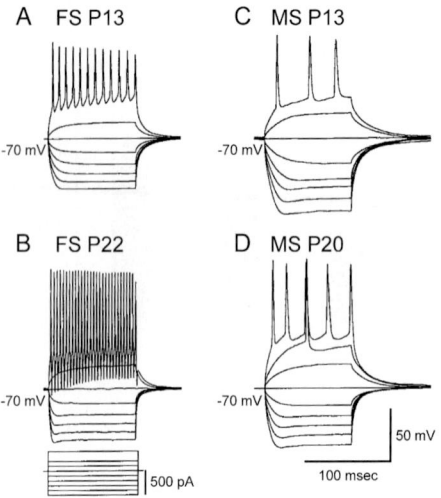

A FS P13

−70 mV

B FS P22

−70 mV

C MS P13

−70 mV

D MS P20

−70 mV

50 mV

500 pA

100 msec

Fig. 2. Examples of responses of FS and MS to depolarizing and hyperpolarizing current injections (100 ms duration and varying amplitude). (A, B) Whole cell patch current clamp responses of FS from the two developmental age groups. Note the rapid rate of firing at both ages. (C, D) Whole cell patch current clamp responses of MS from the two developmental age groups. Note the ramping depolarization and delayed first AP in medium-sized spiny neurons, particularly evident at the lower amplitude depolarizing current injection at P20 in panel D. Also note the slower firing frequencies compared to (A, B). All AP trains represent the maximum firing rates of each type of cell to injected current. Baseline membrane potential is −70 mV for all cells. Calibration refers to all cells. FS = fast-spiking interneuron; MS = medium-sized spiny neuron. Figure modified from Plotkin et al. (2005).

rats, respectively. At the end of the experiment, biocytin labeling was performed to confirm the identity of the recorded fast-spiking interneurons and medium-sized spiny neurons. Microscopic examination of recovered P19–23 fast-spiking neurons confirmed the smoothness of their dendrites, except for occasional protrusions in the distal dendritic regions in some of the neurons (Fig. 3). Overall, we observed morphological characteristics that were consistent with those previously reported (Kawaguchi, 1993).

It was much more difficult to recover the younger neurons. The single P12–14 fast-spiking interneuron recovered had 5–6 primary processes emanating from the soma and smooth, spineless dendrites. The non-fast-spiking neurons recovered from both P12–14 and P19–23 age groups had moderate spine density (except for one young

neuron with smooth dendrites), which was expected considering that spine density continues to increase until the end of the fourth postnatal week in efferent striatal neurons (Tepper et al., 1998). As previously observed (Tepper and Trent, 1993; Tepper et al., 1998), most of the non-fast-spiking neurons recovered in the younger age group had varicose dendrites (Fig. 3).

The electrophysiological analysis confirmed the hypothesis that functional properties of both fast-spiking and medium-sized spiny efferent neurons mature considerably between P12–14 and P19–23. In addition, the data demonstrated that marked differences exist between these two types of cells (Plotkin et al., 2005). Briefly, input resistance decreased whereas AP amplitude increased in both types of neurons between P12–14 and P19–23. Only fast-spiking interneurons, however, showed a significant increase in AHP amplitude, and significant decreases in AP duration at half maximum amplitude and time to peak AHP during the same developmental period (Table 1). Furthermore, the relative developmental rates of many electrophysiological properties were greater in fast-spiking interneurons than medium-sized spiny neurons during this time. When P12–14 values were expressed as the percent of their value at P19–23, the AP duration at half maximum amplitude and time to peak AHP decreased and AHP amplitude and maximum firing frequency increased at significantly greater rates in fast-spiking interneurons than in medium-sized spiny neurons. Input resistance and AP amplitude changed at similar rates in both cell types (Table 2).

In response to the maximum depolarizing current injection that does not induce AP failures, the maximum firing frequencies of both fast-spiking interneurons and medium-sized spiny neurons significantly increased from P12–14 to P19–23 (Fig. 2; Table 1). The firing pattern of fast-spiking interneurons, however, was similar in the two age groups. In response to depolarizing current injections near threshold strength for AP initiation, fast-spiking interneurons in both age groups responded with interrupted bursts of APs and membrane potential oscillations, although the frequency within each burst was markedly lower in the younger age group (Fig. 4; Plotkin et al., 2005).

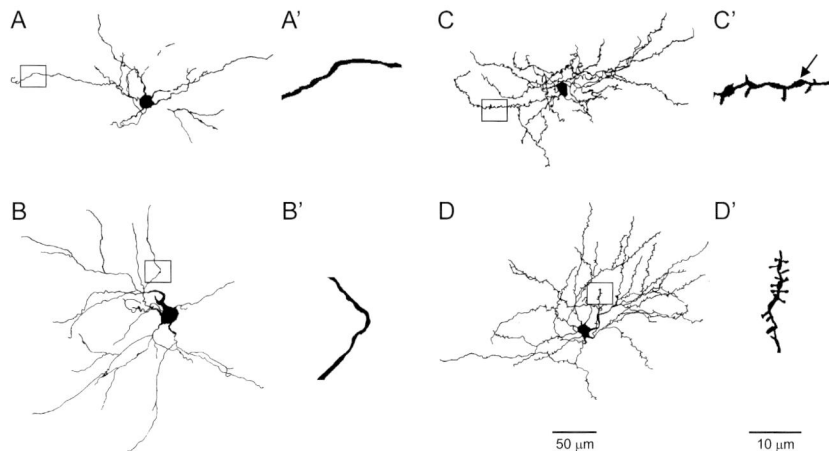

Fig. 3. Partial reconstructions of biocytin-filled FS and MS. On the left of each pair are camera lucida drawings of (A) a P14 FS (B) a P21 FS, (C) a P14 MS and (D) a P22 MS. On the right (A′–D′) are high power camera lucida drawings of the dendrites of (A–D). Note the presence of spines on MS from each age group (C, D), which are slightly more dense in the older neuron (D). Also note the smoothness of FS dendrites from each age group (A, B). Boxes in the camera lucida reconstructions in (A–D) depict the regions from which the high power camera lucida drawings in (A′)–(D′) were made. Arrow in (C′) indicates a dendritic varicosity. Scale bar in (D) corresponds to (A)–(D); scale bar in (D′) corresponds to (A′)–(D′). FS = fast-spiking interneuron; MS = medium-sized spiny neuron. Figure taken from Plotkin et al. (2005).

Table 1. Membrane properties of fast-spiking interneurons and medium-sized spiny neurons

	FS P12–14 ($n = 10$)	FS P19–23 ($n = 11$)	MS P12–14 ($n = 12$)	MS P19–23 ($n = 9$–11)
Input resistance (MΩ)	117.42 ± 10.12	91.76 ± 5.89*	120.47 ± 6.23	92.41 ± 8.84[†]
AP frequency (Hz)	133 ± 12.02	284.54 ± 13.24**	39.17 ± 3.13	52.73 ± 3.33[†]
AP threshold current injection (pA)	200 ± 21.08	236.36 ± 62.19	175 ± 25	200 ± 33.03
Amplitude of AP (mV)	62.23 ± 3.46	74.03 ± 2.77*	79.43 ± 2.32	95.56 ± 3.34[†]
AP duration at 1/2 amplitude (ms)	0.91 ± 0.06	0.58 ± 0.02**	1.7 ± 0.06	1.55 ± 0.16
Amplitude of AHP (mV)	11.81 ± 1.8	35.18 ± 1.42**	6.38 ± 0.52	5.99 ± 1.22
Time to peak of AHP (ms)	3.11 ± 0.31	1.6 ± 0.04**	9.66 ± 0.73	12.35 ± 1.84

Note: All AP properties were measured from the first AP in a train resulting from a depolarizing current injection of threshold strength for inducing APs. The frequency was measured as the maximum frequency at which the cell fired before failures were observed, in response to a 100 ms pulse. Significance was determined with t-tests. For MS P19–23 $n = 11$ for all properties except input resistance ($n = 9$). FS = fast-spiking interneuron; MS = medium-sized spiny neuron. Modified from Plotkin et al. (2005).
*Indicates a significant difference ($p < 0.05$) from FS P12–14.
**Indicates a significant difference ($p < 0.01$) from FS P12–14.
[†]Indicates a significant difference ($p < 0.01$) from MS P12–14.

This bursting pattern is consistent with previous reports in mature fast-spiking interneurons (Koos and Tepper, 1999; Bracci et al., 2002), and was not seen in medium-sized spiny neurons in either age group (Fig. 4). Medium-sized spiny neurons in both age groups showed a ramping depolarization and noticeable delay to the first AP in response to threshold strength depolarizing current injections (Fig. 2D), consistent with previous observations (Tepper et al., 1998).

As described above, some properties of mature fast-spiking interneurons are already observed at P12, such as their bursting pattern of firing. Our data (Plotkin et al., 2005) show that, in some aspects, fast-spiking interneurons display mature properties earlier than medium-sized spiny neurons. For example, another mature trait seen in fast-spiking interneurons at P12–14 is their lack of AP train adaptation. AP train adaptation was measured every 50 ms over a 500 ms period in response to depolarizing current injections of two different intensities: (1) the threshold current injection that induced an AP train and; (2) the maximum current injection that produced an AP train without inducing AP failures (Plotkin et al., 2005). Fast-spiking interneurons in both age groups showed little to no adaptation in response to threshold or maximum current injections, consistent with previous reports showing little to no adaptation in mature fast-spiking interneurons (Kawaguchi, 1993; Koos and Tepper, 1999).

Table 2. Rates of development of membrane properties in fast-spiking interneurons and medium-sized spiny neurons

	FS ($n = 10$)	MS ($n = 12$)
Input resistance	128 ± 11	130 ± 7
AP frequency	47 ± 4	$74 \pm 6^*$
AP threshold current injection	85 ± 9	88 ± 13
Amplitude of AP	84 ± 5	83 ± 2
AP duration at 1/2 amplitude	156 ± 10	$110 \pm 4^*$
Amplitude of AHP	34 ± 5	$107 \pm 9^*$
Time to peak of AHP	194 ± 19	$78 \pm 6^*$

Note: Values at P12–14 represented as the percent of the P19–23 value. Note the greater developmental change in AP frequency, AP duration at half maximum amplitude, AHP amplitude and time to peak AHP in FS compared to MS. FS = fast-spiking interneuron; MS = medium-sized spiny neuron. Modified from Plotkin et al. (2005).
*Indicates a significant difference ($p < 0.01$) with t-tests.

Medium-sized spiny neurons in the older group showed marked adaptation in response to threshold or maximum current injections, whereas medium-sized spiny neurons in the younger group showed little to no adaptation in response to either maximum or threshold current injections (Fig. 5). Thus, medium-sized spiny neurons acquire the tendency to adapt between P12–14 and P19–23, representing an aspect of maturation that is prolonged in medium-sized spiny neurons and not required in fast-spiking interneurons.

We addressed the question of how GABAergic fast-spiking interneurons and medium-sized spiny neurons respond to cortical inputs during development because each neuronal type receives prominent cortical innervation when mature but there are profound different consequences that excitation of each cell type will have in the striatum (Plotkin et al., 2005). Stimulation of the corpus callosum overlying the dorsolateral striatum permitted examination of cortical inputs to fast-spiking interneurons and medium-sized spiny neurons. Both cell types showed increasing amplitude excitatory postsynaptic potentials (EPSPs) in response to increasing stimulus intensities at each age group. Stimulus intensities required to evoke relatively low amplitude EPSPs (20% of the

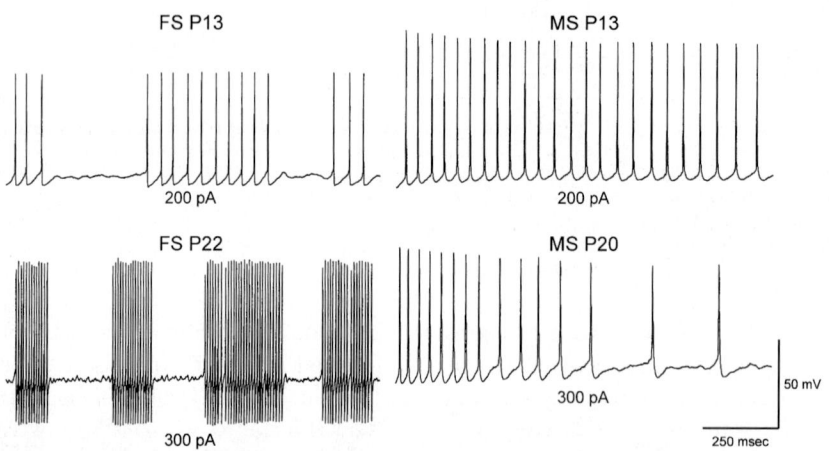

Fig. 4. Whole cell patch clamp responses of FS and MS to depolarizing current injections of near-threshold strength for action potential (AP) trains. Note the interrupted bursts of APs exhibited by FS at each age, with membrane potential oscillations between bursts. Also note the lack of bursts in MS, the marked adaptation of the MS in the older age group and minimum adaptation in the MS in the younger age group. FS P13 and MS P13 are in response to 200 pA current injections. FS P22 and MS P20 are in response to 300 pA current injections. Baseline membrane potential of all cells is –70 mV. Calibration refers to all cells. FS = fast-spiking interneuron; MS = medium-sized spiny neuron. Figure taken from Plotkin et al. (2005).

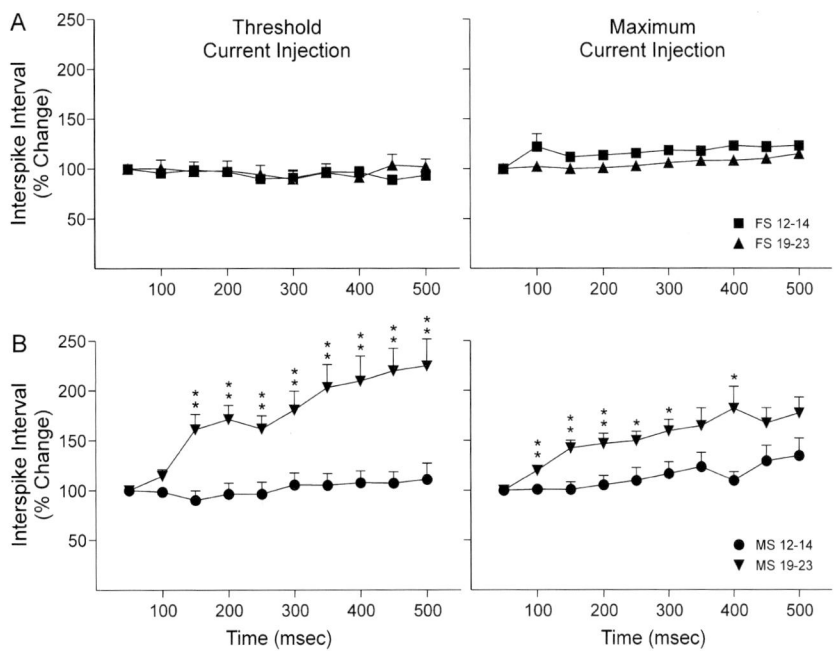

Fig. 5. Adaptation of action potential (AP) trains in FS and MS. Adaptation curves for (A) FS and (B) MS showing the instantaneous interspike interval every 50 ms for a 500 ms AP train initiated by a depolarizing current injection of threshold strength or the maximum depolarizing current injection that does not induce AP failures. To facilitate comparisons the interspike intervals of all cells were normalized to the first (50 ms) interspike interval measured. (A) FS in both age groups show minimum adaptation in response to threshold or maximum current injections. Interspike intervals are not significantly different between age groups over time at either current injection intensity (two way ANOVA). (B) MS in the P19–23 age group show prominent adaptation at both threshold and maximum current injections, while P12–14 MS show minimum adaptation in response to threshold current injections and significantly less adaptation than P19–23 MS in response to maximum current injections. Interspike intervals are significantly different between age groups over time at both current injection intensities ($p < 0.01$, two way ANOVA). * indicates a significant difference ($p < 0.05$) and ** indicates a significant difference ($p < 0.01$) using t-tests. FS = fast-spiking interneuron; MS = medium-sized spiny neuron. Figure taken from Plotkin et al. (2005).

maximum response) and maximum EPSPs (resulting in APs) increased with age in both cell types. There was no significant difference in the stimulus intensities required to evoke EPSPs at 20% of the maximum response between fast-spiking interneurons and medium-sized spiny neurons of the same age. In the P12–14 age group, significantly lower stimulation was required to evoke maximum EPSPs, resulting in APs, in fast-spiking interneurons compared to medium-sized spiny neurons. There was a non-significant trend for lower stimulation to be required to evoke maximum EPSPs in fast-spiking interneurons compared to medium-sized spiny neurons in the P19–23 age group (Fig. 6). Thus, in the developing striatum, weaker cortical inputs are likely to be perceived more intensely by fast-spiking interneurons. Furthermore,

the more gradual slopes of the input-output curves of both cell types at P19–23 (Fig. 6B) demonstrate that fast-spiking interneurons respond with larger subthreshold EPSPs than medium-sized spiny neurons to a broader range of stimulus intensities. This may suggest greater integration of cortical inputs by the two cell types in the older age group.

As another means of studying excitatory afferents over development, spontaneous depolarizations were examined in fast-spiking interneurons and medium-sized spiny neurons in the two age groups. Such spontaneous depolarizations are glutamatergic in origin, as they can be blocked by the non-N-methyl-D-aspartate (NMDA) glutamate receptor antagonist 6-cyano-7-nitroquinoxaline (CNQX) (Plotkin et al., 2005). The average number of events/min between 0.5–1 mV was

268

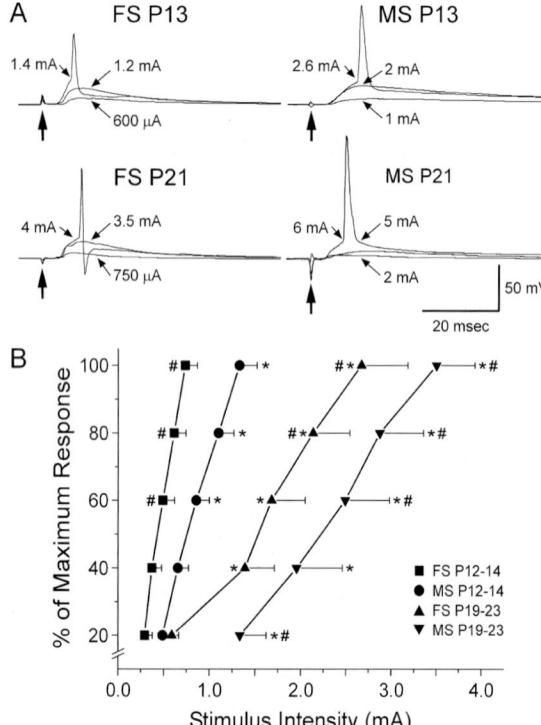

Fig. 6. Postsynaptic responses to stimulation of corticostriatal inputs. (A) Responses to stimulation of the overlying corpus callosum with increasing current intensities. Each group consists of three traces of increasing stimulation intensities taken from the same cell. Traces are overlaid and aligned to the stimulus artifact (vertical arrow). Membrane potential is −70 mV. Calibration refers to all traces. (B) Input-output curves comparing the responses of striatal FS and MS at P12–14 and P19–23 to increasing stimulation intensities. Proportionately larger responses were evoked by lower stimulation intensities in P12–14 compared to P19–23 groups in both types of neurons. In addition, at P12–14 greater responses occurred to lower stimulation intensities in FS compared to MS. * indicates a significant difference ($p < 0.05$) from FS P12–14 and # indicates a significant difference ($p < 0.05$) from MS P12–14 ($n = 6$–11 per group) using Mann-Whitney U nonparametric tests. FS = fast-spiking interneuron; MS = medium-sized spiny neuron. Figure taken from Plotkin et al. (2005).

significantly greater in fast-spiking interneurons compared to medium-sized spiny neurons at each age group. There was no significant change in the average number of spontaneous depolarizations between age groups for either cell type, although it should be noted that fast-spiking interneurons show a rather large variation in both age groups (Fig. 7). Thus, fast-spiking interneurons receive

more spontaneous small amplitude depolarizing inputs than medium-sized spiny neurons at each age group examined, and fast-spiking interneurons already receive more spontaneous depolarizing inputs than medium-sized spiny neurons by P12–14. Taken together, the data suggest that fast-spiking interneurons are under stronger cortical control than medium-sized spiny neurons by P12, a time when the striatum is still actively maturing.

Neurochemical control of striatal fast-spiking GABAergic interneuron development

The molecular and electrophysiological findings clearly show that fast-spiking GABAergic interneurons undergo important maturational changes during the third postnatal week in rat pups. Based on ultrastructural studies, this period corresponds to a time of major increase and then pruning of corticostriatal synapses (Hattori and McGeer, 1973; Sharpe and Tepper, 1998; Uryu et al., 1999). However, the published studies have essentially sampled the development of contacts between cortical inputs and the abundant striatal efferent neurons. Because interneurons represent such a small percent of the population of cells sampled, little is known of the development of cortical innervation of fast-spiking interneurons. Interestingly, the electrophysiological data we have obtained in striatal slices from P12–14 rat pups show that at this early age, fast-spiking GABAergic interneurons strongly respond to cortical stimulation, suggesting that they receive substantial cortical innervation at a time when asymmetric synapses, the type made by cortical inputs, are still very rare in the rat striatum (Hattori and McGeer, 1973; Sharpe and Tepper, 1998; Uryu et al., 1999). It is unknown, however, whether cortical inputs are required for the development of the molecular characteristics that mature after birth and distinguish the fast-spiking interneurons from other striatal cells, in particular the presence of parvalbumin (Schlosser et al., 1999). To begin addressing this question, we have examined parvalbumin immunohistochemistry in organotypic cultures of rat striatum, grown in the absence of cerebral cortex.

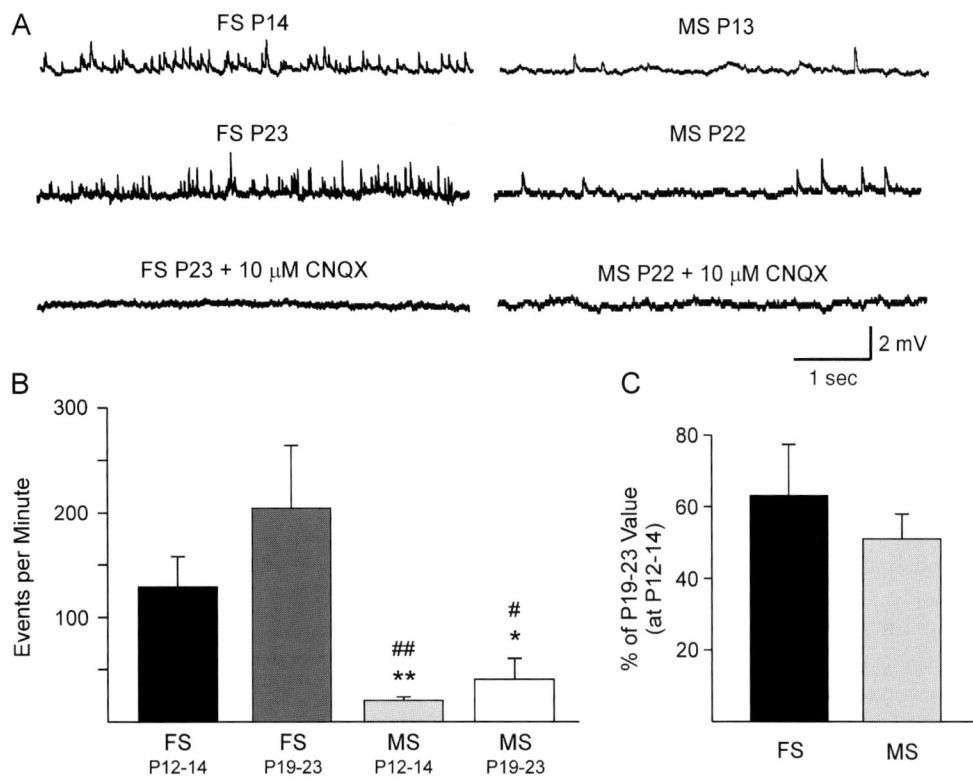

Fig. 7. Spontaneous depolarizing potentials in FS and MS. (A) Sample traces showing spontaneous depolarizing potentials in FS and MS. Bottom traces show the same P23 FS (left) and P22 MS (right), in the presence of the non-NMDA glutamate receptor antagonist CNQX (10 μM). Note that these potentials no longer occur indicating they are mediated by non-NMDA ionotropic glutamate receptors. All recordings were done in current clamp conditions in the presence of 20 μM bicuculline methiodide. Baseline membrane potential of all traces is −70 mV. Calibration refers to all traces. (B) Bar graphs depicting the frequency of spontaneous depolarizations between 0.5–1.0 mV in amplitude for each neuron type. Note the increased frequencies of spontaneous depolarizations in FS at each developmental age, compared to the relatively low frequencies seen in MS. (C) Histogram depicting the frequency of spontaneous depolarizations at P12–14 as the percent of the P19–23 value. Note the similar percentage in both cell types. * indicates a significant difference ($p < 0.05$) from FS P12–14; ** indicates a significant difference ($p < 0.01$) from FS P12–14; # indicates a significant difference ($p < 0.05$) from FS P19–23; ## indicates a significant difference ($p < 0.01$) from FS P19–23 ($n = 5$–6 per group) using Mann-Whitney U nonparametric tests. FS = fast-spiking interneuron; MS = medium-sized spiny neuron. Figure taken from Plotkin et al. (2005).

Striatal organotypic cultures were prepared by the roller tube method (Plenz and Aertsen, 1996) from brains isolated from P1 Sprague-Dawley rat pups. The cultures were grown for two weeks and processed for parvalbumin immunohistochemistry following routine protocols (Soghomonian et al., 1992) with slight modifications (Plotkin and Chesselet, in preparation). Numbers of parvalbumin-positive neurons were analyzed from camera lucida drawings. A first observation was that some 14–15-day-old striatal organotypic slice cultures contained a substantial

number of parvalbumin-positive cells, many of them with numerous process and distinct neuronal morphology. This indicates that cortical inputs are not needed after birth for the expression of parvalbumin in striatal fast-firing neurons and their morphological maturation. Determining whether corticostriatal inputs play a role in regulating the level of expression of parvalbumin and the degree of morphological differentiation, however, will require the use of corticostriatal co-cultures.

In vivo data (Schlosser et al., 1999) indicate that parvalbumin expression is very low at birth in striatal neurons. This suggests that, in our cultures, parvalbumin expression is developed without glutamatergic influences from striatal inputs, either cortical or thalamic, although we cannot exclude a role of glutamate from other, i.e., intrastriatal, sources. Given that no intrinsic striatal glutamatergic neurons have been described, the source of such glutamate would most likely be metabolic, either neuronal or glial.

Since parvalbumin expression can develop in the absence of striatal inputs, we examined a potential role for intrinsic striatal neurotransmitters in controlling interneuron development. We first focused on a role for acetylcholine mediated by nicotinic receptors. Indeed, in prior in vivo experiments, we have found that administration of 2 mg/kg injections of nicotine hydrogen tartrate twice a day for the first two postnatal weeks to rat pups decreased the number of parvalbumin-positive cells in the lateral striatum at P15. To further confirm a role of endogenous acetylcholine in the regulation of parvalbumin expression by way of nicotinic receptors, we exposed cultures to the nicotinic antagonist mecamylamine, which non-selectively blocks most subtypes of nicotinic receptors (Connolly et al., 1992). As previously reported (Plenz and Aertzen, 1996), cultures contained a wide range of parvalbumin cells. Therefore, the distribution of the number of parvalbumin cells per culture was compared between experimental groups. In the presence of mecamylamine, we observed a significant increase in the proportion of cultures that contained a high number of parvalbumin-positive cells, suggesting that by acting on nicotinic receptors, acetylcholine decreases parvalbumin expression, a result compatible with our in vivo data with the nicotinic agonist nicotine.

Nicotinic acetylcholine receptors have been shown to modulate GABA release in the striatum (Limberger et al., 1986; Radcliffe et al., 1999; Guo and Chiappinelli, 2000; Li et al., 2001). Therefore, we examined whether the effects observed after manipulation of nicotinic receptors both in vivo and in vitro could be mediated by alteration of GABAergic mechanisms. Cultures were grown for 14 days in the presence or absence of the $GABA_A$ receptor antagonist bicuculline methiodide (20 μM), and processed for parvalbumin immunoreactivity and quantitative analysis. Similar to the treatment with mecamylamine, we observed an increase in the number of cultures expressing high numbers of parvalbumin-positive cells. To explore the possibility of a mechanistic link between the two effects, cultures were grown for 14 days in the presence of mecamylamine hydrochloride (10 μM) plus the $GABA_A$ receptor agonist muscimol (10 μM), or without either drug. No significant difference in the number of cultures expressing either high or low numbers of parvalbumin-positive cells was observed, suggesting that continuous infusion of muscimol prevented mecamylamine's effect. Taken together, these data suggest that parvalbumin expression in the developing striatum is under the control of both cholinergic and GABAergic neurons, and that acetylcholine's effects may be mediated by its ability to increase GABA release by acting on nicotinic receptors.

Conclusions

Much remains to be elucidated about the timing and molecular mechanisms of striatal fast-spiking GABAergic interneuron development. The development of their electrophysiological characteristics suggests that they may mature earlier than the medium-sized spiny striatal efferent neurons. An intriguing possibility is that these early maturing neurons could then influence the later maturation of the striatal efferents. This hypothesis could be tested by examining the development of medium-sized spiny neurons in mice lacking molecules that are critical for the functioning of the fast-spiking interneurons, such as Kv3.1 potassium channels. A potentially important consequence of the asynchronous maturation of different classes of striatal neurons is that insults occurring at a given time during postnatal development may differentially affect each class of neuron and have different functional consequences on the striatum in adults. For example, we have shown that peripheral administration of nicotine during postnatal development produces a regionally specific decrease in parvalbumin expression in the striatum, suggesting that nicotine exposure, as caused by exposure to cigarette smoke, could affect the maturation of the fast-spiking GABAergic interneurons. Unfortunately, the behavioral consequences of dysfunctions of these striatal interneurons remain to be determined. Given their recognized importance in the shaping of striatal responses to cortical stimulation (Koos and Tepper, 1999), this should be an important avenue for research in the future. One may expect that results from such studies would even more strongly establish the significance of this important source of GABA in the striatum.

Abbreviations

AHP	afterhyperpolarization
AP	action potential
CNQX	6-cyano-7-nitroquinoxaline
EPSP	excitatory postsynaptic potential
GABA	gamma-aminobutyric acid
GAD-67	glutamic acid decarboxylase, Mr 67,000
IR-DIC	infrared differential interference contrast
NADPH-diaphorase	nicotinamide adenine dinucleotide phosphate-diaphorase
NMDA	N-methyl-D-aspartate
NOS	nitric oxide synthase
P	postnatal

Acknowledgments

We thank members of the Chesselet and Levine laboratories for stimulating discussions and assistance. Funded by Public Health Service grants R37-MH44894 (M-FC) and NS-33538 (MSL).

References

Aaron, L.I. and Chesselet, M.F. (1989) Heterogeneous distribution of polysialylated neuronal-cell adhesion molecule during post-natal development and in the adult: an immunohistochemical study in the rat brain. Neuroscience, 28: 701–710.

Augood, S.J., Herbison, A.E. and Emson, P.C. (1995) Localization of GAT-1 GABA transporter mRNA in rat striatum: cellular coexpression with GAD67 mRNA, GAD67 immunoreactivity, and parvalbumin mRNA. J. Neurosci., 15: 865–874.

Bayer, S. (1984) Neurogenesis in the Rat Neostriatum. Int. J. Devl. Neurosci., 2: 163–175.

Belluardo, N., Mudo, G., Trovato-Salinaro, A., Le Gurun, S., Charollais, A., Serre-Beinier, V., Amato, G., Haefliger, J.A., Meda, P. and Condorelli, D.F. (2000) Expression of connexin36 in the adult and developing rat brain. Brain Res., 865: 121–138.

Bennett, B.D. and Bolam, J.P. (1994) Synaptic input and output of parvalbumin-immunoreactive neurons in the neostriatum of the rat. Neuroscience, 62: 707–719.

Bracci, E., Centonze, D., Bernardi, G. and Calabresi, P. (2002) Dopamine excites fast-spiking interneurons in the striatum. J. Neurophysiol., 87: 2190–2194.

Butler, A.K., Uryu, K. and Chesselet, M.F. (1998) A role for N-methyl-D-aspartate receptors in the regulation of synaptogenesis and expression of the polysialylated form of the neural cell adhesion molecule in the developing striatum. Dev. Neurosci., 20: 253–262.

Chesselet, M.F. (1996) Quantitative analysis and interpretation of data for in situ hybridization at the cellular level. In: Henderson Z. (Ed.), In Situ Hybridization Techniques for the Brain. Wiley, pp. 141–149.

Chesselet, M.F. and Robbins, E. (1989) Characterization of striatal neurons expressing high levels of glutamic acid decarboxylase messenger RNA. Brain Res., 492: 237–244.

Chesselet, M.F., Weiss, L., Wuenschell, C., Tobin, A.J. and Affolter, H.U. (1987) Comparative distribution of mRNAs for glutamic acid decarboxylase, tyrosine hydroxylase, and tachykinins in the basal ganglia: an in situ hybridization study in the rodent brain. J. Comp. Neurol., 262: 125–140.

Colwell, C.S., Cepeda, C., Crawford, C. and Levine, M.S. (1998) Postnatal development of glutamate receptor-mediated responses in the neostriatum. Dev. Neurosci., 20: 154–163.

Connolly, J., Boulter, J. and Heinemann, S.F. (1992) Alpha 4-2 beta 2 and other nicotinic acetylcholine receptor subtypes as targets of psychoactive and addictive drugs. Br. J. Pharmacol., 105: 657–666.

Cowan, R.L., Wilson, C.J., Emson, P.C. and Heizmann, C.W. (1990) Parvalbumin-containing GABAergic interneurons in the rat neostriatum. J. Comp. Neurol., 302: 197–205.

Guo, J.Z. and Chiappinelli, V.A. (2000) Muscarinic receptors mediate enhancement of spontaneous GABA release in the chick brain. Neuroscience, 95: 273–282.

Hattori, T. and McGeer, P.L. (1973) Synaptogenesis in the corpus striatum of infant rat. Exp. Neurol., 38: 70–79.

Hurst, R.S., Cepeda, C., Shumate, L.W. and Levine, M.S. (2001) Delayed postnatal development of NMDA receptor function in medium-sized neurons of the rat striatum. Dev. Neurosci., 23: 122–134.

Kanemasa, T., Gan, L., Perney, T.M., Wang, L.Y. and Kaczmarek, L.K. (1995) Electrophysiological and pharmacological characterization of a mammalian Shaw channel expressed in NIH 3T3 fibroblasts. J. Neurophysiol., 74: 207–217.

Kawaguchi, Y. (1993) Physiological, morphological, and histochemical characterization of three classes of interneurons in rat neostriatum. J. Neurosci., 13: 4908–4923.

Kawaguchi, Y., Katsumaru, H., Kosaka, T., Heizmann, C.W. and Hama, K. (1987) Fast spiking cells in rat hippocampus (CA1 region) contain the calcium-binding protein parvalbumin. Brain Res., 416: 369–374.

Kawaguchi, Y. and Kubota, Y. (1993) Correlation of physiological subgroupings of nonpyramidal cells with parvalbumin- and calbindinD28k-immunoreactive neurons in layer V of rat frontal cortex. J. Neurophysiol., 70: 387–396.

Kawaguchi, Y., Wilson, C.J., Augood, S.J. and Emson, P.C. (1995) Striatal interneurones: chemical, physiological and morphological characterization. Trends Neurosci., 18: 527–535.

Kita, H., Kosaka, T. and Heizmann, C.W. (1990) Parvalbumin-immunoreactive neurons in the rat neostriatum: a light and electron microscopic study. Brain Res., 536: 1–15.

Koos, T. and Tepper, J.M. (1999) Inhibitory control of neostriatal projection neurons by GABAergic interneurons. Nat. Neurosci., 2: 467–472.

Lenz, S., Perney, T.M., Qin, Y., Robbins, E. and Chesselet, M.F. (1994) GABA-ergic interneurons of the striatum express the Shaw-like potassium channel Kv3.1. Synapse, 18: 55–66.

Li, D.P., Pan, Y.Z. and Pan, H.L. (2001) Acetylcholine attenuates synaptic GABA release to supraoptic neurons through presynaptic nicotinic receptors. Brain Res., 920: 151–158.

Limberger, N., Spath, L. and Starke, K. (1986) A search for receptors modulating the release of gamma-[3 H]aminobutyric acid in rabbit caudate nucleus slices. J. Neurochem., 46: 1109–1117.

Luk, K.C. and Sadikot, A.F. (2001) GABA promotes survival but not proliferation of parvalbumin-immunoreactive interneurons in rodent neostriatum: an in vivo study with stereology. Neuroscience, 104: 93–103.

Perney, T.M. and Kaczmarek, L.K. (1997) Localization of a high threshold potassium channel in the rat cochlear nucleus. J. Comp. Neurol., 386: 178–202.

Plenz, D. and Aertsen, A. (1996) Neural dynamics in cortex-striatum co-cultures — I. anatomy and electrophysiology of neuronal cell types. Neuroscience, 70: 861–891.

Plenz, D. and Kitai, S.T. (1998) Up and down states in striatal medium spiny neurons simultaneously recorded with spontaneous activity in fast-spiking interneurons studied in cortex-striatum-substantia nigra organotypic cultures. J. Neurosci., 18: 266–283.

Plotkin, J.L., Wu, N., Chesselet, M.F. and Levine, M.S. (2005) Functional and molecular development of striatal fast-spiking

GABAergic interneurons and their cortical inputs. Eur. J. Neurosci., 22: 1097–1108.

Radcliffe, K.A., Fisher, J.L., Gray, R. and Dani, J.A. (1999) Nicotinic modulation of glutamate and GABA synaptic transmission of hippocampal neurons. Ann. N. Y. Acad. Sci., 868: 591–610.

Sadikot, A.F. and Sasseville, R. (1997) Neurogenesis in the mammalian neostriatum and nucleus accumbens: parvalbumin-immunoreactive GABAergic interneurons. J. Comp. Neurol., 389: 193–211.

Schlosser, B., Klausa, G., Prime, G. and Ten Bruggencate, G. (1999) Postnatal development of calretinin- and parvalbumin-positive interneurons in the rat neostriatum: an immunohistochemical study. J. Comp. Neurol., 405: 185–198.

Sharpe, N.A. and Tepper, J.M. (1998) Postnatal development of excitatory synaptic input to the rat neostriatum: an electron microscopic study. Neuroscience, 84: 1163–1175.

Soghomonian, J.J., Gonzales, C. and Chesselet, M.F. (1992) Messenger RNAs encoding glutamate-decarboxylases are differentially affected by nigrostriatal lesions in subpopulations of striatal neurons. Brain Res., 576: 68–79.

Szele, F.G., Dowling, J.J., Gonzales, C., Theveniau, M., Rougon, G. and Chesselet, M.F. (1994) Pattern of expression of highly polysialylated neural cell adhesion molecule in the developing and adult rat striatum. Neuroscience, 60: 133–144.

Tepper, J.M., Koos, T. and Wilson, C.J. (2004) GABAergic microcircuits in the neostriatum. Trends Neurosci., 27: 662–669.

Tepper, J.M., Sharpe, N.A., Koos, T.Z. and Trent, F. (1998) Postnatal development of the rat neostriatum: electrophysiological, light- and electron-microscopic studies. Dev. Neurosci., 20: 125–145.

Tepper, J.M. and Trent, F. (1993) In vivo studies of the postnatal development of rat neostriatal neurons. Prog. Brain Res., 99: 35–50.

Uryu, K., Butler, A.K. and Chesselet, M.F. (1999) Synaptogenesis and ultrastructural localization of the polysialylated neural cell adhesion molecule in the developing striatum. J. Comp. Neurol., 405: 216–232.

Wang, L.Y., Gan, L., Forsythe, I.D. and Kaczmarek, L.K. (1998) Contribution of the Kv3.1 potassium channel to high-frequency firing in mouse auditory neurones. J. Physiol., 509(Pt 1): 183–194.

Yokoyama, S., Imoto, K., Kawamura, T., Higashida, H., Iwabe, N., Miyata, T. and Numa, S. (1989) Potassium channels from NG108-15 neuroblastoma-glioma hybrid cells. Primary structure and functional expression from cDNAs. FEBS Lett., 259: 37–42.

Tepper, Abercrombie & Bolam (Eds.)
Progress in Brain Research, Vol. 160
ISSN 0079-6123

CHAPTER 16

Co-localization of GABA with other neuroactive substances in the basal ganglia

Jacqueline F. McGinty*

Department of Neurosciences, Medical University of South Carolina, Charleston, SC 29425, USA

Abstract: The dorsal striatum (caudate putamen) contains two types of GABAergic medium spiny neurons (MSNs) that are distinguished by the expression of either the opioid peptide, enkephalin, or the opioid peptide, dynorphin, as well as the tachykinin substance P. Pharmacological studies suggest that these peptides modulate local neurotransmission in the striatum in response to direct and indirect dopamine agonists. In contrast, GABA appears to have minimal impact within the striatum under these conditions. The actions of the peptide cocktail are dependent on the cellular distribution of their receptors in the striatal network. The net result of their actions is a homeostatic response that regulates striatal output and balances dopamine and glutamate receptor stimulation.

Keywords: BDNF; dynorphin; enkephalin; substance P; striatum

The dorsal striatum (caudate putamen) is a critical area for the integration and regulation of motor and cognitive functions that are mediated by interactions involving dopaminergic and glutamatergic afferents and intrinsic neurons that reside in patch (striosome) or matrix compartments (Gerfen, 1985; Parent and Hazrati, 1995; Calabresi et al., 1997). Both compartments contain GABAergic medium spiny neurons (MSNs), as well as cholinergic and GABAergic interneurons, that express different neuromodulators and receptors (Angulo and McEwen, 1994; Holt et al., 1997). The regulation of the local striatal network by MSN neuropeptides is a good illustration of the functional significance of neuropeptides co-localized with GABA in the CNS. This review will focus on the expression and modulatory actions of neuropeptides expressed by MSNs in the intact striatum of rodents.

———————
*Corresponding author. Tel.: +843-792-9036;
Fax: +843-792-4423; E-mail: mcginty@musc.edu

Location of neuropeptides and receptors in MSNs

Figure 1 illustrates the circuitry and cellular location of some of the major peptides and receptors that will be discussed in this review. MSNs in both patch and matrix compartments contribute to two distinct projection pathways: striatonigral and striatopallidal (Gerfen and Young, 1988). The striatonigral pathway projects from the striatum to the internal segment of the globus pallidus (rodent entopeduncular nucleus) and the substantia nigra; these neurons are distinguished by the expression of D1 receptors, dynorphin, and substance P. In contrast, the striatopallidal pathway projects from the striatum to the external segment of the globus pallidus. The striatopallidal pathway is distinguished by the expression of D2 receptors and enkephalin. For reviews of the literature on peptide co-localization in MSNs prior to the last decade, see Afifi, 1994; Angulo and McEwen, 1994; Graybiel, 1986;

DOI: 10.1016/S0079-6123(06)60016-2

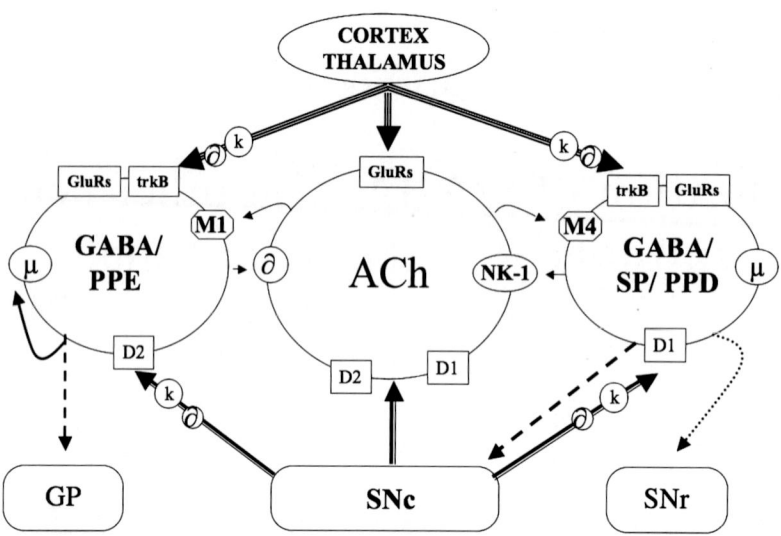

Fig. 1. In the striatum, dopamine and glutamate afferents converge on striatonigral (containing PPD/SP) and striatopallidal (containing PPE) MSNs that express D1 or D2 receptors respectively. Striped lines indicate dopamine and glutamate afferents that contain BDNF. In addition to dopamine and glutamate, stimulant-induced MSN gene expression is modulated by local muscarinic cholinergic, GABA, and neuropeptide receptors in the striatum. Ach = acetylcholine, δ = delta opioid receptor, D = dopaminergic receptors, GluR = glutamate receptor classes, k = kappa opioid receptor, M = muscarinic receptors, μ = mu opioid receptor, NK-1 = SP receptors, PPD = preprodynorphin, PPE = preproenkephalin, SP = Substance P, trkB = tyrosine kinase receptor B. [*Source*: Modified from Gonzalez-Nicolini et al., 2003 with permission.]

Anderson and Reiner, 1990. For reviews of neuroactive substances co-localized with GABA in striatal interneurons, see Aoki and Pickel, 1990; Kubota and Kawaguchi, 1993.

The pre- and postsynaptic localization of neuropeptide receptors in the striatum has helped to clarify the actions of neuropeptide receptor agonists and antagonists that modulate the response to dopamine receptor stimulation (see Fig. 1). Kappa opioid receptors (KORs) are synthesized in cortical pyramidal neurons and in dopamine neurons in the ventral mesencephalon (Mansour et al., 1987) and are transported to a presynaptic position in terminals of excitatory (glutamate) and inhibitory (dopamine) afferents (Meshul and McGinty, 2000; Svingos et al., 2001), where they decrease dopamine and glutamate release (Rawls and McGinty, 1997, 1999; Gray et al., 1999). In contrast, delta opioid receptors (DORs) are synthesized by cholinergic interneurons (Le Moine et al., 1994) and are detected in morphologically heterogeneous terminals in the striatum (Svingos et al., 1999; Wang and Pickel, 2001). Moreover, a DOR agonist decreases

Ach release (Mulder et al., 1984) whereas a DOR antagonist decreases AMPH-stimulated extracellular levels of dopamine and glutamate (Schad et al., 1996; Rawls and McGinty, 2000). Mμ-opioid receptors (MOR) are predominantly expressed by MSNs primarily in striatal patches (striosomes) that express D1 receptors, substance P (SP), and dynorphin (Guggenheim et al., 1996; Georges et al., 1999), but little is known about their function. The differential cellular localization of the opioid receptors suggests partially overlapping but distinct roles for these receptors in the development and maintenance of opioid-mediated behaviors (Shippenberg et al., 1987; Bals-Kubik et al., 1990; Meyer and Meyer, 1993). With regard to the receptors that bind SP, neurokinin-1 (NK-1) receptors, expressed most abundantly by cholinergic (and somatostatin/GABA-containing) interneurons in the striatum (Kaneko et al., 1993), have been reported to mediate D1 receptor-stimulated release of ACh and dopamine (Anderson et al., 1994; Kraft et al., 2001b). Finally, all MSNs express the tyrosine kinase B receptor that binds BDNF and NT-4.

Both cortical glutamatergic and nigral dopaminergic inputs express and release BDNF (Fig. 1). Exogenous BDNF administration enhances MSN neuropeptide expression (Arenas et al., 1996), whereas partial depletion of the BDNF gene in 3-month-old BDNF heterozygous mice results in lower striatal PPE and PPD, but not GAD_{67}, mRNA levels (Saylor et al., 2006, Fig. 2). These data indicate that endogenous BDNF/trkB signaling is required to maintain the normal phenotype and functioning of striatal neuropeptide expression in MSNs.

Dopamine agonists increase neuropeptide expression in MSNs

Direct and indirect dopamine agonists induce behavioral activity in rodents that is positively correlated with neuropeptide gene expression in the striatum. Induction of striatal neuropeptides, preprodynorphin (PPD), preproenkephalin (PPE), neurotensin/neuromedin N (NT), and the SP precursor, preprotachykinin (PPT) in MSNs occurs after acute or repeated administration of the psychostimulants, cocaine, amphetamine, or methamphetamine (Castel et al., 1994; Merchant et al., 1994; Smith and McGinty, 1994; Daunais and McGinty, 1994b; Daunais et al., 1995; Hanson et al., 1988, 1989; Steiner and Gerfen, 1998; Wang and McGinty, 1998; Zahm et al., 1998; Adams et al., 2001). (Because neurotensin is expressed by both types of MSNs, it will not be discussed further here. For a review of neurotensin-dopamine interactions, see Binder et al., 2001.) The induction of neuropeptide expression by psychostimulants is universally blocked by

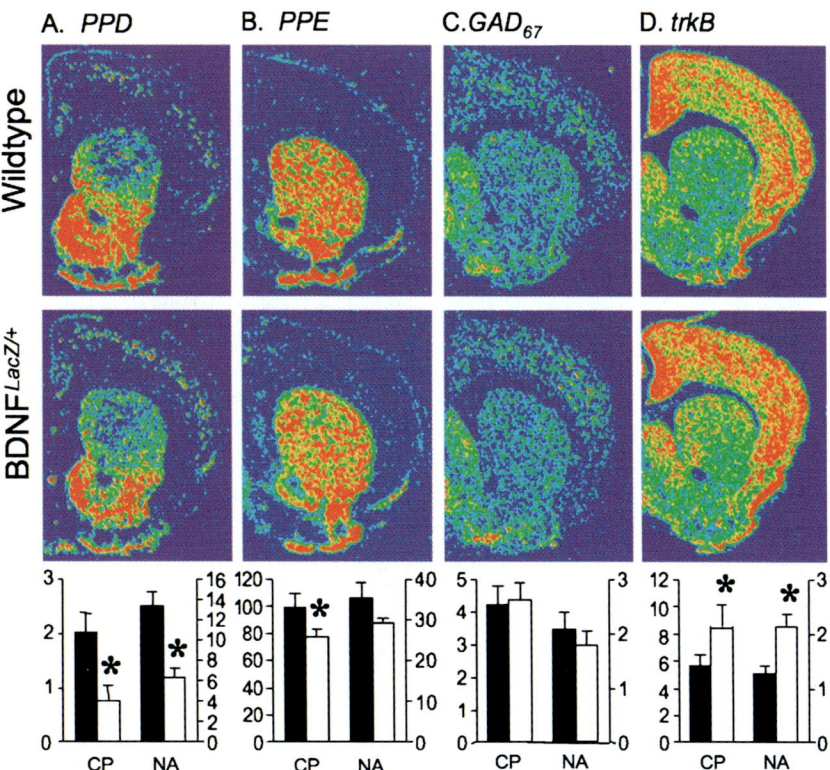

Fig. 2. Digitized photomicrographs and semiquantitative image analysis of the mRNA expression of (A) PPD, (B) PPE, (C) GAD_{67}, and (D) trkB in the striatum of 3-month-old wild type and $BDNF^{LacZ/+}$ mice. Note that caudate-putamen (CPu) data are associated with the left y-axis and nucleus accumbens (NA) data are associated with the right y-axis. Black bars denote wild type, white bars denote $BDNF^{LacZ/+}$; mice. *$p < .05$ versus wild type as determined by unpaired Student's t-test. [*Source*: Modified from Saylor et al. (2006) with permission.]

276

dopamine D1 receptor and ionotropic and metabotropic glutamate receptor antagonists (Hanson
et al., 1987; Singh et al., 1991; Daunais and
McGinty, 1994a, b; Wang and McGinty, 1996a, c).
Conversely, pharmacological studies have suggested that after binding to their corresponding
receptors, striatal neuropeptides activated by
psychostimulants are able to modulate the release
of DA, glutamate, and ACh in the striatum
(Anderson et al., 1994; Gray et al., 1999; Rawls
and McGinty, 1997, 2000). Thus, neuropeptides
expressed by MSNs modify dopamine agonist-
induced changes in striatal neurotransmission.

Neuropeptide regulation of local striatal circuitry

It often goes unrecognized that local regulation
within the striatum has a major impact on MSN
output and that endogenous inhibitory or homeostatic systems also are activated by dopamine
agonists. Two of the inhibitory substrates that are
critical to controlling excessive dopamine and
glutamate neurotransmission in the striatum are
the muscarinic cholinergic (Wang and McGinty,
1996b, d; 1997a, b; 1998) and neuropeptide systems
(Gray et al., 1999; Rawls and McGinty, 1998, 1999;
Tzaferis and McGinty, 2001; Gonzalez-Nicolini
and McGinty, 2002, Gonzalez-Nicolini et al., 2003).
Intrastriatal infusion of selective antagonists for
MOR, DOR, or NK-1 receptors or selective KOR
agonists regulate AMPH-stimulated motor behaviors and MSN gene expression based on the
cellular locations of their respective receptors in the
striatum (see above). The differential localization of
each receptor subtype suggests partially overlapping but distinct roles for these receptors in the
expression of dopamine-mediated behaviors
(Shippenberg et al., 1987; Meyer and Meyer, 1993).

**Stimulation of kappa opioid receptors decreases
amphetamine-induced behavior and striatal
neuropeptide mRNA**

Induction of striatal PPD expression after psychostimulant administration may reflect a compensatory increase in synthesis due to excessive

stimulation of D1 receptors. This hypothesis is
supported by evidence that D1 agonists increase
extracellular levels of dynorphin in the striatum
and substantia nigra (You et al., 1994). Increased
dynorphin release may inhibit the effects of
stimulants by activating KORs on dopaminergic
and glutamatergic terminals in the dorsal and
ventral striatum (Meshul and McGinty, 2000;
Svingos et al., 2001, see Fig. 1).

Indeed, KOR stimulation blocks basal and
stimulated increases in extracellular dopamine in
the striatum (Mulder et al., 1984; Di Chiara and
Imperato, 1988; Werling et al., 1988; Spanagel
et al., 1992; Maisonneuve et al., 1994). In contrast,
the selective KOR antagonist, norbinaltorphimine,
increases basal dopamine and glutamate levels
(Spanagel et al., 1992; Rawls and McGinty, 1998).
Moreover, the selective KOR agonist, U69,593,
decreases the acute amphetamine-induced increase
in extracellular levels of dopamine and glutamate
(Gray et al., 1999) as well as evoked striatal
glutamate levels in striatal synaptosomes (Rawls
and McGinty, 1999). Several studies have demonstrated that selective KOR agonists inhibit
stimulant-induced alterations in behavior and
neuropeptide mRNA in the striatum of rodents
(Ukai et al., 1992; Heidbreder et al., 1993;
Shippenberg et al., 1996; Gray et al., 1999; Tzaferis
and McGinty, 2001). Figure 3 illustrates the
suppressive effects of U69,593 (0.32 mg/kg, i.p.)
on amphetamine-induced PPD and PPE gene
expression (Tzaferis and McGinty, 2001). These
data suggest that excessive dopaminergic and
glutamatergic activity induces the striatal dynor-
phin/kappa receptor system to exert a compensatory, inhibitory influence on the behavioral and
neurochemical effects of receptor overstimulation.

**Blockade of mu or delta opioid receptors decreases
amphetamine-induced behavior and striatal
neuropeptide mRNA**

In contrast to the inhibitory effects of KOR
stimulation, DOR agonists potentiate the release
of newly synthesized dopamine (Lubetski et al.,
1982; Petit et al., 1986; Pentney and Gratton, 1995)
and inhibit ACh release and D1 receptor-stimulated

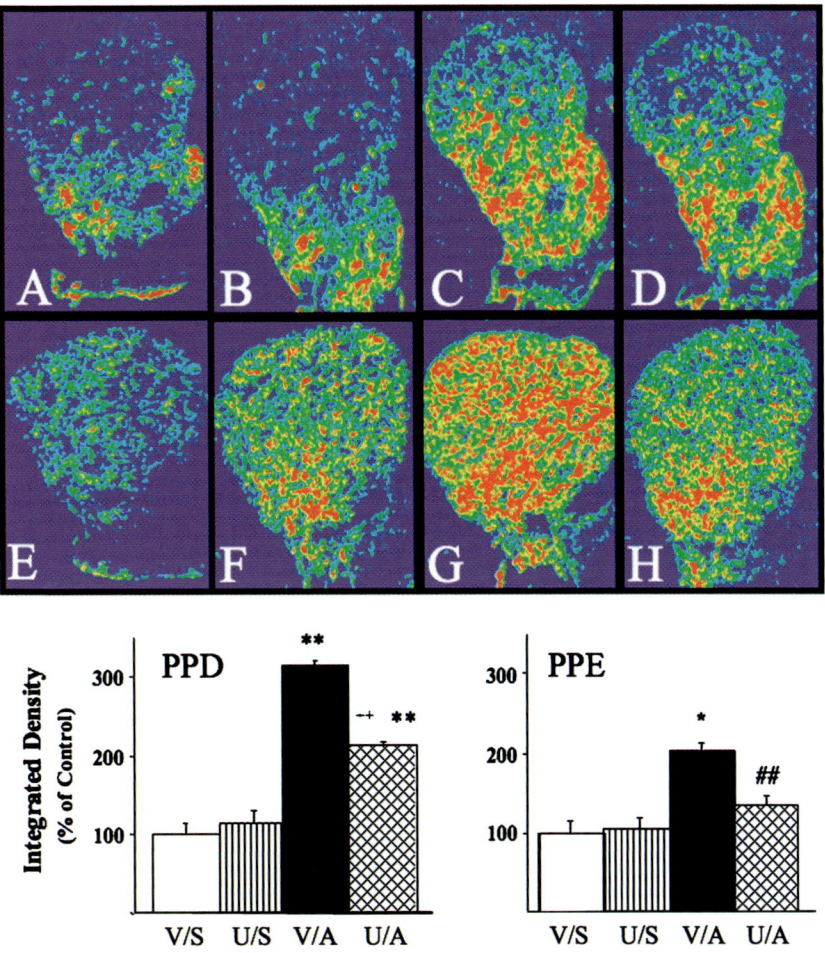

Fig. 3. The kappa opioid receptor agonist, U-69593, decreases AMPH-induced PPD and PPE mRNA in the striatum.Top: A–D: PPD mRNA; E–H: PPE mRNA; A, E: Vehicle/Saline; B, F: U69,593/Saline; C, G: Vehicle/AMPH; D, H: U69,593/AMPH. Bottom left: U69,593 significantly decreased AMPH-induced PPD mRNA. **$p < .005$ vs. V/S; $^{++}p < .0002$ vs. U/S; $^{\#\#}p < .004$ vs. V/A. Bottom right: U69,593 significantly decreased AMPH-induced PPE mRNA. *$p < .05$ vs. V/S; $^{\#}p < .05$ vs. V/A. V/S = Vehicle/Saline; U/S = U69,593/Saline; V/A = Vehicle/AMPH; U/A = U69,593/AMPH. [*Source*: Modified from Tzaferis and McGinty, 2001 with permission.]

adenylate cyclase (Heijna et al., 1990; Sandor et al., 1992) in the striatum. However, local infusion of selective DOR agonists decrease the ability of the D2 antagonist, eticlopride, to increase immediate early gene expression in striatopallidal neurons (Steiner and Gerfen, 1999). In addition, low doses (0.03–0.1 mg/kg, s.c.) of the DOR antagonist, naltrindole, block sensitization to the locomotor activating and conditioned reinforcing effects, without affecting the acute behavioral effects, of cocaine (Shippenberg and Heidbreder, 1995; Heidbreder

et al., 1996). Similarly, Jones and colleagues reported that naltrindole had no effect on acute cocaine-induced behaviors but intracisternal infusion did attenuate behavioral activity induced by AMPH (Jones and Holtzman, 1992; Jones et al., 1993). In addition, naltrindole attenuated AMPH-induced release of dopamine in the dorsal striatum but not AMPH- or cocaine-evoked dopamine release in the nucleus accumbens (Schad et al., 1996). Furthermore, intrastriatal naltrindole decreased an AMPH-induced increase in extracellular levels of glutamate in the striatum (Rawls and McGinty, 2000). Thus, it appears that KOR and DOR exert opposite effects on stimulant-induced behavioral and neurochemical changes and may differentially affect different dopamine pools in striatal neurons in response to cocaine and AMPH.

In an attempt to investigate the role of striatal MOR and DOR receptors, the selective mu opioid antagonists, CTOP or CTAP, or the selective delta opioid antagonists, naltrindole or TIPPψ, were infused into the striatum prior to systemic AMPH (Gonzalez-Nicolini et al., 2003). Both DOR and MOR antagonists significantly decreased AMPH-induced vertical activity. However, only MOR antagonists reduced AMPH-induced distance traveled. Quantitative in situ hybridization histochemistry revealed that the DOR antagonists significantly decreased AMPH-induced mRNA expression of PPD, PPE, and PPT. In contrast, both MOR antagonists blocked AMPH-induced PPD and PPT mRNA increases but PPE mRNA levels in the dorsal striatum were increased to the same extent by MOR antagonists, AMPH, or a combination of MOR antagonist and AMPH. The effects of the MOR antagonists were not due to an alteration in basal or AMPH-stimulated dopamine release (unpublished observations). However, a previous study from our lab demonstrated that delta receptor blockade decreased AMPH-stimulated dopamine release in the striatum (Rawls and McGinty, 2000), suggesting activation of presynaptic heteroreceptors on dopamine terminals (Svingos et al., 1999). These data and those described above indicate that striatal MOR, DOR, and KOR differentially regulate AMPH-induced behavior and neuropeptide gene expression in the rat striatum most likely due to the different cellular localization of these receptors in the local network.

Neurokinin-1 receptor blockade decreases AMPH-induced behavior and neuropeptide mRNA expression in the striatum

Substance P is the endogenous ligand for NK-1 receptors that are expressed by ACh terminals and possibly, dopamine terminals, in the striatum. Blockade of NK-1 receptors decreased a cocaine-induced increase in locomotor activity and extra-cellular dopamine release in the striatum (Kraft et al., 2001a, b). Furthermore, when the selective NK-1 receptor antagonist, LY306740, was infused intrastriatally, there was a significant decrease in AMPH-induced vertical activity (Gonzalez-Nicolini and McGinty, 2002). Furthermore, semi-quantitative in situ hybridization histochemistry revealed that LY306740 significantly decreased AMPH-induced mRNA expression of all three MSN neuropeptides (Fig. 4). These data indicate that striatal NK-1 receptors modulate AMPH-induced behavior and mRNA expression of striatal neuropeptides possibly through presynaptic inhibition of dopamine (Kraft et al., 2001b) or ACh (Anderson et al., 1994) release.

D1 receptor-mediated effects on PPE and D2 receptor-mediated effects on PPD are mediated indirectly through muscarinic, but not GABA$_A$ receptors in the striatum

The nonselective muscarinic agonist, oxotremorine, decreases D1 dopamine agonist-induced behavior and neuropeptide gene expression in PPD/SP neurons, whereas it enhances PPE gene expression (Wang and McGinty, 1996b). In contrast, the nonselective antagonist, scopolamine, blocks the D1 agonist-induced increase in PPE mRNA whereas it augments the increase in PPD and SP mRNA (Wang and McGinty, 1996b, d, 1997b). Furthermore, these actions are mediated by muscarinic receptors in the striatum because intrastriatal infusion of oxotremorine and scopolamine is as effective as systemic injections (Wang and McGinty, 1997a). These data indicate that muscarinic receptor activity positively modulates D1 receptor regulation of enkephalinergic striato-pallidal neurons and negatively modulates SP/PPD striatonigral neurons. Thus, PPE- and PPD/SP-containing neurons, which are regulated in opposite directions by D2 and D1 receptors, respectively, are also modulated in opposite directions by muscarinic receptors in the striatum. These conclusions, which are based on pharmacological

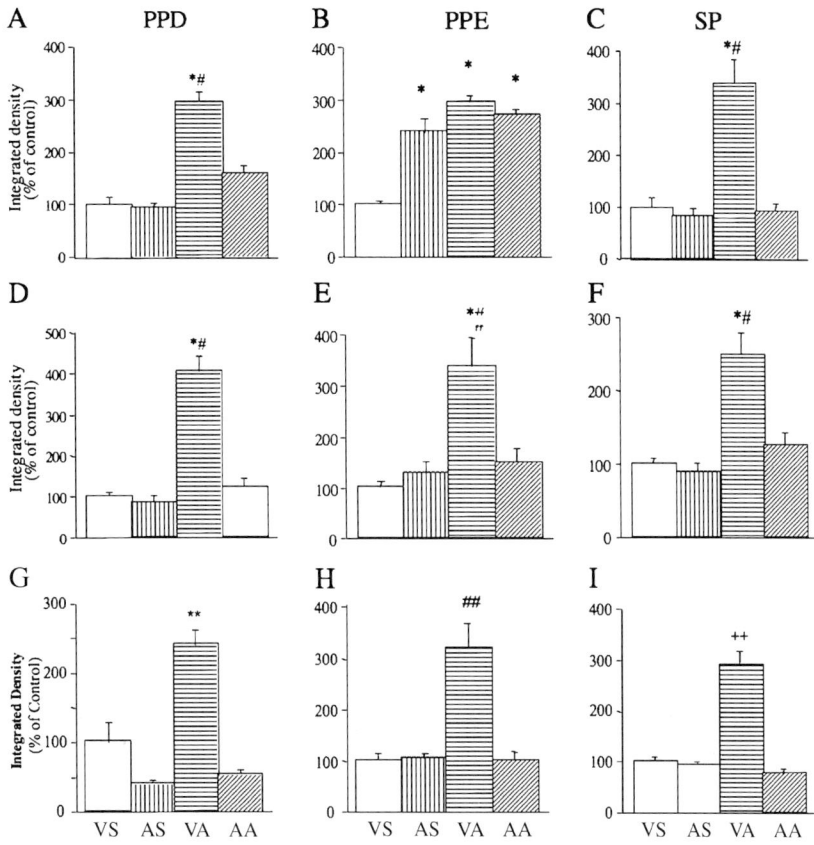

Fig. 4. Quantitative analysis of the effect of 1 µg/µl CTAP (A–C), 2 µg/µl TIPPψ (D–F), or 35 nmoles of LY3340760 (G–I) on the integrated density values for AMPH-induced PPD (A, C, G) PPE (B, D, H), or SP (C, E, I) mRNA expression in the dorsal striatum. VS = vehicle + saline; VA = vehicle + amphetamine; AS = antagonist + saline; AA = antagonist + amphetamine. *$p < .001$ vs. vehicle + saline, #$p < .0003$ vs. TIPPψ + amphetamine or CTAP + amphetamine. **$p < .01$ vs. vehicle + saline, ##$p < .01$ vs. LY306740 + amphetamine. [*Source*: Redrawn from Gonzalez-Nicolini and McGinty (2002) and Gonzalez-Nicolini et al. (2003) with permission.]

data, are supported by data from mice in which striatal cholinergic interneurons were ablated by immunotoxin-mediated cell targeting (Kaneko et al., 2000). However, presynaptic inhibitory muscarinic heteroceptors on glutamatergic terminals (Rawls and McGinty, 1998, 1999) may also contribute to the regulation of MSN responsiveness to AMPH or D1 agonists.

Because D2 receptor antagonists activate striatopallidal neurons, the possibility that increased GABA release from local medium spiny axon collaterals also contributed to the ability of a D2 antagonist to block the effects of D1 receptor stimulation was considered. However, intrastriatal infusion of the GABA$_A$ antagonist, bicuculline,

did not significantly affect basal or SKF82958-stimulated PPD or SP mRNA expression in striatonigral neurons (Jones et al., 1999). These data indicated that GABA$_A$ receptors in the rat striatum are not involved in the regulation of basal or dopamine agonist stimulated PPD/SP mRNA expression. Similarly, in contrast to the actions of the muscarinic antagonist, scopolamine, bicuculline did not affect the increase in behaviors induced by SKF-82958 or the ability of eticlopride to block them (Jones et al., 2001). Quantitative in situ hybridization demonstrated that bicuculline did not affect basal PPD mRNA nor did it affect the ability of eticlopride to decrease SKF-82958-induced PPD mRNA. As expected, bicuculline, eticlopride, or SKF-82958 independently increased basal PPE mRNA. However, there was no significant interaction among bicuculline and eticlopride and SKF-82958 on PPE mRNA levels. These data indicate that blockade of striatal GABA$_A$ receptors does not affect stimulant-induced behavior and gene expression in the same robust way that blockade of muscarinic receptors does and the effects of D2 receptor blockade on D1 agonist-induced PPD/SP mRNA are not mediated by GABA$_A$ receptors.

Extrastriatal GABA$_B$ receptor blockade decreases AMPH-induced behavior, dialysate dopamine levels, and neuropeptide gene expression in the striatum

GABA$_B$ receptors are abundant in basal ganglia circuitry (Chu et al., 1990; Boyes and Bolam, 2003; Lacey et al., 2005) and their stimulation at presynaptic sites has been reported to antagonize the actions of cortically evoked excitatory post-synaptic potentials in MSNs (Nisenbaum et al., 1993) and cocaine self-administration in animal models as well as in humans (Brebner et al., 2000, 2002). Therefore, it was predictable that systemic administration of the GABA$_B$ receptor agonist, (+)-baclofen, blocked AMPH-induced vertical activity, decreased the peak level of striatal DA release, and blocked mRNA expression of medium spiny peptides in the striatum (Zhou et al., 2004). Surprisingly, however, the infusion of (+)-baclofen, into the dorsal striatum did not affect stimulant-induced behavior or neuropeptide mRNA in the striatum (Zhou et al., 2005). Instead, intra-VTA infusion of baclofen completely blocked, whereas intraaccumbal or intranigral infusion of baclofen attenuated, AMPH-induced vertical activity (without affecting AMPH-induced total distance traveled) and MSN peptide gene expression in the dorsal striatum (Zhou et al., 2005). Thus, although the neural substrates that mediate AMPH-induced rearing and gene expression are incompletely understood, this study suggests that stimulation of pre- and postsynaptic GABA$_B$ receptors in the VTA, SN, and NA, but not in the caudate-putamen, are differentially involved. Alternatively,

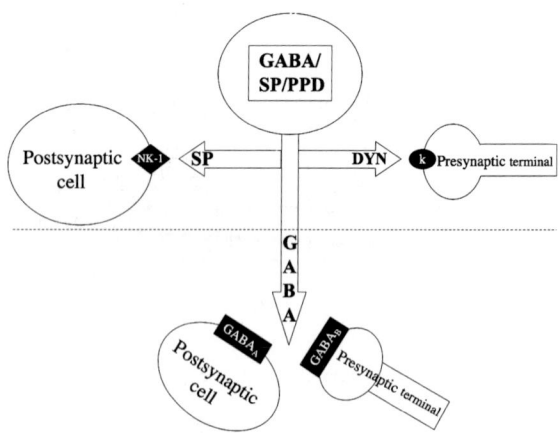

Fig. 5. Schematic illustration of a GABAergic medium spiny neuron that expresses D1 receptors, dynorphin, and substance P. When a dopamine agonist stimulates D1 receptors on striatonigral MSNs, SP and dynorphin are released from local collaterals. SP stimulates NK-1 receptors, predominantly expressed by cholinergic interneurons, and stimulates ACh release. Dynorphin stimulates KORs that presynaptically inhibit dopamine and glutamate release. In contrast, dopamine agonists do not appear to evoke local GABA release that affects GABA$_A$ receptors enough to substantially contribute to intrastriatal regulation of behavior and gene expression. However, extrastriatal GABA$_A$ and GABA$_B$ receptors are stimulated and inhibit dopamine agonist-induced behavior and striatal gene expression.

GABA$_B$ receptors may already have been saturated by GABA release after AMPH and further stimulation with baclofen was inconsequential. The intrastriatal infusion of a GABA$_B$ antagonist to test whether AMPH responses are augmented when local GABA$_B$ receptors are blocked may answer this question.

Conclusion

The studies described above suggest a model for local striatal interactions when peptides are released from MSNs (Fig. 5). When dopamine stimulates D1 receptors on striatonigral MSNs, for example, SP and dynorphin are released from local colla-terals. SP stimulates NK-1 receptors, predominantly expressed by cholinergic interneurons, and stimulates ACh release. Dynorphin stimulates KORs that presynaptically inhibit dopamine and glutamate release. The net effect is a local regulation of striatal output that compensates for overstimulation of dopamine and glutamate receptors. In contrast, although tonic GABA tone on PPE expression is attenuated by intrastriatal GABA$_A$ receptor blockade, dopamine agonists do not appear to evoke local GABA release that affects GABA$_A$ receptors enough to substantially contribute to intrastriatal regulation of behavior and gene expression. However, dopamine-induced GABA released in the NA, VTA, and SN stimulates GABA$_B$ (and possibly GABA$_A$) receptors that inhibit behavior and striatal gene expression. In this way, colocalized GABA and neuropeptides synergize their actions in the basal ganglia network.

Abbreviations

ACh	acetylcholine
AMPH	amphetamine
BDNF	brain-derived neurotrophic factor
DA	dopamine
DOR	delta opioid receptor
KOR	kappa opioid receptor
MSN	medium spiny neuron
MOR	mu opioid receptor
NA	nucleus accumbens
NK-1	neurokinin-1
PPD	preprodynorphin
PPE	preproenkephalin
SN	substantia nigra
SP	substance P
VTA	ventral tegmental area

Acknowledgment

The studies from the author's laboratory were supported by NIH DA03982.

References

Adams, D.H., Hanson, G.R. and Keefe, K.A. (2001) Differential effects of cocaine and methamphetamine on neurotensin/neuromedin N and preprotachykiniin messenger RNA expression in unique regions of the striatum. Neuroscience, 102: 843–851.

Afifi, A.K. (1994) The basal ganglia: functional anatomy and physiology, Part 1. J. Child Neurol., 9: 249–260.

Anderson, J.J., Kuo, S., Chase, T.N. and Engber, T.M. (1994) Dopamine D1 receptor-stimulated release of acetylcholine in rat striatum is mediated by activavtion of striatal neurokinin1 receptors. J. Pharmacol. Exper. Therap., 269: 1144–1151.

Anderson, K.D. and Reiner, A. (1990) Extensive co-occurrence of substance P and dynorphin in striatal projection neurons: an evolutionarily conserved feature of basal ganglia organization. J. Comp. Neurol., 295: 339–369.

Angulo, J.A. and McEwen, B.S. (1994) Molecular aspects of neuropeptide regulation and function in the corpus striatum and nucleus accumbens. Brain Res. Rev., 19: 1–28.

Aoki, C. and Pickel, V.M. (1990) Neuropeptide Y in cortex and striatum. Ultrastructural distribution and coexistence with classical neurotransmitters and neuropeptides. Ann. N. Y. Acad. Sci., 611: 186–205.

Arenas, E., Akerud, P., Wong, V., Boylan, C., Persson, H., Lindsay, R.M. and Altar, C.A. (1996) Effects of BDNF and NY-4/5 on striatonigral neuropeptides or nigral GABA neurons In Vivo. Eur. J. Neurosci., 8: 1707–1717.

Bals-Kubik, R., Shippenberg, T.S. and Herz, A. (1990) Involvement of central mu and delta opioid receptors in mediating the reinforcing effects of beta-endorphin in the rat. Eur. J. Pharmacol., 175: 63–69.

Binder, E.B., Kinkead, B., Owens, M.J. and Nemeroff, C.B. (2001) Neurotensin and dopamine interactions. Pharmacol. Rev., 53: 453–486.

Boyes, J. and Bolam, J.P. (2003) The subcellular localization of GABA(B) receptor subunits in rat substantia nigra. Eur. J. Neurosci., 18: 3279–3293.

Brebner, K., Childress, A.R. and Roberts, D.C. (2002) A potential role for GABA(B) agonists in the treatment of psychostimulant addiction. Alcohol & Alcoholism, 37: 478–484.

Brebner, K., Phelan, R. and Roberts, D.C. (2000) Effect of baclofen on cocaine self-administration in rats reinforced under fixed-ratio 1 and progressive ratio schedules. Psychopharmacology, 143: 209–214.

Calabresi, P., De Murtas, M. and Bernardi, G. (1997) The neostriatum beyond the motor function: experimental and clinical evidence. Neuroscience, 78: 39–60.

Castel, M.N., Morino, P., Nylander, I., Terenius, L. and Hokfelt, T. (1994) Differential dopaminergic regulation of the neurotensin striatonigral and striatopallidal pathways in the rat. Eur. J. Pharmacol., 262: 1–10.

Chu, D.C., Albin, R.L., Young, A.B. and Penny, J.B. (1990) Distribution and kinetics of GABAB binding sites in rat central nervous system: a quantitative autoradiographic study. Neuroscience, 34: 341–357.

Daunais, J.B. and McGinty, J.F. (1994a) The effects of dopamine receptor blockade on zif/268 and preprodynorphin gene expression in rat forebrain following a short-term cocaine binge. Mol. Brain Res., 35: 237–248.

Daunais, J.B. and McGinty, J.F. (1994b) Acute and chronic cocaine administration differentially alters striatal opioid and nuclear transcription factor mRNA. Synapse, 18: 35–45.

Daunais, J.B., Roberts, D.C.S. and McGinty, J.F. (1995) Short-term cocaine self administration increases preprodynorphin, but not preproenkephalin, *c-fos* or *zif/268* mRNAs in rat forebrain. Brain Res. Bull., 37: 523–527.

Di Chiara, G. and Imperato, A. (1988) Drugs abused by humans preferentially increase synaptic dopamine concemtrations in the mesolimbic system of freely moving rats. Proc. Natl. Acad. Sci. USA, 85: 4274–4278.

Georges, F., Stinus, L., Bloch, B. and Le Moine, C. (1999) Chronic morphine exposure and spontaneous withdrawal are associated with modifications of dopamine receptor and neuropeptide gene expression in the rat striatum. Eur. J. Neurosci., 11: 481–490.

Gerfen, C.R. (1985) The neostriatal mosaic. I. Compartmental organization of projections from the striatum to the substantia nigra in the rat. J. Comp. Neurol., 236: 454–456.

Gerfen, C.R. and Young III, W.S. (1988) Distribution of striatonigral and striatopallidal peptidergic neurons in both patch and matrix compartments: an in situ hybridization histochemistry and fluorescent retrograde tracing study. Brain Res., 460: 161–167.

Gonzalez-Nicolini, M.V., Berglind, W., Cole, K.S., Keogh, C.L. and McGinty, J.F. (2003) Local mu and delta opioid receptors regulate amphetamine-induced behavior and neuropeptide mRNA in the striatum. Neuroscience, 121: 387–398.

Gonzalez-Nicolini, V. and McGinty, J.F. (2002) NK-1 receptor blockade decreases amphetamine-induced behavior and neuropeptide mRNA expression in the striatum. Brain Res., 931: 41–49.

Gray, A.M., Rawls, S.M., Shippenberg, T.S. and McGinty, J.F. (1999) The kappa opioid agonist, U-69593, decreases acute amphetamine-evoked behaviors and calcium-dependentdialysate levels of dopamine and glutamate in the ventral striatum. J. Neurochem., 73: 1066–1074.

Graybiel, A.M. (1986) Neuropeptides in the basal ganglia. Res. Publ. Assoc. Res. Nerv. Ment. Dis., 64: 135–161.

Guggenheim, N.D., Klop, H., Minami, M., Satoh, M. and Voorn, P. (1996) Co-localization of μ receptor is greater with dynorphin than enkephalin in rat striatum. NeuroReport, 7: 2119–2124.

Hanson, G.R., Letter, A.A., Bush, L. and Gibb, J.W. (1987) Methamphetamine-induced changes in the striatal-nigral dynorphin system: role of D-1 and D-2 receptors. Eur. J. Pharmacol., 144: 245–246.

Hanson, G.R., Letter, A.A., Bush, L. and Gibb, J.W. (1988) Characterization of methamphetamine effects on the striatalnigral dynorphin system. Eur. J. Pharmacol., 155: 11–18.

Hanson, G.R., Smiley, P., Johnson, M., Letter, A., Bush, L. and Gibb, J.W. (1989) Response by the neurotensin systems of the basal ganglia to cocaine treatment. Eur. J. Pharmacol., 160: 23–30.

Heidbreder, C., Goldberg, S.R. and Shippenberg, T.S. (1993) The kappa-opioid receptor agonist U-69593 attenuates cocaine-induced behavioral sensitization in the rat. Brain Res., 616: 335–338.

Heidbreder, C., Shoaib, M. and Shippenberg, T.S. (1996) Differential role of delta receptors in the development and expression of behavioral sensitization to cocaine. Eur. J. Pharmacol., 298: 207–216.

Heijna, M.H., Padt, M., Hogenboom, F., Portoghese, P.S., Mulder, A.H. and Schoffelmeer, A.N.M. (1990) Opioid receptor-mediated inhibition of dopamine and acetylcholine release from slices of rat nucleus accumbens, olfactory tubercle and frontal cortex. Eur. J. Pharmacol., 18: 267–278.

?A3B2 tvjmline = 1.2h? > Holt, D.J., Graybiel, A.M. and Saper, C.B. (1997) Neurochemical architecture of the human striatum. J. Comp. Neurol., 384: 1–25.

Jones, D.N.C. and Holtzman, S.G. (1992) Interaction between opioid antagonists and amphetamine: evidence for mediation by central delta opioid receptors. J. Pharmacol. Exper. Therap., 262: 638–645.

Jones, D.N.C., Bowen, W.D., Portoghese, P.S. and Holtzman, S.G. (1993) Delta-opioid receptor antagonists attenuate motor activity induced by amphetamine but not cocaine. Eur. J. Pharmacol., 249: 167–177.

Jones, E.A., Wang, J.Q., Mayer, D.C. and McGinty, J.F. (1999) The role of striatal GABA$_A$ receptors in dopamine agonist-induced behavior and gene expression. Brain Res., 836: 99–109.

Jones, E.A., Wang, J.Q. and McGinty, J.F. (2001) Intrastriatal GABA$_A$ receptor blockade does not alter D1/D2 interactions in the intact rat striatum. Neuroscience, 102: 381–389.

Kaneko, S., Hikidi, T., Watanabe, D., Ichinose, H., Nagatsu, T., Kreitman, R.J., Postan, I. and Nakanishi, S. (2000) Synaptic integration mediated by striatal cholinergic interneurons in basal ganglia function. Science, 289: 633–637.

Kaneko, T., Shigemoto, R., Nakanishi, S. and Mizuno, N. (1993) Substance P receptor-immunoreactive neurons in the rat striatum are segregated into somatostatinergic and cholinergic aspiny neurons. Brain Res., 631: 297–303.

Kraft, M., Ahluwahlia, S. and Angulo, J.A. (2001a) Neurokini-1 receptor antagonists block acute cocaine-induced horizontal locomotion. Ann. N.Y. Acad. Sci., 977: 121–131.

Kraft, M., Noailles, P. and Angulo, J.A. (2001b) Substance P modulates cocaine-induced dopamine overflow in the striatum of rat brain. Ann. N.Y. Acad. Sci., 977: 132–139.

Kubota, Y. and Kawaguchi, Y. (1993) Spatial distributions of chemically identified intrinsic neurons in relation to patch and matrix compartments of rat neostriatum. J. Comp. Neurol., 332: 499–513.

Lacey, C.J., Boyes, J., Gerlach, O., Chen, L., Magill, P.J. and Bolam, J.P. (2005) GABA$_B$ receptors at glutamatergic synapses in the rat striatum. Neuroscience, 136: 1083–1095.

Le Moine, C., Gaveriaux-Ruff, C., Befort, K. and Bloch, B. (1994) Delta opioid receptor gene expression in the mouse forebrain: localization in cholinergic neurons of the striatum. Neuroscience, 62: 635–640.

Lubetski, C., Chesselet, M.-F. and Glowinski, J. (1982) Modulation of dopamine release in striatal slices by delta opioid agonists. J. Pharmacol. Exper. Therap., 222: 435–440.

Maisonneuve, I.M., Archer, S. and Glick, S.D. (1994) U50,488, a kappa opioid receptor agonist, attenuates cocaine-induced increases in dialysate dopamine in the nucleus accumbens of rats. Neurosci. Lett., 181: 57–60.

Mansour, A., Khachaturian, H., Lewis, M.E., Akil, H. and Watson, S.J. (1987) Autoradiographic differentiation of mu, delta, and kappa opioid receptors in the rat forebrain and midbrain. J. Neurosci., 7: 2445–2464.

Meshul, C.K. and McGinty, J.F. (2000) Kappa opioid receptor immunoreactivity in nucleus accumbens shell and caudate-putamen is primarily associated with synaptic vesicles in axons. Neuroscience, 96: 91–99.

Merchant, K.M., Hanson, G.R. and Dorsa, D.M. (1994) Induction of neurotensin and c-fos mRNA in distinct subregions of rat neostriatum after acute methamphetamine: comparison with acute haloperidol effects. J. Pharmacol. Exper. Therap., 269: 806–812.

Meyer, M.E. and Meyer, M.E. (1993) Behavioral effects of the mu-opioid peptide agonists DAMGO, DALDA, and PL017 on locomotor activities. Pharmacol. Biochem. Behav., 46: 391–395.

Mulder, A.H., Wardeh, G., Hogenboom, F. and Frankhuyzen, A.L. (1984) k- and d-opioid receptor agonists differentially inhibit striatal dopamine and acetylcholine release. Nature, 308: 278–280.

Nisenbaum, E.S., Berger, T.W. and Grace, A.A. (1993) Depression of glutamatergic and GABAergic synaptic responses in striatal medium spiny neurons by stimulation of presynaptic $GABA_B$ receptors. Synapse, 14: 221–242.

Parent, A. and Hazrati, L.N. (1995) Functional anatomy of the basal ganglia. I. The cortico-basal ganglia-thalamo-cortical loop. Brain Res. Rev., 20: 91–127.

Pentney, R.J.W. and Gratton, A. (1995) Effects of local delta and mu opioid receptor activation of basal and stimulated dopamine release in striatum and nucleus accumbens of rat: an in vivo electrochemical study. Neuroscience, 45: 95–102.

Petit, F., Hamon, M., Fournie-Zaluski, M.C. and Roques, B.P. (1986) Further evidence for a role of delta opiate receptor in the presynaptic regulation of newly synthesized dopamine release. Eur. J. Pharmacol., 126: 1–9.

Rawls, S.M. and McGinty, J.F. (1997) L-trans-PDC-evoked striatal glutamate levels are attenuated by extracellular calcium reduction and glutamate receptor blockade. J. Neurochem., 68: 1553–1563.

Rawls, S.M. and McGinty, J.F. (1998) Muscarinic receptors regulate extracellular glutamate levels in the rat striatum: an in vivo microdialysis study. J. Pharmacol. Exper. Therap., 286: 91–98.

Rawls, S.M. and McGinty, J.F. (1999) Presynaptic kappa opioid and muscarinic receptors inhibit the calcium-dependent component of evoked glutamate release from striatal synaptosomes. J. Neurochem., 73: 1058–1065.

Rawls, S.M. and McGinty, J.F. (2000) Delta opioid receptors regulate calcium-dependent, amphetamine-evoked glutamate levels in the striatum: an in vivo microdialysis study. Brain Res., 861: 296–304.

Sandor, N.T., Lendvai, B. and Vizi, E.S. (1992) Effect of opiate antagonists on striatal acetylcholine and dopamine release. Brain Res. Bull., 29: 369–373.

Saylor, A.J., Meredith, G.E., Vercillo, M.S., Zahm, D.S. and McGinty, J.F. (2006) BDNF heterozygous mice demonstrate age-related changes in striatal and nigral gene expression. Exp. Neurol., 199: 362–372.

Schad, C.A., Justice, J.B. and Holtzmann, S.G. (1996) Differential effects of delta- and mu-opioid receptor antagonists on the amphetamine-induced increase in extracellular dopamine in striatum and nucleus accumbens. J. Neurochem., 67: 2292–2299.

Shippenberg, T.S. and Heidbreder, C. (1995) The delta-opioid receptor antagonist naltrindole prevents sensitization to the conditioned rewarding effects of cocaine. Eur. J. Pharmacol., 280: 55–61.

Shippenberg, T.S., Bals-Kubik, R. and Herz, A. (1987) Motivational properties of opioids: evidence that an activation of delta-receptors mediates reinforcement processes. Brain Res., 436: 234–239.

Shippenberg, T.S., LeFevour, A. and Heidbreder, C. (1996) Kappa opioid receptor agonists prevent sensitization to the conditioned rewarding effects of cocaine. J. Pharmacol. Exp. Ther., 276: 545–554.

Singh, N.A., Midgley, L.P., Bush, L.G., Gibb, J.W. and Hanson, G.R. (1991) N-Methyl-D-aspartate receptors mediate dopamine-induced changes in extrapyramidal and limbic dynorphin systems. Brain Res., 555: 233–238.

Smith, A.J.W. and McGinty, J.F. (1994) Acute amphetamine and methamphetamine alter opioid peptide mRNA expression in the rat striatum. Mol. Brain Res., 21: 359–362.

Spanagel, R., Herz, A. and Shippenberg, T.S. (1992) Opposing tonically active endogenous opioid systems modulate the mesolimbic dopaminergic pathway. Proc. Natl. Acad. Sci. USA, 89: 2046–2050.

Steiner, H. and Gerfen, C.R. (1998) Role of dynorphin and enkephalin in the regulation of straital output pathways and behavior. Exp. Brain Res., 123: 60–73.

Steiner, H. and Gerfen, C.R. (1999) Enkephalin regulates acute D2 receptor antagonist-induced immediate-early gene expression in striatal neurons. Neurosci., 88: 795–810.

Svingos, A.L., Clarke, C.L. and Pickel, V.M. (1999) Localization of the delta-opioid receptor and dopamine transporter in the nucleus accumbens shell: implications for opiate and psychostimulant cross-sensitization. Synapse, 34: 1–10.

Svingos, A.L., Chavkin, C., Colago, E.E. and Pickel, V.M. (2001) Major coexpression of kappa-opioid receptors and the dopamine transporter in nucleus accumbens axonal profiles. Synapse, 42: 185–192.

Tzaferis, J.A. and McGinty, J.F. (2001) Kappa opioid receptor stimulation decreases amphetamine-induced behavior and neuropeptide mRNA expression in the striatum. Mol. Brain Res., 93: 27–35.

Ukai, M., Kamiya, T., Toyoshi, T. and Kameyama, T. (1992) Systemic administration of dynorphin A (1–13) markedly inhibits different behavioural responses induced by cocaine in the mouse. Neuropharmacol., 31: 843–849.

Wang, H. and Pickel, V.M. (2001) Preferential cytoplasmic localization of delta-opioid receptors in rat striatal patches: comparison with plasmalemmal μ-opioid receptors. J. Neurosci., 21: 3242–3250.

Wang, J.Q. and McGinty, J.F. (1996a) D1 and D2 receptor regulation of preproenkephalin and preprodynorphin mRNA

284

in rat striatum following acute injection of amphetamine or methamphetamine. Synapse, 22: 114–122.

Wang, J.Q. and McGinty, J.F. (1996b) Scopolamine augments c-fos and zif/268 mRNA expression induced by the full D1 dopamine receptor agonist SKF-82958 in the intact rat striatum. Neuroscience, 72: 601–616.

Wang, J.Q. and McGinty, J.F. (1996c) Intrastriatal injection of the metabotropic glutamate receptor antagonist MCPG attenuates acute amphetamine-stimulated neuropeptide mRNA expression in rat striatum. Neurosci. Lett., 218: 13–16.

Wang, J.Q. and McGinty, J.F. (1996d) Muscarinic receptors regulate striatal neuropeptide gene expression in normal and amphetamine-treated rats. Neuroscience, 75: 43–56.

Wang, J.Q. and McGinty, J.F. (1997a) The full D1 dopamine receptor agonist SKF-82958 induces neuropeptide mRNA in the normosensitive striatum of rats: Regulation of D1/D2 interactions by muscarinic receptors. J. Pharmacol. Exper. Therap., 281: 972–982.

Wang, J.Q. and McGinty, J.F. (1997b) Intrastriatal injection of a muscarinic receptor agonist and antagonist regulates striatal neuropeptide mRNA expression in normal and amphetamine-treated rats. Brain Res., 748: 62–70.

Wang, J.Q. and McGinty, J.F. (1998) Glutamatergic and cholinergic regulation of immediate early gene and neuropeptide gene expression in the striatum. In: Merchant K. (Ed.), Pharmacological Regulation of Gene Expression in the CNS. CRC Press, Boca Raton, pp. 81–113.

Werling, L., Frattali, A., Portoghese, P., Takemori, A. and Cox, B. (1988) Kappa receptor regulation of dopamine release from striatum and cortex of rats and guinea pigs. J. Pharmacol. Exper. Therap., 246: 282–286.

You, Z.-B., Herrera-Marschitz, M., Nylander, I., Goiny, M., O'Connor, W.T., Ungerstedt, U. and Terenius, L. (1994) The striatonigral dynorphin pathway of the rat studied with in vivo microdialysis–II. Effects of dopamine D1 and D2 receptor agonists. Neuroscience, 63: 427–434.

Zahm, D.S., Krause, J.E., Welch, M.A. and Grosu, D.S. (1998) Distinct and interactive effects of d-amphetamine and haloperidol in levels of neurotensin and its mRNA in subterritories in the dorsal and ventral striatum of the rat. J. Comp. Neurol., 400: 487–503.

Zhou, W., Mailloux, A., Jung, B.J., Edmunds, H.S. and McGinty, J.F. (2004) GABA$_B$ receptor stimulation decreases amphetamine-induced neuropeptide mRNA expression in the striatum of rats. Brain Res., 1004: 18–28.

Zhou, W., Mailloux, A. and McGinty, J.F. (2005) Intracerebral baclofen administration decreases amphetamine-induced behavior and neuropeptide gene expression in the striatum. Neuropsychopharmacology, 30: 880–890.

Systems Level Aspects of GABA in the Basal Ganglia

Tepper, Abercrombie & Bolam (Eds.)
Progress in Brain Research, Vol. 160
ISSN 0079-6123

CHAPTER 17

GABAergic circuits in the basal ganglia and movement disorders

Adriana Galvan* and Thomas Wichmann

Department of Neurology, School of Medicine and Yerkes National Primate Research Center, Emory University, Atlanta, GA 30322, USA

Abstract: GABA is the major inhibitory neurotransmitter in the basal ganglia, and GABAergic pathways dominate information processing in most areas of these structures. It is therefore not surprising that abnormalities of GABAergic transmission are key elements in pathophysiologic models of movement disorders involving the basal ganglia. These include hypokinetic diseases such as Parkinson's disease, and hyperkinetic diseases, such as Huntington's disease or hemiballism. In this chapter, we will briefly review the major anatomic features of the GABAergic pathways in the basal ganglia, and then describe in greater detail the changes of GABAergic transmission, which are known to occur in movement disorders.

Keywords: Parkinson's disease; Huntington's disease; Hemiballism; GABA receptors; GABAergic neurons

Anatomy and function of GABAergic circuits in the basal ganglia

The basal ganglia comprise the neostriatum (caudate nucleus and putamen), the ventral striatum, the external and internal pallidal segment (GP$_e$ and GP$_i$, respectively), the subthalamic nucleus (STN), and the substantia nigra with its pars reticulata and pars compacta (SNr and SNc, respectively). These structures are components of larger circuits that involve thalamus and cortex (Fig. 1, Normal). The striatum and STN are the main entry points for inputs from the cerebral cortex and thalamus. Basal ganglia outflow arises from GP$_i$ and SNr and is directed toward frontal areas of the cerebral cortex (via the thalamus) and toward various brainstem structures (superior colliculus, pedunculopontine nucleus (PPN), and parvicellular reticular formation).

With the exception of a small number of cholinergic (Kawaguchi et al., 1995) and dopaminergic neurons (e.g., Betarbet et al., 1997), the striatum is composed of GABAergic interneurons and output neurons. The majority of these are medium-sized spiny projection neurons that receive intrinsic GABAergic input from collaterals of other medium-sized spiny neurons (Wilson and Groves, 1980), as well as interneurons (Tepper et al., 2004).

Medium-sized spiny neurons can be grossly divided into those that give rise to the 'direct' pathway and those that give rise to the 'indirect' pathway. The direct pathway is a monosynaptic connection that links the striatum to the basal ganglia output nuclei, GP$_i$ and SNr, while the indirect pathway is a polysynaptic connection to the same targets, and involves synaptic interactions in GP$_e$ and the STN (Albin et al., 1989; DeLong, 1990).

*Corresponding author. Tel.: (404) 712 8841;
Fax: (404) 727 9294; E-mail: agalvan@emory.edu

DOI: 10.1016/S0079-6123(06)60017-4

287

288

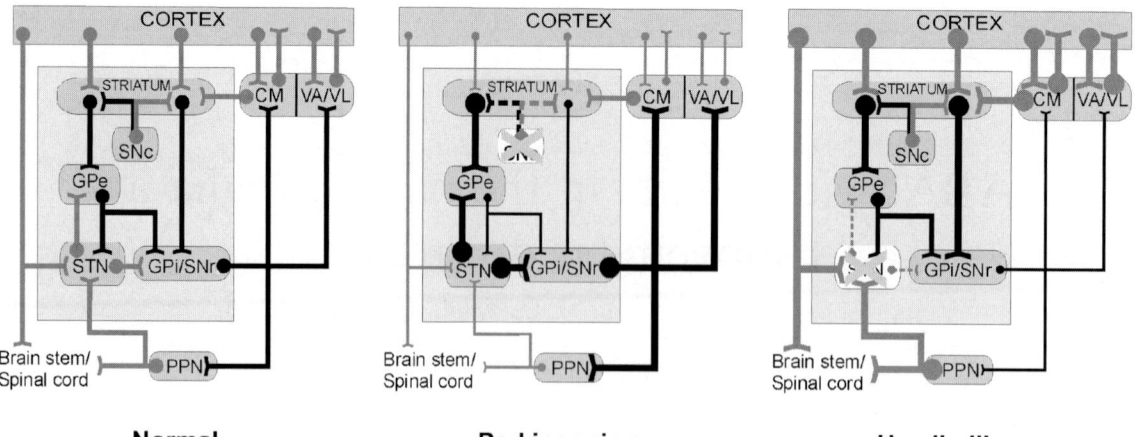

Normal   **Parkinsonism**   **Hemiballism**

Fig. 1. Traditional schematic diagram of the basal ganglia–thalamocortical circuits, under normal conditions (left), in parkinsonism (center) and in hemiballism (right). Black connections are inhibitory; gray connections are excitatory. The striatum and the STN are the principal input nuclei of the basal ganglia, and they are connected to the output nuclei, GP$_i$ and SNr, through a direct and an indirect pathway. Basal ganglia output is directed toward several thalamic nuclei (VA/VL and CM), and toward brain stem nuclei. In parkinsonism, degeneration of SNc dopaminergic cells produce a cascade of changes in the basal ganglia, which ultimately results in increased output from GP$_i$ and SNr (illustrated by thicker connections in this diagram). This leads to inhibition of thalamic and cortical neurons. In hemiballism, by contrast, lesions of the STN lead to reduced basal ganglia output (thinner connections), and increased activity of related thalamic and cortical cells. Abbreviations: GP$_e$ and GP$_i$: Globus pallidus external and internal segments, respectively; STN, subthalamic nucleus; SNc and SNr, substantia nigra compacta and reticulata, respectively; PPN, pedunculopontine nucleus; CM, centromedian nucleus; VA, ventral anterior nucleus; VL, ventrolateral nucleus.

The two groups of striatal source neurons differ substantially, with regard to the type of dopamine receptors found on them, and with regard to peptides that are co-expressed with gamma-aminobutyric acid (GABA) in them. Thus, medium-sized spiny neurons giving rise to the direct pathway carry dopamine D1-receptors and express substance P, while those neurons that give rise to the indirect pathway carry D2-receptors, and express enkephalin (for a summary, see e.g., Wichmann and DeLong, 2003). In addition, it is likely that different classes of cortical neurons project to the striatal medium-sized spiny neurons that give rise to the direct and indirect pathways (Lei et al., 2004), but the functional significance of this finding is unclear. One of the principal functions of dopamine release in the striatum is the gating of corticostriatal transmission. It is an important component of the traditional models of basal ganglia anatomy and function that D1 and D2 receptors mediate different effects in this regard. Thus, corticostriatal transmission onto striatal neurons that give rise to the indirect pathway is

thought to be inhibited by dopamine via D2 receptor activation, while corticostriatal transmission onto neurons in the direct pathway is thought to be facilitated by dopamine via D1 receptor activation. Of course, dopamine release also affects other aspects of striatal function (such as plasticity) that are beyond the scope of this chapter.

The anatomic concept of the direct and indirect GABAergic connection systems and their modulation through dopaminergic inputs have undergone significant revisions since its original description in the 1980s. For instance, it is now clear that these pathways are not as separate as initially proposed. The experimental evidence indicates that many striatal output neurons branch to both segments of the globus pallidus (thus contributing to direct and indirect routes (see e.g., Kawaguchi et al., 1990; Wu et al., 2000), and that a subpopulation of medium-sized spiny neurons carry both D1 and D2 receptors (Surmeier et al., 1992).

GABAergic transmission in the other basal ganglia nuclei is anatomically less complicated than that found in the striatum. Cells in GP_e, GP_i, or SNr receive substantial GABAergic input from the striatum. In addition, GP_i and SNr also receive GABAergic afferents from the GP_e as part of the indirect pathway (Parent et al., 2000). These GP_e terminals preferentially contact proximal regions of the dendritic arbors in SNr and GP_i, while striatal terminals are found more distally (Smith et al., 1998). Cells in GP_e, GP_i, and SNr are known to extend local axon collaterals that represent another source of GABA in these nuclei (Grofova et al., 1982; Francois et al., 1984; Kita and Kitai, 1994). In the STN, a nucleus that primarily contains glutamatergic projection neurons, the primary sources of GABA are the terminals of GP_e efferents, although a small number of GABAergic local interneurons may also contribute (Levesque and Parent, 2005).

Current models of the role of the basal ganglia in movement consider the potential interactions between direct and indirect pathways, but do not (yet) take into account local GABAergic interactions via collaterals or interneurons within these nuclei. Such models are discussed here because an understanding of the normal function of these pathways may lead to a better understanding of the abnormalities associated with movement disorders. Activity initiated in cortical motor areas is thought to be projected onto the striatal neurons that give rise to the direct and indirect pathways. Direct pathway activation results in inhibition of neurons in GP_i and SNr with subsequent disinhibition of related thalamocortical neurons (Inase et al., 1996), thereby facilitating movement. By contrast, activation of the striatal neurons that give rise to the indirect pathway would lead to increased basal ganglia output onto thalamocortical neurons, thereby suppressing movement. Interactions between direct and indirect pathways may serve to scale individual movement parameters, such as the amplitude or velocity of movements, or to focus entire movements, by facilitating intended movements, and suppressing unintended ones (Mink, 1996). The basal ganglia clearly have functions that go beyond simple scaling or focusing, such as roles in movement preparation, in self-initiated (internally generated) movements, as well as motor (especially

procedural) learning and movement sequencing (e.g., Graybiel, 1995; Schultz, 1998).

Despite the fact that many recent experimental findings qualify or even contradict specific aspects of this model for the role of interactions between the direct and indirect pathways in the control of movement (see critique in Wichmann and DeLong, 2003), the principal insight that movement tends to be facilitated by direct pathway activation and inhibited by indirect pathway activation is generally accepted. There is also little doubt that globally increased basal ganglia output is associated with a reduction of movement (as is seen in parkinsonism). Similarly, globally reduced basal ganglia output is associated with the production of excessive movements, as seen in hyperkinetic disorders. Many facets and distinguishing features of these disorders are, of course, not fully explained in these general terms, and more detailed analyses of patterns of neurons or neuronal assemblies must be considered. It is also increasingly recognized that many factors other than tonic activity may influence the impact of GABA released from GABAergic terminals, including the density, type, and subcellular localization of GABA receptors, the intracellular effects triggered by receptor activation, the rate of GABA uptake, and the pattern of electrical activity mediated by GABAergic transmission.

Hypokinetic disorders

Parkinson's disease is the most prominent hypokinetic movement disorder. It is characterized by progressive degeneration of brain cells, including areas in the brainstem, midbrain, olfactory tubercle and cortex (Braak et al., 2003). The disease is clinically dominated by the progressive appearance of motor signs such as poverty of movement (akinesia), muscle stiffness (rigidity) and tremor at rest. These signs are thought to be related to the profound loss of dopaminergic neurons in the SNc. Additional symptoms include a myriad of cognitive and mood abnormalities, as well as autonomic and balance problems. While the non-motor symptoms of the disease are poorly understood at this time, dopamine loss has been linked to the development of the motor symptoms by the neurochemical and pathophysiologic mechanisms outlined in this section.

The study of pathophysiologic and neurochemical changes occurring in Parkinson's disease has been facilitated by the availability of animal models that faithfully replicate many of the behavioral, biochemical, and electrophysiologic features of the human disorder. These models of parkinsonism include rodents treated with 1-methyl-4-phenyl-1,2,3,6-tetrahydropyridine (MPTP) or 6-hydroxydopamine (6-OHDA), and primates treated with MPTP (Burns et al., 1983; Forno et al., 1993). The latter animal model remains the gold standard for changes in activity in basal ganglia pathways in response to dopamine depletion. Studied with a variety of techniques (see below), these animal models have led to the development and subsequent refinement of a model of the pathway abnormalities in Parkinson's disease. According to this model (Fig. 1, Parkinsonism), the disorder is the result of a combination of increased indirect pathway activity and decreased direct pathway activity, both resulting from loss of dopamine in the striatum (Albin et al., 1989; DeLong, 1990), and leading to increased GP_i/SNr activity. GABAergic dysfunction is thus a key component of the pathophysiologic and neurochemical abnormalities that underlie Parkinson's disease. These dysfunctions have been extensively documented, as will be shown in the following paragraphs, but it is important to recognize that many of the changes proposed by the model to occur outside of the basal ganglia remain hypothetical. For instance, the effects of increased basal ganglia output on thalamic or PPN activity, and the changes in cortical activity in parkinsonism, have not been investigated to the same extent as the abnormalities that occur in the basal ganglia.

Changes in GABAergic transmission in Parkinson's disease

Early studies in MPTP-treated primates have indicated that metabolic activity (as measured with the 2-deoxy-glucose technique) is increased in both pallidal segments (e.g., Crossman et al., 1985;

Schwartzman and Alexander, 1985; Mitchell et al., 1989). Because metabolic activity measured by this technique appears to parallel the level of synaptic activity, the findings were interpreted as indicative of increased activity along the first portion of the indirect pathway (the striatum-GP_e connection) and along the STN-GP_i pathway, or, alternatively, as evidence for increased activity along the projections from the STN to GP_i and GP_e. An alternative metabolic marker, the subunit I of cytochrome oxidase (CO-I), was also found to be increased in the STN and in the output basal ganglia nuclei of parkinsonian monkeys, but no changes were observed in GP_e (Hirsch et al., 2000). Taken together, these general metabolic markers indicate that dopaminergic depletion leads to increased activity in the output nuclei. The studies also provide some (albeit inconsistent) evidence for increased activity along the indirect pathway in parkinsonism.

The specific involvement of GABAergic neurons in these changes was further evaluated by measurement of mRNA or protein levels of the GABA-synthesizing enzyme glutamate decarboxylase (GAD). This enzyme exists in two isoforms, GAD65 and GAD67, which are expressed ubiquitously in GABAergic cells (Pinal and Tobin, 1998). GAD67 is present in terminals and cell bodies, and may preferentially synthesize cytoplasmic GABA. GAD65 is mostly found in nerve terminals, and may be primarily involved in synthesizing GABA for vesicular release (Soghomonian and Martin, 1998). GAD mRNA appears to be selectively increased in the source neurons of the striatopallidal projection system after dopaminergic degeneration (Soghomonian and Laprade, 1997; Laprade and Soghomonian, 1999). This finding is in line with the notion that the indirect pathway is hyperactive in parkinsonism (Soghomonian and Laprade, 1997), resulting in increased GP_i and SNr activity. The corresponding increase in GAD mRNA levels in GP_i and SNr has indeed been found (e.g., Kincaid et al., 1992; Soghomonian et al., 1994; Salin et al., 2002). However, GAD mRNA levels in GP_e (Pedneault and Soghomonian, 1994; Schneider and Wade, 2003; Soares et al., 2004) and GAD protein levels in the STN (Soares et al., 2004), both presumably reflecting the activity of GABAergic GP_e cells and their projections, are not altered in parkinsonian monkeys. These findings are inconsistent with some of the metabolic and electrophysiologic results (see also below) and may indicate that GAD activity does not necessarily reflect the overall activity of GABAergic neurons. There is, in fact, evidence that the increased expression of GAD in the output nuclei and GP_e may result from increased STN 'drive' rather than from increased pallidal activity. For instance, lesions of the STN reverse the increased levels of GAD mRNA in the pallidum (Delfs et al., 1995; Herrero et al., 1996; Billings and Marshall, 2004).

Another approach to examining the GABAergic changes in the basal ganglia in Parkinson's disease has been the study of GABA levels or GABA-receptor binding in postmortem tissue from parkinsonian patients or animals. These studies have found no changes in levels of GABA in most of the basal ganglia (Kish et al., 1986; Gerlach et al., 1996; Calon et al., 1999), with the exception of the striatum, where GABA levels are increased (Perry et al., 1983; Kish et al., 1986; Molina-Holgado et al., 1993). These results are in contrast with the evidence obtained by in vivo microdialysis in parkinsonian animals, in which GABA levels were increased in GP_e (Robertson et al., 1991; Schroeder and Schneider, 2002; Galeffi et al., 2003), and decreased in the STN (Soares et al., 2004). These microdialysis observations are in line with the aforementioned general model of the pathophysiology of Parkinson's disease. In the SNr, levels of GABA have been reported to be increased (Windels et al., 2005) or unchanged (Galeffi et al., 2003; Ochi et al., 2004).

GABA exerts its effects through ionotropic (GABA-A) and metabotropic (GABA-B) receptors. There are many reports of changes in expression levels of both types of GABA receptors in parkinsonism. For example, GABA-A receptor binding has been found to be decreased in GP_e and increased in GP_i and SNr of parkinsonian patients (Griffiths et al., 1990) or animals (e.g., Pan et al., 1985; Robertson et al., 1990; Gnanalingham and Robertson, 1993; Calon et al., 1999; Chadha et al., 2000b). These changes are likely to be compensatory in nature, counteracting changes in GABA levels (see above).

The changes in GABA-A receptor binding are also accompanied by changes in the mRNA levels for specific GABA-A receptor subunits. After dopamine depletion, the mRNA for α1, α2, β2, and γ2 subunits is decreased in the rat globus pallidus (equivalent to the primate GP_e), and increased in the SNr (Chadha et al., 2000a; Yu et al., 2001; Katz et al., 2005). Interestingly, even though binding studies show little or no change in striatal GABA-A receptors (Griffiths et al., 1990; Robertson et al., 1990; Calon et al., 2003), in situ hybridization reveals increased expression of α4 and γ2 in the striatum of 6-OHDA-treated rats (Chadha et al., 2000a). To date, the functions of α4 and γ2-containing GABA-A receptors have not been studied in the basal ganglia, but these receptors are known to mediate tonic inhibition in the hippocampus and cerebellum (Semyanov et al., 2004).

Similar changes have been observed in metabotropic GABA-B receptors. Thus, in patients with Parkinson's disease and in MPTP–treated monkeys, GABA-B receptor levels are decreased in GP_e, but increased in GP_i or SNr, as measured in binding studies (Calon et al., 2000; Calon et al., 2003), or by using mRNA in situ hybridization in dopamine-depleted rats (Johnston and Duty, 2003). Striatal GABA-B receptor levels are not significantly changed in 6-OHDA-treated rats (de Groote et al., 1999), or in MPTP-treated mice or monkeys (Calon et al., 2001). In the striatum, GABA-B receptor binding is decreased (Calon et al., 2003), and increased (Johnston and Duty, 2003) or unchanged (Calon et al., 2001) in the STN.

The ligand binding and mRNA studies, however, do not provide the spatial resolution needed to describe changes in the localization of these receptors at the subcellular or subsynaptic levels. In fact, preliminary evidence suggests that the subcellular pattern of localization of GABA-B receptors in GP_e and GP_i is altered in MPTP-treated monkeys (Galvan et al., 2005). After dopamine depletion, the expression of presynaptic GABA-B receptors in unmyelinated axons is increased (Fig. 2), suggesting that these receptors could have a more significant function modulating neurotransmitters release in the parkinsonian state than

292

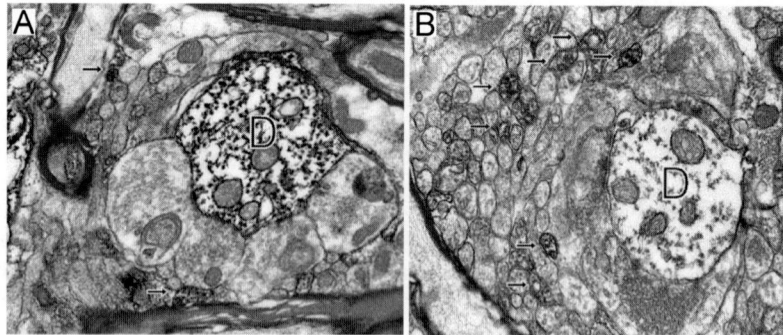

Fig. 2. GABA-BR1-immunoreactive dendrites (D) and unmyelinated axons (arrows) in GP$_e$ of normal (A) and MPTP-treated (B) monkeys. Note the increased number of labeled axons in the MPTP monkey. Despite the intense dendritic labeling in (A) in the example presented, the overall number of labeled dendrites was not different between the normal and the MPTP-treated monkey.

in normal conditions. Furthermore, it has not yet been analyzed whether the changes in expression are associated with changes in affinity or in coupling to second messenger systems. GABAergic transmission and receptor activation may also critically depend on GABA re-uptake. This process has thus far been studied only in normal basal ganglia (Bahena-Trujillo and Arias-Montano, 1999; Chen and Yung, 2003; Galvan et al., 2005) (Fig. 3).

Electrophysiologic in vivo studies have also yielded substantial evidence about the activity of individual (presumably GABAergic) neurons in the basal ganglia. These studies have shown that the neuronal activity in the primate MPTP-model of parkinsonism is reduced overall in GP$_e$, and increased in the STN and GP$_i$, as compared to normal controls (Miller and DeLong, 1987; Filion et al., 1988; Bergman et al., 1994) (Fig. 4). Some of these findings have also been supported by electrophysiologic recording studies in parkinsonian patients undergoing neurosurgical interventions guided by microelectrode recording (e.g., Vitek et al., 1993; Lozano et al., 1996). More recently, experiments in MPTP-treated monkeys have suggested that SNr neurons are also hyperactive in parkinsonism (Wichmann et al., 1999).

The general pathophysiologic model in which overactivity along the indirect pathway is a major contributor to the development of parkinsonism is also supported by the demonstration that lesions of STN, GP$_i$, or SNr in MPTP-treated primates reverse some or all signs of parkinsonism

(Bergman et al., 1990b; Lieberman et al., 1999; Wichmann et al., 2001). These findings have led to a resurgence of basal ganglia lesioning as treatment for patients with Parkinson's disease. Although the experience with procedures such as pallidotomies or thalamotomies also generally supports the anatomy- and rate-based pathophysiologic model for Parkinson's disease, several findings in patients undergoing such lesion procedures are incompatible with the model. For instance, lesions of the VA/VL nuclei of the thalamus do not lead to parkinsonism, as would be predicted by the models, and are, in fact, beneficial in the treatment of tremor and rigidity (see, e.g., Tasker et al., 1997; Giller et al., 1998). Similarly, lesions of GP$_i$ improve all aspects of Parkinson's disease without producing dyskinesias or other obvious detrimental effects. In fact, these procedures are highly effective in reducing drug-induced dyskinesias (see, e.g., Dogali et al., 1995; Rabey et al., 1995; Baron et al., 1996). In contrast to the hypokinetic features of parkinsonism, dyskinesias appear to arise from pathologic reduction in basal ganglia outflow (see below, and Papa et al., 1999), and thus would be predicted by the model to be made worse by further reduction of pallidal outflow (see for example, Marsden and Obeso, 1994).

These findings may reflect a dependence of parkinsonism on changes in basal ganglia activity other than alterations in discharge rates (Fig. 4). Such other changes may include altered processing of proprioceptive inputs, as well as abnormal

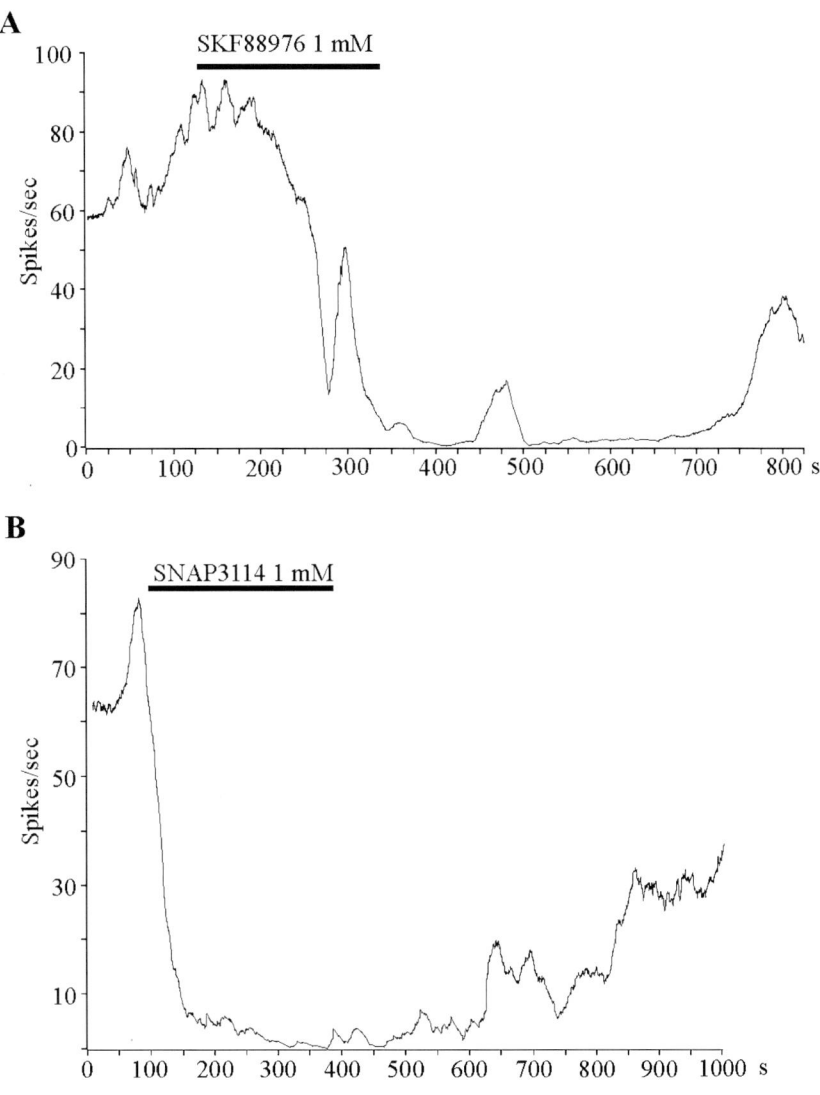

Fig. 3. Blocking GABA transport in monkey GP$_e$ or GP$_i$ decreased pallidal firing rate, in vivo. Shown are extracellular recording of pallidal cells after local microinjections of GAT-1 or GAT-3 blockers. (**A**) The GAT-1 blocker, SKF-89976A, decreased the firing of a GP$_e$ cell. (**B**) The GAT-3 blocker, SNAP-5114, decreased the firing rate in a pallidal cell. Horizontal bars indicate the duration of drug infusions. Dashed lines represent the mean discharge rate ± 2 standard deviations. For more details see Galvan et al. (2005).

timing, patterning, or synchronization of discharge that introduces errors and non-specific noise into the thalamocortical signal. Altered sensory processing in the basal ganglia has been extensively documented in parkinsonian monkeys and patients. For instance, neuronal responses to passive limb manipulations in STN, GP$_i$, and thalamus (e.g., Miller and DeLong, 1987; Filion et al., 1988; Bergman et al., 1994) have been shown to occur more often, to be more pronounced, and to have widened receptive fields after treatment with MPTP. In addition, the proportion of neurons showing increased discharge in response to somatosensory inputs increases, supporting the view that somatosensory processing is fundamentally altered in this disease (Boraud et al., 2000). It is also remarkable that in contrast to the virtual absence of synchronized discharge of basal ganglia

294

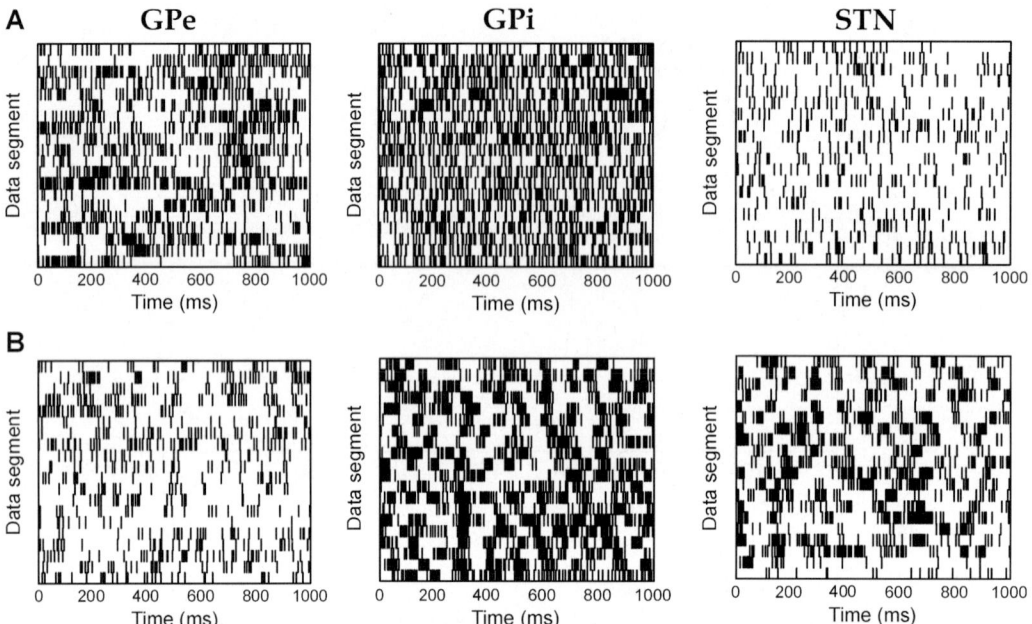

Fig. 4. Spontaneous neuronal activity, shown as raster displays, recorded in GP_e, GP_i, or STN, in normal (A) and parkinsonian (B) primates. Shown are 20 consecutive 1 s segments of data. In GP_e, the neuronal activity is reduced, whereas in STN and GP_i it is increased. In addition to the changes in firing rate, there are changes in the firing patterns of neurons with a marked increase of burstiness and oscillatory discharge patterns in the parkinsonian state.

neurons in normal monkeys (e.g., Wichmann et al., 1994a), a substantial proportion of neighboring neurons in globus pallidus and STN discharge in unison in parkinsonian primates (Bergman et al., 1994). Finally, the proportion of cells in STN, GP_i, and SNr that discharge in (synchronized) oscillatory or non-oscillatory bursts is greatly increased in parkinsonism (see Gatev et al., 2006).

The development of pathologic coherent oscillatory activity in the basal ganglia–thalamocortical network has taken center stage in recent work regarding the pathophysiology of Parkinson's disease. These oscillations may involve not only single cells but also larger portions of the network. Oscillations are not only recordable in animals but also in patients with Parkinson's disease, using intraoperative single-cell recordings, or postoperative local field potential (LFP) recordings from implanted deep brain stimulation (DBS) electrodes. Together with single-cell recording studies, the LFP studies have shown that oscillations in STN and GP_i at low frequencies (below 30 Hz) are

disruptive for movement ('antikinetic', see Brown, 2003). The processes leading to the generation of these (synchronized) oscillations are still unclear, but several different possible explanations have been offered. One is the existence of local pacemaker circuits, consisting of re-entry loops linking the GABAergic GP_e neurons and the glutamatergic STN neurons. In co-culture, GP_e and STN neurons form connections that support the development of spontaneously occurring slow synchronized oscillatory bursts of neuronal spikes (Plenz and Kitai, 1999). The oscillatory bursts are dependent on the tendency of STN neurons to respond to strong inhibitory inputs with a long inhibitory post-synaptic potential (IPSP) followed by a rebound burst (e.g., Nakanishi et al., 1987; Nambu and Llinas, 1994; Beurrier et al., 1999; Bevan et al., 2002), which may then promote subsequent pallidal activity via the glutamatergic STN–GP_e connection. The long IPSP leading to the rebound burst is, at least in part, mediated by GABA-B receptors (Hallworth and Bevan, 2005). The emergence of oscillatory bursts in the STN is also promoted by the fact that GABA-receptor dependent IPSPs in the STN appear to transiently enhance the postsynaptic excitability of STN neurons (Baufreton et al., 2005). Another prerequisite for the production of synchronized GP$_e$–STN bursting is that functionally related areas of STN and GP$_e$ are interconnected. Anatomic studies in primates (Shink et al., 1996) suggest that this is the case, but it is not clear whether these circuits in vivo are closed at the level of individual cells that are reciprocally connected to one another, or involve larger groups of neurons.

The coherence in the GP$_e$–STN network may also be modulated by striatal inputs. The potential relevance of these inputs was highlighted in recent neural network modeling studies of the striato-GP$_e$–STN interaction, where the normally irregular activity in the pallido–STN network was converted into synchronized oscillatory activity by changes in the extent of striatal inhibition of GP$_e$ (or through changes of the extent of inhibition within GP$_e$, see Bevan et al., 2002; Terman et al., 2002). Brain slice recordings in rats have also shown the potential for striatal inputs to induce coherent firing of otherwise independently oscillating GP$_e$ neurons (Stanford, 2003).

Finally, the occurrence of oscillations in the GP$_e$–STN network, and in the basal ganglia output nuclei may be strongly influenced by cortical inputs to the STN (Hartmann-von Monakow et al., 1978; Nambu et al., 2002). It has been shown in rodents, primates, and humans that cortical activity (as measured with electroencephalography (EEG)) is closely linked to single-neuron or LFP activity in the STN (Magill et al., 2000; Wichmann et al., 2002; Gatev and Wichmann, 2003). The apparent link between basal ganglia activity and cortical activity persist and may even strengthen in parkinsonism.

Drugs aimed at GABAergic transmission in the treatment of Parkinson's disease

As expected, interventions that restore dopaminergic transmission have been shown to be highly effective treatments for the motor signs in Parkinson's disease, and to 'normalize' GABAergic transmission or receptor levels. For instance,

dopamine replacement therapy with levodopa or dopaminergic agonists partially reverse the changes in GABA-A receptors in GP$_e$ (Calon et al., 1995), SNr, and GP$_i$ (Gnanalingham and Robertson, 1993; Calon et al., 1999; Katz et al., 2005). Similarly, increased GABA-B receptor binding in GP$_i$ is partially reversed by treatment with dopamine D2-receptor agonists (Calon et al., 2000), and fetal mesencephalic tissue grafts into the striatum have been shown to reverse GABA-A receptor changes in striatum, GP, and SNr of rodent models of Parkinson's disease (Stasi et al., 1999; Yu et al., 2001).

Experimental local activation of GABA-A receptors in SNr, GP$_i$, STN, or the striatum, also alleviates motor deficits. For instance, infusions in the GP$_i$ of the GABA-A receptor agonist, muscimol, ameliorates motor symptoms in parkinsonian patients (Penn et al., 1998), and in MPTP–treated monkeys (Baron et al., 2002). Similar results have also been obtained in the SNr with injections of muscimol in parkinsonian monkeys (Wichmann et al., 2001), or with the implantation of genetically engineered GABA-releasing cells in the SNr of 6-OHDA–lesioned rats (Carlson et al., 2003). Activation of GABA-A receptors in the STN also results in dramatic improvement of parkinsonian symptoms in animal models and in patients with Parkinson's disease (Wichmann et al., 1994b; Levy et al., 2001b; Baron et al., 2002; Mehta and Chesselet, 2005). In the striatum, injections of benzodiazepines, i.e., compounds that activate GABA-A receptors (Tenn and Niles, 1997), or transplantation of GABA–rich grafts (Winkler et al., 1999) induce behavioral recovery in 6-OHDA-treated rats. A problem with all of these GABA receptor agonist approaches is that GABA-receptor activation often amounts to a total inactivation of the treated tissue, similar to a (transient) lesion. The impressive effects of these interventions may therefore be rather nonspecific, and may simply act by interrupting the flow of abnormal basal ganglia firing patterns to cortex.

Treatment trials in parkinsonian patients with systemically applied GABAergic compounds have had only limited success in relieving the motor symptoms (Nutt et al., 1979; Marjama-Lyons and

Koller, 2000), and produce sedation and other side effects, probably due to the fact that they act at multiple brain sites. On the other hand, the GABA-A receptor antagonist flumazenil or the GABA release inhibitor budipin (Przuntek, 2000; Ondo and Hunter, 2003) appear to have anti-parkinsonian properties.

Another group of drugs, adenosine A2a receptor antagonists also alter the activity of GABA-ergic neurons in Parkinson's disease. Adenosine A2a receptors are primarily expressed on striatal medium-sized spiny neurons that give rise to the indirect pathway. Monotherapy with the specific adenosine A2a receptor antagonist istradefylline (KW-6002) ameliorates parkinsonism without inducing dyskinesias (Jenner, 2005; Pinna et al., 2005), and, when co-administered with levodopa, potentiates the latter's antiparkinsonian effects without producing dyskinesias in parkinsonian rodents (e.g., Lundblad et al., 2003), primates (Bibbiani et al., 2003), and patients (Bara-Jimenez et al., 2003; Hauser et al., 2003).

Procedures such as lesioning or high-frequency electrical stimulation of GP_i or STN (deep brain stimulation, DBS) are highly effective non-pharmacologic treatments for Parkinson's disease. Both of these strongly affect GABAergic functions in the basal ganglia. While lesioning may simply interrupt abnormal activity in the basal ganglia, the mechanisms underlying DBS effects are not clearly understood (Dostrovsky et al., 2002). Stimulation of the STN may result in an inhibition of neuronal activity in the STN itself (Magarinos-Ascone et al., 2002), and possibly also in those brain regions that receive, and are driven by, the glutamatergic outputs of this nucleus (Windels et al., 2000). These inhibitory effects are most likely due to stimulation of local GABAergic axon terminals, and of fibers of passage. However, DBS of the STN or GP_i may also result in a net activation of the output projections from these nuclei (Hashimoto et al., 2003), perhaps via direct activation of axons emanating from STN neurons (McIntyre et al., 2004). These stimulatory effects may act to impose a highly abnormal firing pattern (i.e., high-frequency activity) on basal ganglia output that is easier to tolerate than the one found in the untreated parkinsonian state.

Hyperkinetic disorders

Disordered function of GABAergic neurons in the basal ganglia frequently result in involuntary movements, the hallmark of hyperkinetic disorders. Pathophysiologically, these disorders are all characterized by reduced basal ganglia output, which may result in disinhibition of the thalamo-cortical system(s), releasing cortical motor areas to allow movements that would normally be suppressed (Fig. 1). The cause of reduced $GP_i\backslash SNr$ activity differs between different diseases (see below). Furthermore, the specific manifestations of the hyperkinetic diseases may in large part be determined by the presence or absence of the other abnormalities, such as changes in the patterning or the level of synchronization of basal ganglia output.

Drug-induced dyskinesias

Long-term treatment with levodopa is associated with the development of dyskinesias in the majority of parkinsonian patients (Obeso et al., 2000). The severity of dopamine depletion, the length of levodopa exposure, and the pulsatile application of the drug all contribute to the development of dyskinesias. While levodopa-induced dyskinesias are often portrayed as severely disruptive to behavior, they are often surprisingly well tolerated.

Many of the proposed mechanisms underlying dyskinesias involve altered signaling at glutamatergic synapses on GABAergic neurons in the striatum (Chase and Oh, 2000), changes in the distribution of striatal dopamine receptors or the uneven distribution of dopaminergic terminals in the striatum. These changes are thought to change the activity of the GABAergic medium-sized spiny output neurons. Levodopa-induced dyskinesias are associated with changes in GAD mRNA levels in striatal projection neurons in dopamine-depleted rats (Cenci et al., 1998; Katz et al., 2005) and monkeys (Soghomonian et al., 1996), perhaps specifically affecting indirect pathway neurons (Cenci et al., 1998; Carta et al., 2001). The indirect pathway is further implicated by the fact that D2 receptor activation (hence, inhibition of the indirect pathway) is a strong stimulus for the development of drug-induced dyskinesias. This is also supported by the finding that GAD mRNA levels in the globus pallidus of dopamine-depleted rats (equivalent to GP_e in monkeys) are increased (Cenci et al., 1998), possibly indicating enhanced activity of GP_e neurons. Increased GP_e activation may lead to inhibition of the excitatory connection linking STN and GP_i, and, thus, to a reduction of GP_i output.

Changes in GABA receptor levels in GP_i may also contribute to the development of dyskinesias. In non-dyskinetic parkinsonian animals and patients, levodopa reverses the abnormal increase in GABA-A receptor binding that accompanies the dopamine-depleted state (see above). However, GABA-A receptor levels remain high in GP_i of dyskinetic parkinsonian patients or in MPTP-treated monkeys (Calon et al., 1995; Calon and Di Paolo, 2002; Calon et al., 2003). Similarly, GABA-B receptors may be upregulated in dyskinetic subjects (Calon et al., 2000).

Electrophysiologic recording studies in monkeys and human patients with Parkinson's disease have confirmed that levodopa-induced dyskinesias are associated with abnormally low background activity in GP_i (Papa et al., 1999; Levy et al., 2001a). However, it is also clear that reduced basal ganglia output per se does not induce dyskinesias (Mink and Thach, 1991; Wenger et al., 1999; Baron et al., 2002), suggesting that alterations in the patterning of pallidal firing may also be important for dyskinesias (see, for instance, Levy et al., 2001a).

The insight that GABAergic basal ganglia pathways are involved in the development of involuntary movements associated with levodopa treatment has led to early attempts to treat dyskinesias with drugs aimed at enhancing GABAergic transmission (e.g., Lees et al., 1978; Nutt et al., 1979; Morselli et al., 1985). The results of these trials were generally disappointing. However, more recent pharmacologic experiments aimed at modulating the activity of specific GABAergic pathways in the basal ganglia have been more promising. A relatively well-developed example of these approaches is the therapeutic use of cannabinoid receptor agonists. These compounds are considered to be potentially useful treatments of levodopa-induced dyskinesias because activation of cannabinoid receptors in GP_e may reduce GABA reuptake and enhance GABA transmission in this nucleus, which may reduce GP_e activity, and may act to reduce the excessive inhibition of STN and GP_i to more normal levels. Small trials, using such agents in dyskinetic patients (Sieradzan et al., 2001) and monkeys (Fox et al., 2002) have indeed been promising. Interestingly, a selective cannabinoid receptor antagonist also reduced dyskinesias in monkeys, perhaps through a different mechanism (van der Stelt et al., 2005).

Hemiballismus

Hemiballismus is a hyperkinetic movement disorder, characterized by proximal large-amplitude limb movements that are usually of sudden onset, and are in many cases caused by vascular lesions involving the (glutamatergic) STN. Although the primary deficit in these cases is a loss of excitatory subthalamic 'drive' onto GP_e or GP_i, the STN lesion has prominent secondary effects on the balance of GABAergic transmission in the basal ganglia as a whole. Thus, STN lesions reduce firing rates in GP_e and GP_i (Hamada and DeLong, 1992; Suarez et al., 1997; Vitek et al., 1999), and disrupt the cortico-subthalamic pathway, which is a likely reason for the reduced sensitivity of pallidal cells to cortical inputs in individuals with hemiballism (Hamada and DeLong, 1992; Nambu et al., 2000; Nambu et al., 2002; Kita et al., 2005). STN lesions also (partially) disrupt the function of the indirect pathway of the basal ganglia. Together, these changes would result in lowered firing rates in the basal ganglia output nuclei, and an emphasis on direct pathway (with its facilitatory effect on movement, Fig. 1, Hemiballism).

In some respect, hemiballism and parkinsonism are conditions on opposite ends of the spectrum of basal ganglia disorders. This point is perhaps best demonstrated by the fact that STN lesions are effective treatments for parkinsonian symptoms in both animals (Bergman et al., 1990a; Aziz et al., 1991; Wichmann et al., 1994b; Guridi et al., 1996) and humans (Guridi and Obeso, 2001; Chen et al., 2002a; Keep et al., 2002; Su et al., 2003). There are also reports of antiparkinsonian effects of spontaneously occurring STN lesions in humans (Scoditti et al., 1989; Sellal et al., 1992). Interestingly, parkinsonian patients rarely develop severe hemiballistic movements after STN lesions (Guridi and Obeso, 2001). This may be due to the inadvertent surgical damage to pallidal outflow fibers that pass close to the STN's dorsal borders, or may be due to the fact that the direct pathway is underactive in Parkinson's disease, which may partially protect against hemiballism (Guridi and Obeso, 2001).

Systemic GABAergic drugs, such as GABA/benzodiazepine receptor agonists have limited effects in hemiballism (Becker and Lal, 1983), while GABA/benzodiazepine-receptor blockers are without clinical benefit (Kulisevsky et al., 1994). Other drugs that act on the GABA system, such as gabapentin (Kothare et al., 2000) sodium valproate (Lenton et al., 1981; Chandra et al., 1982; Kouyoumdjian, 1984), or topiramate (Gatto et al., 2004; Driver-Dunckley and Evidente, 2005) have also been occasionally used with success. Of course, the more commonly used treatment with neuroleptic drugs (see, e.g., Vidakovic et al., 1994; Postuma and Lang, 2003) may also act through modulation of GABAergic pathways in the basal ganglia, through reduction of the activity along the 'direct' pathway.

It may seem paradoxical that surgical pallidotomies are also useful treatments in severe, drug-resistant hemiballism, both in animals (Whittier and Mettler, 1949; Carpenter et al., 1950), and humans (Suarez et al., 1997; Vitek et al., 1999). In this case, the rate-based model of pathway abnormalities in the basal ganglia does not explain the beneficial effects, because a pallidotomy procedure would induce a further reduction of basal ganglia output. The effects of pallidotomies on hemiballism may, however, be explainable through the reduction of abnormal basal ganglia output patterns to thalamus and brainstem. The basal ganglia discharge patterns associated with hemiballism have not been explored in detail.

Huntington's disease

Patients with Huntington's disease present a combination of motor and non-motor signs. One of the primary and relatively early clinical signs of the disease is chorea, i.e., the occurrence of involuntary arrhythmic, jerky movements of limbs and face. Later in the disease, these involuntary movements are often replaced by bradykinesia and rigidity, i.e., a parkinsonian appearing phenotype. In addition to these motor disturbances, non-motor signs such as depression, behavioral disturbances, and cognitive impairments also occur, and may represent the most significant source of disability for these patients.

The disorder results from an autosomal dominant trinuclotide repeat mutation of the huntingtin gene on the short arm of chromosome 4 (e.g., Group, 1993; Reddy et al., 1999). Huntingtin is a cytosolic and nuclear protein, which is associated with microtubules and synaptic vesicles, and is widely expressed throughout the nervous system, and in non-neuronal tissues. Expansion of polyglutamine sections within the huntingtin protein may result in processes such as abnormal protein aggregation, increased expression of caspase-1, impairment of proteasomal function, and problems with mitochondrial function.

The relationship between these changes and the clinical manifestations of the disease are unclear. It is known, however, that at least in early stages of the disease, changes in GABAergic transmission in the basal ganglia contribute significantly to the motor signs of the disease. Studies of the neuropathology and GABAergic function in the basal ganglia have been carried out both in humans with the disease, and in animal models, including striatal lesion models, induced by quinolinic acid and other neurotoxins (Ferrante et al., 1993; Brickell et al., 1999; Brouillet et al., 1999), or injections of 3-nitropropionic acid, an irreversible inhibitor of complex II of the mitochondrial respiratory chain (Beal et al., 1993; Brouillet et al., 1999), or in transgenic mouse models, such as the R6/1 and R6/2 lines (Dunnett and Rosser, 2004).

The pathologic studies have shown widespread neuronal degeneration throughout cortex, basal ganglia, thalamus, and brainstem. Of these, the GABAergic striatal output neurons appear to be particularly affected (Reiner et al., 1988), while some of the striatal interneurons (specifically, calretinin-positive interneurons) are selectively spared (Cicchetti et al., 2000). Striatal degeneration follows a fairly predictable pattern, first involving neurons of the indirect pathway, which may result in reduced inhibition of GP_e neurons and, secondarily, in increased inhibition of STN neurons, decreased facilitation of GP_i output and disinhibition of thalamocortical neurons. In later stages of Huntington's disease, inhibitory striatal output neurons to GP_i degenerate as well, resulting in disinhibition of GP_i neurons and development of hypokinetic (parkinsonian) features.

Striatal degeneration in Huntington's disease goes far beyond the motor portion of this structure. In fact, the earliest and most prominent anatomic and radiologic features of Huntington's disease occur in the caudate nucleus, i.e., the non-motor portion of the striatum. The involvement of basal ganglia–thalamocortical associative and limbic circuits may contribute to the prominent psychiatric and cognitive abnormalities in this disease.

Biochemical studies have largely supported the anatomic findings mentioned above. Thus, measurements of GABA levels in postmortem studies have suggested that choreic Huntington's disease patients show a substantial loss of GABA in the striatum (Ellison et al., 1987; Bonilla et al., 1988; Reynolds et al., 1990), and a greater reduction of GABA in GP_e than in GP_i (Spokes, 1980; Reynolds et al., 1990; Storey and Beal, 1993), along with an increase in GABA levels in the STN (but see Spokes, 1980; Storey and Beal, 1993), and in the VA nucleus of thalamus (Storey and Beal, 1993). Levels of GAD are also reduced (Spokes, 1980). Animals with striatal quinolinic acid lesions also show substantial striatal GABA depletion (Beal et al., 1986).

Receptor binding studies have confirmed the prominent involvement of GABAergic mechanisms in the pathology and pathophysiology of Huntington's disease. Striatal GABA receptors were found to be reduced in patients with Huntington's disease (Mohler and Okada, 1978; Reisine et al., 1980; Van Ness et al., 1982), likely as the result of the loss of striatal neurons, and the concomitant loss of postsynaptic GABA receptors. In the regions most strongly affected by loss of striatal efferents, GP_e and GP_i/SNr, GABA-A receptor immunoreactivity (Reisine et al., 1980;

Thompson-Vest et al., 2003), or binding is increased (Cross and Waddington, 1981; Van Ness et al., 1982; Walker et al., 1984; Glass et al., 2000; Kunig et al., 2000). In accordance with the model that the indirect pathway is affected before the direct pathway, increased GABA-A receptor binding in GP_e predates that in GP_i or SNr (Walker et al., 1984; Glass et al., 2000). Similar patterns of changes in receptor binding were also demonstrated in animal models of the disease (Nicholson et al., 1995; Fujiyama et al., 2002; Bauer et al., 2005), although others have shown increased expression of the $\alpha 1$ subunit of GABA-A receptors in the striatum of transgenic animals (Cepeda et al., 2004). Electrophysiologic recordings in these animals have suggested that striatal GABAergic transmission onto striatal medium-sized spiny neurons is, in fact, hyperactive (Cepeda et al., 2004), perhaps in part due to changes in the interaction between GABA and cannabinoid receptors in the striatum (Centonze et al., 2005).

The finding of GABAergic dysfunction in Huntington's disease has motivated clinical trials directed at the GABAergic system in these patients. These treatment attempts have been largely ineffective. Systemic pharmacologic treatments with GABA transaminase inhibitors, aimed at enhancing brain GABA levels (Perry et al., 1979; Manyam et al., 1981; Tell et al., 1981; Scigliano et al., 1984), have resulted in increased CSF levels of GABA, but in little or no clinical improvements. Similarly, treatments with the GABA-A receptor agonist muscimol (Shoulson et al., 1978), and other GABAergic agents, such as progabide (Mondrup et al., 1985), sodium valproate (Symington et al., 1978) or dipropylacetic acid (Shoulson et al., 1976), were also without effect.

Cell transplantation into the striatum is a more specific method to partially restore striatal anatomy and possibly 'normalize' GABAergic function in the striatum. Transplantation of rat or human fetal striatal tissue (containing GABAergic neurons, among other elements) has beneficial behavioral effects in rats with excitotoxin-induced lesions of the striatum (Isacson et al., 1984; Giordano et al., 1990; Pundt et al., 1996), and, at least in some studies, improves the biochemical and electrophysiologic abnormalities seen in this

300

model in the striatum (Isacson et al., 1984; Chen et al., 2002b), and in GP$_e$ and SNr (Isacson et al., 1985; Sanberg et al., 1986; Nakao et al., 1999). In the same rodent model, transplantation of GABAergic cells derived from mouse stem cells (Dinsmore et al., 1996), or of striatum-like cells derived from human stem cells (Hurlbert et al., 1999; McBride et al., 2004) was also effective. Intravenous injection of human stem cells appears to also improve lesion-induced deficits (Lee et al., 2005). Fetal tissue grafting (Pearlman et al., 1991) or transplantation of neurturin or brain derived neurotrophic factor (BDNF) producing cells (Perez-Navarro et al., 2000a, b) reduces the impact of quinolinic acid lesioning, an effect that may specifically spare those striatal cells that project to GP$_e$ (Marco et al., 2002). Thus far, grafting studies in transgenic animals have resulted in only small behavioral improvements (Dunnett et al., 1998). Several authors have also documented successful neuronal replacement and some behavioral recovery in primates with striatal lesions (Isacson et al., 1989; Schumacher et al., 1992). Small pilot series of human patients in whom fetal striatal tissue was transplanted showed that the transplantation approach itself is safe, but has little or no beneficial effect (Madrazo et al., 1995; Philpott et al., 1997; Bachoud-Levi et al., 2000a, b; Furtado et al., 2005). As in other applications of neural transplantation, many practical issues of the transplantation procedure need to be resolved in order to optimize the treatment outcome. Such factors include finding the selection of the source of fetal tissue, the number and location of implants, and the use of immunosuppressive therapy.

Dystonia

In dystonia, voluntary movements are severely disrupted by the occurrence of abnormal postures and slow involuntary movements, along with agonist–antagonist co-contraction and excessive 'overflow' activation of inappropriate musculature. Dystonia can be the result of a variety of genetic and non-genetic diseases. Many of these involve the basal ganglia, but parts of the nervous system outside of the basal ganglia may also be affected.

In many cases, there is no obvious pathology (so-called 'primary' dystonia), while in 'secondary' forms of dystonia structural or biochemical abnormalities such as strokes or other focal lesions in the basal ganglia, thalamus, or cerebellum may occur. Dystonia is also seen in patients with abnormalities of dopaminergic transmission, for instance in the context of parkinsonism, either as an early parkinsonian sign or in response to dopaminergic drugs, or in diseases of dopamine metabolism, such as levodopa-responsive dystonia (Segawa et al., 1976; Segawa, 2000).

Research into the pathophysiology of primary dystonia has demonstrated that this disorder is associated with widespread loss of GABAergic inhibition and a loss of specificity in sensorimotor maps at the cortical level (Chen and Hallett, 1998; Levy and Hallett, 2002). The notion that dystonia may result from aberrant cortical or subcortical plasticity in susceptible individuals is also supported by the fact that dystonia due to sudden insults such as strokes or injuries usually does not develop abruptly, but progresses slowly, over the course of weeks or months, after the lesion. The beneficial effects of neurosurgical interventions for dystonia, such as pallidotomy or DBS (see below), are also typically gradual and delayed by several weeks or months.

Our knowledge regarding the pathophysiology of dystonia remains very limited. In most cases, dystonia of basal ganglia origin is thought to start with a change in striatal activity. For instance striatal lesions may affect the affinity or the number of dopamine receptors in the unlesioned portion of the striatum, or may lead to reorganization of striatal topography, which may then affect the balance of activity along the direct and indirect pathways. Recent positron emission tomography (PET) studies have demonstrated global hypermetabolism in the striatum of patients with dystonia both at rest (Eidelberg et al., 1995; Galardi et al., 1996; Kumar et al., 1999) and with activation (Brooks, 1995; Playford et al., 1998), and have suggested a combination of activity changes in the direct and indirect pathways. Single-cell recording studies in humans and animals (Segawa, 2000; Friedman et al., 2002; Vitek, 2002; Starr et al., 2005) have placed particular emphasis on increased activity along the GABAergic 'direct' pathway of the basal ganglia in dystonia. As in the other hyperkinetic disorders, this may result in reduced basal ganglia output.

This is further supported by pharmacologic studies. Use of D2 receptor antagonists is associated with the development of dystonia, presumably by increasing striatal outflow to GP_e via the indirect pathway, whereas D1-receptors antagonists may be beneficial in this regard, presumably by reducing striatal outflow to GP_i along the direct pathway (see also Segawa, 2000). Increased activity along the direct pathway may not only be due to intrastriatal mechanisms but may also involve other mechanism, such as changes in GABAergic transmission in GP_i, or activity changes in transthalamic feedback loops (Smith et al., 2004), which may regulate GP_i activity.

Abnormal discharge patterns of basal ganglia neurons may also contribute to dystonia. For instance, several studies have reported abnormal oscillatory activity in the basal ganglia of dystonic patients. LFP recordings in dystonic patients, using implanted DBS electrodes for recording purposes, have revealed increased spectral power in the 4–10 Hz band, particularly in recordings from GP_e (Silberstein et al., 2003). Single-cell recording studies in GP_i have suggested the presence of beta-band oscillations, similar to those seen in Parkinson's disease (Starr et al., 2005; see also discussion in Gatev et al., 2006), but, of course, superimposed on relatively low discharge rates in dystonia, contrasting with the abnormally high rates in Parkinson's disease. Agonist–antagonist co-activation in dystonia may reflect an increased degree of neuronal synchronization that may be affected by striatal dopaminergic transmission. An extra-basal ganglia origin of the presumed synchrony between neighboring cells, e.g., through the absence of GABAergic inhibition at the cortical level, can also not be excluded.

Studies into the pathway abnormalities leading to dystonia of basal ganglia origin are hampered by the fact that there are far few animal models available for the motor disorder. The best-studied model is a strain of Syrian hamsters that develop dystonia (dtsz hamsters). These animals show attacks of sustained posturing of limbs and trunk upon environmental stimuli such as being handled or their exposure to a new environment. In these animals, there are significant changes in the concentration of GABA, glutamate, and other amino acids in multiple brain areas, including the striatum and thalamus (Loscher and Horstermann, 1992), with significant increases in the binding of GABA receptors in thalamus, and STN (among other areas), perhaps indicating reduced GABAergic transmission in these areas (Nobrega et al., 1995; Pratt et al., 1995). The movement disorder is age dependent with a gradual increase to a maximum between the 30th and 40th day of life, followed by a continuous decline in severity until complete remission at about 70 days. This time course is perhaps explained by the findings that the retarded development of GABAergic inhibition plays a critical role in the pathogenesis (Kreil and Richter, 2005). In anesthetized (and, thus, not dystonic) animals, there is an increased firing rate in the striatum (Gernert et al., 1999a), and significant firing rate or pattern changes in GP (Gernert et al., 1999b), but a reduction of spontaneous discharge rate and a highly irregular pattern in the entopeduncular nucleus (the rodent analogue of the primate GP_i) (Gernert et al., 2002).

Dystonia can also be pharmacologically induced in cats or monkeys. For instance, in monkeys, acute dystonic reactions can be induced by systemic neuroleptic exposure in a variety of monkey species (Rupniak et al., 1986). These forms of dystonia typically respond to anticholinergic and GABAergic agents (Casey et al., 1980). Furthermore, focal striatal injections of GABA-A receptor antagonists in cat were shown to lead to dystonia of neck and trunk muscles (Yoshida et al., 1991). Inactivation of the medial SNr is also known to result in abnormalities of axial posture, which may represent elements of dystonia (Burbaud et al., 1998; Wichmann et al., 2001). Bicuculline injections into the ventrolateral thalamus are also known to produce dystonic limb and trunk posturing (Guehl et al., 2000; Macia et al., 2002). These injections were shown to enhance discharge rates of thalamic neurons, and to increase their responsiveness to sensory stimulation or microstimulation.

Except in the rare cases in which known mechanisms are present and specific therapies are

302

available, the treatment of dystonia is symptomatic. The pharmacotherapy of focal or generalized dystonia is at best only partially effective. Drugs that either increase GABA concentrations, or directly act at GABA-A receptors show some benefit (Korein et al., 1981; Thaker et al., 1990; Marino Junior et al., 1993; Chudnow et al., 1997; Guimaraes and Vale Santos, 2000; Pina and Modrego, 2005).

The GABA-B receptor agonist baclofen is also effective for treating focal or generalized dystonia. Systemic treatment with this drug has been used in a variety of dystonic conditions (e.g., Sandyk, 1984; Rosse et al., 1986; Fahn, 1987; Kyriagis et al., 2004). In addition, intrathecal application of baclofen may be helpful in severe forms of dystonia (e.g., Silbert and Stewart-Wynne, 1992; Penn et al., 1995; Albright et al., 1996; Ford et al., 1996; Awaad et al., 1999; Meythaler et al., 1999; Kyriagis et al., 2004). While some of these effects may be mediated within the basal ganglia–thalamocortical circuitry, the ubiquitous distribution of GABA receptors make other locations of action also a possibility, specifically in cortical and spinal areas. Stereotactic lesioning or DBS of the pallidum or thalamus is also increasingly used to treat severe dystonia, often with pronounced (albeit delayed) benefits. As in other applications of these procedures in the treatment of movement disorders, they likely work by releasing the thalamocortical network from abnormally patterned basal ganglia output.

Conclusion

Changes in GABAergic transmission are important components of all movement disorders. The development of functional models of hyperkinetic and hypokinetic disorders in the late 1980s has prompted a large number of studies into the changes in the complex network of GABAergic and other connections in the basal ganglia in these diseases. These studies have clearly shown that movement disorders such as Parkinson's disease or Huntington's chorea cannot be simply characterized by global changes in the release of neurotransmitters like GABA. With regard to GABAergic transmission, it is now clear that many additional neurochemical changes affecting the composition, levels, and subcellular localization of receptors, as well as the uptake and metabolism of GABA need to be considered. At this moment we are only beginning to understand the complex interactions between these changes and the rate and pattern abnormalities described in GABAergic neurons. With hindsight, many of the previous treatment attempts with systemically acting GABAergic agents appear premature, given the complexity of the functional changes that are now known to occur in the basal ganglia. However, the outlook for control of the motor symptoms of these diseases is promising. As discussed for some of the individual diseases mentioned above, there are now several pharmacologic and non-pharmacologic techniques available that may facilitate the modulation of GABAergic mechanisms at specific sites within the basal ganglia.

Abbreviations

6-OHDA	6-hydroxydopamine
BDNF	brain-derived neurotrophic factor
CM	centromedian nucleus
CSF	cerebro spinal fluid
DBS	deep brain stimulation
GABA	gamma-aminobutyric acid
GABA$_A$	GABA receptor subtype A
GABA$_B$	GABA receptor subtype B
GABA$_B$ R1	GABA$_B$ receptor subunit R1
GAD	glutamate decarboxylase
GAT	GABA transporter
GP$_e$	globus pallidus, external segment
GP$_i$	globus pallidus, internal segment
LFP	local field potential
MPTP	1-methyl-4-phenyl-1,2,3,6-tetrahydropyridine
PET	positron emission tomography
PPN	pedunculopontine nucleus
SNc	substantia nigra pars compacta
SNr	substantia nigra pars reticulata
STN	subthalamic nucleus
VA	ventral anterior nucleus
VL	ventrolateral nucleus

References

Albin, R.L., Young, A.B. and Penney, J.B. (1989) The functional anatomy of basal ganglia disorders. Trends Neurosci., 12: 366–375.

Albright, A.L., Barry, M.J., Fasick, P., Barron, W. and Shultz, B. (1996) Continuous intrathecal baclofen infusion for symptomatic generalized dystonia. Neurosurgery, 38: 934–938 (discussion 938–939).

Awaad, Y., Munoz, S. and Nigro, M. (1999) Progressive dystonia in a child with chromosome 18p deletion, treated with intrathecal baclofen. J. Child Neurol., 14: 75–77.

Aziz, T.Z., Peggs, D., Sambrook, M.A. and Crossman, A.R. (1991) Lesion of the subthalamic nucleus for the alleviation of 1-methyl-4-phenyl-1,2,3,6-tetrahydropyridine (MPTP)-induced parkinsonism in the primate. Mov. Disord., 6: 288–292.

Bachoud-Levi, A., Bourdet, C., Brugieres, P., Nguyen, J.P., Grandmougin, T., Haddad, B., Jeny, R., Bartolomeo, P., Boisse, M.F., Barba, G.D., Degos, J.D., Ergis, A.M., Lefaucheur, J.P., Lisovoski, F., Pailhous, E., Remy, P., Palfi, S., Defer, G.L., Cesaro, P., Hantraye, P. and Peschanski, M. (2000a) Safety and tolerability assessment of intrastriatal neural allografts in five patients with Huntington's disease. Exp. Neurol., 161: 194–202.

Bachoud-Levi, A.C., Remy, P., Nguyen, J.P., Brugieres, P., Lefaucheur, J.P., Bourdet, C., Baudic, S., Gaura, V., Maison, P., Haddad, B., Boisse, M.F., Grandmougin, T., Jeny, R., Bartolomeo, P., Dalla Barba, G., Degos, J.D., Lisovoski, F., Ergis, A.M., Pailhous, E., Cesaro, P., Hantraye, P. and Peschanski, M. (2000b) Motor and cognitive improvements in patients with Huntington's disease after neural transplantation. Lancet, 356: 1975–1979.

Bahena-Trujillo, R. and Arias-Montano, J.A. (1999) [3H] gamma-aminobutyric acid transport in rat substantia nigra pars reticulata synaptosomes: pharmacological characterization and phorbol ester-induced inhibition. Neurosci. Lett., 274: 119–122.

Bara-Jimenez, W., Sherzai, A., Dimitrova, T., Favit, A., Bibbiani, F., Gillespie, M., Morris, M.J., Mouradian, M.M. and Chase, T.N. (2003) Adenosine A(2A) receptor antagonist treatment of Parkinson's disease.[see comment]. Neurology, 61: 293–396.

Baron, M.S., Vitek, J.L., Bakay, R.A.E., Green, J., Kaneoke, Y., Hashimoto, T., Turner, R.S., Woodard, J.L., Cole, S.A., McDonald, W.M. and DeLong, M.R. (1996) Treatment of advanced Parkinson's disease by GPi pallidotomy: 1 year pilot-study results. Ann. Neurol., 40: 355–366.

Baron, M.S., Wichmann, T., Ma, D. and DeLong, M.R. (2002) Effects of transient focal inactivation of the basal ganglia in parkinsonian primates. J. Neurosci., 22: 299–592.

Bauer, A., Zilles, K., Matusch, A., Holzmann, C., Riess, O. and von Horsten, S. (2005) Regional and subtype selective changes of neurotransmitter receptor density in a rat transgenic for the Huntington's disease mutation. J. Neurochem., 94: 639–650.

Baufreton, J., Atherton, J.F., Surmeier, D.J. and Bevan, M.D. (2005) Enhancement of excitatory synaptic integration by GABAergic inhibition in the subthalamic nucleus. J. Neurosci., 25: 8505–8517.

Beal, M.F., Brouillet, E., Jenkins, B.G., Ferrante, R.J., Kowall, N.W., Miller, J.M., Storey, E., Srivastava, R., Rosen, B.R. and Hyman, B.T. (1993) Neurochemical and histologic characterization of striatal excitotoxic lesions produced by the mitochondrial toxin 3-nitropropionic acid. J. Neurosci., 13: 4181–4192.

Beal, M.F., Kowall, N.W., Ellison, D.W., Mazurek, M.F., Swartz, K.J. and Martin, J.B. (1986) Replication of the neurochemical characteristics of Huntington's disease by quinolinic acid. Nature, 321: 168–171.

Becker, R.E. and Lal, H. (1983) Pharmacological approaches to treatment of hemiballism and hemichorea. Brain Res. Bull., 11: 187–189.

Bergman, H., Wichmann, T. and DeLong, M.R. (1990a) Amelioration of parkinsonian symptoms by inactivation of the subthalamic nucleus (STN) in MPTP treated green monkeys. Mov. Disord., 5(Suppl. 1): 79.

Bergman, H., Wichmann, T. and DeLong, M.R. (1990b) Reversal of experimental parkinsonism by lesions of the subthalamic nucleus. Science, 249: 1436–1438.

Bergman, H., Wichmann, T., Karmon, B. and DeLong, M.R. (1994) The primate subthalamic nucleus. II. Neuronal activity in the MPTP model of parkinsonism. J. Neurophysiol., 72: 507–520.

Betarbet, R., Turner, R., Chockkan, V., DeLong, M.R., Allers, K.A., Walters, J., Levey, A.I. and Greenamyre, J.T. (1997) Dopaminergic neurons intrinsic to the primate striatum. J. Neurosci., 17: 6761–6768.

Beurrier, C., Congar, P., Bioulac, B. and Hammond, C. (1999) Subthalamic nucleus neurons switch from single-spike activity to burst-firing mode. J. Neurosci., 19: 599–609.

Bevan, M.D., Magill, P.J., Terman, D., Bolam, J.P. and Wilson, C.J. (2002) Move to the rhythm: oscillations in the subthalamic nucleus-external globus pallidus network. Trends Neurosci., 25: 525–531.

Bibbiani, F., Oh, J.D., Petzer, J.P., Castagnoli Jr., N., Chen, J.F., Schwarzschild, M.A. and Chase, T.N. (2003) A2A antagonist prevents dopamine agonist-induced motor complications in animal models of Parkinson's disease [see comment]. Exp. Neurol., 184: 285–294.

Billings, L.M. and Marshall, J.F. (2004) Glutamic acid decarboxylase 67 mRNA regulation in two globus pallidus neuron populations by dopamine and the subthalamic nucleus. J. Neurosci., 24: 3094–3103.

Bonilla, E., Prasad, A.L. and Arrieta, A. (1988) Huntington's disease: studies on brain free amino acids. Life Science, 42: 1153–1158.

Boraud, T., Bezard, E., Bioulac, B. and Gross, C.E. (2000) Ratio of inhibited-to-activated pallidal neurons decreases dramatically during passive limb movement in the MPTP-treated monkey. J. Neurophysiol., 83: 1760–1763.

Braak, H., Del Tredici, K., Rub, U., de Vos, R.A., Jansen Steur, E.N. and Braak, E. (2003) Staging of brain pathology related to sporadic Parkinson's disease. Neurobiol. Aging, 24: 197–211.

Brickell, K.L., Nicholson, L.F., Waldvogel, H.J. and Faull, R.L. (1999) Chemical and anatomical changes in the striatum and substantia nigra following quinolinic acid lesions in the striatum of the rat: a detailed time course of the cellular and GABA(A) receptor changes. J. Chem. Neuroanat., 17: 75–97.

Brooks, D.J. (1995) The role of the basal ganglia in motor control: contributions from PET. J. Neurol. Sci., 128: 1–13.

Brouillet, E., Conde, F., Beal, M.F. and Hantraye, P. (1999) Replicating Huntington's disease phenotype in experimental animals. Prog. Neurobiol., 59: 427–468.

Brown, P. (2003) Oscillatory nature of human basal ganglia activity: relationship to the pathophysiology of Parkinson's disease. Mov. Disord., 18: 357–363.

Burbaud, P., Bonnet, B., Guehl, D., Lagueny, A. and Bioulac, B. (1998) Movement disorders induced by gamma-aminobutyric agonist and antagonist injections into the internal globus pallidus and substantia nigra pars reticulata of the monkey. Brain Res., 780: 102–107.

Burns, R.S., Chiueh, C.C., Markey, S.P., Ebert, M.H., Jacobowitz, D.M. and Kopin, I.J. (1983) A primate model of parkinsonism: selective destruction of dopaminergic neurons in the pars compacta of the substantia nigra by N-methyl-4-phenyl-1,2,3,6-tetrahydropyridine. Proc. Natl. Acad. Sci. USA, 80: 4546–4550.

Calon, F. and Di Paolo, T. (2002) Levodopa response motor complications—GABA receptors and preproenkephalin expression in human brain. Parkinsonism Relat. Disord., 8: 449–454.

Calon, F., Goulet, M., Blanchet, P.J., Martel, J.C., Piercey, M.F., Bédard, P.J. and Di Paolo, T. (1995) Levodopa or D2 agonist induced dyskinesia in MPTP monkeys: correlation with changes in dopamine and GABAA receptors in the striatopallidal complex. Brain Res., 680: 43–52.

Calon, F., Lavertu, N., Lemieux, A.M., Morissette, M., Goulet, M., Grondin, R., Blanchet, P.J., Bedard, P.J. and Di Paolo, T. (2001) Effect of MPTP-induced denervation on basal ganglia GABA(B) receptors: correlation with dopamine concentrations and dopamine transporter. Synapse, 40: 225–234.

Calon, F., Morissette, M., Goulet, M., Grondin, R., Blanchet, P.J., Bedard, P.J. and Di Paolo, T. (1999) Chronic D1 and D2 dopaminomimetic treatment of MPTP-denervated monkeys: effects on basal ganglia GABA(A)/benzodiazepine receptor complex and GABA content. Neurochem. Int., 35: 81–91.

Calon, F., Morissette, M., Goulet, M., Grondin, R., Blanchet, P.J., Bedard, P.J. and Di Paolo, T. (2000) 125I-CGP 64213 binding to GABA(B) receptors in the brain of monkeys: effect of MPTP and dopaminomimetic treatments. Exp. Neurol., 163: 191–199.

Calon, F., Morissette, M., Rajput, A.H., Hornykiewicz, O., Bedard, P.J. and Di Paolo, T. (2003) Changes of GABA receptors and dopamine turnover in the postmortem brains of parkinsonians with levodopa-induced motor complications. Mov. Disord., 18: 241–253.

Carlson, B.B., Behrstock, S., Tobin, A.J. and Salamone, J.D. (2003) Brain implantations of engineered GABA-releasing cells suppress tremor in an animal model of Parkinsonism. Neuroscience, 119: 927–932.

Carpenter, M.B., Whittier, J.R. and Mettler, F.A. (1950) Analysis of choreoid hyperkinesia in the rhesus monkey: surgical and pharmacological analysis of hyperkinesia resulting from lesions in the subthalamic nucleus of Luys. J. Comp. Neurol., 92: 293–332.

Carta, A., Fenu, S. and Morelli, M. (2001) Alterations in GAD67, dynorphin and enkephalin mRNA in striatal output neurons following priming in the 6-OHDA model of Parkinson's disease. Neurol. Sci., 22: 59–60.

Casey, D.E., Gerlach, J. and Christensson, E. (1980) Dopamine, acetylcholine, and GABA effects in acute dystonia in primates. Psychopharmacology (Berl.), 70: 83–87.

Cenci, M.A., Lee, C.S. and Bjorklund, A. (1998) L-DOPA-induced dyskinesia in the rat is associated with striatal overexpression of prodynorphin- and glutamic acid decarboxylase mRNA. Eur. J. Neurosci., 10: 2694–2706.

Centonze, D., Rossi, S., Prosperetti, C., Tscherter, A., Bernardi, G., Maccarrone, M. and Calabresi, P. (2005) Abnormal sensitivity to cannabinoid receptor stimulation might contribute to altered gamma-aminobutyric acid transmission in the striatum of R6/2 Huntington's disease mice. Biol. Psychiatry, 57: 1583–1589.

Cepeda, C., Starling, A.J., Wu, N., Nguyen, O.K., Uzgil, B., Soda, T., Andre, V.M., Ariano, M.A. and Levine, M.S. (2004) Increased GABAergic function in mouse models of Huntington's disease: reversal by BDNF. J. Neurosci. Res., 78: 855–867.

Chadha, A., Dawson, L.G., Jenner, P.G. and Duty, S. (2000a) Effect of unilateral 6-hydroxydopamine lesions of the nigrostriatal pathway on GABA(A) receptor subunit gene expression in the rodent basal ganglia and thalamus. Neuroscience, 95: 119–126.

Chadha, A., Howell, O., Atack, J.R., Sur, C. and Duty, S. (2000b) Changes in [³H]zolpidem and [³H]Ro 15-1788 binding in rat globus pallidus and substantia nigra pars reticulata following a nigrostriatal tract lesion. Brain Res., 862: 280–283.

Chandra, V., Wharton, S. and Spunt, A.L. (1982) Amelioration of hemiballismus with sodium valproate. Ann. Neurol., 12: 407.

Chase, T.N. and Oh, J.D. (2000) Striatal dopamine- and glutamate-mediated dysregulation in experimental parkinsonism. Trends Neurosci., 23: S86–S91.

Chen, C.C., Lee, S.T., Wu, T., Chen, C.J., Huang, C.C. and Lu, C.S. (2002a) Hemiballism after subthalamotomy in patients with Parkinson's disease: report of 2 cases. Mov. Disord., 17: 1367–1371.

Chen, G.J., Jeng, C.H., Lin, S.Z., Tsai, S.H., Wang, Y. and Chiang, Y.H. (2002b) Fetal striatal transplants restore electrophysiological sensitivity to dopamine in the lesioned striatum of rats with experimental Huntington's disease. J. Biomed. Sci., 9: 303–310.

Chen, L. and Yung, W.H. (2003) Effects of the GABA-uptake inhibitor tiagabine in rat globus pallidus. Exp. Brain Res., 152: 263–269.

Chen, R. and Hallett, M. (1998) Focal dystonia and repetitive motion disorders. Clin. Orthop. Relat. Res., 351: 102–106.

Chudnow, R.S., Mimbela, R.A., Owen, D.B. and Roach, E.S. (1997) Gabapentin for familial paroxysmal dystonic choreoathetosis. Neurology, 49: 1441–1442.

Cicchetti, F., Prensa, L., Wu, Y. and Parent, A. (2000) Chemical anatomy of striatal interneurons in normal individuals and in patients with Huntington's disease. Brain Res. Brain Res. Rev., 34: 80–101.

Cross, A.J. and Waddington, J.L. (1981) Substantia nigra gamma-aminobutyric acid receptors in Huntington's disease. J. Neurochem., 37: 321–324.

Crossman, A.R., Mitchell, I.J. and Sambrook, M.A. (1985) Regional brain uptake of 2-deoxyglucose in N-methyl-4-phenyl-1,2,3,6-tetrahydropyridine (MPTP)-induced parkinsonism in the macaque monkey. Neuropharmacology, 24: 587–591.

de Groote, C., Wullner, U., Lochmann, P.A., Luiten, P.G. and Klockgether, T. (1999) Functional characterization and expression of thalamic GABA(B) receptors in a rodent model of Parkinson's disease. Neuropharmacology, 38: 1683–1689.

Delfs, J.M., Anegawa, N.J. and Chesselet, M.F. (1995) Glutamate decarboxylase messenger RNA in rat pallidum: comparison of the effects of haloperidol, clozapine and combined haloperidol-scopolamine treatments. Neuroscience, 66: 67–80.

DeLong, M.R. (1990) Primate models of movement disorders of basal ganglia origin. Trends Neurosci., 13: 281–285.

Dinsmore, J., Ratliff, J., Deacon, T., Pakzaban, P., Jacoby, D., Galpern, W. and Isacson, O. (1996) Embryonic stem cells differentiated in vitro as a novel source of cells for transplantation. Cell Transplant, 5: 131–143.

Dogali, M., Fazzini, E., Kolodny, E., Eidelberg, D., Sterio, D., Devinsky, O. and Beric, A. (1995) Stereotactic ventral pallidotomy for Pakinson's disease. Neurology, 45: 753–761.

Dostrovsky, J.O., Hutchison, W.D. and Lozano, A.M. (2002) The globus pallidus, deep brain stimulation, and Parkinson's disease. Neuroscientist, 8: 284–290.

Driver-Dunckley, E. and Evidente, V.G. (2005) Hemichorea-hemiballismus may respond to topiramate. Clin. Neuropharmacol., 28: 142–144.

Dunnett, S.B., Carter, R.J., Watts, C., Torres, E.M., Mahal, A., Mangiarini, L., Bates, G. and Morton, A.J. (1998) Striatal transplantation in a transgenic mouse model of Huntington's disease. Exp. Neurol., 154: 31–40.

Dunnett, S.B. and Rosser, A.E. (2004) Cell therapy in Huntington's disease. NeuroRx, 1: 394–405.

Eidelberg, D., Moeller, J.R., Ishikawa, T., Dhawan, V., Spetsieris, P., Przedborski, S. and Fahn, S. (1995) The metabolic topography of idiopathic torsion dystonia. Brain, 118: 1473–1484.

Ellison, D.W., Beal, M.F., Mazurek, M.F., Malloy, J.R., Bird, E.D. and Martin, J.B. (1987) Amino acid neurotransmitter abnormalities in Huntington's disease and the quinolinic acid animal model of Huntington's disease. Brain, 110(Pt 6): 1657–1673.

Fahn, S. (1987) Systemic therapy of dystonia. Can. J. Neurol. Sci., 14: 528–532.

Ferrante, R.J., Kowall, N.W., Cipolloni, P.B., Storey, E. and Beal, M.F. (1993) Excitotoxin lesions in primates as a model for Huntington's disease: histopathologic and neurochemical characterization. Exp. Neurol., 119: 46–71.

Filion, M., Tremblay, L. and Bedard, P.J. (1988) Abnormal influences of passive limb movement on the activity of globus pallidus neurons in parkinsonian monkeys. Brain Res., 444: 165–176.

Ford, B., Greene, P., Louis, E.D., Petzinger, G., Bressman, S.B., Goodman, R., Brin, M.F., Sadiq, S. and Fahn, S. (1996) Use of intrathecal baclofen in the treatment of patients with dystonia. Arch. Neurol., 53: 1241–1246.

Forno, L.S., DeLanney, L.E., Irwin, I. and Langston, J.W. (1993) Similarities and differences between MPTP-induced parkinsonism and Parkinson's disease. Adv. Neurol., 60: 600–608.

Fox, S.H., Henry, B., Hill, M., Crossman, A. and Brotchie, J. (2002) Stimulation of cannabinoid receptors reduces levodopa-induced dyskinesia in the MPTP-lesioned nonhuman primate model of Parkinson's disease. Mov. Disord., 17: 1180–1187.

Francois, C., Percheron, G., Yelnik, J. and Heyner, S. (1984) A Golgi analysis of the primate globus pallidus. I. Inconstant processes of large neurons, other neuronal types, and afferent axons. J. Comp. Neurol., 227: 182–199.

Friedman, Y., Richter, A., Raymond, R., Loscher, W. and Nobrega, J.N. (2002) Regional decreases in NK-3, but not NK-1 tachykinin receptor binding in dystonic hamster (dt(sz)) brains. Neuroscience, 112: 639–645.

Fujiyama, F., Stephenson, F.A. and Bolam, J.P. (2002) Synaptic localization of GABA(A) receptor subunits in the substantia nigra of the rat: effects of quinolinic acid lesions of the striatum. Eur. J. Neurosci., 15: 1961–1975.

Furtado, S., Sossi, V., Hauser, R.A., Samii, A., Schulzer, M., Murphy, C.B., Freeman, T.B. and Stoessl, A.J. (2005) Positron emission tomography after fetal transplantation in Huntington's disease. Ann. Neurol., 58: 331–337.

Galardi, G., Perani, D., Grassi, F., Bressi, S., Amadio, S., Antoni, M., Comi, G.C., Canal, N. and Fazio, F. (1996) Basal ganglia and thalamo-cortical hypermetabolism in patients with spasmodic torticollis. Acta Neurol. Scand., 94: 172–176.

Galeffi, F., Bianchi, L., Bolam, J.P. and Della Corte, L. (2003) The effect of 6-hydroxydopamine lesions on the release of amino acids in the direct and indirect pathways of the basal ganglia: a dual microdialysis probe analysis. Eur. J. Neurosci., 18: 856–868.

Galvan, A., Villalba, R.M., West, S.M., Maidment, N.T., Ackerson, L.C., Smith, Y. and Wichmann, T. (2005) GABAergic modulation of the activity of globus pallidus neurons in primates: in vivo analysis of the functions of GABA receptors and GABA transporters. J. Neurophysiol., 94: 990–1000.

Gatev, P., Darbin, O. and Wichmann, T. (2006) Oscillations in the basal ganglia under normal conditions and in movement disorders. Mov. Disord., 21: 1566–1577.

Gatev, P.G. and Wichmann, T. (2003) Changes in arousal alter neuronal activity in primate basal ganglia. Soc. Neurosci. Abstr., 29.

Gatto, E.M., Uribe, R.C., Raina, G., Gorja, M., Folgar, S. and Micheli, F.E. (2004) Vascular hemichorea/hemiballism and topiramate. Mov. Disord., 19: 836–838.

Gerlach, M., Gsell, W., Kornhuber, J., Jellinger, K., Krieger, V., Pantucek, F., Vock, R. and Riederer, P. (1996) A post mortem study on neurochemical markers of dopaminergic, GABA-ergic and glutamatergic neurons in basal ganglia-thalamocortical circuits in Parkinson syndrome. Brain Res., 741: 142–152.

Gernert, M., Bennay, M., Fedrowitz, M., Rehders, J.H. and Richter, A. (2002) Altered discharge pattern of basal ganglia output neurons in an animal model of idiopathic dystonia. J. Neurosci., 22: 7244–7253.

Gernert, M., Richter, A. and Loscher, W. (1999a) Alterations in spontaneous single unit activity of striatal subdivisions during ontogenesis in mutant dystonic hamsters. Brain Res., 821: 277–285.

Gernert, M., Richter, A. and Loscher, W. (1999b) In vivo extracellular electrophysiology of pallidal neurons in dystonic and nondystonic hamsters. J. Neurosci. Res., 57: 894–905.

Giller, C.A., Dewey, R.B., Ginsburg, M.I., Mendelsohn, D.B. and Berk, A.M. (1998) Stereotactic pallidotomy and thalamotomy using individual variations of anatomic landmarks for localization. Neurosurgery, 42: 56–62.

Giordano, M., Ford, L.M., Shipley, M.T. and Sanberg, P.R. (1990) Neural grafts and pharmacological intervention in a model of Huntington's disease. Brain Res. Bull., 25: 453–465.

Glass, M., Dragunow, M. and Faull, R.L. (2000) The pattern of neurodegeneration in Huntington's disease: a comparative study of cannabinoid, dopamine, adenosine and GABA(A) receptor alterations in the human basal ganglia in Huntington's disease. Neuroscience, 97: 505–519.

Gnanalingham, K.K. and Robertson, R.G. (1993) Chronic continuous and intermittent L-3,4-dihydroxyphenylalanine treatments differentially affect basal ganglia function in 6-hydroxydopamine lesioned rats—an autoradiographic study using [3 H]flunitrazepam. Neuroscience, 57: 673–681.

Graybiel, A.M. (1995) Building action repertoires: memory and learning functions of the basal ganglia. Curr. Op. Neurobiol., 5: 733–741.

Griffiths, P.D., Sambrook, M.A., Perry, R. and Crossman, A.R. (1990) Changes in benzodiazepine and acetylcholine receptors in the globus pallidus in Parkinson's disease. J. Neurol. Sci., 100: 131–136.

Grofova, I., Deniau, J.M. and Kitai, S.T. (1982) Morphology of the substantia nigra pars reticulata projection neurons intracellularly labelled with HRP. J. Comp. Neurol., 208: 352–368.

Group, H.s.D.C.R. (1993) A novel gene containing a trinucleotide repeat that is expanded and unstable on Huntington's disease chromosomes. Cell, 72: 971–983.

Guehl, D., Burbaud, P., Boraud, T. and Bioulac, B. (2000) Bicuculline injections into the rostral and caudal motor thalamus of the monkey induce different types of dystonia. Eur. J. Neurosci., 12: 1033–1037.

Guimaraes, J. and Vale Santos, J. (2000) Paroxysmal dystonia induced by exercise and acetazolamide. Eur. J. Neurol., 7: 237–240.

Guridi, J., Herrero, M.T., Luquin, M.R., Guillen, J., Ruberg, M., Laguna, J., Vila, M., Javoy-Agid, F., Agid, Y., Hirsch, E. and Obeso, J.A. (1996) Subthalamotomy in parkinsonian monkeys. Behavioural and biochemical analysis. Brain, 119: 1717–1727.

Guridi, J. and Obeso, J.A. (2001) The subthalamic nucleus, hemiballismus and Parkinson's disease: reappraisal of a neurosurgical dogma. Brain, 124: 5–19.

Hallworth, N.E. and Bevan, M.D. (2005) Globus pallidus neurons dynamically regulate the activity pattern of subthalamic nucleus neurons through the frequency-dependent activation of postsynaptic GABAA and GABAB receptors. J. Neurosci., 25: 6304–6315.

Hamada, I. and DeLong, M.R. (1992) Excitotoxic acid lesions of the primate subthalamic nucleus result in reduced pallidal neuronal activity during active holding. J. Neurophysiol., 68: 1859–1866.

Hartmann-von Monakow, K., Akert, K. and Kunzle, H. (1978) Projections of the precentral motor cortex and other cortical areas of the frontal lobe to the subthalamic nucleus in the monkey. Exp. Brain Res., 33: 395–403.

Hashimoto, T., Elder, C.M., Okun, M.S., Patrick, S.K. and Vitek, J.L. (2003) Stimulation of the subthalamic nucleus changes the firing pattern of pallidal neurons. J. Neurosci., 23: 1916–1923.

Hauser, R.A., Hubble, J.P. and Truong, D.D. (2003) Randomized trial of the adenosine A(2A) receptor antagonist istradefylline in advanced PD. Neurology, 61: 297–303.

Herrero, M.T., Levy, R., Ruberg, M., Javoy-Agid, F., Luquin, M.R., Agid, Y., Hirsch, E.C. and Obeso, J.A. (1996) Glutamic acid decarboxylase mRNA expression in medial and lateral pallidal neurons in the MPTP-treated monkey and patients with Parkinson's disease. Adv. Neurol., 69: 209–216.

Hirsch, E.C., Perier, C., Orieux, G., Francois, C., Feger, J., Yelnik, J., Vila, M., Levy, R., Tolosa, E.S., Marin, C., Trinidad Herrero, M., Obeso, J.A. and Agid, Y. (2000) Metabolic effects of nigrostriatal denervation in basal ganglia. Trends Neurosci., 23: S78–S85.

Hurlbert, M.S., Gianani, R.I., Hutt, C., Freed, C.R. and Kaddis, F.G. (1999) Neural transplantation of hNT neurons for Huntington's disease. Cell Transplant, 8: 143–151.

Inase, M., Buford, J.A. and Anderson, M.E. (1996) Changes in the control of arm position, movement, and thalamic discharge during local inactivation in the globus pallidus of the monkey. J. Neurophysiol., 75: 1087–1104.

Isacson, O., Brundin, P., Gage, F.H. and Bjorklund, A. (1985) Neural grafting in a rat model of Huntington's disease: progressive neurochemical changes after neostriatal ibotenate lesions and striatal tissue grafting. Neuroscience, 16: 799–817.

Isacson, O., Brundin, P., Kelly, P.A., Gage, F.H. and Bjorklund, A. (1984) Functional neuronal replacement by grafted striatal neurones in the ibotenic acid-lesioned rat striatum. Nature, 311: 458–460.

Isacson, O., Riche, D., Hantraye, P., Sofroniew, M.V. and Maziere, M. (1989) A primate model of Huntington's disease: cross-species implantation of striatal precursor cells to the excitotoxically lesioned baboon caudate-putamen. Exp. Brain Res., 75: 213–220.

Jenner, P. (2005) Istradefylline, a novel adenosine A2A receptor antagonist, for the treatment of Parkinson's disease. Expert Opin. Investig. Drugs, 14: 729–738.

Johnston, T. and Duty, S. (2003) Changes in GABA(B) receptor mRNA expression in the rodent basal ganglia and thalamus following lesion of the nigrostriatal pathway. Neuroscience, 120: 1027–1035.

Katz, J., Nielsen, K.M. and Soghomonian, J.J. (2005) Comparative effects of acute or chronic administration of levodopa to 6-hydroxydopamine-lesioned rats on the expression of glutamic acid decarboxylase in the neostriatum and GABAA receptors subunits in the substantia nigra, pars reticulata. Neuroscience, 132: 833–842.

Kawaguchi, Y., Wilson, C.J., Augood, S.J. and Emson, P.C. (1995) Striatal interneurones: chemical, physiological and morphological characterization. Trends Neurosci., 18: 527–535.

Kawaguchi, Y., Wilson, C.J. and Emson, P.C. (1990) Projection subtypes of rat neostriatal matrix cells revealed by intracellular injection of biocytin. J. Neurosci., 10: 3421–3438.

Keep, M.F., Mastrofrancesco, L., Erdman, D., Murphy, B. and Ashby, L.S. (2002) Gamma knife subthalamotomy for Parkinson disease: the subthalamic nucleus as a new radiosurgical target. Case report. J. Neurosurg., 97: 592–599.

Kincaid, A.E., Albin, R.L., Newman, S.W., Penney, J.B. and Young, A.B. (1992) 6-Hydroxydopamine lesions of the nigrostriatal pathway alter the expression of glutamate decarboxylase messenger RNA in rat globus pallidus projection neurons. Neuroscience, 51: 705–718.

Kish, S.J., Rajput, A., Gilbert, J., Rozdilsky, B., Chang, L.J., Shannak, K. and Hornykiewicz, O. (1986) Elevated gamma-aminobutyric acid level in striatal but not extrastriatal brain regions in Parkinson's disease: correlation with striatal dopamine loss. Ann. Neurol., 20: 26–31.

Kita, H. and Kitai, S.T. (1994) The morphology of globus pallidus projection neurons in the rat: an intracellular staining study. Brain Res., 636: 308–319.

Kita, H., Tachibana, Y., Nambu, A. and Chiken, S. (2005) Balance of monosynaptic excitatory and disynaptic inhibitory responses of the globus pallidus induced after stimulation of the subthalamic nucleus in the monkey. J. Neurosci., 25: 8611–8619.

Korein, J., Lieberman, A., Kupersmith, M. and Levidow, L. (1981) Effect of L-glutamine and isoniazid on torticollis and segmental dystonia. Ann. Neurol., 10: 247–250.

Kothare, S.V., Pollack, P., Kulberg, A.G. and Ravin, P.D. (2000) Gabapentin treatment in a child with delayed-onset hemichorea/hemiballismus. Pediatr. Neurol., 22: 68–71.

Kouyoumdjian, J.A. (1984) [Hemiballism treated with valproic acid: report of 2 cases]. Arq. Neuropsiquiatr., 42: 274–276.

Kreil, A. and Richter, A. (2005) Antidystonic efficacy of gamma-aminobutyric acid uptake inhibitors in the dtsz mutant. Eur. J. Pharmacol., 521: 95–98.

Kulisevsky, J., Avila, A., Berthier, M.L. and Barbanoj, M. (1994) Double-blind trial of flumazenil in hemiballismus. Clin. Neuropharmacol., 17: 470–472.

Kumar, R., Dagher, A., Hutchison, W.D., Lang, A.E. and Lozano, A.M. (1999) Globus pallidus deep brain stimulation for generalized dystonia: clinical and PET investigation. Neurology, 53: 871–874.

Kunig, G., Leenders, K.L., Sanchez-Pernaute, R., Antonini, A., Vontobel, P., Verhagen, A. and Gunther, I. (2000) Benzodiazepine receptor binding in Huntington's disease: [11C]flumazenil uptake measured using positron emission tomography. Ann. Neurol., 47: 644–648.

Kyriagis, M., Grattan-Smith, P., Scheinberg, A., Teo, C., Nakaji, N. and Waugh, M. (2004) Status dystonicus and Hallervorden-Spatz disease: treatment with intrathecal baclofen and pallidotomy. J. Paediatr. Child Health, 40: 322–325.

Laprade, N. and Soghomonian, J.J. (1999) Gene expression of the GAD67 and GAD65 isoforms of glutamate decarboxylase is differentially altered in subpopulations of striatal neurons in adult rats lesioned with 6-OHDA as neonates. Synapse, 33: 36–48.

Lee, S.T., Chu, K., Park, J.E., Lee, K., Kang, L., Kim, S.U. and Kim, M. (2005) Intravenous administration of human neural stem cells induces functional recovery in Huntington's disease rat model. Neurosci. Res., 52: 243–249.

Lees, A.J., Shaw, K.M. and Stern, G.M. (1978) Baclofen in Parkinson's disease. J. Neurol. Neurosurg. Psychiatry, 41: 707–708.

Lei, W., Jiao, Y., Del Mar, N. and Reiner, A. (2004) Evidence for differential cortical input to direct pathway versus indirect pathway striatal projection neurons in rats. J. Neurosci., 24: 8289–8299.

Lenton, R.J., Copti, M. and Smith, R.G. (1981) Hemiballismus treated with sodium valproate. Br. Med. J. (Clin. Res. Ed.), 283: 17–18.

Levesque, J.C. and Parent, A. (2005) GABAergic interneurons in human subthalamic nucleus. Mov. Disord., 20: 574–584.

Levy, L.M. and Hallett, M. (2002) Impaired brain GABA in focal dystonia. Ann. Neurol., 51: 93–101.

Levy, R., Dostrovsky, J.O., Lang, A.E., Sime, E., Hutchison, W.D. and Lozano, A.M. (2001a) Effects of apomorphine on subthalamic nucleus and globus pallidus internus neurons in patients with Parkinson's disease. J. Neurophysiol., 86: 249–260.

Levy, R., Lang, A.E., Dostrovsky, J.O., Pahapill, P., Romas, J., Saint-Cyr, J., Hutchison, W.D. and Lozano, A.M. (2001b) Lidocaine and muscimol microinjections in subthalamic nucleus reverse parkinsonian symptoms. Brain, 124: 2105–2118.

Lieberman, D.M., Corthesy, M.E., Cummins, A. and Oldfield, E.H. (1999) Reversal of experimental parkinsonism by using selective chemical ablation of the medial globus pallidus. J. Neurosurg., 90: 928–934.

Loscher, W. and Horstermann, D. (1992) Abnormalities in amino acid neurotransmitters in discrete brain regions of genetically dystonic hamsters. J. Neurochem., 59: 689–694.

Lozano, A., Hutchison, W., Kiss, Z., Tasker, R., Davis, K. and Dostrovsky, J. (1996) Methods for microelectrode-guided posteroventral pallidotomy. J. Neurosurg., 84: 194–202.

Lundblad, M., Vaudano, E. and Cenci, M.A. (2003) Cellular and behavioural effects of the adenosine A2a receptor antagonist KW-6002 in a rat model of l-DOPA-induced dyskinesia. J. Neurochem., 84: 1398–1410.

Macia, F., Escola, L., Guehl, D., Michelet, T., Bioulac, B. and Burbaud, P. (2002) Neuronal activity in the monkey motor thalamus during bicuculline-induced dystonia. Eur. J. Neurosci., 15: 1353–1362.

Madrazo, I., Franco-Bourland, R.E., Castrejon, H., Cuevas, C. and Ostrosky-Solis, F. (1995) Fetal striatal homotransplantation for Huntington's disease: first two case reports. Neurol. Res., 17: 312–315.

Magarinos-Ascone, C., Pazo, J.H., Macadar, O. and Buno, W. (2002) High-frequency stimulation of the subthalamic nucleus silences subthalamic neurons: a possible cellular mechanism in Parkinson's disease. Neuroscience, 115: 1109–1117.

Magill, P.J., Bolam, J.P. and Bevan, M.D. (2000) Relationship of activity in the subthalamic nucleus-globus pallidus network to cortical electroencephalogram. J. Neurosci., 20: 820–833.

Manyam, B.V., Katz, L., Hare, T.A., Kaniefski, K. and Tremblay, R.D. (1981) Isoniazid-induced elevation of CSF GABA levels and effects on chorea in Huntington's disease. Ann. Neurol., 10: 35–37.

Marco, S., Perez-Navarro, E., Tolosa, E., Arenas, E. and Alberch, J. (2002) Striatopallidal neurons are selectively protected by neurturin in an excitotoxic model of Huntington's disease. J. Neurobiol., 50: 323–332.

Marino Junior, R., Benabou, R. and Benabou, S. (1993) Therapeutic effects of flunitrazepam in dystonias and torticollis. Preliminary communication. Arq. Neuropsiquiatr., 51: 285–286.

Marjama-Lyons, J. and Koller, W. (2000) Tremor-predominant Parkinson's disease. Approaches to treatment. Drugs Aging, 16: 273–278.

Marsden, C.D. and Obeso, J.A. (1994) The functions of the basal ganglia and the paradox of stereotaxic surgery in Parkinson's disease. Brain, 117: 877–897.

McBride, J.L., Behrstock, S.P., Chen, E.Y., Jakel, R.J., Siegel, I., Svendsen, C.N. and Kordower, J.H. (2004) Human neural stem cell transplants improve motor function in a rat model of Huntington's disease. J. Comp. Neurol., 475: 211–219.

McIntyre, C.C., Savasta, M., Walter, B.L. and Vitek, J.L. (2004) How does deep brain stimulation work? Present understanding and future questions. J. Clin. Neurophysiol., 21: 40–50.

Mehta, A. and Chesselet, M.F. (2005) Effect of GABA(A) receptor stimulation in the subthalamic nucleus on motor deficits induced by nigrostriatal lesions in the rat. Exp. Neurol., 193: 110–117.

Meythaler, J.M., Guin-Renfroe, S., Grabb, P. and Hadley, M.N. (1999) Long-term continuously infused intrathecal baclofen for spastic-dystonic hypertonia in traumatic brain injury: 1-year experience. Arch. Phys. Med. Rehabil., 80: 13–19.

Miller, W.C. and DeLong, M.R. (1987) Altered tonic activity of neurons in the globus pallidus and subthalamic nucleus in the

primate MPTP model of parkinsonism. In: Carpenter M.B. and Jayaraman A. (Eds.), The Basal Ganglia II. New York, Plenum Press, pp. 415–427.

Mink, J.W. (1996) The basal ganglia: focused selection and inhibition of competing motor programs. Prog. Neurobiol., 50: 381–425.

Mink, J.W. and Thach, W.T. (1991) Basal ganglia motor control. III. Pallidal ablation: normal reaction time, muscle cocontraction, and slow movement. J. Neurophysiol., 65: 330–351.

Mitchell, I.J., Clarke, C.E., Boyce, S., Robertson, R.G., Peggs, D., Sambrook, M.A. and Crossman, A.R. (1989) Neural mechanisms underlying parkinsonian symptoms based upon regional uptake of 2-deoxyglucose in monkeys exposed to 1-methyl-4-phenyl-1,2,3,6-tetrahydropyridine. Neuroscience, 32: 213–226.

Mohler, H. and Okada, T. (1978) The benzodiazepine receptor in normal and pathological human brain. Br. J. Psychiatry, 133: 261–268.

Molina-Holgado, E., Dewar, K.M., Grondin, L., van Gelder, N.M. and Reader, T.A. (1993) Amino acid levels and gamma-aminobutyric acidA receptors in rat neostriatum, cortex, and thalamus after neonatal 6-hydroxydopamine lesion. J. Neurochem., 60: 936–945.

Mondrup, K., Dupont, E. and Braendgaard, H. (1985) Progabide in the treatment of hyperkinetic extrapyramidal movement disorders. Acta Neurol. Scand., 72: 341–343.

Morselli, P.L., Fournier, V., Bossi, L. and Musch, B. (1985) Clinical activity of GABA agonists in neuroleptic- and L-dopa-induced dyskinesia. Psychopharmacology Suppl., 2: 128–136.

Nakanishi, H., Kita, H. and Kitai, S.T. (1987) Electrical membrane properties of rat subthalamic neurons in an in vitro slice preparation. Brain Res., 437: 35–44.

Nakao, N., Ogura, M., Nakai, K. and Itakura, T. (1999) Embryonic striatal grafts restore neuronal activity of the globus pallidus in a rodent model of Huntington's disease. Neuroscience, 88: 469–477.

Nambu, A. and Llinas, R. (1994) Electrophysiology of globus pallidus neurons in vitro. J. Neurophysiol., 72: 1127–1139.

Nambu, A., Tokuno, H., Hamada, I., Kita, H., Imanishi, M., Akazawa, T., Ikeuchi, Y. and Hasegawa, N. (2000) Excitatory cortical inputs to pallidal neurons via the subthalamic nucleus in the monkey. J. Neurophysiol., 84: 289–300.

Nambu, A., Tokuno, H. and Takada, M. (2002) Functional significance of the cortico-subthalamo-pallidal 'hyperdirect' pathway. Neurosci. Res., 43: 111–117.

Nicholson, L.F., Faull, R.L., Waldvogel, H.J. and Dragunow, M. (1995) GABA and GABAA receptor changes in the substantia nigra of the rat following quinolinic acid lesions in the striatum closely resemble Huntington's disease. Neuroscience, 66: 507–521.

Nobrega, J.N., Richter, A., Burnham, W.M. and Loscher, W. (1995) Alterations in the brain GABAA/benzodiazepine receptor-chloride ionophore complex in a genetic model of paroxysmal dystonia: a quantitative autoradiographic analysis. Neuroscience, 64: 229–239.

Nutt, J., Williams, A., Plotkin, C., Eng, N., Ziegler, M. and Calne, D.B. (1979) Treatment of Parkinson's disease with sodium valproate: clinical, pharmacological, and biochemical observations. Can. J. Neurol. Sci., 6: 337–343.

Obeso, J.A., Olanow, C.W. and Nutt, J.G. (2000) Levodopa motor complications in Parkinson's disease. Trends Neurosci., 23: S2–S7.

Ochi, M., Shiozaki, S. and Kase, H. (2004) Adenosine A(2A) receptor-mediated modulation of GABA and glutamate release in the output regions of the basal ganglia in a rodent model of Parkinson's disease. Neuroscience, 127: 223–231.

Ondo, W.G. and Hunter, C. (2003) Flumazenil, a GABA antagonist, may improve features of Parkinson's disease. Mov. Disord., 18: 683–685.

Pan, H.S., Penney, J.B. and Young, A.B. (1985) Gamma-aminobutyric acid and benzodiazepine receptor changes induced by unilateral 6-hydroxydopamine lesions of the medial forebrain bundle. J. Neurochem., 45: 1396–1404.

Papa, S.M., Desimone, R., Fiorani, M. and Oldfield, E.H. (1999) Internal globus pallidus discharge is nearly suppressed during levodopa-induced dyskinesias. Ann. Neurol., 46: 732–738.

Parent, A., Sato, F., Wu, Y., Gauthier, J., Levesque, M. and Parent, M. (2000) Organization of the basal ganglia: the importance of axonal collateralization. Trends Neurosci., 23: S20–S27.

Pearlman, S.H., Levivier, M., Collier, T.J., Sladek Jr., J.R. and Gash, D.M. (1991) Striatal implants protect the host striatum against quinolinic acid toxicity. Exp. Brain Res., 84: 303–310.

Pedneault, S. and Soghomonian, J.J. (1994) Glutamate decarboxylase (GAD65) mRNA levels in the striatum and pallidum of MPTP-treated monkeys. Brain Res. Mol. Brain Res., 25: 351–354.

Penn, R.D., Gianino, J.M. and York, M.M. (1995) Intrathecal baclofen for motor disorders. Mov. Disord., 10: 675–677.

Penn, R.D., Kroin, J.S., Reinkensmeyer, A. and Corcos, D.M. (1998) Injection of GABA-agonist into globus pallidus in patient with Parkinson's disease. Lancet, 351: 340–341.

Perez-Navarro, E., Akerud, P., Marco, S., Canals, J.M., Tolosa, E., Arenas, E. and Alberch, J. (2000a) Neurturin protects striatal projection neurons but not interneurons in a rat model of Huntington's disease. Neuroscience, 98: 89–96.

Perez-Navarro, E., Canudas, A.M., Akerund, P., Alberch, J. and Arenas, E. (2000b) Brain-derived neurotrophic factor, neurotrophin-3, and neurotrophin-4/5 prevent the death of striatal projection neurons in a rodent model of Huntington's disease. J. Neurochem., 75: 2190–2199.

Perry, T.L., Javoy-Agid, F., Agid, Y. and Fibiger, H.C. (1983) Striatal GABAergic neuronal activity is not reduced in Parkinson's disease. J. Neurochem., 40: 1120–1123.

Perry, T.L., Wright, J.M., Hansen, S. and MacLeod, P.M. (1979) Isoniazid therapy of Huntington disease. Neurology, 29: 370–375.

Philpott, L.M., Kopyov, O.V., Lee, A.J., Jacques, S., Duma, C.M., Caine, S., Yang, M. and Eagle, K.S. (1997) Neuropsychological functioning following fetal striatal transplantation in Huntington's chorea: three case presentations. Cell Transplant, 6: 203–212.

Pina, M.A. and Modrego, P.J. (2005) Dystonia induced by gabapentin. Ann. Pharmacother., 39: 380–382.

Pinal, C.S. and Tobin, A.J. (1998) Uniqueness and redundancy in GABA production. Perspect. Dev. Neurobiol., 5: 109–118.

Pinna, A., Wardas, J., Simola, N. and Morelli, M. (2005) New therapies for the treatment of Parkinson's disease: adenosine A2A receptor antagonists. Life Sciences, 77: 3259–3267.

Playford, E.D., Passingham, R.E., Marsden, C.D. and Brooks, D.J. (1998) Increased activation of frontal areas during arm movement in idiopathic torsion dystonia. Mov. Disord., 13: 309–318.

Plenz, D. and Kitai, S. (1999) A basal ganglia pacemaker formed by the subthalamic nucleus and external globus pallidus. Nature, 400: 677–682.

Postuma, R.B. and Lang, A.E. (2003) Hemiballism: revisiting a classic disorder. Lancet Neurol., 2: 661–668.

Pratt, G.D., Richter, A., Mohler, H. and Loscher, W. (1995) Regionally selective and age-dependent alterations in benzodiazepine receptor binding in the genetically dystonic hamster. J. Neurochem., 64: 2153–2158.

Przuntek, H. (2000) Non-dopaminergic therapy in Parkinson's disease. J. Neurol., 247(Suppl. 2): II19–II24.

Pundt, L.L., Kondoh, T., Conrad, J.A. and Low, W.C. (1996) Transplantation of human striatal tissue into a rodent model of Huntington's disease: phenotypic expression of transplanted neurons and host-to-graft innervation. Brain Res. Bull., 39: 23–32.

Rabey, J.M., Orlov, E. and Spiegelman, R. (1995) Levodopa-induced dyskinesias are the main feature improved by contralateral pallidotomy in Parkinson's disease. Neurology, 45: A377.

Reddy, P.H., Williams, M. and Tagle, D.A. (1999) Recent advances in understanding the pathogenesis of Huntington's disease. Trends Neurosci., 22: 248–255.

Reiner, A., Albin, R.L., Anderson, K.D., D'Amato, C.J., Penney, J.B. and Young, A.B. (1988) Differential loss of striatal projection neurons in Huntington disease. Proc. Natl. Acad. Sci. USA, 85: 5733–5737.

Reisine, T.D., Overstreet, D., Gale, K., Rossor, M., Iversen, L. and Yamamura, H.I. (1980) Benzodiazepine receptors: the effect of GABA on their characteristics in human brain and their alteration in Huntington's disease. Brain Res., 199: 79–88.

Reynolds, G.P., Pearson, S.J. and Heathfield, K.W. (1990) Dementia in Huntington's disease is associated with neurochemical deficits in the caudate nucleus, not the cerebral cortex. Neurosci. Lett., 113: 95–100.

Robertson, R.G., Clarke, C.A., Boyce, S., Sambrook, M.A. and Crossman, A.R. (1990) The role of striatopallidal neurones utilizing gamma-aminobutyric acid in the pathophysiology of MPTP-induced parkinsonism in the primate: evidence from [3 H]flunitrazepam autoradiography. Brain Res., 531: 95–104.

Robertson, R.G., Graham, W.C., Sambrook, M.A. and Crossman, A.R. (1991) Further investigations into the pathophysiology of MPTP-induced parkinsonism in the primate: an intracerebral microdialysis study of gamma-aminobutyric

310

acid in the lateral segment of the globus pallidus. Brain Res., 563: 278–280.

Rosse, R.B., Allen, A. and Lux, W.E. (1986) Baclofen treatment in a patient with tardive dystonia. J. Clin. Psychiatry, 47: 474–475.

Rupniak, N.M., Jenner, P. and Marsden, C.D. (1986) Acute dystonia induced by neuroleptic drugs. Psychopharmacology (Berl.), 88: 403–419.

Salin, P., Manrique, C., Forni, C. and Kerkerian-Le Goff, L. (2002) High-frequency stimulation of the subthalamic nucleus selectively reverses dopamine denervation-induced cellular defects in the output structures of the basal ganglia in the rat. J. Neurosci., 22: 5137–5148.

Sanberg, P.R., Henault, M.A. and Deckel, A.W. (1986) Locomotor hyperactivity: effects of multiple striatal transplants in an animal model of Huntington's disease. Pharmacol. Biochem. Behav., 25: 297–300.

Sandyk, R. (1984) Beneficial effect of sodium valproate and baclofen in spasmodic torticollis. A case report. S. Afr. Med. J., 65: 62–63.

Schneider, J.S. and Wade, T.V. (2003) Experimental parkinsonism is associated with increased pallidal GAD gene expression and is reversed by site-directed antisense gene therapy. Mov. Disord., 18: 32–40.

Schroeder, J.A. and Schneider, J.S. (2002) GABA-opioid interactions in the globus pallidus: [D-Ala2]-Met-enkephalinamide attenuates potassium-evoked GABA release after nigrostriatal lesion. J. Neurochem., 82: 666–673.

Schultz, W. (1998) The phasic reward signal of primate dopamine neurons. Adv. Pharmacol., 42: 686–690.

Schumacher, J.M., Hantraye, P., Brownell, A.L., Riche, D., Madras, B.K., Davenport, P.D., Maziere, M., Elmaleh, D.R., Brownell, G.L. and Isacson, O. (1992) A primate model of Huntington's disease: functional neural transplantation and CT-guided stereotactic procedures. Cell Transplant, 1: 313–322.

Schwartzman, R.J. and Alexander, G.M. (1985) Changes in the local cerebral metabolic rate for glucose in the 1-methyl-4-phenyl-1,2,3,6-tetrahydropyridine (MPTP) primate model of Parkinson's disease. Brain Res., 358: 137–143.

Scigliano, G., Giovannini, P., Girotti, F., Grassi, M.P., Caraceni, T. and Schechter, P.J. (1984) Gamma-vinyl GABA treatment of Huntington's disease. Neurology, 34: 94–96.

Scoditti, U., Rustichelli, P. and Calzetti, S. (1989) Spontaneous hemiballism and disappearance of parkinsonism following contralateral lenticular lacunar infarct. Ital. J. Neurol. Sci., 10: 575–577.

Segawa, M. (2000) Hereditary progressive dystonia with marked diurnal fluctuation. Brain Dev., 22(Suppl. 1): S65–S80.

Segawa, M., Hosaka, A., Miyagawa, F., Nomura, Y. and Imai, H. (1976) Hereditary progressive dystonia with marked diurnal fluctuation. In: Eldridge R. and Fahn S. (Eds.), Adv. Neurol. Raven Press, New York, pp. 215–233.

Sellal, F., Lisovoski, F., Hirsch, E., Mutschler, V., Collard, M. and Marescaux, C. (1992) Contralateral disappearance of parkinsonian signs after subthalamic hematoma. Neurology, 42: 255.

Semyanov, A., Walker, M.C., Kullmann, D.M. and Silver, R.A. (2004) Tonically active GABAA receptors: modulating gain and maintaining the tone. Trends Neurosci., 27: 262–269.

Shink, E., Bevan, M.D., Bolam, J.P. and Smith, Y. (1996) The subthalamic nucleus and the external pallidum: two tightly interconnected structures that control the output of the basal ganglia in the monkey. Neuroscience, 73: 335–357.

Shoulson, I., Goldblatt, D., Charlton, M. and Joynt, R.J. (1978) Huntington's disease: treatment with muscimol, a GABA-mimetic drug. Ann. Neurol., 4: 279–284.

Shoulson, I., Kartzinel, R. and Chase, T.N. (1976) Huntington's disease: treatment with dipropylacetic acid and gamma-aminobutyric acid. Neurology, 26: 61–63.

Sieradzan, K.A., Fox, S.H., Hill, M., Dick, J.P., Crossman, A.R. and Brotchie, J.M. (2001) Cannabinoids reduce levodopa-induced dyskinesia in Parkinson's disease: a pilot study. Neurology, 57: 2108–2111.

Silberstein, P., Kuhn, A.A., Kupsch, A., Trottenberg, T., Krauss, J.K., Wohrle, J.C., Mazzone, P., Insola, A., Di Lazzaro, V., Oliviero, A., Aziz, T. and Brown, P. (2003) Patterning of globus pallidus local field potentials differs between Parkinson's disease and dystonia. Brain, 126: 2597–2608.

Silbert, P.L. and Stewart-Wynne, E.G. (1992) Increased dystonia after intrathecal baclofen. Neurology, 42: 1639–1640.

Smith, Y., Bevan, M.D., Shink, E. and Bolam, J.P. (1998) Microcircuitry of the direct and indirect pathways of the basal ganglia. Neuroscience, 86: 353–387.

Smith, Y., Raju, D.V., Pare, J.F. and Sidibe, M. (2004) The thalamostriatal system: a highly specific network of the basal ganglia circuitry. Trends Neurosci., 27: 520–527.

Soares, J., Kliem, M.A., Betarbet, R., Greenamyre, J.T., Yamamoto, B. and Wichmann, T. (2004) Role of external pallidal segment in primate parkinsonism: comparison of the effects of 1-methyl-4-phenyl-1,2,3,6-tetrahydropyridine-induced parkinsonism and lesions of the external pallidal segment. J. Neurosci., 24: 6417–6426.

Soghomonian, J.J. and Laprade, N. (1997) Glutamate decarboxylase (GAD67 and GAD65) gene expression is increased in a subpopulation of neurons in the putamen of parkinsonian monkeys. Synapse, 27: 122–132.

Soghomonian, J.J. and Martin, D.L. (1998) Two isoforms of glutamate decarboxylase: why? Trends Pharmacol. Sci., 19: 500–505.

Soghomonian, J.J., Pedneault, S., Audet, G. and Parent, A. (1994) Increased glutamate decarboxylase mRNA levels in the striatum and pallidum of MPTP-treated primates. J. Neurosci., 14: 6256–6265.

Soghomonian, J.J., Pedneault, S., Blanchet, P.J., Goulet, M., Di Paolo, T. and Bedard, P.J. (1996) L-DOPA regulates glutamate decarboxylases mRNA levels in MPTP-treated monkeys. Brain Res. Mol. Brain Res., 39: 237–240.

Spokes, E.G. (1980) Neurochemical alterations in Huntington's chorea: a study of post-mortem brain tissue. Brain, 103: 179–210.

Stanford, I.M. (2003) Independent neuronal oscillators of the rat globus pallidus. J. Neurophysiol., 89: 1713–1717.

Starr, P.A., Rau, G.M., Davis, V., Marks Jr., W.J., Ostrem, J.L., Simmons, D., Lindsey, N. and Turner, R.S. (2005) Spontaneous pallidal neuronal activity in human dystonia: comparison with Parkinson's disease and normal macaque. J. Neurophysiol., 93: 3165–3176.

Stasi, K., Mitsacos, A., Giompres, P., Kouvelas, E.D. and Triarhou, L.C. (1999) Partial restoration of striatal GABAA receptor balance by functional mesencephalic dopaminergic grafts in mice with hereditary parkinsonism. Exp. Neurol., 157: 259–267.

Storey, E. and Beal, M.F. (1993) Neurochemical substrates of rigidity and chorea in Huntington's disease. Brain, 116: 1201–1222.

Su, P.C., Tseng, H.M., Liu, H.M., Yen, R.F. and Liou, H.H. (2003) Treatment of advanced Parkinson's disease by subthalamotomy: one-year results. Mov. Disord., 18: 531–538.

Suarez, J.I., Metman, L.V., Reich, S.G., Dougherty, P.M., Hallett, M. and Lenz, F.A. (1997) Pallidotomy for hemiballismus: efficacy and characteristics of neuronal activity. Ann. Neurol., 42: 807–811.

Surmeier, D.J., Eberwine, J., Wilson, C.J., Cao, Y., Stefani, A. and Kitai, S.T. (1992) Dopamine receptor subtypes colocalize in rat striatonigral neurons. Proc. Natl. Acad. Sci. USA, 89: 10178–10182.

Symington, G.R., Leonard, D.P., Shannon, P.J. and Vajda, F.J. (1978) Sodium valproate in Huntington's disease. Am. J. Psychiatry, 135: 352–354.

Tasker, R.R., Lang, A.E. and Lozano, A.M. (1997) Pallidal and thalamic surgery for Parkinson's disease. Exp. Neurol., 144: 35–40.

Tell, G., Bohlen, P., Schechter, P.J., Koch-Weser, J., Agid, Y., Bonnet, A.M., Coquillat, G., Chazot, G. and Fischer, C. (1981) Treatment of Huntington disease with gamma-acetylenic GABA an irreversible inhibitor of GABA-transaminase: increased CSF GABA and homocarnosine without clinical amelioration. Neurology, 31: 207–211.

Tenn, C.C. and Niles, L.P. (1997) Mechanisms underlying the antidopaminergic effect of clonazepam and melatonin in striatum. Neuropharmacology, 36: 1659–1663.

Tepper, J.M., Koos, T. and Wilson, C.J. (2004) GABAergic microcircuits in the neostriatum. Trends Neurosci., 27: 662–669.

Terman, D., Rubin, J.E., Yew, A.C. and Wilson, C.J. (2002) Activity patterns in a model for the subthalamopallidal network of the basal ganglia. J. Neurosci., 22: 2963–6976.

Thaker, G.K., Nguyen, J.A., Strauss, M.E., Jacobson, R., Kaup, B.A. and Tamminga, C.A. (1990) Clonazepam treatment of tardive dyskinesia: a practical GABAmimetic strategy. Am. J. Psychiatry, 147: 445–451.

Thompson-Vest, N.M., Waldvogel, H.J., Rees, M.I. and Faull, R.L. (2003) GABA(A) receptor subunit and gephyrin protein changes differ in the globus pallidus in Huntington's diseased brain. Brain Res., 994: 265–270.

van der Stelt, M., Fox, S.H., Hill, M., Crossman, A.R., Petrosino, S., Di Marzo, V. and Brotchie, J.M. (2005) A role for endocannabinoids in the generation of parkinsonism and levodopa-induced dyskinesia in MPTP-lesioned non-human primate models of Parkinson's disease. FASEB J., 19: 1140–1142.

Van Ness, P.C., Watkins, A.E., Bergman, M.O., Tourtellotte, W.W. and Olsen, R.W. (1982) gamma-Aminobutyric acid receptors in normal human brain and Huntington disease. Neurology, 32: 63–68.

Vidakovic, A., Dragasevic, N. and Kostic, V.S. (1994) Hemiballism: report of 25 cases. J. Neurol. Neurosurg. Psychiatry, 57: 945–949.

Vitek, J.L. (2002) Pathophysiology of dystonia: a neuronal model. Mov. Disord., 17(Suppl. 3): S49–S62.

Vitek, J.L., Chockkan, V., Zhang, J.Y., Kaneoke, Y., Evatt, M., DeLong, M.R., Triche, S., Mewes, K., Hashimoto, T. and Bakay, R.A. (1999) Neuronal activity in the basal ganglia in patients with generalized dystonia and hemiballismus. Ann. Neurol., 46: 22–35.

Vitek, J.L., Kaneoke, Y., Turner, R., Baron, M., Bakay, R. and DeLong, M. (1993) Neuronal activity in the internal (GP$_i$) and external (GP$_e$) segments of the globus pallidus (GP) of parkinsonian patients is similar to that in the MPTP-treated primate model of parkinsonism. Soc. Neurosci. Abstr., 19: 1584.

Walker, F.O., Young, A.B., Penney, J.B., Dovorini-Zis, K. and Shoulson, I. (1984) Benzodiazepine and GABA receptors in early Huntington's disease. Neurology, 34: 1237–1240.

Wenger, K.K., Musch, K.L. and Mink, J.W. (1999) Impaired reaching and grasping after focal inactivation of globus pallidus pars interna in the monkey. J. Neurophysiol., 82: 2049–2060.

Whittier, J.R. and Mettler, F.A. (1949) Studies of the subthalamus of the rhesus monkey. II. Hyperkinesia and other physiologic effects of subthalamic lesions with special references to the subthalamic nucleus of Luys. J. Comp. Neurol., 90: 319–372.

Wichmann, T., Bergman, H. and DeLong, M.R. (1994a) The primate subthalamic nucleus. I. Functional properties in intact animals. J. Neurophysiol., 72: 494–506.

Wichmann, T., Bergman, H. and DeLong, M.R. (1994b) The primate subthalamic nucleus. III. Changes in motor behavior and neuronal activity in the internal pallidum induced by subthalamic inactivation in the MPTP model of parkinsonism. J. Neurophysiol., 72: 521–530.

Wichmann, T., Bergman, H., Starr, P.A., Subramanian, T., Watts, R.L. and DeLong, M.R. (1999) Comparison of MPTP-induced changes in spontaneous neuronal discharge in the internal pallidal segment and in the substantia nigra pars reticulata in primates. Exp. Brain Res., 125: 397–409.

Wichmann, T. and DeLong, M.R. (2003) Functional neuroanatomy of the basal ganglia in Parkinson's disease. Adv. Neurol., 91: 9–18.

Wichmann, T., Gatev, P.G. and Kliem, M.A. (2002) Neuronal discharge in the basal ganglia is correlated with EEG in normal and parkinsonina primates. Mov. Disord. (Abstract Volume), 17(Suppl. 5): S155.

312

Wichmann, T., Kliem, M.A. and DeLong, M.R. (2001) Anti-parkinsonian and behavioral effects of inactivation of the substantia nigra pars reticulata in hemiparkinsonian primates. Exp. Neurol., 167: 410–424.

Wilson, C.J. and Groves, P.M. (1980) Fine structure and synaptic connections of the common spiny neuron of the rat neostriatum: a study employing intracellular inject of horseradish peroxidase. J. Comp. Neurol., 194: 599–615.

Windels, F., Bruet, N., Poupard, A., Urbain, N., Chouvet, G., Feuerstein, C. and Savasta, M. (2000) Effects of high frequency stimulation of subthalamic nucleus on extracellular glutamate and GABA in substantia nigra and globus pallidus in the normal rat. Eur. J. Neurosci., 12: 4141–4146.

Windels, F., Carcenac, C., Poupard, A. and Savasta, M. (2005) Pallidal origin of GABA release within the substantia nigra pars reticulata during high-frequency stimulation of the subthalamic nucleus. J. Neurosci., 25: 5079–5086.

Winkler, C., Bentlage, C., Nikkhah, G., Samii, M. and Bjorklund, A. (1999) Intranigral transplants of GABA-rich striatal tissue induce behavioral recovery in the rat Parkinson model and promote the effects obtained by intrastriatal dopaminergic transplants. Exp. Neurol., 155: 165–186.

Wu, Y., Richard, S. and Parent, A. (2000) The organization of the striatal output system: a single-cell juxtacellular labeling study in the rat. Neurosci. Res., 38: 49–62.

Yoshida, M., Nagatsuka, Y., Muramatsu, S. and Niijima, K. (1991) Differential roles of the caudate nucleus and putamen in motor behavior of the cat as investigated by local injection of GABA antagonists. Neurosci. Res., 10: 34–51.

Yu, T.S., Wang, S.D., Liu, J.C. and Yin, H.S. (2001) Changes in the gene expression of GABA(A) receptor alpha1 and alpha2 subunits and metabotropic glutamate receptor 5 in the basal ganglia of the rats with unilateral 6-hydroxydopamine lesion and embryonic mesencephalic grafts. Exp. Neurol., 168: 231–241.

Tepper, Abercrombie & Bolam (Eds.)
Progress in Brain Research, Vol. 160
ISSN 0079-6123

CHAPTER 18

Simulation of GABA function in the basal ganglia: computational models of GABAergic mechanisms in basal ganglia function

Jeffery R. Wickens*, Gordon W. Arbuthnott and Tomomi Shindou

Basal Ganglia Research Group, School of Medical Sciences, University of Otago, Dunedin, New Zealand

Abstract: This chapter outlines current interpretation of computational aspects of GABAergic circuits of the striatum. Recent hypotheses and controversial matters are reviewed. Quantitative aspects of striatal synaptology relevant to computational models are considered, with estimates of the connectivity of the spiny projection neurons and fast-spiking interneurons. Against this background, insights into the computational properties of inhibitory circuits based on analysis and simulation of simple models are discussed. The paper concludes with suggestions for further theoretical and experimental studies.

Keywords: Lateral inhibition; GABA; computation; dynamics; competition; striatum; spiny projection neuron

Introduction

This chapter considers the computational operations of GABAergic circuits in the basal ganglia with a particular focus on the striatum. Although recognised as playing a key role in movement, behaviour, and learning, the function of the striatum remains enigmatic. This is a problem for 'top-down' modelling efforts, which must start with assumptions about the computational operation the structure is performing.

We advocate the use of 'bottom-up' modelling — using models based on the actual anatomy and physiology — to constrain our assumptions. The neural network of the striatum provides a special opportunity to analyse the computational operations of a brain structure. Much of the information required to quantify the GABAergic circuits of the striatum has become available recently, and formal computational models are beginning to emerge. These models have stimulated sharp debates and experimental tests, which are already leading to their refinement.

Computational models are needed to define the operations of the striatum and to understand its contribution to information processing in the basal ganglia. We use the term 'operations' in a mathematical sense, defined as the transformations applied to inputs to produce output. These operations are determined by the properties of the neural network that makes up the striatum. They cannot be deduced from the effects of lesions on behaviour.

Optimistically, there may be a common and mathematically definable input–output operation performed by the striatum, and reiterated throughout its extent. Even so, striatal subregions may

---
*Corresponding author. Tel.: +64 3 379 7373; Fax: +64 3 479 7254; E-mail: jeff.wickens@stonebow.otago.ac.nz

DOI: 10.1016/S0079-6123(06)60018-6

314

have specific functions on account of their different input sources and output targets. Alternatively, striatal subregions may show regional specialisation of function due to regional differences in the input–output operations. To date, however, few regional differences in cellular properties or connectivity have been defined, with most data suggesting similarities in these features (Kawaguchi et al., 1989; Taverna et al., 2004). On the other hand, regional differences in the input sources and output targets are well established (Alexander et al., 1986; Hoover and Strick, 1993). Thus, it is reasonable to pursue the hypothesis of a common input–output operation.

If a common input–output operation could be defined, such a definition would extend current theories of basal ganglia function based on the effects of clinical–pathological correlations. For example, current 'box and arrow' models of the basal ganglia stop short of defining the computational operations performed in individual structures. Such models serve as valuable heuristic devices in the development of theories of basal ganglia function, but require specification of the input–output transformations represented by the boxes and connecting arrows. While such specification has been accepted as a laudable aim, previous computational models have been strongly challenged by experimental findings, and are regarded as in need of substantial revision.

Anatomical descriptions of the spiny projection neurons of the striatum suggest a particular type of circuitry, namely a network of projection neurons interconnected by local axon collaterals (Wilson and Groves, 1980). Based on the GABA-ergic nature of these neurons and their synaptic contacts with other spiny neurons, several authors have proposed that the spiny projection neurons form a lateral or feedback inhibitory network (Groves, 1983; Rolls and Williams, 1986; Wickens et al., 1991). In this idea of a lateral inhibition network, each output neuron makes inhibitory synaptic contact with its neighbours, as shown in Fig. 1A. Because of obvious physical limitations, lateral inhibition is usually thought of in terms of a local domain of inhibition, extending over the dimensions of the axonal and dendritic arborisations of neighbouring cells. Such local domains need not

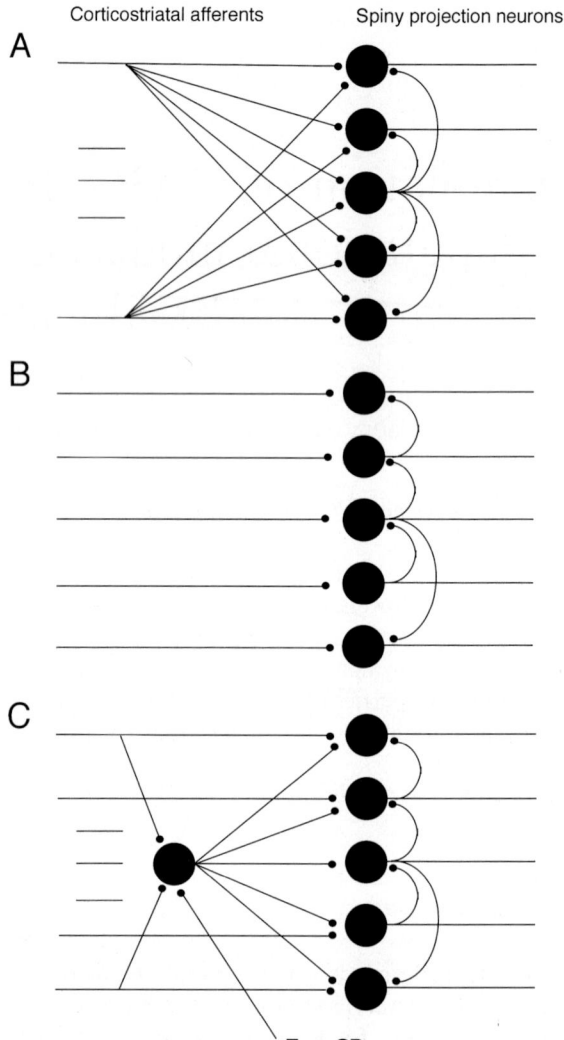

Fig. 1. Changing concepts of striatal organisation. In all versions, the topology is such that the striatum is a single layer of spiny projection neurons with one synapse between cortical inputs and striatal outputs. The connectivity among spiny projection cells has changed over time. (A) High connectivity model in which every neuron within a domain inhibits every other neuron, and all cells are reciprocally connected. (B) Low connectivity model in which connectivity is sparse and asymmetric. (C) Low connectivity model showing contribution of feedforward interneurons. These constitute about 1% of the total number of cells but are thought to make about 10% of the total number of inhibitory synapses. Their influence may be as great as the spiny projection neurons if their synapses are made predominantly on the soma.

be physically compartmentalised, but may arise as a dynamic property of local inhibitory interactions across a homogeneous field (Wilson, 2000).

Recently, our concept of the striatal circuit has been modified in the light of new findings. The anatomical connectivity among spiny projection neurons has been recognised to be quite sparse, so that not every neuron contacts every other, even within spatially limited domains (Oorschot et al., 2002; Wickens, 2002; Wickens and Oorschot, 2000). As a result connections between pairs of neurons are not in general reciprocal or symmetrical (Kotter and Wickens, 1995; Wickens et al., 1995; Kotter and Wickens, 1998; Plenz and Kitai, 1999). Furthermore, the cortical input has also been shown to be sparsely distributed, with adjacent neurons receiving few inputs in common (Kincaid et al., 1998). This modified circuit is shown in Fig. 1B. This represents the revised model of the striatal spiny cell network.

In addition to changes in our concept of the local interactions among the spiny projection neurons, it has also become evident that the feedforward interneurons of the striatum, although relatively few in number, have a disproportionately large influence. These are represented in Fig. 1C. The number of synaptic contacts that a feedforward interneuron makes is many times that of the local synaptic contacts of a single projection neuron. However, the feedforward interneurons are outnumbered almost 100 to 1 by the spiny projection neurons. Even if they made 10 times as many connections per cell, the feedforward interneurons would account for only 10% of the total GABAergic input. On the other hand they are thought to make the majority of their synaptic connections on the soma of the spiny projection neurons, and to be relatively more excitable, so it is likely they have as much influence over the striatal network as the spiny projection neuron collaterals.

We are now at the stage where the first generation of computational models is being replaced by models that take into account our updated concepts of the striatal circuit. The more recent findings have shown that many of the assumptions made in the earlier models overestimated the strength of feedback inhibition in the striatum,

and hence these models require revision. On the other hand, there have so far been few models presented which incorporate the new data into the network structure of the striatum. New computer simulation models are needed in order to understand how the measurements obtained from paired recordings translate into effects on the overall network dynamics in which many neurons are active.

Insights into inhibitory circuits from artificial neural networks

Before considering the debate over the significance of feedforward and feedback inhibition in the striatum, it is useful to consider what is at stake. Inhibitory neural networks have powerful computational properties, some of which are understood in mathematical terms. If inhibitory interactions are a central organising principle of the striatal network, then the theoretical tools developed for studying abstract inhibitory networks can be brought to bear on the operations performed by the striatum.

Studies of artificial neural networks provide valuable insights into the computational power of inhibitory circuits. For example, it is sometimes assumed that competitive dynamics can only arise when neurons do not inhibit themselves (or the members of the same group). It has been argued that such networks would only be effective in producing competition if spiny neurons formed groups that did not inhibit each other. If this is true then these cannot give rise to winner-take-all (WTA) groups, or 'k-WTA' dynamics (where k is a set of active neurons). Any group of active cells that grew large enough to inhibit its neighbours — it is argued — would inhibit its own members as much as its competitors (Koos et al., 2004). As noted below, however, the ratio of self-inhibition to lateral inhibition may produce competitive dynamics even when this ratio is less than unity (Fukai and Tanaka, 1997). Computationally useful competitive dynamics can be shown to arise when self-inhibition is as much or even more than inhibition of neighbours, leading to 'winner-share-all' type of dynamics (Fukai and Tanaka, 1997). This softer competitive dynamic has greater computational power than the single WTA type of interaction (Maass, 2000a, 2000b).

GABAergic interactions may have different functions in different parts of the brain. A brief survey of neural network models illustrates some of the functions proposed for GABA interactions. In networks that contain both excitatory and inhibitory populations of neurons, inhibition is important to maintain activity at an intermediate level between maximum and minimum, in other words to protect against explosive type dynamics. Beurle (1956) showed that a mass of purely excitatory cells has inherently unstable activity, such that if started from an intermediate state of activity the mass would rapidly become either completely quiescent or completely active. Griffith (1963) and Wilson and Cowen (1972) considered the dynamics of networks containing both excitatory and inhibitory types of neurons. The stabilising function of inhibition is unlikely to be the major role of inhibition in the striatum, however, because explosive increases of activity do not occur when there are no recurrent excitatory connections.

Contour enhancement is another proposed function of recurrent networks. Grossberg (1973, 1988) notes that such networks can store a particular feature of a spatial pattern, such as the boundary of a solid figure, while other parts are quenched. Feedback inhibition can also lead to networks in which a pattern of activity can reverberate. This has usually been studied in an anatomy where connections with other cells are inhibitory, but the connections with self are excitatory. Amari (1977a, 1977b) and Kishimoto and Amari (1979) considered the dynamics of pattern formation in lateral-inhibition type one-dimensional homogenous neural fields. In the fields studied, excitatory connections dominate for proximate neurons and inhibitory connections dominate at greater distances. The dynamics of these fields depend on the general level of excitation, the threshold, and how the synaptic weights vary as a function of distance. Several different types of dynamics have been characterised. A localised excitation may die out or if it is above a certain level it may grow until all neurons are maximally active. Conditions also exist under which the dynamics

are such that a field could remain quiescent but retain a localised excitation that had been briefly applied. Under certain conditions a periodic solution arises, consisting of regularly spaced localised excitations. Although many of these studies in the artificial neural network literature are based on networks that include local excitatory interconnections and further reaching inhibitory connections (Amari, 1977a, 1977b; Kishimoto and Amari, 1979), some purely inhibitory networks have been studied, and these are especially relevant to the current topic.

Mathematical analysis of purely inhibitory feedback networks has provided important insight into the dynamics of networks with a structure similar to the striatal spiny cell network. Fukai and Tanaka (1997) analysed a simple model in which there were excitatory inputs and all-to-all recurrent inhibitory connections, including self-inhibition. They showed that the dynamics of such a network depend on the strength of lateral inhibition relative to self-inhibition. When lateral inhibition is weaker than self-inhibition, a certain number of more strongly excited neurons are active. This was called a 'winners-share-all' type of dynamics. Thus, instead of only the most strongly excited neuron becoming active and suppressing the rest, the 2nd, 3rd, etc., most strongly excited neurons can become active.

Fukai and Tanaka (1997) also showed that when self-inhibition is weaker than lateral inhibition, only one neuron is activated, but it is not necessarily the one receiving the largest input. This is a kind of WTA dynamic, but the winner depends on the initial conditions as well as the current input. In such networks the first-activated neuron is able to suppress other cells, even if they receive more excitation on average. Thus, although it is often stated that competitive dynamics require neurons to inhibit themselves less than they inhibit others, this is not actually the case, and when it is, the dynamics are not necessarily the most useful for computation. On the other hand, when the strength of the lateral inhibition is equal to self-inhibition, the dynamics of the network are such that the most active neuron suppresses activity in the others. In this case we have the commonly understood form of WTA dynamic, in which the most strongly excited neuron becomes the most active cell and suppresses the others.

Despite their apparent simplicity, pure inhibitory networks have a number of interesting and important dynamical properties that are useful for information processing (Majani et al., 1989; Wolfe et al., 1991). One operation that WTA dynamics can provide is a rapid readout of the neuron that receives the greatest net excitation. This may be important, for example, in voluntary movement, where the activation of a relatively small number of neurons is necessary to select a particular action from the wide range of possible movements. A role for WTA dynamics in the striatum in such action selection was suggested by Wickens (1993). This dynamic may be necessary to prevent cocontraction of groups of muscular antagonists in movement. In Parkinson's disease, in which cocontraction replaces reciprocal inhibition of muscles at the onset of movements, this dynamic may break down (Wickens et al., 1991; Alexander and Wickens, 1993).

The computational operation performed by WTA type networks appears deceptively simple, when in fact it is a computationally demanding operation. In terms of an algorithm, it translates into the operation of finding the most depolarised neuron. To perform this operation using a serial computer, it would be necessary to use an algorithm that made pairwise comparisons between the largest stored value, and every other value, i.e., for an n-neuron network it would be necessary to make at least n comparisons and store the identify of the neuron with the greater value in memory between each comparison. To compute k-winners is obviously even more involved. A lateral inhibitory network can compute this operation in essentially one time-step, which implies that it might be a very useful architecture that would be conserved by evolution and employed extensively in the brain.

Computational analysis of WTA dynamics shows the potential of neural computation based on inhibitory networks (Maass, 2000a). Analogue circuits, which can be fabricated on chips, are able to compute WTA operations with very few transistors and short connection lengths (Lazzaro et al., 1989). Theoretical analysis indicates that WTA modules have surprisingly powerful computational capabilities in comparison to threshold gates and sigmoidal gates. Arbitrary functions can be approximated by circuits employing lateral inhibition. For example, any Boolean function can be computed by a single k-WTA unit applied to weighted sums of the input variables (Maass, 2000b). Furthermore, it can be computed using biological neurons in which only positive weights are employed on the inputs and only the positive weights are subjected to plasticity (Maass, 2000a).

Challenges to the concept of lateral inhibition

While the utility of lateral inhibition is well recognised, whether lateral inhibition is in reality a central organising principle of striatal function has been challenged on several grounds. Initially, there was great difficulty demonstrating functional inhibitory interactions between spiny projection neurons, and doubts were raised about their existence. Functional GABA interactions between spiny projection neurons were finally demonstrated using paired intracellular recording techniques together with spike-triggered averaging (Oorschot et al., 2002; Tunstall et al., 2002). With improvements in technique, such as the use of dual whole-cell recordings with patch electrodes, the existence of such interactions has been repeatedly confirmed (Czubayko and Plenz, 2002; Koos et al., 2004; Taverna et al., 2004; Venance et al., 2004; Gustafson et al., 2006). Therefore, the first challenge to the reality of lateral inhibition was overturned by unequivocal proof of functional GABAergic synapses between spiny projection neurons.

However, a second and more difficult challenge is that the inhibitory interactions among the spiny neurons are too sparse and weak to play the role suggested by early models of lateral inhibition. These models assumed high conductance values for inhibitory synapses, giving rise to a dynamic 'WTA' network, in which the most active neurons suppressed activity in the less active neurons (Wickens et al., 1991; Alexander and Wickens, 1993; Wickens, 1993; Wickens and Arbuthnott, 1993; Wickens et al., 1995; Plenz et al., 1996; Fukai, 1999). In contrast to these assumptions, measurements of the magnitude of the synaptic currents produced by single spiny neurons on other spiny neurons revealed that these are an order of magnitude less than was assumed in the earlier models (Koos et al., 2004; Tepper et al., 2004). The frequency of detection of connections was also low, suggesting a sparse connectivity in the inhibitory network. In particular, there was a lack of directly reciprocal connections, which had been assumed in models based on mutual inhibition.

With these facts now widely acknowledged, it is appropriate to consider whether the concept of lateral inhibition in the striatum is still viable. This is a more difficult question than it may seem at first. Analyses of the effects of one single spiny neuron on another provide a crucial piece of information. However, other data are also required. In the striatal network, the effects of individual synaptic interactions summate, according to the population activity and the connectivity among different groups of neurons. Relating the results of pairwise studies to the activity of the striatal network in an awake, behaving animal is not trivial.

An important variable in determining the network-level effects of lateral inhibition is the fraction of spiny cells that are active in a given period. The great majority of the spiny projection neurons do not fire action potentials when the animal is awake but at rest (Sandstrom and Rebec, 2003). It is not known how much inhibition is produced by this active subset. Moreover, it is unclear what fraction of cells becomes active when the animal is moving. A further difficulty is that the spatial distribution of the active neurons is unknown, probably depending on the relationship of activity of specific cells to the task. For this reason it is helpful to consider the effects of realistic lateral inhibition on different scenarios of population activity.

In order to resolve these issues, some quantification of the effects of lateral inhibition in the striatum is needed at the network level in awake animals. The direct experimental measures needed to provide such quantification in vitro are not yet available for populations in awake, behaving animals. Single unit recording provides important information about the active cells but is hard to

use to provide information about cells that only become active in relation to specific behaviours. Hence, it is useful to explore the implications of the measures obtained from pairs of cells for the activity of larger networks with anatomically realistic connections. In the following section the functional properties of lateral inhibition in the striatum are considered. The key anatomical and physiological parameters of inhibitory interactions among spiny projection neurons are reviewed. A simple model of a small block of striatal tissue is developed, and studied by computer simulation. This provides an indication of the effect that lateral inhibition has on overall activity dynamics of a striatal network. The results suggest that under natural firing conditions in the striatal network as a whole, lateral inhibition produces strong effects on the overall activity levels.

Anatomy of spiny projection neuron interconnections in the striatum

The striatum is the major input structure in the basal ganglia. The principal cells of the striatum are the spiny projection neurons (also known as medium-spiny neurons). They receive excitatory synaptic input from the cerebral cortex (Somogyi et al., 1981), and thalamus (Xu et al., 1991) and project to output structures such as the globus pallidus, the substantia nigra and the entopeduncular nucleus (Somogyi et al., 1981; Dube et al., 1988). The spiny projection neurons comprise 97% of the total number of striatal cells as determined by stereology in the rat (Oorschot, 1996). Since the excitatory input synapses terminate directly on the spiny projection neurons, the equivalent neural network circuit of the spiny projection neuron skeleton of the striatum is a single layer network of spiny neurons, with only one synapse between cortical input and striatal output (Fig. 1A, B).

Lateral inhibitory interactions among spiny neurons were suggested in the earliest descriptions of the anatomy of single, injected neurons (Wilson and Groves, 1980). Spiny projection neurons have a relatively uniform somatodendritic architecture. There are usually several primary dendrites radiating from the soma, which divide three or four times to form a dendritic tree that occupies a spherical space of radius, ca. 200 μm. Each neuron gives rise to a main axon that projects to target structures. From the main axon, multiple collateral branches arise. These divide repeatedly to form a network of local axon collaterals that overlaps extensively with the dendritic tree of the parent neuron (Wilson and Groves, 1980). On the basis of the overlap of axonal and dendritic trees of adjacent neurons, synaptic interactions seem very probable.

Electron microscopic studies confirm that the local axon collaterals of spiny projection neurons form synapses with other spiny projection neurons (Somogyi et al., 1981). These synapses have the appearance of inhibitory synaptic contacts, with symmetrical synaptic densities and large pleomorphic vesicles (Wilson and Groves, 1980). More direct evidence comes from immunohistochemical techniques, which show that a proportion of striatal neurons stain positively for glutamate decarboxylase (GAD), the synthesising enzyme for GABA (Oertel and Mufnini, 1984; Bolam et al., 1985; Kubota et al., 1987; Kita and Kitai, 1988). When conditions are optimised for detection of GAD, the great majority (80%) of neurons with the morphological characteristics of spiny projection neurons stain positively for GAD (Kita, 1993). It has also been shown that GAD-positive boutons form synapses with the cell body and dendritic shafts of neurones identified as projection neurones by retrograde labelling from the substantia nigra (Aronin et al., 1986). Immunohistochemical staining for GABA has identified numerous synapses between GABA-positive boutons and similarly staining dendrites (Pasik et al., 1988). The final definitive proof was provided by Wilson (1994) and Oorschot et al. (2002), who showed synaptic contact between identified spiny projection neurons. Thus, the spiny projection neurons both produce and receive GABAergic synaptic connections with other spiny projection neurons.

As noted above, the anatomical description of inhibitory collaterals led to the concept of a domain of inhibition shown in Fig. 1A. This concept assumes a high degree of connectivity between nearby neurons, leading to mutual inhibition. However, when the probability of a synaptic connection is quantified, it turns out to be much lower than one might expect. Our expectations have been strongly influenced by the two-dimensional projections of axonal and dendritic arborisations, which suggest that overlap of axonal and dendritic trees is very probable. In three-dimensional reality, however, these arborisations occupy only a small fraction of the volume of potential overlap, and the probability of a connection is in fact quite low, as illustrated in Fig. 1B. We consider estimates of this probability in detail below.

Synaptic organisation of the neural network of the striatum: spiny–spiny connections

The probability of a synapse between the local axonal collaterals of one spiny neuron and the dendrites of another spiny neuron located at a certain distance from the first can be estimated using statistical arguments. We assume that each receiving neuron has a number of potential sites for symmetrical synapses, which are intermingled with postsynaptic sites belonging to many other spiny neurons. The postsynaptic sites on a given cell can be visualised as a subset of a cloud-like distribution of postsynaptic targets of all cells, as illustrated in Fig. 2. The probability of the connection between a pair of cells depends on the number of postsynaptic sites in that volume that belong to a given postsynaptic cell, relative to the total number of postsynaptic sites to choose from in the volume. Each presynaptic cell makes a number of 'choices' represented by the number of synapses it makes within the volume of the overlapping region. This is analogous to choosing balls of one colour from an urn containing balls of different colours (sampling without replacement) and is described statistically by the hypergeometric distribution.

Neuroanatomical studies of the rat striatum provide values for the number of synaptic sites of a given neuron, and the total number of synaptic sites of all neurons in the volume in question. Ingham et al. (1998) give the density of asymmetrical synapses in the rat striatum as 0.91 μm^{-3} and

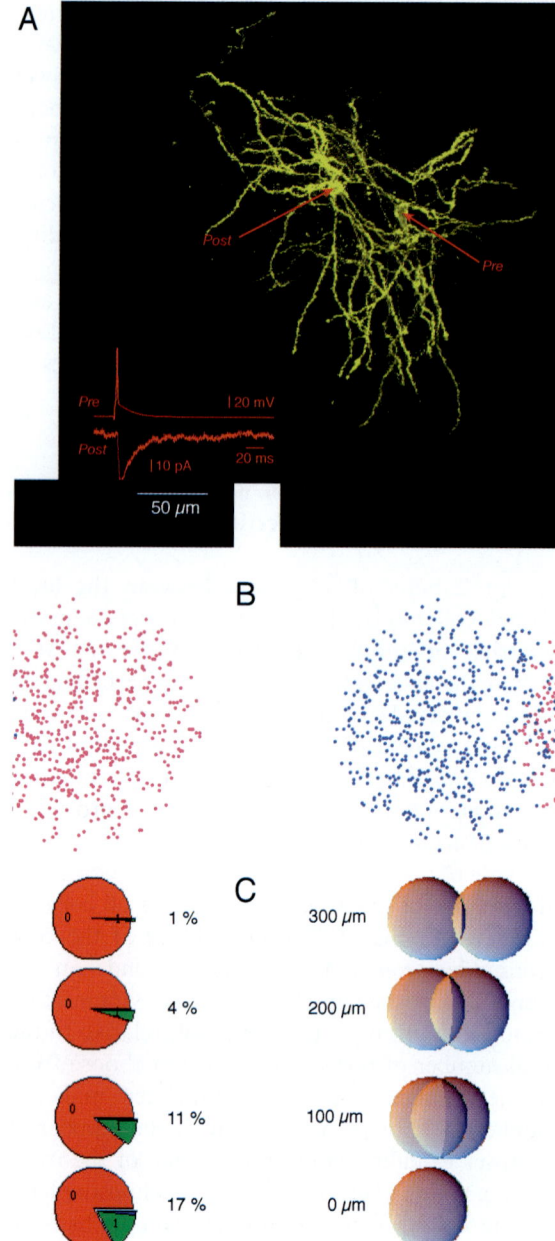

Fig. 2. Estimate of probability of synaptic contact. (A) A pair of filled spiny projection neurons showing their overlapping axonal and dendritic trees. Inset shows the response evoked in the postsynaptic cell by firing of the presynaptic cell. (B) Model of a pair of cells as 'clouds' over overlapping dendritic and axonal sites. (C) Results of using hypergeometric distribution and data reviewed in text to calculate the probability of a connection at distances ranging from 0 μm to 300 μm between centres of the cells.

estimate that symmetrical synapses account for about 20% of the total number of synapses. The other sources of symmetrical synapses include GABA/parvalbumin, somatostatin and cholinergic interneurons, and dopaminergic afferents, which are also predominantly of the symmetrical type. The proportion of symmetrical synapses that come from spiny projection neurons can be estimated to be about 1 in 6 (Wilson, 2000). Using these numbers, the number of synapses of spiny projection neurons per unit volume is on the order of $0.038\,\mu m^{-3}$. The proportion of the synapses that belong to each individual neuron, on average, can be calculated from the density of medium-sized somata. According to Oorschot (1996) this number is $84{,}900\,mm^{-3}$. From these values the average number of postsynaptic sites on one spiny projection neuron is estimated as 448.

Independent experimental studies have produced results in agreement with the results of the statistical argument given above. Lee et al. (1997) counted varicosities of identified striatonigral neurons in brain slices and found on average 749 varicosities per cell (in a sample of 5 cells). Boutons were labelled as synaptic by synaptophysin, a presynaptic vesicle protein. Of 39 varicosities studied by electron microscopy, 31 were presynaptic terminals as defined by synaptophysin labelling (see Table 1). Based on this proportion, we can estimate 595 synapses per striatal cell, in good agreement with the estimate derived from quantitative neuroanatomy, above.

The number of synapses per striatal cell can be used in computational estimates of the probability of a synaptic connection between pairs of spiny projection neurons. The probability of synaptic contact as a function of distance between somata can be calculated from the volume of the solid formed by the intersection of two spheres representing the dendritic and axonal arborisations of the respective neurons (Braitenberg and Schüz, 1991; Wickens and Oorschot, 2000). The number of postsynaptic sites in the volume that belong to the neuron in question (j), and the total number of synaptic sites in the volume (n) are calculated, and from this the fraction of the postsynaptic sites belonging to the receiving neuron is calculated. This gives the probability (p) that a postsynaptic site

Table 1. Estimate of number of GABA synapses per spiny neuron based on individually labelled spiny neurons

	Quantity	Source	Estimate
I	Mean number of varicosities per striatonigral cell	1	749
II	Fraction of varicosities with synaptophysin	1	0.79
III	Mean number of synapses per striatal neuron	ii × iii	592

Source: 1, Lee et al. (1997).

Table 2. Estimate of number of GABA synapses per spiny neuron based on total synaptic numbers

	Quantity	Source	Estimate
I	Total number of spiny cells in rat striatum	1	2.8×10^6
II	Total number of symmetric synapses in rat striatum	2	4.1×10^9
III	Total number of asymmetric synapses in rat striatum	2	16.1×10^9
IV	Total number of all synapses in rat striatum	ii + iii	20.2×10^9
V	Fraction of all synapses in striatum that are GABA-like	3	7.5%
VI	Total number of GABA-like synapses in striatum	iv × v	1.52×10^9
VII	Number of GABA-like synapses per striatal neuron	vi/i	541

Sources: 1, Oorschot (1996); 2, Ingham et al. (1998); 3, Groves et al. (1994).

chosen at random will belong to the postsynaptic neuron ($p = j/n$). The number of contacts made in the same volume by the presynaptic neuron (k) is similarly determined. Finally, the probability of the presynaptic neuron making one, two, or more synapses with the postsynaptic neuron is calculated from the cumulative hypergeometric distribution with parameters j, k, and n (Wickens and Miller, 1997). Using this method (see Table 2), the estimated probability of a synapse between the axons and dendrites of immediately adjacent neurons is $p = 0.146$ (assuming 448 synapses per cell, and the diameter of the dendritic and axonal arborisations both to be 400 μm). It is important to note, however, that this method is sensitive to the assumptions. A higher probability applies if the axonal and dendritic diameter is smaller, e.g., if a diameter of 300 μm is assumed then $p = 0.310$; or if the number of synapses per cell is larger, e.g., if a total number of 595 is assumed (with diameter 400 μm), then $p = 0.243$.

Paired recordings provide an independent experimental estimate of the probability of a connection. Using dual intracellular recording, Tunstall et al. (2002) found 9 connected cells in a sample of 45 pairs of spiny projection neurons,

corresponding to a probability of 0.1. With improved sensitivity for detecting a connection, we have recently detected 56 connected cells in a sample of 194 pairs ($p = 0.14$) of spiny projection neurons of which 4 were bidirectional (Shindou et al., 2005). These measures are in agreement with those obtained using similar techniques in other labs. In the ventral striatum, Taverna et al. (2004) found 13 connections in a sample of 38 pairs, corresponding to a probability of $p = 0.17$. Koos et al. (2004) found 39 connections in a sample of 325 pairs that were studied in one direction only, corresponding to a probability of $p = 0.12$. Venance et al. (2004) detected connections in 5/50 pairs in horizontally cut brain slices, and 7/22 pairs in saggital slices, corresponding to unidirectional probabilities of $p = 0.05$ and $p = 0.16$, respectively. Of course, the probabilities cited above are likely to be an underestimate, due to the proximity of the cells to the cut surface, and that on average about 50% of the axonal and dendritic arborisations will be superficial to the recorded cell. A much higher connection probability ($p = 0.25$) has been detected in organotypic slice cultures (Czubayko and Plenz, 2002; Gustafson et al., 2006) and this may be closer to the situation in the intact animal.

Towards a realistic computational model of the spiny cell network

As a device to muster the relevant data and make quantitative interpretations, a model of the spiny cell network was studied by computer simulation. The model is based on the relevant physiology of the component neurons of the striatum and the anatomy of their interconnections. A two-compartment model of a single striatal projection neuron was used. This incorporated excitatory and inhibitory synaptic conductances and two forms of a depolarisation-activated potassium conductance. Parameters were chosen to represent the UP state of striatal cell activity (i.e., depolarised close to threshold by afferent activity). The neurons were interconnected into a network with realistic connection probabilities, synaptic numbers, and cell activities. The activity patterns produced in a network model consisting of a number of interconnected neurons, both with and without feedback inhibition. These computer simulations indicated that feedback inhibition plays a significant role in population activity dynamics.

In formalising the model of the spiny cell network, the values of a number of key parameters need to be chosen (see Table 3). These include the probability of synaptic contact between nearby pairs of spiny neurons, the conductance and reversal potential of the inhibitory and excitatory synapses, and the pattern and efficacy of cortical synaptic input. These parameters determine the proportion of spiny neurons that fire under baseline and active conditions, and the distribution of firing activity among them. The baseline condition approximates the state of the spiny projection neurons during a barrage of afferent activity from the cerebral cortex, i.e., the UP state of the spiny projection neuron. The active condition approximates the conditions required for action potential firing in the UP state.

The model was implemented using the general simulation system (GENESIS) interpreter developed by Wilson and Bower (1989). Integration was done at 0.1 ms time-steps. The physical dimensions of the dendrite and soma determined the passive electrical membrane properties of spiny striatal neurons. The dendritic tree was reduced to

Table 3. Assumptions of the model for spiny–spiny interactions

Symbol	Quantity	Estimate
n	Total number of spiny cells in local network	2500
p	Probability of local synaptic connection	0.20
Gi	Conductance of inhibitory synaptic connection	0.6 nS
f	Frequency of cortical input (total rate over all synapses)	3000 Hz
Ge	Conductance of excitatory synaptic connection	2–3 nS
T_1e	Time constant first term, excitatory synapse	2.0 ms
T_2e	Time constant second term, excitatory synapse	2.0 ms
T_1i	Time constant first term, inhibitory synapse	2.9 ms
T_2i	Time constant second term, inhibitory synapse	8.9 ms
Ga	Peak conductance of after hyperpolarisation	50 nS
T_1a	Time constant first term, after hyperpolarisation	1.0 ms
T_2a	Time constant second term, after hyperpolarisation	5.0 ms
R	Refractory period	5.0 ms
T	Threshold	−45 mV

a single compartment and modelled as a cylinder of diameter 2 μm and length 400 μm. Assuming a membrane resistivity of 10 KΩcm^2 gives a membrane resistance for the dendritic compartment of 398 MΩ. If the membrane capacitivity is assumed to be 1 μF/cm^2, the membrane time constant is then approximately 10 ms. The soma was modelled as a cylinder of diameter 15 μm and length 15 μm.

The dendritic compartment was connected to the soma compartment by an axial resistance, which was determined by the cord conductance of the dendritic cylinder. A cytoplasmic resistance of 300 Ω-cm was assumed, giving a total axial resistance of 382 MΩs from the soma to the dendrite.

The corticostriatal connections were modelled in a way that represented the highly specific sparse coding of the corticostriatal pathway. Kincaid et al. (1998) estimated that about 380,000 cortical axons pass through the volume of striatum occupied by a single spiny projection neuron. Based on interbouton distances of filled corticostriatal axons, an individual axon gives rise to a maximum of 40 synapses in the same volume of striatum. Since there are about 2840 spiny neurons overlapping in the same volume, then a single axon can only contact less than 1.4% of MSNs. Thus striatal spiny neurons with overlapping dendritic volumes have few cortical axons in common and cortical axons few MSN in common. Therefore, we modelled corticostriatal input lines as independent inputs, although with the caveat that their firing patterns may not be so independent if subsets of these inputs are associated into cell assemblies in the cerebral cortex.

In the model, separate input lines were used to represent the corticostriatal afferent axons. Each afferent axon innervated only one striatal neuron in the model, terminating on the distal end of the dendritic compartment. The inputs were statistically independent, in accordance with the sparse corticostriatal connectivity proposed by Kincaid et al. (1998).

Afferent rates of corticostriatal neurons in the monkey have median rate of 1.4 spikes/s (Turner and DeLong, 2000), see also (Bauswein et al., 1989). If we assumed that each spiny cell received 5000 corticostriatal inputs, this would give an upper limit of 7000 synaptic events/s. This does not allow for the fraction of corticostriatal neurons that are silent in a particular context, or for the fraction of synapses that are silent. An alternative estimate based on measurements obtained from organotypic slice cultures suggests the synaptic frequency in the UP state is on the order of 800 synaptic events/s (Blackwell et al., 2003). We chose rates that resulted in UP state membrane potentials in the neurons in the simulated network.

The conductance and reversal potential of the excitatory corticostriatal synapse was also determined from the literature. Individual spiny projection neurons receive a large number of corticostriatal inputs. Both 'small' and 'huge' unitary glutamatergic EPSCs have been described in striatal slices (Mori et al., 1994a). The small type is most common, with a mean amplitude of about 2.0 pA, and is thought to represent the monosynaptic input from the cerebral cortex (Mori et al., 1994b). These display a single peak amplitude, suggesting that each afferent fibre contacts a given

spiny cell at few, or only one, release site, consistent with the assumptions made above.

The strength of the spiny cell–spiny cell inhibitory synapse has been measured in several labs. Using mean-variance analysis Koos et al. (2004) determined a peak quantal conductance change at spiny–spiny synapses of from 0.3 nS to 0.9 nS. They reported also that the use of intracellular Cs^+ increased their estimate of the mean synaptic conduction for this connection by a factor of 3. Allowing for the effects of attenuation of apparent synaptic currents at distant synapses by the electronic properties of the spiny cell dendrites, a conservative estimate for the effectiveness of this connection is $G_i = 0.6$ nS.

To determine the current flowing through this conductance under natural conditions, we also need to know the reversal potential for these synapses. This has been variously reported as -62 mV (Tunstall et al., 2002), -76 mV (Koos et al., 2004), and $-66-$mV (Venance et al., 2004). In order to ensure a conservative estimate of the effects of the inhibitory synapses, the model assumed a reversal potential of $E_i = -60$ mV. That is, the inhibitory synapses were depolarising with a reversal potential near threshold.

Results of the computer simulation

Simulations were conducted on networks that ranged in size up to 2500 cells, representing the number of spiny neurons overlapping in the same volume. The results reported were not sensitive to variations on the numbers of cells. One aim of the simulation was to investigate how the presence or absence of inhibition affected the difference in firing rate produced by differences in excitation, i.e., contrast. To produce a controlled range of synaptic conductance, the values were assigned stepwise in a gradient from 2.0 nS to 5.0 nS so that no two cells received exactly the same excitatory synaptic input and the differences between cells were graded across the network. These inputs had no spatial organisation, with the location of individual cells effectively randomised within the network.

The rate of afferent activity was set to 3000 synaptic events/s, to represent the UP state during rest and during activity, respectively. Accurately modelling the firing patterns of spiny neurons in awake animals is crucial for understanding the effects of their inhibitory synapses. In awake animals, the large majority (72.7%) of striatal neurons are silent in the absence of overt movement (Sandstrom and Rebec, 2003). The presumed spiny projection neurons that do show spontaneous activity (i.e., activity while the animal is resting quietly) discharge at low rates (<5 spikes/s) (Sandstrom and Rebec, 2003). The rates of firing increase dramatically when the animal makes a movement; however, the firing rate under those conditions is movement- and context-dependent. The afferent activity rates were chosen to produce activity in about 25% of cells when all parameters were at their final physiological values.

Figure 3 shows that inhibition via local axon collaterals of spiny projection neurons exerts a major effect, when realistic assumptions about connectivity and conductance of inhibitory synapses are made. The firing activity across the network in the absence of feedback inhibition is compared with that occurring when feedback inhibition is turned on. Feedback inhibition using synapses of 0.6 nS produced a 10-fold reduction in the firing rate, as can be seen by comparing Fig. 3A with Fig. 3B. Importantly, this simulation shows that this 10-fold reduction in firing rate can be produced by a minority of active spiny cells, of which only about 20% fire >2 spikes/s.

The effect of feedback inhibition on the overall activity pattern of the whole network is not only to reduce firing rates in all cells, but also to increase the difference in firing rates between more and less active cells. This effect can be seen in Fig. 3C, where the individual neurons have been arranged in order from the least excited to the most excited. This graph shows the network input–output function. In the absence of inhibition, increasing excitation increases cell-firing rates until saturation occurs, and differences between less excited and more excited cells become vanishingly small. In the presence of inhibition, the upward slope of the right-hand curve on the network input–output function is steep, reflecting competitive interactions.

These results show that using realistic assumptions about the probability and conductance of

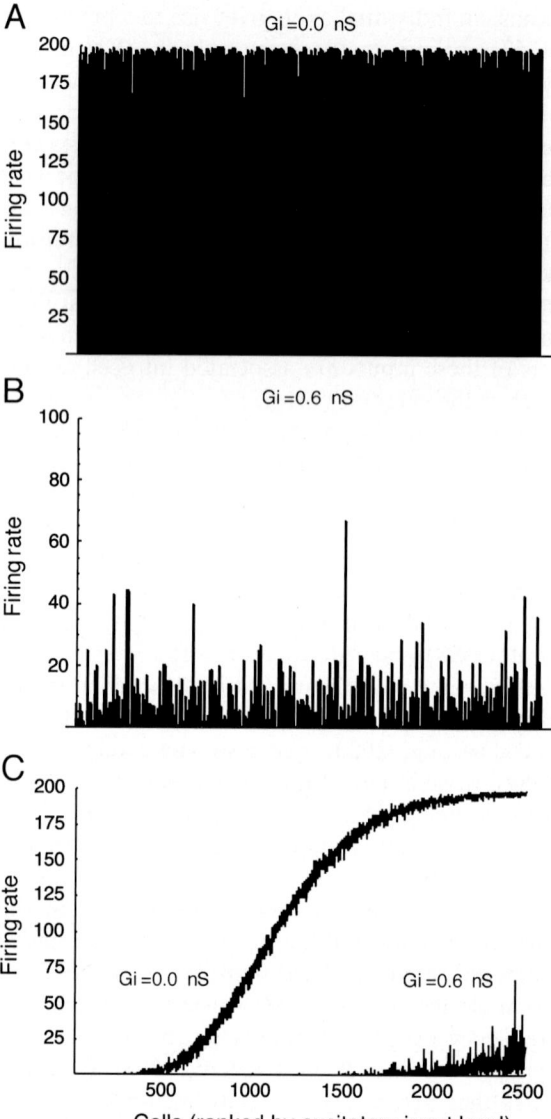

Fig. 3. Results of computer simulaton, based on realistic estimates of connectivity and synaptic conductance. (A) Average firing rates of all cells in a 2500 cell network, with zero inhibitory interactions and an average excitatory afferent rate of 3000 Hz.(B) Firing rates in same network as in A, but with inhibitory synapses ($G_i = 0.6\,nS$) formed with probability of $p = 0.2$, giving about 500 inhibitory inputs per cell. (C). Comparison of distribution of firing activity (cells arranged in order of excitatory input) showing effects of inhibition as shown in (A) and (B). Note that the presence of physiological levels of feedback inhibition, the difference between less and more active cells is increased, showing functional competitive dynamics in the network.

inhibitory interactions among spiny projection neurons results in a network in which inhibition plays a significant and important role. In addition to setting the overall activity level of the network, the feedback nature of the interactions is evident in the competitive dynamics. In particular, feedback inhibition amplifies the differences in output activity between cells that receive different levels of excitation. Although this is not as strong as the single-WTA dynamics formerly proposed, it is nevertheless a strong effect of the sum of many weak interactions.

Other GABAergic computations in the striatum

Although beyond the scope of the formal computational modelling described above, there are several other important GABAergic mechanisms. These include the feedforward interneurons (Kita, 1993) and their GABAergic inputs from the globus pallidus (Bevan et al., 1998) as depicted in Fig. 1C.

The feedforward interneurons are a subpopulation of parvalbumin and GABA-positive cells that are also known as fast-spiking interneurons. These interneurons are numerically in the minority, numbering about 16,900 in total (Luk and Sadikot, 2001), in contrast to spiny projection neurons, which number about 2,791,000 (Oorschot, 1996). Parvalbumin-positive interneurons therefore account for less than 1% of the total population of GABAergic cells in the striatum.

The GABA/parvalbumin neurons account for only a small minority of the total number of neurons in the striatum. However, their axonal arborisation is very dense in the area surrounding their cell body. They make symmetrical synapses on spiny projection neurons (Bennett and Bolam, 1994). Individual GABA/parvalbumin-containing interneurons exert a relatively strong inhibitory effect on the spiny projection neurons (Koos and Tepper, 1999; Koos et al., 2004; Tepper et al., 2004). Comparisons of feedforward and feedback interactions suggest that on average the feedforward inhibitory effects of one single neuron on another are several times greater than the feedback effects (Koos et al., 2004), although some individual spiny-to-spiny cell interactions are just as strong (Gustafson et al., 2006).

Cortical afferents make asymmetric synapses on the GABA/parvalbumin interneurons (Kita et al., 1990; Lapper et al., 1992). The pattern of cortical innervation of these interneurons is such that individual cortical axons frequently form multiple synaptic contacts with an individual parvalbumin-positive neuron (Ramanathan et al., 2002). This may account for ease with which the interneurons are activated following cortical stimulation (Kita, 1993). In the awake animal, presumed feedforward interneurons fire mostly single spikes, but at a relatively high mean rate of 15 Hz (Berke et al., 2004).

As with the spiny projection neurons, understanding the computational role of the GABA/parvalbumin interneurons requires an analysis of their population level effects. Up to now very few computational models have incorporated feedforward interneurons. Wickens and Arbuthnott (1993) proposed that feedforward interneurons played a reset or blanking role, at the onset of a behavioural sequence, to improve the reliability and increase the length of sequences produced by the spiny cell network. Another, more general property of feedforward inhibitory networks is normalisation (Grossberg, 1988). The dynamics of a feedforward network may ensure that the total activity is independent of the number of active cells. Such dynamics may be important for explaining brightness contrast in visual processing because they cause a network which receives a fixed total brightness to respond to increased brightness in one part of the field with increased darkness in another part of the field (Grossberg, 1988). These dynamical properties of networks that include feedforward inhibition may complement the feedback competition mediated by the spiny projection neurons.

A number of crucial quantitative neuroanatomical details are needed to develop computational models incorporating the feedforward interneurons. These include: the fraction of GABA/parvalbumin terminals that are on the soma of the spiny projection neurons, versus the dendrites; the fraction of GABA synapses that are parvalbumin positive; and also the efficacy of individual corticostriatal inputs to the GABA/parvalbumin interneurons. With these data it would be possible to undertake a detailed computer simulation of both feedforward and feedback circuits interacting.

Discussion

As noted by Wilson (2000), anatomy has inspired modelling efforts in the basal ganglia. This has been particularly true in the striatum. The existence of synaptic connections among spiny projection neurons led to the concept of the striatum as a lateral inhibitory network, in which the prevailing dynamic is one of competition. The effect of competition in the proposed network was that the neurons that receive the greatest excitation should inhibit less strongly excited cells, thus increasing the contrast between 'winners' and 'losers' in the competitive dynamic.

Formal models of the striatal network were initially based on the assumption of high connection probability among spiny neurons, inspired by the two-dimensional projection of the three-dimensional structure. Thus, a 'domain' of competition was defined in terms of mutual interaction among all members of the domain. The quantitative estimates of connectivity based on anatomy and physiology reviewed suggest that within the volume of arborisation of the axon of a spiny cell, the average probability of a connection is much less than 1. Moreover, each connection also probably involves very few synaptic contacts. Such constraints limit the degree of competition that is possible.

What are the computational properties of a block of striatal tissue that includes realistic inhibitory connectivity and efficacy? Obviously, the extreme form of competition in which a single 'winner' suppresses the activity in all other neurons is not possible. Does it follow that lateral inhibition has no impact on the dynamics of the striatal network?

A crucial consideration in the dynamics of lateral inhibitory networks is the relative weight of self-inhibition to other inhibition. This is especially important when we assume that strong inhibitory effects will require cooperation among multiple spiny cells. In such a case the spiny cells in an active group will be inhibiting each other as much as they inhibit non-group members. Alternatively, the active group could be selected so as to ensure minimal interaction among members of the group, by a simple activity-dependent plasticity rule.

In the case where self-inhibition is as strong as other inhibition, we reach a winner-share-all situation (Fukai and Tanaka, 1997). As well as being more consistent with the known anatomy and physiology, this model is also advantageous for information processing. Winner-share-all dynamics converge on the actual winner and are less influenced by initial conditions than WTA networks. Allowing multiple winners also permits a more efficient coding of information than the extreme WTA situation. For example, if k out of n winners is allowed, then many more combinations are possible than if only one out of n winners is allowed.

As has been shown in other brain regions, analyses of artificial neural networks may offer some insight into the role of a variety of GABAergic circuits in the basal ganglia. On the other hand, the interpretation of GABAergic circuits in the basal ganglia is especially challenging, because the organisation of the basal ganglia GABAergic circuits is unlike any existing artificial neural network model. The anatomy and physiology of GABAergic circuitry in the basal ganglia may, therefore, suggest novel mechanisms of computation that have the potential to contribute to new types of artificial neural networks.

Acknowledgments

This work was supported by the Health Research Council of New Zealand.

References

Alexander, G.E., DeLong, M.R. and Strick, P.L. (1986) Parallel organization of functionally segregated circuits linking basal ganglia and cortex. Ann. Rev. Neurosci., 9: 357–381.

Alexander, M.E. and Wickens, J.R. (1993) Analysis of striatal dynamics: the existence of two modes of behaviour. J. Theor. Biol., 163: 413–438.

Amari, S.I. (1977a) Competition and cooperation in neural nets. In: Metzler J. (Ed.), Systems Neuroscience. Academic Press, New York, pp. 119–165.

Amari, S.I. (1977b) Dynamics of pattern formation in lateral-inhibition type neural fields. Biol. Cybern., 27: 77–87.

Aronin, N., Chase, K. and DiFiglia, M. (1986) Glutamic acid decarboxylase and enkephalin immunoreactive axon terminals in the rat neostriatum synpase with striatonigral neurons. Brain Res, 365: 151–158.

Bauswein, E., Fromm, C. and Preuss, A. (1989) Corticostriatal cells in comparison with pyramidal tract neurons: contrasting properties in the behaving monkey. Brain Res, 493: 198–203.

Bennett, B.D. and Bolam, J.P. (1994) Synaptic input and output of parvalbumin-immunoreactive neurons in the neostriatum of the rat. Neuroscience, 62: 707–719.

Berke, J.D., Okatan, M., Skurski, J. and Eichenbaum, H.B. (2004) Oscillatory entrainment of striatal neurons in freely moving rats. Neuron, 43: 883–896.

Beurle, R.L. (1956) Properties of a mass of cells capable of regenerating pulses. Phil. Trans. R. Soc. London [Biol.], 240: 55–94.

Bevan, M.D., Booth, P.A., Eaton, S.A. and Bolam, J.P. (1998) Selective innervation of neostriatal interneurons by a subclass of neuron in the globus pallidus of the rat. J. Neurosci., 18: 9438–9452.

Blackwell, K.T., Czubayko, U. and Plenz, D. (2003) Quantitative estimate of synaptic inputs to striatal neurons during up and down states in vitro. J. Neurosci., 23: 9123–9132.

Bolam, J.P., Powell, J.F., Wu, J.-Y. and Smith, A.D. (1985) Glutamate decarboxylase-immunoreactive structures in the rat neostriatum: a correlated light and electron microscopic study including a combination of Golgi-impregnation with immunocytochemistry. J. Comp. Neurol., 237: 1–20.

Braitenberg, V. and Schüz, A. (1991) Anatomy of the cortex: statistics and geometry. Springer, Berlin.

Czubayko, U. and Plenz, D. (2002) Fast synaptic transmission between striatal spiny projection neurons. Proc. Natl. Acad. Sci. U S A, 99: 15764–15769.

Dube, L., Smith, A.D. and Bolam, J.P. (1988) Identification of synaptic terminals of thalamic or cortical origin in contact with distinct medium-size spiny neurons in the rat neostriatum. J. Comp. Neurol., 267: 455–471.

Fukai, T. (1999) Sequence generation in arbitrary temporal patterns from theta-nested gamma oscillations: a model of the basal ganglia-thalamo-cortical loops. Neural Netw, 12: 975–987.

Fukai, T. and Tanaka, S. (1997) A simple neural network exhibiting selective activation of neuronal ensembles: from winner-take-all to winners-share-all. Neural Comput, 9: 77–97.

Griffith, J.S. (1963) On the stability of brain-like structures. Biophys J, 3: 299–308.

Grossberg, S. (1973) Contour enhancement, short term memory and constancies in reverberating neural networks. Stud. Appl. Math., 52: 213–257.

Grossberg, S. (1988) Nonlinear neural networks: principles, mechanisms and architectures. Neural Netw., 1: 17–61.

Groves, P.M. (1983) A theory of the functional organisation of the neostriatum and the neostriatal control of voluntary movement. Brain Res. Rev., 5: 109–132.

Groves, P.M., Linder, J.C. and Young, S.J. (1994) 5-hydroxydopamine-labeled dopaminergic axons: three-dimensional reconstructions of axons, synapses and postsynaptic targets in rat neostriatum. Neuroscience, 58: 593–604.

Gustafson, N., Gireesh-Dharmaraj, E., Czubayko, U., Blackwell, K.T. and Plenz, D. (2006) A comparative voltage and current-clamp analysis of feedback and feedforward synaptic transmission in the striatal microcircuit in vitro. J. Neurophysiol., 95: 737–752.

Hoover, J.E. and Strick, P.L. (1993) Multiple output channels in the basal ganglia. Science, 259: 819–821.

Ingham, C.A., Hood, S.H., Taggart, P. and Arbuthnott, G.W. (1998) Plasticity of synapses in the rat neostriatum after unilateral lesion of the nigrostriatal dopaminergic pathway. J. Neurosci., 18: 4732–4743.

Kawaguchi, Y., Wilson, C.J. and Emson, P.C. (1989) Intracellular recording of identified neostriatal patch and matrix spiny cells in a slice preparation preserving cortical inputs. J. Neurophys., 62: 1052–1068.

Kincaid, A.E., Zheng, T. and Wilson, C.J. (1998) Connectivity and convergence of single corticostriatal axons. J. Neurosci., 18: 4722–4731.

Kishimoto, K. and Amari, S.I. (1979) Existence and stability of local excitations in homogenous neural fields. J. Math. Biol., 7: 303–318.

Kita, H. (1993) GABAergic circuits of the striatum. Prog. Brain Res., 90: 51–72.

Kita, H. and Kitai, S.T. (1988) Glutamate decarboxylase immunoreactive neurons in cat neostriatum: Their morphological types and populations. Brain Res, 447: 346–352.

Kita, H., Kosaka, T. and Heizmann, C.W. (1990) Parvalbumin-immunoreactive neurons in the rat neostriatum: a light and electron microscopic study. Brain Res, 536: 1–15.

Koos, T. and Tepper, J.M. (1999) Inhibitory control of neostriatal projection neurons by GABAergic interneurons. Nat. Neurosci., 2: 467–472.

Koos, T., Tepper, J.M. and Wilson, C.J. (2004) Comparison of IPSCs evoked by spiny and fast-spiking neurons in the neostriatum. J. Neurosci., 24: 7916–7922.

Kotter, R. and Wickens, J.R. (1995) Interactions of glutamate and dopamine in a computational model of the striatum. J. Comput. Neurosci., 2: 195–214.

Kotter, R. and Wickens, J.R. (1998) Striatal modeling and Parkinson's disease. Art. Intel. Med., 13: 37–55.

Kubota, Y., Inagaki, S., Shimada, S., Kito, S. and Wu J, Y. (1987) Glutamate decarboxylase-like immunoreactive neurons in the rat caudate putamen. Brain Res. Bull., 18: 687–697.

Lapper, S.R., Smith, Y., Sadikot, A.F., Parent, A. and Bolam, J.P. (1992) Cortical input to parvalbumin-immunoreactive neurones in the putamen of the squirrel monkey. Brain Res, 580: 215–224.

Lazzaro, J., Ryckebusch, S., Mahowald, M. and Mead, C. (1989) Winner-take-all networks of O(n) complexity. In: Touretzky D.S. (Ed.), Adv Neur Inf Proc Sys, Vol. 2. Morgan Kauffmann, San Mateo, CA, pp. 703–711.

Lee, T., Kaneko, T., Shigemoto, R., Nomura, S. and Mizuno, N. (1997) Collateral projections from striatonigral neurons to substance P receptor-expressing intrinsic neurons in the striatum of the rat. J. Comp. Neurol., 388: 250–264.

Luk, K.C. and Sadikot, A.F. (2001) GABA promotes survival but not proliferation of parvalbumin-immunoreactive neurons in rodent neostriatum: An in vivo study with stereology. Neuroscience, 104: 93–103.

Maass, W. (2000a) Neural computation with winner-take-all as the only nonlinear operation. Adv. Neur. Inf. Proc. Sys., 12: 293–299.

Maass, W. (2000b) On the computational power of winner-take-all. Neural Comput, 12: 2519–2535.

Majani, E., Erlanson, R. and Abu-Mostafa, Y. (1989) On the K-winners-take-all-network. In: Touretzky D.S. (Ed.), Adv. Neur. Inf. Proc. Sys., Vol. 1. Morgan Kaufmann, San Francisco, CA, pp. 634–642.

Mori, A., Takahashi, T., Miyashita, Y. and Kasai, H. (1994a) Two distinct glutamatergic synaptic inputs to striatal medium spiny neurones of neonatal rats and paired-pulse depression. J. Physiol., 476: 217–228.

Mori, A., Takahashi, T., Miyashita, Y. and Kasai, H. (1994b) Quantal properties of S-type glutamatergic synaptic input to the striatal medium spiny neuron from neonate rat. Neurosci Lett, 169: 199–202.

Oertel, W.H. and Mufnini, E. (1984) Immunocytochemical studies of GABAergic neurons in rat basal ganglia and their relations to other neuronal systems. Neurosci. Lets., 47: 233–238.

Oorschot, D.E. (1996) Total number of neurons in the neostriatal, pallidal, subthalamic, and substantia nigral nuclei of the rat basal ganglia: a stereological study using the Cavalieri and optical disector methods. J. Comp. Neurol., 366: 580–599.

Oorschot, D.E., Tunstall, M.J. and Wickens, J.R. (2002) Local connectivity between striatal spiny projection neurons: a re-evaluation. In: Nicholson L. and Faull R.L.M. (Eds.), Basal Ganglia VII. Plenum Press, New York, pp. 421–434.

Pasik, P., Pasik, T., Holstein, G. and Hamori, J. (1988) GABAergic elements in the neuronal circuits of the monkey neostriatum: a light and electron microscopic immunocytochemical study. J. Comp. Neurol., 270: 157–170.

Plenz, D. and Kitai, S.T. (1999) Adaptive classification of cortical input to the striatum by competitive learning. In: Miller R. and Wickens J.R. (Eds.), Brain Dynamics and the Striatal Complex. Harwood Academic.

Plenz, D., Wickens, J. and Kitai, S.T. (1996) Basal ganglia control of sequential activity in the cerebral cortex: A model. In: Bower J.M. (Ed.), Computational Neuroscience. Academic Press, San Diego, CA, pp. 397–402.

Ramanathan, S., Hanley, J.J., Deniau, J.M. and Bolam, J.P. (2002) Synaptic convergence of motor and somatosensory cortical afferents onto GABAergic interneurons in the rat striatum. J. Neurosci., 22: 8158–8169.

Rolls, E.T. and Williams, G.V. (1986) Sensory and movement-related activity in different regions of the primate striatum. In: Schneider J.S. and Lidsky T.I. (Eds.), Basal Ganglia and Behavior. Hans Huber, Stuttgart, pp. 37–60.

Sandstrom, M.I. and Rebec, G.V. (2003) Characterization of striatal activity in conscious rats: contribution of NMDA and AMPA/kainate receptors to both spontaneous and glutamate-driven firing. Synapse, 47: 91–100.

Shindou, T., Ochi-Shindou, M., Arbuthnott, G.W. and Wickens, J.R. (2005) Adenosine A2A receptor-mediated modulation of fast-spiking interneurons in the striatum. Program No. 298.17, Society for Neuroscience, 2005. Washington, DC. Online.

Somogyi, J.P., Bolam, J.P. and Smith, A.D. (1981) Monosynaptic cortical input and local axon collaterals of identified striatonigral neurons. A light and electron microscope study using the Golgi-peroxidase transport degeneration procedure. J. Comp. Neurol., 195: 567–584.

Taverna, S., van Dongen, Y.C., Groenewegen, H.J. and Pennartz, C.M. (2004) Direct physiological evidence for synaptic connectivity between medium-sized spiny neurons in rat nucleus accumbens in situ. J. Neurophysiol., 91: 1111–1121.

Tepper, J.M., Koos, T. and Wilson, C.J. (2004) GABAergic microcircuits in the neostriatum. Trends Neurosci, 27: 662–669.

Tunstall, M.J., Oorschot, D.E., Kean, A. and Wickens, J.R. (2002) Inhibitory interactions between spiny projection neurons in the rat striatum. J. Neurophys., 88: 1263–1269.

Turner, R.S. and DeLong, M.R. (2000) Corticostriatal activity in primary motor cortex of the macaque. J. Neurosci., 20: 7096–7108.

Venance, L., Glowinski, J. and Giaume, C. (2004) Electrical and chemical transmission between striatal GABAergic output neurones in rat brain slices. J. Physiol., 559: 215–230.

Wickens, J.R. (1993) A Theory of the Striatum. Pergamon Press, Oxford.

Wickens, J.R. (2002) Surround inhibition in the basal ganglia. In: DeLong M. and Graybiel A.M. (Eds.), The Basal Ganglia VI. Plenum/Kluwer Press, pp. 187–197.

Wickens, J.R. and Arbuthnott, G.W. (1993) The corticostriatal system on computer simulation: an intermediate mechanism for sequencing of actions. Prog. Brain Res., 99: 325–339.

Wickens, J.R. and Miller, R. (1997) A formalisation of the neural assembly concept: 1. Constraints on neural assembly size. Biol. Cybern., 77: 351–358.

Wickens, J.R. and Oorschot, D.E. (2000) Neural dynamics and surround inhibition in the neostriatum: a possible connection. In: Miller R. and Wickens J.R. (Eds.), Brain Dynamics and the Striatal Complex. Gordon and Breach, Reading, U.K., pp. 141–150.

Wickens, J.R., Alexander, M.E. and Miller, R. (1991) Two dynamic modes of striatal function under dopaminergic-cholinergic control: simulation and analysis of a model. Synapse, 8: 1–12.

Wickens, J.R., Kotter, R. and Alexander, M.E. (1995) Effects of local connectivity on striatal function: simulation and analysis of a model. Synapse, 20: 281–298.

Wilson, C.J. (1994) Understanding the neostriatal microcircuitry: high-voltage electron microscopy. Microscopy Res. Tech., 29: 368–380.

Wilson, C.J. (2000) Striatal circuitry: Categorically selective, or selectively categorical? In: Miller R. and Wickens J. (Eds.), Brain Dynamics and the Striatal Complex, pp. 289–305.

Wilson, C.J. and Groves, P.M. (1980) Fine structure and synaptic connection of the common spiny neuron of the rat neostriatum: a study employing intracellular injection of horseradish peroxidase. J. Comp. Neurol., 194: 599–615.

Wilson, H.R. and Cowen, J.D. (1972) Excitatory and inhibitory interactions in localized populations of model neurons. Biophys. J., 12: 1–24.

Wilson, M.A. and Bower, J.M. (1989) The simulation of large-scale neural networks. In: a.I. Segev C.K. (Ed.), Methods in Neuronal Modeling: From Synapses to Networks. MIT Press, Cambridge, pp. 291–334.

Wolfe, W.J., Mathis, D., Anderson, C.M., Rothman, J., Gottler, M., Brady, G., Walker, R., Dunae, G. and Alaghband, G. (1991) K-Winner networks. IEEE Trans. Neur. Net., 2: 310–315.

Xu, Z.C., Wilson, C.J. and Emson, P.C. (1991) Restoration of thalamostriatal projections in rat neostriatal grafts: an electron microscopic analysis. J. Comp. Neurol., 303: 22–34.

Subject Index